32nd International Symposium on Space Terahertz Technology (ISSTT 2022)

Baeza, Spain
16-20 October 2022

ISBN: 978-1-7138-8988-5

Printed from e-media with permission by:

Curran Associates, Inc.
57 Morehouse Lane
Red Hook, NY 12571

Some format issues inherent in the e-media version may also appear in this print version.

International Symposium on Space Terahertz Technology
c/o NRAO Headquarters
520 Edgemont Road
Charlottesville, VA 22903-2475
USA

Phone: (434) 296-0254
Fax: (434) 296-0278

mbishop@nrao.edu

Additional copies of this publication are available from:

Curran Associates, Inc.
57 Morehouse Lane
Red Hook, NY 12571 USA
Phone: 845-758-0400
Fax: 845-758-2633
Email: curran@proceedings.com
Web: www.proceedings.com

October 17, 2022 (Monday)

Venue: Ruinas de San Francisco Auditorium (Hall)

08:15	*Coffee & Registration (30 mins)*

Venue: Ruinas de San Francisco Auditorium

8:45		**ISSTT 2022 Opening Remarks & Information**

Venue: Ruinas de San Francisco Auditorium

9:15	Invited Talk I **- Dr. Kartik Seth, NASA** Chair: *Jose V. Siles (NASA Jet Propulsion Laboratory)*

Venue: Ruinas de San Francisco Auditorium

Session I – Instruments & Systems for Future Missions I
Chair: *Victor Belitski (Chalmers University of Technology)*

09:50	S1.1	"A Direct Up-Conversion G-band Radar Prototype for Atmospheric Measurements"...1 **Raquel R. Monje,** *NASA Jet Propulsion Laboratory*
10:10	S1.2	"MetOp-SG Ice Cloud Imager 183-664 GHz Front-End Receivers Proto-Flight Model Qualification and Acceptance Test Results"...2 **Bertrand Thomas**, *Radiometer Physics GmbH*
10:30	S1.3	"On-Ground Calibration Targets for the Ice Cloud Imager Instrument on the MetOP Second Generation Satellite"...3 **Axel Murk**, *University of Bern*
10:50	S1.4	"Highly-Compact Terahertz Spectrometers on Ultra-Small Platforms"...4 **Goutam Chattopadhyay**, *NASA Jet Propulsion Laboratory*

Venue: Ruinas de San Francisco Auditorium (Patio)

11:10	*Coffee Break (30 mins)*

Venue: Ruinas de San Francisco Auditorium

Session II – Best Student Paper Competition Finalists Chair: *Jeffrey Hesler (Virginia Diodes, Inc.)*		
11:40	S2.1	"Optimization of a PID controlled TES detector array for use in an FIR double-Fourier interferometer"...88 **Chris Benson**, *University of Lethbridge*
11:50	S2.2	"The effect of complex dispersion and impedance in the gain of superconducting traveling-wave kinetic inductance parametric amplifiers"...92 **Javier Carrasco**, *University of Chile*
12:00	S2.3	"Gas Analysis with High-Resolution Fabry-Perot spectrometry in the 550-650 GHz Range"...94 **Coralie Elmaleh**, *Université du Littoral Côte d'Opale*
12:10	S2.4	"Design and Characterisation of a 3.5-THz Fundamental Schottky Mixer"...102 **Divya Jayasankar**, *Chalmers University of Technology*
12:20	S2.5	"Experimental characterisation of titanium nitride transmission lines for applications as kinetic inductance travelling wave parametric amplifiers"...104 **Joseph Longden**, *Chalmers University of Technology*
12:30	S2.6	"A Cryogenic Scalable Small-Signal & Noise Model of GaN HEMTs"...113 **Mohamed Aniss Mebarki**, *University of Chile*
12:40	S2.7	"Development of an instantaneous multiband digital 2SB receiver for the 67–116-GHz band"...117 **David Monasterio**, *University of Chile*
12:50	S2.8	"Exploring the Limits of the Tunnel Junction Fabrication Technique for Josephson Junctions TWPA and the Preliminary Characterisation Results"...120 **Javier Navarro Montilla**, *University of Oxford*
13:00	S2.9	"Design, Operation, and Characterization of a Laboratory Spatial-Spectral Fourier Transform Interferometer"...126 **Jeremy Scott**, *University of Lethbridge*
13:10	S2.10	"Preliminary Characterisation of a Compact 240 GHz SIS Dual-Polarisation Receiver for Large Array Applications"...136 **Jakob Wenninger**, *University of Oxford*

Venue: Ruinas de San Francisco Auditorium

Diamond Sponsors Session Chair: *Imran Mehdi (NASA Jet Propulsion Laboratory)*		
13:20	DS1	Virginia Diodes, Inc.
13:30	DS2	Genera Tecnologías, S.A.

13:40 - 15:00	*Launch Break (90 mins)*

Venue: Ruinas de San Francisco Auditorium

		Session III – Terahertz Sources I Chair: *Alain Maestrini (NASA Jet Propulsion Laboratory)*
15:00	S3.1	"An Upgrade to the 1.9THz Local Oscillator for the SOFIA GREAT Instrument"...5 **Thomas Crowe**, *Virginia Diodes, Inc.*
15:20	S3.2	"Electron Speed Saturation Effects in the Schottky Barrier Diodes Implemented in an 800 – 900 GHz Frequency Tripler"...6 **Diego Moro Melgar**, *ACST GmbH*
15:40	S3.3	"High-Power Compact Sources in the 200 - 300 GHz Frequency Range"...7 **Verónica Laín Rubio**, *ACST GmbH*
16:00	S3.4	"Ultra-broadband integrated room-temperature Schottky diode based local oscillators for line surveys in the 400-2070 GHz range"...8 **José V. Siles**, *NASA Jet Propulsion Laboratory*

Venue: Ruinas de San Francisco Auditorium (Patio)

16:20	*Coffee Break (30 mins)*

Venue: Ruinas de San Francisco Auditorium

		Session IV – Solid State Receivers Chair: Berhanu Bulcha (NASA Goddard Space Flight Center)
16:50	S4.1	"Terahertz Heterodyne Receiver Frontend Based on SiGe BiCMOS Technology for Space Applications"...9 **Nick Rothbart**, *German Aerospace Center (DLR)*
17:10	S4.2	"A Solid-State, Non-Cryogenic Receiver Operating at 2.5THz"...10 **Theodore Reck**, *Virginia Diodes, Inc.*
17:30	S4.3	"THz Receivers for Thermospheric Science"...12 **Alain Maestrini**, *NASA Jet Propulsion Laboratory*

Venue: Teatro Montemar

18:15	Plenary Talk – **Dr. John Mather, 2006 Nobel Prize of Physics, NASA** **"Opening the Infrared Treasure Chest with JWST"...N/A** Chair: *Paul Goldsmith (NASA Jet Propulsion Laboratory)*

20:00	*Ruta de la Tapa (Scavenger Tapas Night)*

Venue: Café Bar Central

22:00	*A Night of Magic* **Javi Benítez, 2018 FISM World Champion**

October 18, 2022 (Tuesday)

Venue: Ruinas de San Francisco Auditorium (Hall)

08:30	*Coffee & Registration (30 mins)*

Venue: Ruinas de San Francisco Auditorium

9:00	**Roundtable | Women on Space Terahertz Research** *Chairs: Martina Wiedner (Observatory of Paris), Kartik Sheth (NASA)*

Venue: Ruinas de San Francisco Auditorium

Platinum Sponsors Session Chair: *Imran Mehdi (NASA Jet Propulsion Laboratory)*		
09:30	PS1	ACST
09:35	PS2	Thomas Keating
09:40	PS3	RPG / Rohde & Schwarz

Venue: Ruinas de San Francisco Auditorium

Session V – Novel Devices, Technologies & Components Chair: Hiroshi Matsuo (*National Astronomical Observatory of Japan, NAOJ)*		
09:50	S5.1	"Measurement of Transmission Losses of Superconducting Coplanar Waveguide and Microstrip Lines with On-chip Resonators at 2 mm Wavelength"...13 **Wenlei Shan**, *National Astronomical Observatory of Japan (NAOJ)*
10:10	S5.2	"A Micromachined 1.37 THz Waveguide-based 2X2 Beam Divider for HEB Detectors"...14 **Haotian Zhu**, *National Space Science Center (CAS)*
10:30	S5.3	"A Turnstile Quad-Ridge OMT for Full Octave-Bandwidth Receivers"...16 **Doug Henke**, *NRC Herzberg Astronomy and Astrophysics*
10:50	S5.4	"Recent research on (sub)mm-wave OMTs at NAOJ"...18 **Alvaro González**, *National Astronomical Observatory of Japan (NAOJ)*

Venue: Ruinas de San Francisco Auditorium (Patio)

11:10	*Coffee Break (30 mins)*

Venue: Ruinas de San Francisco Auditorium

Session VI – Instruments & Systems for Ground Telescopes
Chair: Patricio Mena (NRAO)

11:40	S6.1	"Results from ALMA Band 2 cryogenic LNA pre-production run"...20 **Sener Türk**, *Max Planck Institute for Radio Astronomy*
12:00	S6.2	"ALMA Band 2 Cold Cartridge Assembly Design"...21 **Victor Belitsky**, *Chalmers University*
12:20	S6.3	"ALMA Band 9 Sideband Separating Upgrade"...25 **Ronald Hesper**, *Kapteyn Astronomical Institute - University of Groningen*
12:40	S6.4	"Status of ALMA Band 6v2 Receiver Development"...31 **Joseph Lambert**, *National Radio Astronomy Observatory (NRAO)*
13:00	S6.5	"16-pixel, 3 mm band, Cryogenic Array Receiver for Users of the Sardinia Observatory (CARUSO)"...34 **Hui Wang**, *STFC – RAL Space*

13:20 - 15:00	*Launch Break / LOC & SOC Lunch (100 mins)*

Venue: Ruinas de San Francisco Auditorium

Session VII – Terahertz Antennas & Optical Components
Chair: Heinz-Wilhelm Hübers (German Aerospace Center, DLR)

15:00	S7.1	"Algorithmic Design of Terahertz Silicon Metaoptics"...35 **Conner Ballew**, *NASA Jet Propulsion Laboratory*
15:20	S7.2	"Optics for the TEMPERA-C polarimetric middle atmosphere temperature sounder"...36 **Roland Albers**, *Bern University*
15:40	S7.3	"Monolithic focal plane array concept for space and ground applications"...39 **Andrey Baryshev**, *Kapteyn Astronomical Institute (NOVA, RUG)*
16:00	S7.4	"3D Printed Submillimeter Reflectors: a New Design and Manufacturing Methodology"...40 **Paul Goldsmith**, *NASA Jet Propulsion Laboratory*

Venue: Ruinas de San Francisco Auditorium (Patio)

16:20	*Coffee Break (30 mins)*

Venue: Ruinas de San Francisco Auditorium

		Session VIII – Cryogenic Receivers, Mixers & Back-ends Chair: *Netty Honingh (University of Cologne)*
16:50	S8.1	"Towards the heterodyne receiver for balloons and SmallSats above 2 THz"...44 **Boris Karasik**, *NASA Jet Propulsion Laboratory*
17:10	S8.2	"Performance analysis of superconductor insulator superconductor mixer for 260 GHz atmospheric window"...45 **Andrey Baryshev**, *Kapteyn Astronomical Institute (NOVA, RUG)*
17:30	S8.3	"Low noise, wide band MgB2 hot electron bolometer mixer at 5.3 THz and 20 K"...46 **Behnam Mirzaei**, *Delft University of Technology*
17:50	S8.4	"On embedding of an HEB mixer into a THz photonic integrated circuit"...47 **Alexander Shurakov**, *Moscow Pedagogical State University*
18:10	S8.5	"2-18 GHz Ultra-wideband Cryogenic Amplifier with 4 K Noise Temperature"...48 **Isaac López Fernández**, *Observatorio de Yebes*

18:30	*Baeza Historic City Center Visit* *or Olive Oil Museum Visit*

Venue: Plaza de Santa María (in front of the Cathedral)

19:45	*Flamenco under the stars with* ***Arcángel, 2018 Latin Grammy Award Winner***

Venue: Hotel Puerta de la Luna

21:15	*Conference Banquet*

October 19, 2022 (Wednesday)

Venue: Palacio de Jabalquinto

09:00	*Registration / Poster Session (starts)*

Venue: Palacio de Jabalquinto

Poster Session *Chair: Cecile Jung-Kubiak (NASA Jet Propulsion Laboratory)*	
9:00 10:00	Poster Session Presentations (see complete list of poster presentation in the next pages)
10:00	*Coffee Break*
10:30 11:15	Poster Session Presentations (cont.) (see complete list of poster presentation in the next pages)

Venue: Ruinas de San Francisco Auditorium

Session IX – Measurement Techniques Chair: *Christopher Groppi (Arizona State University)*		
11:30	S9.1	"Versatile Radiometric Testbed for the Submillimeter Wave Instrument"...50 **Mikko Kotiranta**, *University of Bern*
11:50	S9.2	"Harmonic phase and amplitude beam characterization of a wideband on-chip spectrometer"...54 **Stephen Yates**, *SRON*
12:10	S9.3	"Characterization of widefield THz optics using phase shifting interferometry"...55 **Nicolás Reyes**, *Max Planck Institue for Radioastronomy*
12:30	S9.4	"CCAT prime: Multi-map Holographic Measurement for FYST Testbed——near-field beam measurement and data análisis"...57 **Xiadong Ren**, *University of Cologne*
12:50	S9.5	"Introducing LORENTZ: A Novel Low-temperature Near-field Terahertz Chamber for Instrument Characterisation"...61 **Paul Moseley**, *European Space Agency*

Venue: Ruinas de San Francisco Auditorium

13:10	Invited Talk II – **Dr. Paul Hartogh** Chair: *Imran Mehdi (NASA Jet Propulsion Laboratory)*

13:40 - 15:00	*Launch Break (80 min)*

Venue: Ruinas de San Francisco Auditorium

Session X – Far-Infrared Balloon-Borne Missions Chair: *Kartik Sheth (NASA)*		
15:00	S10.1	"GUSTO Payload Design and Performance"...62 **Christopher Walker**, *University of Arizona*
15:20	S10.2	"The Terahertz Intensity Mapper (TIM): design and development status of a far-infrared balloon for spectroscopic galaxy evolution studies"...63 **Christopher Groppi**, *Arizona State University*
15:40	S10.3	"The OSAS-B instrument: a balloon-borne heterodyne spectrometer for atomic oxygen in the mesosphere and lower thermosphere"...65 **Martin Wienold**, *German Aerospace Center (DLR)*
16:00	S10.4	"ASTHROS - Astrophysics Stratospheric Telescope for High-Spectral Resolution Observations at Submillimeter-Wavelengths: Architecture Design and Subsystem Level Integration & Test Status"...66 **Jose V. Siles**, *NASA Jet Propulsion Laboratory*

Venue: Ruinas de San Francisco Auditorium (Hall)

16:20	*Coffee Break (30 mins)*

Venue: Ruinas de San Francisco Auditorium

Session XI – Terahertz Sources II Chair: *Jian-Rong Gao (SRON)*		
16:50	S11.1	"Frequency stabilization of 3.5-THz and 4.7-THz quantum-cascade lasers by a phase-locked loop"...67 **Heiko Richter**, *DLR - Institute of Optical Sensor Systems*
17:10	S11.2	"The ammonia laser: a possible local oscillator for space applications?"...69 **Jean Francois Lampin**, *IEMN CNRS*
17:30	S11.3	"A tunable, high power source for GUSTO's local oscillator at 4.74 THz"...70 **Ali Khalatpour**. *Massachusetts Institute of Technology (MIT)*

Venue: Ruinas de San Francisco Auditorium

		Session XII – Instruments & Systems for Future Missions II Chair: *Martina Wiedner (Observatory of Paris)*
17:50	S12.1	"Kinetic Inductance Detector based focal plane arrays for the Terahertz Intensity Mapper"...71 **Reinier Janssen**, *NASA Jet Propulsion Laboratory*
18:10	S12.2	"Reflector-Based Phased Array for High Power G-band radars"...72 **Alain Maestrini**, *NASA Jet Propulsion Laboratory*
18:30	S12.3	"French contribution on the SWI hardware and flight models acceptance levels"...N/A **Jeanne Treuttel**, *Observatory of Paris*

Venue: Ruinas de San Francisco Auditorium

18:50		**ISSTT 2022 Closing Remarks & Student Award Winner**

Venue: Teatro Montemar

19:30	Public Outreach Talk – **Dr. Paul Goldsmith, NASA** **"The Legacy of the Arecibo Telescope"** Chair: *Jose V. Siles (NASA Jet Propulsion Laboratory)*

Venue: Discoteca Albacara

22:30	*Typical Spanish Late Social*

October 20, 2022 (Thursday)

10:00	*Departure to Granada* **(transit time ~1h30 min)**

11:30	*Short visit to Granada*

Venue: Carmen de los Martires Granada

13:30	Invited Talk III - **Dr. Antxon Alberdi, Director IAA** Chair: *Jose V. Siles (NASA Jet Propulsion Laboratory)*
14:00	*Special reception/lunch offered by Andalusia in Cármen de los Mártires*

16:00	*Visit to Alhambra*

08:30	*Departure to Baeza* **(transit time ~1h30 min)**

Venue: Discoteca Albacara / Bar Central / Burladero de Copas

22:30	*Typical Spanish Late Social*

APPENDIX: LIST OF POSTER PRESENTATIONS

Poster Session Chair: *Cecile Jung-Kubiak (Jet Propulsion Laboratory)*	
P1. INSTRUMENTS AND SYSTEMS FOR FUTURE MISSIONS	
P1.1	"Architecture of ASTHROS' Spectral Data Collection and Onboard Anomaly Detection Pipeline"...73 **Paul Horton**, *Arizona State University*
P1.2	"Atmospheric Phase Monitoring System Evolution for the NOEMA interferometer"...74 **Sylvain Mahieu**, *IRAM*
P1.3	"Heterodyne Array Receiver Study for the FIRSST Space Mission"...75 **Martina Wiedner**, *Observatory of Paris*
P1.4	"Jupiter Icy Moon Explorer, Submillimeter wave Instrument: Delivery status of the 1200 GHz high spectral resolution receiver front end"...76 **Jeanne Treuttel**, *Observatory of Paris*
P1.5	"Cold Optics on the Terahertz Intensity Mapper (TIM)"...77 **Talia Saeid**, *Arizona State University*
P1.6	"Multipixel 2-color Superconducting Receiver for ASTHROS"...78 **Jonathan Kawamura**, *NASA Jet Propulsion Laboratory*
P1.7	"Progress toward Antarctic Terahertz Intensity Interferometry"...79 **Hiroshi Matsuo**, *National Astronomical Observatory of Japan*
P1.8	"Breaking the 10mW/pixel limit for kinetic inductance detector readout electronics"...81 **Adrian STD-0026_ISSTT2022_Elmaleh**, *University of British Columbia*
P1.8	"The Space Interferometer for Cosmic Evolution (SPICE): The Far Infrared Universe at High Spectral Resolution"...84 **Locke Spencer**, *University of Lethbridge*
P2. BEST STUDENT PAPER COMPETITION	
P2.1	"G-Band Metamaterial-Based Circulator for FCC-Compliant Space-to-Earth Communication"...86 **Ali Alqaraghuli**, *Northeastern University*
P2.2	"Optimization of a PID controlled TES detector array for use in an FIR double-Fourier interferometer"...88 **Chris Benson**, *University of Lethbridge*

P4.6	"LO power division circuits for the CCAT-prime Heterodyne Array Instrument (CHAI)"...155 **Ignacio Barrueto**, *University of Cologne*
P4.7	"Characterization of monolithically integrated lithium niobate ring resonator for a high sensitivity room temperature radiometer"...159 **Jessica César Cuello**, *University Carlos III of Madrid*
P4.8	"Characterization of Microwave Properties of Superconducting NbTiN Films using TDS"...162 **Fedor Khan**, *Institute of Radioengineering and Electronics of RAS*
P4.9	"A Low Loss Diplexer for Submillimeter-wave Sideband Separating Receivers"...164 **Subash Khanal**, *NASA Jet Propulsion Laboratory*
P4.10	"Design of RF waveguide structure for 2SB SIS receiver for Millimetron 211-275 GHz VLBI channel"...165 **Andrey Khudchenko**, *Astro Space Center of P.N. Lebedev Physical Institute RAS*
P5. INSTRUMENTS & SYSTEMS FOR GROUND TELESCOPES	
P5.1	"A 67-116 GHz MMIC-based dual-polarized 2SB down-converter for the ALMA Band 2 Warm Cartridge Assembly prototype development"...167 **Bertrand Thomas**, *Radiometer Physics GmbH*
P5.2	"SEPIA345: a dual polarization 2SB cartridge receiver for APEX telescope: Design and Performance"...168 **Denis Meledin**, *Chalmers University of Technology*
P5.3	"ALMA Band 2 Receiver Automated Test System"...172 **Jan Barkhof**, *NOVA*
P5.4	"Development of ALMA Band 1 Corrugated Horns based on Metal 3D Printer"...173 **Keiko Kaneko**, *National Astronomical Observatory of Japan*
P5.5	"Demonstrator of Cryogenic Multibeam Receiver with MMIC LNAs for 75-116 GHz"...174 **Patrice Serres**, *IRAM*
P5.6	"Study of ALMA Band 2 receiver optical design for ACA 7-m antenna"...176 **Hiroaki Imada**, *National Astronomical Observatory of Japan*
P6. TERAHERTZ ANTENNAS & OPTICAL COMPONENTS	
P6.1	"Design of a 480 GHz Metamaterial Flat Lens"...178 **Casssandra Whitton**, *Arizona State University*

P6.2	"A Tunable Linear to Right-Handed Circularly Polarized THz Antenna Based on Graphene Switch"...181 **Mohammad Alibakhshikenari**, *Universidad Carlos III de Madrid*
P6.3	"A broadband and dual-polarization single-layer dichroic filter for applications in Sub-THz Range"...184 **Daniel Montof**, *Chalmers University of Technology*
P6.4	"Stepped Impedance Metal-Mesh Filters for Terahertz Frequencies"...187 **Adhitya B. Sriram**, *Arizona State University*
P6.5	"A Compact Dielectrically Loaded Quad-ridge Feed Horn for Octave Band Radio Astronomy Application"...188 **Sara Salem Hesari**, *National Research Council Canada*
P6.6	"A satellite tracking system at 78GHz using the over-moded TE21 ground-station antenna pattern"...189 **Hugh Gibson**, *Gibson Microwave Design (GMD)*
P6.7	"Diffraction efficiency of reflective metallic gratings operating in the THz range...190 **María Manuela Fernández**, *INTA*
P6.8	"Circularly Polarized Dielectric Resonator Antenna for the Terahertz Band Applications"...192 **Wael Jaafar**, *Carleton University*
P6.9	"Design and Analysis of a Spline-Profile Diagonal Horn Antenna at 104GHz for Microwave Holography"...193 **Daniele Ronso Lima**, *Kapteyn Astronomical Institute*
P7. CRYOGENIC RECEIVERS, MIXERS & BACK-ENDS	
P7.1	"Design of a 350 GHz Circular Waveguide Superconductor-Insulator-Superconductor Mixer for Array Applications"...194 **Boon Kok Tan**, *University of Oxford*
P7.2	"Design of RF waveguide structure for 2SB SIS mixer at 210-280 GHz"...200 **Sabrina Realini**, *Kapteyn Astronomical Institute, University of Groningen*
P7.3	"Towards 100% array yield: understanding (and fixing) the causes of KID array inhomogeneity"...N/A **Eduard Driessen**, *IRAM*
P7.4	"Design and Performance of the Terahertz Photon Counting System: Detectors and Cryogenics"...202 **Hajime Ezawa**, *National Astronomical Observatory of Japan*
P7.5	"Enhanced sensitivity of THz NbN hot electron bolometer mixers"...203 **Behnam Mirzaei**, *Delft University of Technology*
P7.6	"10.7 THz HEB heterodyne mixer designs"...205 **Johanna Böhm**, *University of Cologne*

IMPORTANT NOTE:

Late abstract submissions are still welcome for poster presentations.

*Please visit **isstt2022.com** to submit an abstract*

ADDITIONAL PAPER	

32ND INTERNATIONAL SYMPOSIUM ON SPACE TERAHERTZ TECHNOLOGY

ISSTT 2022

BAEZA

¡MUCHO ESPACIO!

OCTOBER 16 – 20, 2022

www.isstt2022.com

España · Ayuntamiento de BAEZA · Baeza PATRIMONIO MUNDIAL · Andalucía · Junta de Andalucía · Diputación de Jaén · Ciudades Patrimonio de la Humanidad España UNESCO · Universidad Internacional de Andalucía · IEEE Advancing Technology for Humanity · MTT-S

VDI Virginia Diodes, Inc · GENERA TECNOLOGIAS · Radiometer Physics A Rohde & Schwarz Company · ROHDE&SCHWARZ Make ideas real · ACST Technology Solutions for Terahertz Electronics · TK SPACE · TK INSTRUMENTS · Pacific MicroCHIP · Grupo oleícola jaén

ISSTT 2022
Baeza
ESPAÑA
Ayuntamiento de BAEZA
Andalucía
Junta de Andalucía
DIPUTACIÓN DE JAÉN
@isstt2022
@isstt2022
isstt2021
Baeza PATRIMONIOMUNDIAL
Ciudades Patrimonio de la Humanidad
ESPAÑA UNESCO
www.isstt2022.com

Baeza
PATRIMONIOMUNDIAL
www.baeza.es

THE CITY OF BAEZA WELCOMES ISSTT 2022

Located on a hilltop, Baeza offers one of the most singular landscapes in Andalusia: its magnificent 'sea of olive trees'. Its particular location also allows its visitors to appreciate Baeza's beauty even from far away, with its particular skyline dominated by the main tower of its monumental Cathedral.

Baeza is a city that is proud of its past, its present and its future. A charming city that not only perfectly demonstrates the ability of small cities to host top international workshops and conferences, but it also serves as a perfect travel destination due to its 'hidden gems' and exceptional buildings and landscapes.

It is a great pleasure to welcome all the participants of the 32nd International Symposium of Space Terahertz Technology to our city. As Mayor of Baeza, I am immensely proud and honored that our city was selected to host this year's ISSTT symposium, definitely one of the most important scientific meetings in the field for far-infrared space technologies."

María Dolores Marín Torres

Mayor of Baeza

ISSTT 2022 LOCAL ORGANIZING COMMITTEE WELCOME

"On behalf of the Local Organizing Committee, the City of Baeza, and the Tourism Office of Spain it is an immense honor to welcome all of you to Spain, for the first time in more than 30 years of ISSTT, and in particular to my hometown, Baeza. A city full of Renaissance buildings and cobbled streets that you will be surprised to see in the south, and one of the 15 UNESCO World Heritage cities in Spain. Baeza is *definitely a city full of art, culture and diversity, and thus a perfect place to welcome science and technology innovation. All these together represent indeed the best heritage of the human kind. This will be also the first conference after the Covid-19 pandemic, which made the organization more challenging but also more special. We are looking forward to seeing again so many colleagues after all this long and difficult time. We are convinced ISSTT 2022 will be a very unique conference."*

Jose V. Siles, ISSTT 2022 General chair

Local Organizing Committee:

Jose V. Siles (General Chair), Goutam Chattopadhyay, Imran Mehdi, Alain Maestrini, Boris Karasik, and Cecile Jung-Kubiak (from NASA Jet Propulsion Laboratory); F. Javier Rodríguez Mañas (from Tourism Office of Spain in Los Angeles, Spanish Ministry of Tourism); Lola Marín Torres and Ana Rodríguez (from Baeza City Hall).

ORGANIZATION PARTNERS & SPONSORS

The Organizing Committee of the 32nd International Symposium on Space Terahertz Technology (ISSTT 2022) would like to thank the following institutions and companies for they valued support to make this event happen.

In particular, we would like to express our deepest gratitude to the **Ayuntamiento de Baeza (Baeza City Hall)** and to the **International University of Andalusia (UNIA)** for making available all the conference venues at no cost, including free lodging at the campus for more than 30 students attending the conference. Special thanks also to **the Tourism Office of Spain in Los Angeles** for all their support from the very first day, including not only financial support, but also an incredible institutional support that made it possible to bring the ISSTT conference to Spain for the first time ever.

Our immense gratitude as well to the **Andalusia Tourism Board (Ministry of Tourism, Culture and Sports of the Government of Andalusia – Junta de Andalucía) and the cities of Granada and Córdoba**, for sponsoring the visit to Granada and the Alhambra (including a very special reception and evening private visit to the Alhambra palaces), as well as the visit to the city of Córdoba; to the **UNESCO Spanish World Heritage Cities Group (Grupo Ciudades Patrimonio de la Humanidad de España)**, for sponsoring the Welcome Reception; and to **the Government of Jaén (Diputación de Jaén)**, for co-sponsoring the Conference Banquet together with the **Tourism Office of Spain**.

Thanks to our Diamond sponsors for their contribution to the ISSTT 2022 Student Competition Award and the "Flamenco under the stars" performance.

Special mention as well to our technical sponsors: the Institute of Electrical & Electronics Engineers (IEEE) and its Microwave Theory and Techniques Society (MTT-S). Our partnership with IEEE MTT-S is critical to increase even more the value of the symposium.

Last but not least, thanks to our all sponsors (Diamond, Platinum and Gold) and institutional partners, we have been able to provide around 30 full student travel grants, including waived registration fees, free lodging and free access to all social activities for the conference. To our knowledge, this is the first time that travel grants are offered for students at ISTTT. Because of this initiative we have increased the student participation at the ISSTT to more than a 30% of the total number of attendees. This was the main focus of this year's ISSTT: to create a symposium much more attractive and accessible to students and young professionals. Together, we have turned this into a reality!

Thank you!

MAIN ORGANIZERS

ESPAÑA

Ayuntamiento de BAEZA

Baeza PATRIMONIOMUNDIAL

TECHNICAL CO-SPONSORS

IEEE Advancing Technology for Humanity

MTT-S IEEE MICROWAVE THEORY & TECHNOLOGY SOCIETY

ORGANIZATION PARTNERS

Andalucía

Junta de Andalucía

DIPUTACIÓN DE JAÉN

Ciudades Patrimonio de la Humanidad ESPAÑA | UNESCO

unia Universidad Internacional de Andalucía

DIAMOND SPONSORS

VDi Virginia Diodes, Inc.

GENERA TECNOLOGIAS

PLATINUM SPONSORS

ROHDE&SCHWARZ

RPG Radiometer Physics A Rohde & Schwarz Company

TK INSTRUMENTS TK SPACE

ACST Technology Solutions for Terahertz Electronics

GOLD SPONSORS

Pacific MicroCHIP Corp. IC/ASIC Products

GRUPO oleícola jaén

SOCIAL VISITS TO GRANADA AND CÓRDOBA

Andalucía

Ayuntamiento de Granada

AYUNTAMIENTO DE CÓRDOBA

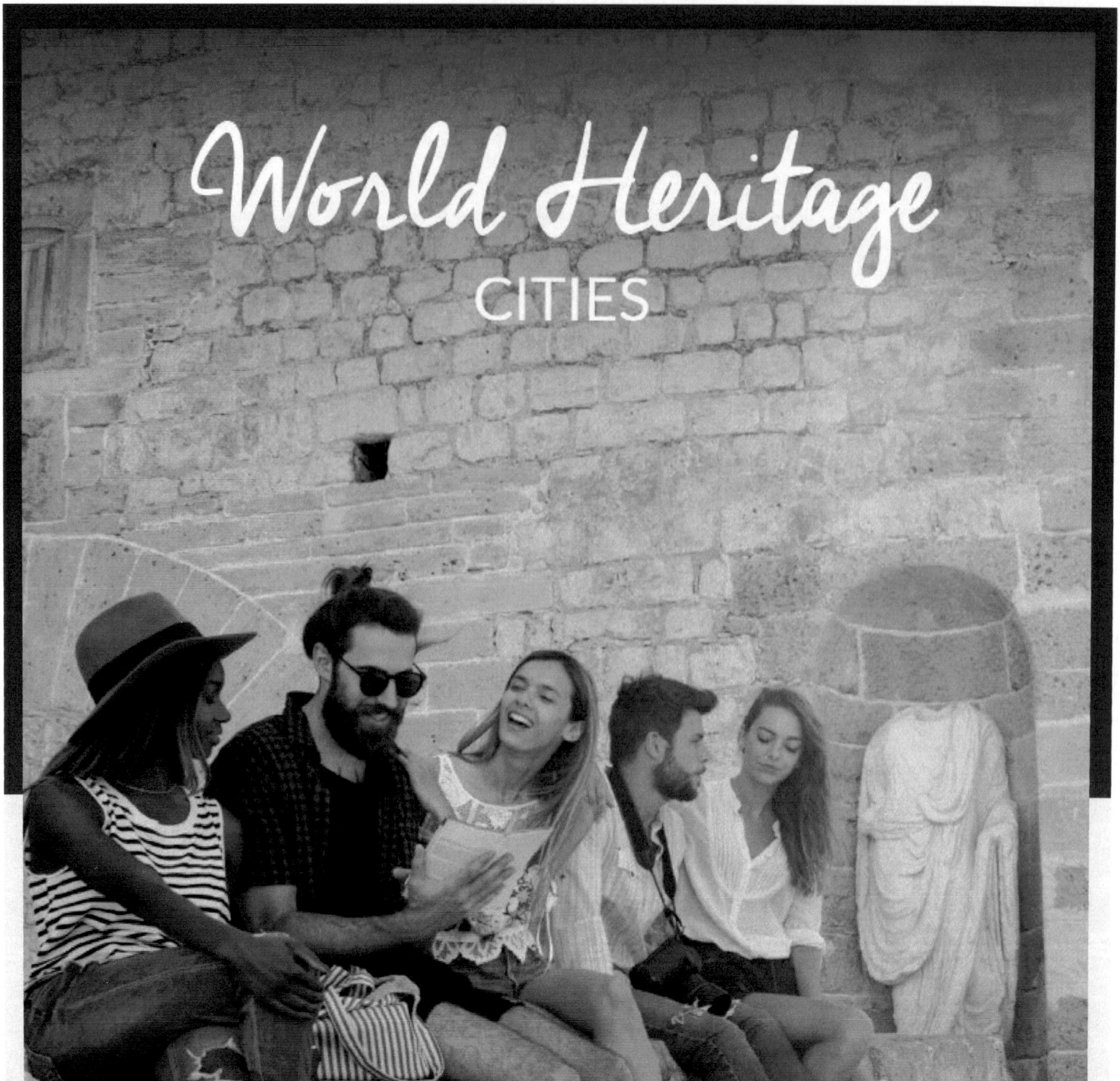
World Heritage
CITIES

ISSTT 2022 SCIENTIFIC ORGANIZING COMMITTEE

The Scientific Organizing Committee (SOC) was responsible for reviewing the abstracts and making recommendations on acceptance and format of presentation for each abstract. This Committee will also decide on the place and dates for the next ISSTT.

Victor Belitsky
Chalmers University of Technology
Sweden

Andrey Baryshev
University of Groningen
The Netherlands

Brian Ellison
Rutherford Appleton Laboratory
United Kingdom

Jian-Rong Gao
SRON
The Netherlands

Gregory Goltsman
Moscow State Pedagogical University
Russia

Christopher Groppi
Arizona State University
USA

Jeffrey Hesler
Virginia Diodes Inc.
USA

Heinz-Wilhelm Hübers
DLR
Germany

Boris Karasik
Jet Propulsion Laboratory
USA

Valery Koshelets
IRE RAS
Russia

Alain Maestrini
Paris Observatory
France

Hiroshi Matsuo
NAOJ
Japan

Imran Mehdi
NASA Jet Propulsion Laboratory
USA

Patricio Mena
University of Chile
Chile

Netty Honingh
University of Cologne
Germany

Christophe Risacher
IRAM
France

Sheng-Cai Shi
Purple Mountain Observatory
China

José V. Siles
NASA Jet Propulsion Laboratory
USA

Jan Stake
Chalmers University of Technology
Sweden

Edward Tong
Smithsonian Astrophysics Observatory
USA

Yoshinori Uzawa
NAOJ
Japan

Ghassan Yassin
University of Oxford
United Kingdom

32nd IEEE International Symposium on Space Terahertz Technology (ISSTT 2022)

October 16-20, 2022, Baeza, Spain

Technical Program

Sunday, October 16, 2022

TIME	MINS	Sunday, October 16, 2022
8:00		
9:00		
10:00	0	Shuttle Madrid Airport - Baeza (transit time ~4h00min)
11:00		
12:00	0	Shuttle Madrid Airport - Baeza (transit time ~4h00min)
13:00		
14:00		
15:00		Shuttle Madrid Airport - Baeza (transit time ~4h00min)
16:00		
17:00	30	Baeza Historic City Center Visit
18:00	0	
19:00		Welcome Reception & Conference Registration *Baeza City Hall*
20:00		
21:00		
22:00		
23:00		

Monday, October 17, 2022

TIME	MINS	Monday, October 17, 2022
8:00	15	Coffee & Registration (30 mins)
	45	Opening Remarks
9:00	15	**Invited Talk I:** Dr. Kartik Seth
	50	**Session I:** Instruments & Systems for Future Missions I *Auditorio Ruinas de San Francisco*
10:00	0	
11:00	10	Coffee Break (30 mins)
	40	**Session II:** Student Competition Finalists *Auditorio Ruinas de San Francisco*
12:00		
13:00	20	Diamond Sponsors Session
	40	Lunch Break (80 mins)
14:00		
15:00	0	**Session III:** THz Sources I *Auditorio Ruinas de San Francisco*
16:00	20	Coffee Break
	50	**Session IV:** Solid-State Receivers *Auditorio Ruinas de San Francisco*
17:00	50	
18:00	15	**Plenary Talk:** Dr. John Mather, 2006 Nobel Prize in Physics *Teatro Montemar*
19:00		
20:00	0	Ruta de la Tapa (Scavenger Tapas Night)
21:00		
22:00	0	A Night of Magic with **Javi Benitez, 2018 FISM World Champion** *Café Bar Central*
23:00		

Tuesday, October 18, 2022

TIME	MINS	Tuesday, October 18, 2022
8:00	30	Coffee & Registration (30 mins)
	0	Roundtable: Women in Space Teraherz
9:00	30	Platinum Sponsors Session
	50	**Session V:** Novel Devices, Technologies & Components *Auditorio Ruinas de San Francisco*
10:00		
11:00	10	Coffee Break (30 mins)
	40	
	0	**Session VI:** Instrument & Systems for Ground Telescopes *Auditorio Ruinas de San Francisco*
12:00		
13:00	20	Lunch Break / LOC & SOC Lunch (100 mins)
	40	
14:00		
15:00	0	**Session VII:** THz Antennas & Optical Components *Auditorio Ruinas de San Francisco*
16:00	20	Coffee Break
	50	**Session VIII:** Cryogenic Receivers, Mixers and Backends *Auditorio Ruinas de San Francisco*
17:00		
18:00	30	Baeza Historic City Center Visit Olive Oil Museum Visit
19:00	45	Flamenco under the stars with **Arcángel, 2018 Latin Grammy Award Winner** *Plaza de Santa María, Baeza Cathedral*
20:00		
21:00	0	Cocktail & Conference Banquet *Hotel Puerta de la Luna*
22:00		
23:00		

Wednesday, October 19, 2022

TIME	MINS	Wednesday, October 19, 2022
	0	
8:00		
	0	**Poster Session** *Palacio de Jabalquinto*
9:00	30	Coffee Break (30 mins)
10:00		**Poster Session** *Palacio de Jabalquinto*
11:00	15	
	30	**Session IX:** Measurement Techniques *Auditorio Ruinas de San Francisco*
12:00		
13:00	10	Invited Talk III: Dr. Paul Hartogh
	40	Lunch Break
14:00		
15:00	0	**Session X:** Far-Infrared Balloon-Borne Missions *Auditorio Ruinas de San Francisco*
16:00	20	Coffee Break (30 mins)
	50	**Session XI:** THz Sources II *Auditorio Ruinas de San Francisco*
17:00	50	**Session XII:** Instruments & Systems for Future Missions II *Auditorio Ruinas de San Francisco*
18:00	50	Closing Remarks / Student Award
19:00	30	**Outreach Talk:** Dr. Paul Goldsmith, NASA/JPL *Teatro Montemar*
20:00		
21:00		
22:00	30	Typical Spanish Late Social *Burladero de Copas / Cafe Bar Central*
23:00		

Thursday, October 20, 2022

TIME	MINS	Thursday, October 20, 2022
8:00	30	
9:00		Departure to Granada (transit time ~1h30min)
10:00		
11:00	30	Visit to Graanda
	0	
12:00		
13:00		Invited Talk III: Dr. Antxon Alberdi
	0	Special lunch/reception offered by Andalusian region *Carmen de los Martires*
14:00		
15:00	30	Visit to the Alhambra Palace & Generalife Gardens
16:00		
17:00		
18:00	0	
19:00		Departure to Baeza (transit time ~1h30min)
20:00		
21:00		
22:00		Typical Spanish Late Social *Burladero de Copas / Cafe Bar Central*
23:00		

32nd IEEE International Symposium on Space Terahertz Technology (ISSTT 2022)

October 16-20, 2022, Baeza, Spain

Conference Proceedings

The 32nd International Symposium on Space Terahertz Technology (ISSTT 2022) was held in Spain for the first time in more than 30 years of history of ISSTT. Baeza is a city full of Renaissance buildings and cobbled streets that you will be surprised to see in the south, and one of the 15 UNESCO World Heritage cities in Spain. Baeza is definitely a city full of art, culture and diversity, and thus a perfect place to welcome science and technology innovation. All these together represent indeed the best heritage of the human kind. This was also the first conference after the Covid-19 pandemic, which made the organization more challenging but also more special. Many colleagues in the filed were able to connect in person again after such long and difficult times. ISSTT 2022 was indeed a unique conference.

In terms of numbers, 126 submissions were accepted for presentation, 3 of which will be Invited Presentations, 2 Plenary Talks, 44 regular contributed Oral Presentations, 10 short Oral Presentations for the Student Competition Finalists, and 67 contributed Poster Presentations (+10 additional poster presentations also from Student Competition Finalists). The review process was carried out by SOC members with each abstract been reviewed by at least three reviewers. For the very first time at ISSTT, the invited plenary talks were open to the general public, including local high schools students, with the goal to get cutting edge science and technology closer to the general public and thus help inspiring new generations to pursue careers in STEM fields. Two very special open plenary talks were presented by Nobel Prize Winner Dr. John Mather from NASA on the discoveries made by the James Webb Space Telescope, and by Dr. Paul Goldsmith of Jet Propulsion Laboratory on the legacy of the Arecibo Radio Observatory in Radio Astronomy. A student competition was also held to identify and recognize outstanding technical contributions from individual students. 10 contributions were selected as finalists (see table below) and had the opportunity to do a ~8 min oral presentation during the conference. Opening remarks were given by the Director of the recently created Spanish Space Agency, Dr. Miguel Belló.

The Local Organizing Committee wanted this conference to also have a very important social agenda to favor interactions and networking between all the attendees after all these difficult times as well as to keep promoting the Andalusian culture. Visits to nearby and historic cities of Córdoba and Granada were offered to the participant and guests, as well as a magic flamenco performance under the stars by Latin Grammy Award Winner "Arcángel" and renowned flamenco dancer "Macarena López", followed by the conference banquet in a historical venue, more than 500 years old. And talking about magic, a network social event was also held for the attendees the first day of the conference featuring a special performance by one of the best magicians in the world: "Javi Benitez", also originary from Andalusia.

The Organizing Committee of the 32nd International Symposium on Space Terahertz Technology (ISSTT 2022) would like to thank the following institutions and companies for they valued support to make this event happen.

In particular, we would like to express our deepest gratitude to the Ayuntamiento de Baeza (Baeza City Hall) and to the International University of Andalusia (UNIA) for making available all the conference venues at no cost, including free lodging at the campus for all students attending the conference. Special thanks also to the Tourism Office of Spain in Los Angeles for all their support from the very first day, including not only financial support, but also an incredible institutional support that made it possible to bring the ISSTT conference to Spain for the first time ever. Our immense gratitude as well to the Andalusia Tourism Board (Ministry of Tourism, Culture and Sports of the Government of Andalusia – Junta de Andalucía) and the Cities of Córdoba and Granada, for sponsoring the visit to Granada and the Alhambra (including a very special reception and evening private visit to the Alhambra palaces), and the visit to Córdoba; to the UNESCO Spanish World Heritage Cities Group (Grupo Ciudades Patrimonio de la Humanidad de España), for sponsoring the Welcome Reception; and to the Government of Jaén (Diputación de Jaén),for co-sponsoring the Conference Banquet together with the Tourism Office of Spain. Thanks to our Diamond sponsors (Virginia Diodes and Genera Tecnologías) for their contribution to the ISSTT 2022 Student Competition Award and the "Flamenco under the stars" performance. Special mention as well to our technical sponsors: the Institute of Electrical & Electronics Engineers (IEEE) and its Microwave Theory and Techniques Society (MTT-S). Our partnership with IEEE MTT-S is critical to increase even more the value of the symposium. Last but not least, thanks to our all sponsors (Virginia Diodes, Genera Tecnologías, Rohde & Schwarz, Radiometer & Physics, Thomas Keating Instruments & Space, ACST, Pacific Microchip Corp. and Oleicola Jaén) and institutional partners (Tourspain, Baeza City Hall, Diputación de Jaén and International University of Andalusia – Baeza Campus), we were able to provide around 40 student travel grants, including waived registration fees, free lodging and free access to all social activities for the conference. To our knowledge, this is the first time that travel grants are offered for students at ISTTT. Because of this initiative we have increased the student participation at the ISSTT to more than a 30% of the total number of attendees. This was the main focus of this year's ISSTT: to create a symposium much more attractive and accessible to students and young professionals. Together, we have turned this into a reality!

LIST OF PARTICIPANTS (as of September 30, 2022)

Last Name	First Name	Institution	E-mail
Ahmad	Sarosh	Universidad Carlos III de Madrid	asarosh@ing.uc3m.es
Al Qaraghuli	Ali	Northeastern University	alqaraghuli.a@northeastern.edu
Albers	Roland	Bern University	roland.albers@unibe.ch
Ali	Muhsin	Universidad Carlos III de Madrid	muali@ing.uc3m.es
Aliaga	Sergi	Northeastern University	sergi.aliaga@gmail.com
Alibakhshikenari	Mohammad	Universidad Carlos III de Madrid, Spain	mohammad.alibakhshikenari@uc3m.es
Amils Sabalot	Ricardo I.	Yebes Observatory	ricardo.amils@uah.es
Auriacombe	Olivier	Omnisys Instruments	olivier.auriacombe@aac-clydespace.com
Ballew	Conner	NASA Jet Propulsion Laboratory	mrconnerballew@gmail.com
Baranauskas	Dalius	Pacific Microchip Corp.	dalius@pacificmicrochip.com
Barkhof	Jan	NOVA	j.barkhof@sron.nl
Barrueto	Ignacio	University of Cologne	barrueto@ph1.uni-koeln.de
Baryshev	Andrey	Kapteyn Astronomical Institute /NOVA	andrey@astro.rug.nl
Bekema	Marielle	NOVA	m.e.bekema@rug.nl
Belitsky	Victor	Chalmers University of Technology	victor.belitsky@chalmers.se
Benson	Chris	University of Lethbridge	chris.benson@uleth.ca
Böhm	Johanna	University of Cologne	boehm@ph1.uni-koeln.de
Cámara Mayorga	Ivan	Max-Planck-Institute for Radio Astronomy	imayorga@mpifr.de
Cardenas	Jorge	Universidad de Antioquia	jorge.cardenas3@udea.edu.co
Carrasco	Javier	University of Chile	javier.carrasco@ug.uchile.cl
César Cuello	Jessica	University Carlos III de Madrid	jecesarc@ing.uc3m.es
Chattopadhyay	Goutam	NASA Jet Propulsion Laboratory	hi.goutam@gmail.com
Chen	Pai-Yen	University of Illinois Chicago	pychen@uic.edu
Cojocari	Oleg	ACST GmbH	oleg.cojocari@acst.de
Crowe	Thomas	Virginia Diodes	crowe@vadiodes.com
Curwen	Christopher	NASA Jet Propulsion Laboratory	chris.a.curwen@jpl.nasa.gov
Desmaris	Vincent	Chalmers University of Technology	vincent.desmaris@chalmers.se
Diez-González	Cármen	Observatorio de Yebes, CDT (IGN)	m.diez@oan.es
Drakinskiy	Vladimir	Chalmers University of Technology	vladimir.drakinskiy@chalmers.se
Driessen	Eduard	IRAM	driessen@iram.fr
Elmaleh	Coralie	Université du Littoral Côte d'Opale	coralie.elmaleh@univ-littoral.fr
Ezawa	Hajime	National Astronomical Observatory of Japan	h.ezawa@nao.ac.jp
Fernández R.	María Manuela	INTA	fernandezrm@inta.es
Gallego Puyol	Juan Daniel	Observatorio de Yebes, CDT (IGN)	jd.gallego@oan.es
Gao	Wenqi	University of Electronic Science and Technology of China	yeleiyao@uestc.edu.cn
Gao	Jian Rong	SRON/TU Delft	j.r.gao@sron.nl
García	Carlos	Rohde & Schwarz	carlos.garcia@rohde-schwarz.com
García Muñoz	Luis Enrique	Universidad Carlos III de Madrid	legarcia@ing.uc3m.es
Gatilova	Lina	Observatoire de Paris	lina.gatilova@obspm.fr
Gibson	Hugh	Gibson Microwave Design EURL	hughrpg@gmail.com
Goldsmith	Paul	NASA Jet Propulsion Laboratory	paul.f.goldsmith@jpl.nasa.gov
González	Alvaro	National Astronomical Observatory of Japan	alvaro.gonzalez@nao.ac.jp
González	Luis M.	INTA	gonzalezfl@inta.es
Groppi	Christopher	Arizona State University	cgroppi@asu.edu
Grutzeck	Gerrit	Max-Planck-Institue for Radioastronomy	ggrutzeck@mpifr-bonn.mpg.de
Hartogh	Paul	Max Planck Institute for Solar System Research (MPS)	hartogh@mps.mpg.de
Hawkins	Bert	NRAO	bhawkins@nrao.edu
Henke	Doug	NRC Herzberg Astronomy and Astrophysics	doug.henke@nrc-cnrc.gc.ca
Hesler	Jeffrey	Virginia Diodes, Inc.	hesler@vadiodes.com
Hesper	Ronald	University of Groningen	r.hesper@astro.rug.nl
Hoh	Jonathan	Arizona State University	jrhoh@asu.edu
Honingh	Netty	Universtiy Cologne	honingh@ph1.uni-koeln.de
Horton	Paul	Arizona State University	pahorton@asu.edu
Hübers	Heinz-Wilhelm	German Aerospace Center (DLR)	heinz-wilhelm.huebers@dlr.de
Imada	Hiroaki	National Astronomical Observatory of Japan	hiroaki.imada@nao.ac.jp
Janssen	Reinier	NASA Jet Propulsion Laboratory	reinier.m.janssen@jpl.nasa.gov
Jayasankar	Divya	Chalmers University of Technology	divyaj@chalmers.se
Jeffers	Sandra	Max-Planck Institute for Solar System Research	jeffers@mps.mpg.de
Jellema	Willem	SRON	w.jellema@sron.nl
Jung-Kubiak	Cecile	NASA Jet Propulsion Laboratory	cecile.d.jung@jpl.nasa.gov
Kaneko	Keiko	NAOJ	keiko.kaneko@nao.ac.jp
Karasik	Boris	NASA Jet Propulsion Laboratory	boris.s.karasik@jpl.nasa.gov
Khalatpour	Ali	Massachusetts Institute of Technology	akhalat@mit.edu
Khan	Fedor	Russian Academy of Sciences	khanfv@hitech.cplire.ru
Khanal	Subash	NASA Jet Propulsion Laboratory	subash.khanal@jpl.nasa.gov
Khudchenko	Andrey	Lebedev Physical Institute RAS	khudchenko@asc.rssi.ru
Klein	Bernd	Max Planck Institute for Radio Astronomy	bklein@mpifr-bonn.mpg.de
Kojima	Takafumi	National Astronomical Observatory of Japan	t.kojima@nao.ac.jp
Kotiranta	Mikko	University of Bern	mikko.kotiranta@unibe.ch
Krieg	Jean Michel	Observatoire de Paris	jean-michel.krieg@obspm.fr
Laín Rubio	Verónica	ACST GmbH	veronica.lain-rubio@acst.de

Lambert	Joseph	National Radio Astronomy Observatory	jlambert@nrao.edu
Lampin	Jean-Francois	IEMN-CNRS	jean-francois.lampin@univ-lille.fr
Lapkin	Igor	Chalmers University of Technology	alpa@chalmers.se
Linden	Emily	Arizona State University	enlinden@asu.edu
Longden	Joseph	Uniersity of Oxford	joeylongden@hotmail.com
Lopez	Cristian	Chalmers University of Technology	cristian.lopez@chalmers.se
López Fernández	Isaac	Observatorio de Yebes, CDT (IGN)	i.lopez@oan.es
Maestrini	Alain	NASA Jet Propulsion Laboratory	alain.e.maestrini@jpl.nasa.gov
Mahdizadeh	Sajjad	University of Cologne	mahdizadeh@ph1.uni-koeln.de
Mahieu	Sylvain	IRAM	mahieu@iram.fr
Malo-Gómez	Inmaculada	Observatorio de Yebes, CDT (IGN)	i.malo@oan.es
Martín	Germán	Rohde & Schwarz	german.martin@rohde-schwarz.com
Martinez Gil	Javier	ACST GmbH	javier.martinez-gil@acst.de
Massingill	Kyle	Arizona State University	kyle.massingill@asu.edu
Masui	Sho	Osaka Metropolitane University/NAOJ	sho.masui@grad.nao.ac.jp
Matsuo	Hiroshi	National Astronomical Observatory of Japan	h.matsuo@nao.ac.jp
Mebarki	Mohamed Aniss	Chalmers University of Technology	aniss@chalmers.se
Mehdi	Imran	NASA Jet Propulsion Laboratory	imran.mehdi@jpl.nasa.gov
Mena	Patricio	NRAO	fpmena@gmail.com
Mercant	Guillermo	INTA	gmerrub@inta.es
Mirzaei	Behnam	Delft University of Technology	b.mirzaei@tudelft.nl
Monasterio	David	University of Chile	david.monasterio@raig.uchile.cl
Mondal	Priyanka	Lytid	priyanka.mondal@obspm.fr
Moro Melgar	Diego	ACST GmbH	diego.moro-melgar@acst.de
Moseley	Paul	European Space Agency	paul.moseley@esa.int
Murk	Axel	University of Bern	axel.murk@unibe.ch
Navarrini	Alessandro	NRAO	anavarri@nrao.edu
Navarro Montilla	Javier	University of Oxford	javier.navarro@physics.ox.ac.uk
Neric	Marko	Arizona State University	mneric@asu.edu
Niwa	Ayako	University of Tsukuba	s2230063@s.tsukuba.ac.jp
Orfao	Beatriz	University of Salamanca	beatrizorfao@usal.es
Pavolotsky	Alexey	Chalmers University of Technology	alexey.pavolotsky@chalmers.se
Pérez Santos	Susana	University of Salamanca	susana@usal.es
Perez-Bosch	Francisco	GENERA Soluciones Tecnológicas, S.L.	fperez@generatecnologias.es
Puf	Teodora	VivaTech TeraHertz SAS	david.vizard@vivatechthz.com
Pütz	Patrick	Max Planck Institute for Radio Astronomy	ppuetz@mpifr-bonn.mpg.de
Realini	Sabrina	University ofGroningen	realini@astro.rug.nl
Reck	Theodore	Virginia Diodes	theodore.reck@vadiodes.com
Ren	Xiadong	University of Cologne	ren@ph1.uni-koeln.de
Reyes	Nicolas	Max Planck Institute for Radioastronomy	nireyes@mpifr-bonn.mpg.de
Richter	Heiko	German Aerospace Center (DLR)	heiko.richter@dlr.de
Rodenhuis	Michiel	NOVA Netherlands Research School for Astronomy	rodenhuis@strw.leidenuniv.nl
Rodriguez Monje	Raquel	NASA Jet Propulsion Laboratory	rmonje@jpl.nasa.gov
Ronso Lima	Daniele	Kapteyn Astronomical Institute	daniele.ronso@usp.br
Saeid	Talia	Arizona State University	tsaeid@asu.edu
Sakai	Ryo	National Astronomical Observatory of Japan	ryo.sakai@nao.ac.jp
Scott	Jeremy	University of Lethbridge	jeremy.scott@uleth.ca
Shan	Wenlei	National Astronomical Observatory of Japan	wenlei.shan@nao.ac.jp
Sheth	Kartik	NASA HQ	kartik.sheth@nasa.gov
Shin	Naeun	Seoul National University / KASI	neshin@kasi.re.kr
Siles	Jose V.	NASA Jet Propulsion Laboratory	jose.v.siles@jpl.nasa.gov
Silva	Jose	SRON	j.r.g.d.silva@sron.nl
Sinclair	Adrian	University of British Columbia	adriankaisinclair@gmail.com
Sirsi	Siddhartha	University of Arizona	ssirsi@arizona.edu
Spencer	Locke	University of Lethbridge	locke.spencer@uleth.ca
Stake	Jan	Chalmers University of Technology	jan.stake@chalmers.se
Tan	Boon Kok	University of Oxford	boonkok.tan@physics.ox.ac.uk
Thomas	Bertrand	Radiometer physics GmbH	bertrand.thomas@radiometer-physics.de
Thuroczy	Tomas	IETR UMR CNRS 6164	tomas.thuroczy@univ-rennes1.fr
María del Carmen	Torquemada	Instituto Nacional de Tecnología Aeropespacial	torquemadavm@inta.es
Türk	Sener	Max Planck Institute for Radio Astronomy	stuerk@mpifr-bonn.mpg.de
Valentin	Jérôme	Observatoire de Paris	jerome.valentin@obspm.fr
Van Berkel	Sven	NASA JetPropulsion Laboratory	sven.l.van.berkel@jpl.nasa.gov
Veenendaal	Ian	Cardiff University / SRON	veenendaali@cardiff.ac.uk
Vizard	David	VivaTech TeraHertz SAS	david.vizard@vivatechthz.com
Watkins	Ryan	NASA Jet Propulsion Laboratory	ryan.t.watkins@jpl.nasa.gov
Wenninger	Jakob	University of Oxford	jakob.wenninger@physics.ox.ac.uk
Whitton	Cassandra	Arizona State University	cnwhitto@asu.edu
Wiedner	Martina	Observatoire de Paris	martina.wiedner@obspm.fr
Wienold	Martin	German Aerospace Center (DLR)	martin.wienold@dlr.de
Wild	Wolfgang	CTAO GmbH	wolfgang.wild@cta-observatory.org
Wylde	Richard	Thomas Keating Ltd	r.wylde@terahertz.co.uk
Yagoubov	Pavel	ESO	pyagoubo@eso.org
Yamasaki	Yasumasa	Osaka Metropolitan University/NAOJ	yasumasa.yamasaki@grad.nao.ac.jp
Yates	Stephen	SRON	s.yates@sron.nl
Yoo	Changyun	University of California Santa Barbara	cyoo@ucsb.edu

A Direct Up-Conversion G-band Radar Prototype for Atmospheric Measurements

Raquel Rodriguez Monje*[1], Ken B. Cooper[2], Robert Beauchamp[1], Srinivas Prasad Mysore Nagaraja[1] and Jose V. Siles[1]

Abstract— **A 240 GHz radar is being built as a part of a multifrequency, compact, low-cost atmospheric radar prototype, called CloudCube. CloudCube instrument consists of three-frequency radar (Ka/W/G-band) modules and uses a radar architecture where the baseband signal is directly upconverted to the RF band without any addition of intermediate frequencies or further multiplication scheme. This architecture, combined with pulse compression techniques, provides a simple and effective solution while achieving the required performance and robustness of an airborne or space flight radar instrument. We present the CloudCube's G-band channel breadboard radar concept and the first-light outdoor observations over clear sky and rain.**

Keywords— Meteorological radar, millimeter wave radar, millimeter wave circuits

I. Introduction

Low-cost orbital radars, compatible with small satellite platforms, are crucial to enable new Earth observing systems capable of providing time evolution observations of atmospheric processes at high resolution to improve the weather and climate models. With RainCube [1] — a Ka-band precipitation radar in a 6U CubeSat — the Jet Propulsion Laboratory demonstrated that disruptive innovation in the areas of deployable antennas, pulse compression, and frequency-modulated waveform generation using offset I/Q up conversion make it possible to build very compact Ka-band radars for CubeSats. Building off the RainCube's technical success, we are developing a new atmospheric multifrequency radar, named CloudCube. CloudCube is intended to be the first, compact, low-cost, pulsed-compression radar that combines three frequency channels Ka-, W- and G-band (35/94/239 GHz) for vertical profiling of clouds, convection and precipitation structures and dynamics from space. Pulse-compression techniques allow the transmission of long-duration, low-power signals that achieve the same total pulse energy as traditional higher-power pulse radar systems. Pulse-compression radars can be built using solid-state technology enabling a compact and high energy efficient radar, providing high quality observations comparable to the current high-power pulsed millimeter-wave cloud and precipitation radars.

[1]Jet Propulsion Laboratory, California Institute of Technology, Pasadena, 91109,

II. Radar Breadboard and Results

CloudCube's G-band breadboard includes several in-house designed components such as the InP LNAs and the 240 GHz high-power handling frequency tripler, as well as the G-band fundamental mixer. For transmit power, a solid-state source based on GaAs Schottky diode frequency-triplers is used [2].

Figure 1 shows the radar's first-light measurements- a range spectra in the presence of rain and no-rain. The radar is operated on a FMCW (Frequency-Modulated Continuous-Wave). The results of Figure 1 demonstrate that CloudCube's first G-band breadboard can successfully capture volumetric scattering targets up to ranges of 2 km.

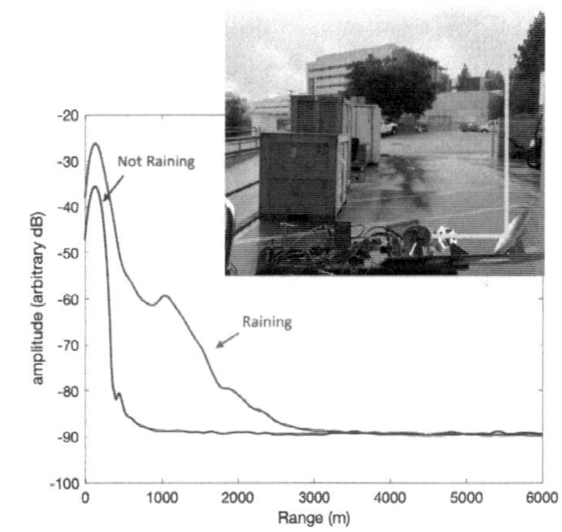

Fig. 3. Range-compressed spectra of the radar in the presence of rain (blue) and without rain (red). Top-right, CloudCube's G-band radar breadboard positioned 60° off zenith outdoors.

References

[1] E. Peral et al., "RainCube: the first ever radar measurements from a CubeSat in space," J. Appl. Rem. Sens. 13(3) 032504 (11 June 2019) https://doi.org/10.1117/1.JRS.13.032504

[2] J. V. Siles et al., "A New Generation of Room-Temperature Frequency Multiplied Sources with up to 10x Higher Output Power in the 160 GHz - 1.6 THz Range," IEEE Transactions on Terahertz Science and Technology, vol. 8, no. 6, pp. 596-604, 2018

NOTES:

MetOp-SG Ice Cloud Imager 183–664 GHz Front-End Receivers Proto-Flight Model Qualification and Acceptance Test Results

B. Thomas[1], A. Kilian[1], G. Sonnabend[1], M. Brandt[1], M. Trasatti[1], T. Stangier[1,6], P. Krause[1], N. Wehres[1], M. Philipp [1], S. Mohr[1], R. Henneberger[1], M. Schmitz[1], S. Rea[2], D. Cuadrado-Calle[2,5], C. Howe[2], K. Parow-Souchon[2], G. Marschall[2], B. Ellison[2], A. Andrés-Beivide[3], M. Bergada[3], N. Alcaraz[3], J. Martinez[3], P. Robustillo[4], M. Gotsmann[4] and U. Klein[5]

***Abstract*—We present the development of the Proto-Flight Model (PFM) receivers covering the frequency range 183-664 GHz as part of the Ice Cloud Imager (ICI) front-end for the next generation of European polar orbiting Meteorological satellite series (MetOp-SG). State-of-the-art radiometric sensitivity results are reported at room temperature. Qualification results from the underlying semi-conductor GaAs Schottky and MMIC technology will also be presented.**

Keywords—Receiver, Schottky, MMIC, Space, Radiometer.

I. Introduction

The next generation of near-earth polar-orbiting satellites (MetOp-SG) is a high priority mission from ESA/ EUMETSAT to ensure continuous data acquisition for weather forecasting and climate monitoring in 2023-2043 period [1]. The prime contractor Airbus is currently developing a series of satellites with a suite of microwave instruments that are to be launched sequentially every 7-8 years to cover at least 20 years of cumulated observation time. The Ice Cloud Imager (ICI) instrument on-board one of the satellite series (Sat-B) is a multi-channel passive imaging radiometer dedicated to the remote sensing of water and ice clouds in the millimetre to submillimetre range [1].

ICI features seven Front-End (FE) receivers designed to operate close to room temperature (28°C) around five frequencies (183 GHz, 243 GHz, 325 GHz, 448 GHz and 664 GHz), two of them being dual-polarised (243 and 664 GHz). These receivers rely on semi-conductor GaAs Schottky and MMIC technology packaged and qualified by RPG[1] and RAL[2], as well as hybrid technology procured from Narda-Miteq (DRO and IF LNAs).

II. Development Status and Results

A major project milestone was achieved with the successful completion of the Proto-Flight Model (PFM) FE test campaign in 2021 [2] (see Fig.1), followed by delivery to ASE[3] for the next level of integration into the ICI instrument. The FM2 receivers are currently under Assembly, Integration and Test (AIT) at RPG.

An extensive qualification test campaign characterises several stringent requirements such as the noise figure, gain and gain flatness, short-term gain stability (1/f noise), sideband imbalance and spurious response over the qualification temperature range (0°C–60°C). Full functional tests on top of specific EMC, vibration as well as thermal vacuum tests will conclude the PFM verification campaign.

Fig. 1. Photo of the ICI receivers fully assembled, including feed cluster components (middle) and LO generation units (left and right towers) prior to final integration into PFM front-end.

A summary of the state-of-the-art results will be presented. We will discuss the lessons learned from the EQM phase and how this has been applied to improve the PFM and FM2 receivers performance and reliability.

[1] Radiometer Physics GmbH, Werner-von-Siemens-Strasse 4, 53340 Meckenheim, Germany

[2] STFC Rutherford Appleton Laboratory, Harwell Campus, Didcot, OX11 0QX, United Kingdom.

[3] Airbus DS Space Systems España, Avda. de Aragón 404, 28022 Madrid, Spain

[4] Airbus DS GmbH, 88039 Friedrichshafen, Germany

[5] ESTEC, European Space Agency, Keplerlaan 1, 2201 AZ Noordwijk, The Netherlands

[6] Deutsches Zentrum für Luft- und Raumfahrt (DLR), Linder Höhe, 51147 Köln

References

[1] V. Kangas et al., "Ice Cloud Imager instrument for MetOp-SG", proceedings of the MicroRad conference, 24-27 March 2014, DOI: 10.1109/MicroRad.2014.6878946

[2] A. Andres et al. "The ICI Instrument proto-flight receiver chain: results and performance prediction", proceedings of the ARSI 2022 conference, 10 – 12 May 2022, ESA/ESTEC, Noordwijk, The Netherlands.

NOTES:

On-Ground Calibration Targets for the Ice Cloud Imager Instrument on the MetOP Second Generation Satellite

Axel Murk*[1], Mikko Kotiranta[1], Tobias Plüss[1], Marc Bergadà[2], Massimo Labriola[2], Ana Andres-Beivide[2], Jaione Martínez[2], David de Castro[2], Daniel Döring[3], Kevin Pike[4], Richard Wylde[4]

Abstract—**This paper presents the optical and thermal design of the On Ground Calibration Targets for the Ice Cloud Imager Instrument for the MetOP-SG satellites, as well as the first radiometric test results of these targets.**

Keywords—radiometric calibration, blackbody target, submillimeter-wave optics,

I. Introduction

The Ice Cloud Imager (ICI) instrument is currently being developed for the second generation of the European operational meteorological satellites (MetOP-SG) [1]. The conically scanning imager includes seven radiometers with different polarizations and frequency channels between 183 GHz and 664 GHz. The radiometer feed horns are clustered in the focal plane of an off-axis parabolic reflector with a projected diameter of 255 mm. The radiometers and the main reflector are mounted under a sun shield on a rotating platform which spins around the nadir axis at a constant rate of 45 rpm. The stationary part of the instrument includes the On-Board Calibration Target (OBCT) [2] and a Cold Sky Reflector (CSR). They block the field-of-view (FOV) of the feed horns at each rotation to provide a hot and cold scene for the radiometric calibration of the instrument. The Earth's atmosphere will be observed over a rotation angle of ±65°, while angular segments of ±2.5° will be used on each calibration scene.

The radiometric performance of the ICI instrument has been verified on ground in a thermal vacuum chamber. The cold space view is simulated by a Fixed Temperature Target (FTT) blackbody cooled to about 77 K, and the Earth view with a Variable Temperature Target (VTT) that is temperature controlled between 80 K and 335 K. An additional liquid nitrogen (LN2) cold target in an ambient temperature and normal pressure environment (CAT) was used for simpler performance checks. The optical and thermal design of these On Ground Calibration Targets (OGCT) are described together with the radiometric model for ICI in [3, 4].

Fig. 1. ICI on-ground calibration targets during integration.

In this contribution, we report on the radiometric performance of the OGCT system which was achieved during the test campaign with the Protoflight Model (PFM) of the ICI instrument. The telemetry data from the radiometers of the instrument will be compared with model results based on Physical Optics simulations and the measured temperature distribution in the blackbody cavities.

References

[1] Ville Kangas et al., "Ice Cloud Imager Instrument for MetOp Second Generation," 13th Specialist Meeting on Microwave Radiometry and Remote Sensing of the Environment, Mar. 2014

[2] A. Schröder et al, "Electromagnetic Design of Calibration Targets for MetOp-SG Microwave Instruments," IEEE Transactions on Terahertz Science and Technology, 7 (6), 2017

[3] M. Bergada et al, "On-ground radiometric calibration of the Ice Cloud Imager for Metop-SG", 6th Workshop on Advanced RF Sensors and Remote Sensing Instruments (ARSI), ESTEC, Nov. 2019

[4] A. Murk et al, "Optical Design and Electromagnetic Verification of the On-Ground Calibration Target for the Ice Cloud Imager on Metop-SG", 6th Workshop on Advanced RF Sensors and Remote Sensing Instruments (ARSI), ESTEC, Nov. 2019

[1]University of Bern, IAP, 3012 Bern, Switzerland, [2]Airbus Defence and Space, 28022 Madrid, Spain, [3]Industrieanlagen-Betriebsgesellschaft mbH, 85521 Ottobrunn, Germany, [4]Thomas Keating Ltd., RH14 9SH, Billingshurst, West Sussex, UK

NOTES:

Highly-Compact Terahertz Spectrometers on Ultra-Small Platforms

Goutam Chattopadhyay, Adrian Tang, Subash Khanal, Jacob Kooi, Jose Siles, Darren Hayton, Cecile Jung-Kubiak, Sven van Berkel, Alain Maestrini, Choonsup Lee, and Imran Mehdi

Jet Propulsion Laboratory, California Institute of Technology, Pasadena, CA 91109, USA

SmallSats and CubeSats are compact and low-volume, sometimes even similar to a shoe box size, satellites with low-mass, low available power, and have somewhat limited communication and instrumentation capability. The size of the CubeSats is referred in the unit of 'U', where one 'U' is a cube with 10 cm x 10 cm x 10 cm dimensions.

Traditionally, SmallSats and CubeSats and their related instrumentation was primarily confined to undergraduate university research. However, in recent years, the international space agencies have been actively looking into CubeSats and SmallSats as useable platforms to supplement main missions as well as using them for standalone scientific missions. In some cases, they are being used to provide a communication relay platform for the missions where direct to earth (DTE) links are not always feasible. A case in point is the Mars Cube One (MarCO) [1] where two CubeSats were used to provide communications with the main space craft and ground control during high-risk maneuvers and critical events during the recent Insight landing.

Although apparently simplistic, development of scientific instruments on SmallSat and CubeSat platform poses a host of challenges: the instrument needs to be highly compact due to the lack of available space, has to be ultra-low power due to the severe restrictions on DC power availability, and one has to be innovative in the design of the antennas as traditional high gain reflector antennas (for scientific payload as well as for data communication) are not practical for such platforms. Design and development of aperture deployable antennas and other innovative structures are gaining a lot of attention in this regard.

We are developing a fully functional submillimeter-wave standalone scientific instrument as technology demonstrations on a CubeSat platform. Specifically, we are developing a low-mass and low-power 500-600 GHz high-resolution spectrometer instrument capable of remotely measuring water isotopes and deuterium/hydrogen (D/H) ratios on comets. The instrument's 18 cm diameter aperture consists of a novel low-profile leaky-wave lens-based antenna with waveguide feed. A microelectromechanical system (MEMS) based calibration switch is integrated with the receiver front-end along with low-power CMOS-based backend spectrometer circuits. This allows the overall instrument mass and power to be in the range ideal for implementation on ultra-small platforms such as CubeSats.

In this presentation, we will provide the design and implementation of the 500-600 GHz spectrometer instrument with details of innovative packaging solutions, antenna technology, and low-power backend solutions suitable for CubeSat and SmallSat platforms.

This work was carried out at Jet Propulsion Laboratory, California Institute of Technology under a contract with the National Aeronautics and Space Administration (NASA).

References

[1] S. Asmar et al., "Mars Cube One (MarCO) Shifting the Paradigm in Relay Deep Space Operations," in the Proceedings of the 14th International Coference on Space Operations (SpaceOps 2016), Daejeon, Korea, May 2016.

An Upgrade to the 1.9THz Local Oscillator for the SOFIA GREAT Instrument

Thomas Crowe[1], Theodore Reck[1], Steven Retzloff[1], and Jeffrey Hesler[1]

***Abstract*—An upgrade to the ~1.9THz channel local oscillator (LO) for SOFIA's GREAT instrument is presented. This LO system produces a minimum of 50uW and up to 120uW with continuous coverage from 1.8 to 2.08THz. A 4-way combined amplifier covering 150-175GHz with over 500mW of power dramatically simplifies the system design by introducing gain at a relatively high frequency in the multiplier chain. In addition, this amplifier reduced the total DC power consumption of each LO chain to 19W. Finally, careful characterization of AM noise was performed to show that the LO source should have minimal impact on the receiver sensitivity.**

Keywords*—*Local oscillator, Schottky diode, multiplier, amplifier

I. Introduction

The initial LO sources for the ~1.9 THz receivers of the for the German Receiver for Astronomy at Terahertz (GREAT, a facility instrument on NASA's Stratospheric Observatory for Infrared Astronomy (SOFIA) aircraft) were developed by VDI in 2013. These relied on very high-power amplifiers at ~26 GHz, followed by a cascade of three high-power varactor doublers and then two broadband varistor triplers. The resulting power was in the range of 10 – 20 uW for room temperature operation; but this was roughly doubled when the triplers were cryogenically cooled during SOFIA missions. Two such sources were used, with each pumping seven elements of the fourteen-pixel mixer array. The performance of these systems was gradually improved over the years to achieve improved power, tuning bandwidth and AMC [1, 2].

More recently, NASA JPL has demonstrated an LO source using higher frequency, W-Band, amplifiers [3]. This effort successfully demonstrated a highly integrated sixteen-pixel source array, and the per-pixel power of 40uW set a new state-of-the-art for power near 2 THz.

Beginning in 2020, NASA funded a significant upgrade to the LO sources for SOFIA. The specific requirements included a minimum power of 50 uW across the band of 1.80 – 2.08 THz with room temperature operation. The high power and broad tuning range of this upgrade presented a significant challenge requiring a complete revisioning of the system. The key to achieving the bandwidth was to introduce a broadband and high-power G-band amplifier to reduce the number of cascaded varactors to reach 1.9 THz. These amplifiers, fabricated by Teledyne Scientific [4] and packaged at Virginia Diodes, produce over 500mW of power from 150-175 GHz when four-way combined. Finally, a broadband, two-way power-combined 325GHz varactor and a 650 GHz varactor were developed to drive the 1.9 THz varistor tripler.

[1]Virginia Diodes, Charlottesville, VA, USA

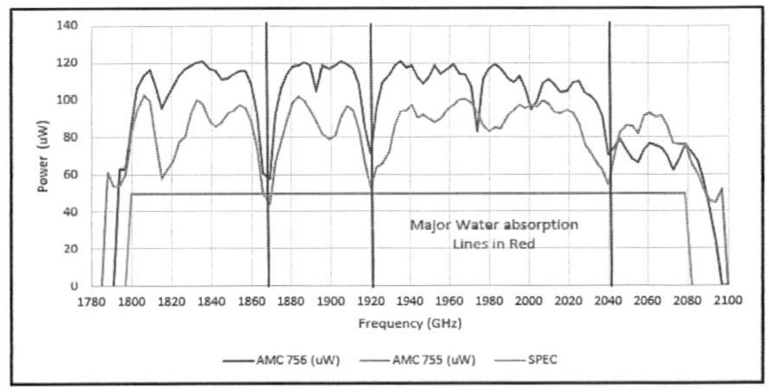

Fig. 1. The output power of the pair of sources, with the water absorbtion lines marked.

II. Results

Figure 1 shows that the 50 uW specification has been exceeded, with one of the two chains delivering as much as 120uW. In total, two pairs of LO chains have now been delivered to NASA, a flight unit and spare. All chains were tested for AM noise, which can contribute significantly to the noise of the single-ended HEB mixers. These measurements were performed at the output of the 650 GHz doublers and yielded AM noise power levels better than -140 dBc/Hz throughout the IF band for all LO frequencies. This result is quite acceptable, especially since the final tripler in the chains is known to be well saturated and thereby further lessens the AM noise at the final output.

The new LO chains have already flown one set of missions on SOFIA and the improved power and LO noise has significantly enhanced the receiver performance. A second set of flights is planned in June 2022.

References

[1] C. Risacher et al., "The upGREAT Dual Frequency Heterodyne Arrays for SOFIA," Journal of Astronomical Instrumentation, Vol. 7, No. 4 (2018).

[2] T.W. Crowe, J.L. Hesler, S.A. Retzloff, D.S. Kurtz, "Higher power multipliers for terahertz sources," 2016 41st Intl. Conf. Infrared, Millimeter, and Terahertz Waves, 2016.

[3] I. Mehdi, J. Siles, C. Lee and R. Lin, "Compact 1.9 THz Multi-Pixel Local Oscillator Chain," 2018 Asia-Pacific Microwave Conference (APMC), 2018, pp. 467-469.

[4] Z. Griffith, M. Urteaga, P. Rowell and L. Tran, "A 150-175-GHz 30-dB S_{21} Power Amplifier With 125-mW Pout and 16.2% PAE Using InP HBT," in IEEE Microwave and Wireless Components Letters.

NOTES:

Electron Speed Saturation Effects in the Schottky Barrier Diodes Implemented in an 800 – 900 GHz Frequency Tripler

D. Moro-Melgar *, O. Cojocari, V. Laín Rubio, I. Oprea, and A. Negrus

***Abstract*— We report the manufacture and testing of a Schottky Barrier Diode (SBD) based frequency tripler able to handle more than 50 mW input power in the 270-300 GHz range with conversion efficiency of 5 %. The tripler integrates a single chip based on discrete SBD diodes developed at ACST. The electron speed saturation in the SBD semiconductor of this tripler can be achieved when increasing the input power. This effect significantly modifies the matching network of the tripler chip and induces strong deviations from expected results. The impact of this effect on the RF performance of the triplers is discussed and compared in this work.**

***Keywords*— Schottky, Tripler, Electron Speed Saturation**

I. Introduction

The implementation of SBD technology for the development of mm-wave and submm-wave modules is a well know option to achieve compact and performant THz sources up to 2.7 THz [1]. SBD-based Frequency triplers are specially interesting to achieve the highest frequencies due to the larger multiplication factor and still reasonable conversion efficiency of up to ~5% at ~1 THz output frequency [2]. The SBD tripler technology developed at ACST can achieve at the moment 20%, 14%, 7% and 5 % at 300 GHz [3], 500 GHz [4], 700 GHz [3] and 900 GHz respectively. The tripler reported in this work uses the same discrete SBD technology described in [4], and it is based on the design architecture of the chip described in [5]. The available input power in WR3.4 for this tripler is much higher (70 mW) than it was possible to optimize for. This is due to the limitations in available space to fit a discrete diode for a 900 GHz tripler. The limitation of space in the chip limited this tripler to a maximum of 2 anodes, compared to other MMIC-based design able to fit even 4 anodes [5]. This tripler also achieved its best performance at an optimum input power of ~25 mW, but it can handle up to 50 mW without failure. The limitation in the number of anodes, the high frequency and the excessive available input power led us to find strong and clear electron speed saturation (ESS) effects in the semiconductor (SC) of the SBD. This is mainly due to the fact that the excessive input power excites a too large voltage amplitude response in the SBDs leading to a too fast voltage change. This effect is experimentally studied in this work and supported by simulations in EMpro/ADS softwares.

[1]ACST GmbH, Hanau, 63457, Germany;

II. Results

The experimental conversion efficiency of one tripler unit for two different levels of input power along its frequency band is plotted in Fig. 1. The solid lines represent a conversion efficiency (red) between 3-5% along the 800-900 GHz range when ~14 dBm input power is applied to the unit. Dashed lines illustrate the drop in conversion efficiency due to ESS when increasing the input power above 15 dBm. This tripler can be considered state-of-the-art in both high power and high efficiency compared to other ~900 GHz reported SBD triplers. The tripler reported in this work can match the similar output power achieved in [2] and [5] while only one chip is used (no power combining of chips) and having only two anodes in the chip instead of four [5]. The conversion efficiency in this work is twice larger than in [5] and as high as in [2] even if discrete SBD instead of MMIC approach.

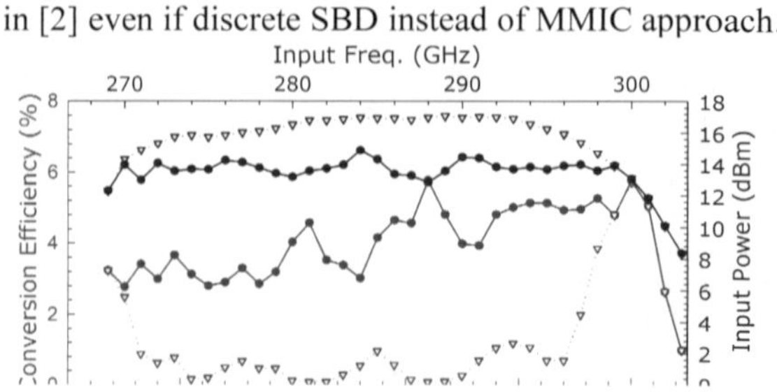

Fig. 1. Experimental efficiency (red) of the 800-900 GHz tripler with respect to the available input power (blue) in the 270-300 GHz range.

References

[1] Maestrini, Alain, and Jose V. Siles. "Signal Generation by Diode Frequency Multiplication." Fundamentals of Terahertz Devices and Applications (2021): 323-381.

[2] Siles, Jose V., et al. "A new generation of room-temperature frequency-multiplied sources with up to 10× higher output power in the 160-GHz–1.6-THz range." *IEEE Transactions on Terahertz Science and Technology 8.6 (2018): 596-604.*

[3] https://acst.de/products/mm-sub-mm-wave-modules/multipliers-2/

[4] Moro-Melgar, D., Cet al., "High Power High Efficiency 475-520 GHz Source Based on Discrete Schottky Diodes", *In 2020 50th European Microwave Conference (EuMC) (pp. 607-610). IEEE*, January 2021.

[5] Maestrini, Alain, et al. "A frequency-multiplied source with more than 1 mW of power across the 840–900-GHz band." *IEEE transactions on microwave theory and techniques 58.7 (2010): 1925-1932.*

NOTES:

High-Power Compact Sources in the 200 - 300 GHz Frequency Range

Veronica Lain-Rubio*[1], Diego Moro-Melgar[1], Oleg Cojocari[1] and Frank Gorski[1]

***Abstract*— We present a set of three high-power compact sources based on ACST high-power multiplier technology providing about 18dBm typical output power covering frequency ranges from 200 - 235 GHz, 230 – 260 GHz and 270-300 GHz. In this work we present the experimental results of this series (1213) of THz sources (1213).**

Keywords—Source, compact, high-power, Schottky, tripler

I. INTRODUCTION

The market for mm-wave sources has highly evolved up to W-band, where 1 W output power can be achieved with MMIC technology [1]. However, in the submm-wave range there still exists a limitation in terms of output power with this technology. Several emerging applications such as THz imaging [2], THz quality management control [3] and wireless THz telecommunication technologies aiming for 6G techniques [4] require new solutions. Therefore, the use of discrete diode structures for frequency multipliers above 100GHz is in comparison of significant advantage.

With our Schottky Diodes-based technology, we have developed a series of three frequency multipliers (213 series [5]) able to efficiently triple the frequency of commercially available E and W-band amplifiers to cover the 200 GHz to 300 GHz frequency range. The 213 series of multipliers have been combined with MMIC technology up to W-Band to produce submm-wave high-power compact sources, fulfilling the requirements of the market in terms of miniaturization and cost reduction.

II. RESULTS

With our Schottky Diode-based technology combined with commercial MMIC technology at ACST GmbH, we have developed a series of compact sources: 1213A, 1213B and 1213C [6] covering the frequency ranges of 200 - 235 GHz, 230 – 260 GHz and 270-300 GHz, respectively. ACST's compact 1213B is shown in Fig. 1 in order to illustrate the level of miniaturization. The 213 series are single chip Schottky based triplers able to efficiently handle 0.5 W output power from commercial sources [1] to achieve the performance depicted in Fig. 1. It is possible to easily increase in 3 dB the available output power of our compact sources implementing power combining of two tripler chips in a single module. This power combining approach allows us to efficiently handle 1 W from the commercially available sources in E-Band and W-Band to deliver up to 160 mW without dramatically increasing the size of the compact source. The development of these multipliers is still on- going.

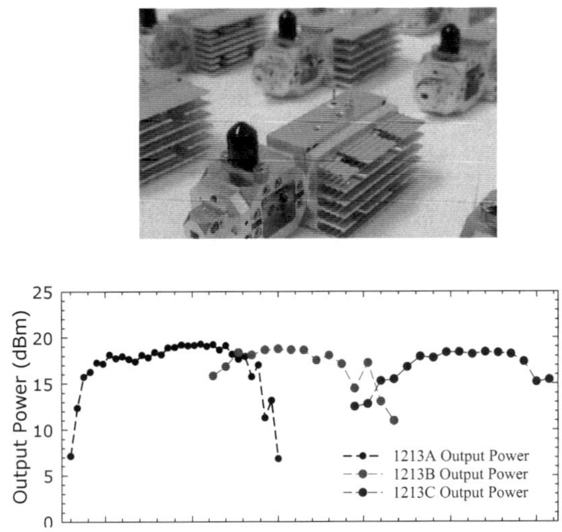

Fig. 1. Image of a set of commercial 1213B sources (Up) and the performance of the 1213 series in the 200-300 GHz range (Bottom).

The 1213 series set the basis of high-power sources required to overcome the THz range, being used in combination with our 229 series of high-power triplers to cover the range of 600-900GHz [7].

REFERENCES

[1] https://www.eravant.com/products/frequency-multipliers/active-frequency-multipliers
[2] Dufour, D., et al., "Review of terahertz technology development at INO" *Journal of Infrared, Millimeter, and Terahertz Waves*, 36(10), 922-946, 2015.
[3] Kahl, M., et al. "3D inspection of fiber-reinforced thermoplastics at THz frequencies." *44th International Conference on Infrared, Millimeter, and Terahertz Waves (IRMMW-THz)*. August, 2019.
[4] Alan, D., et al. "Building an end user focused THz based ultra high bandwidth wireless access network: The TERAPOD approach." *9th International Congress on Ultra Modern Telecommunications and Control Systems and Workshops (ICUMT). IEEE*, 2017.
[5] https://acst.de/products/mm-sub-mm-wave-modules/multipliers-2/
[6] https://acst.de/products/amc-sources/
[7] Moro-Melgar,D., et al. "600 – 900 GHz Frequency Sources with up to 4 mW Output Power based on Flip-Chip Schottky Barrier Diodes Technology", *47th International Conference on Infrared, Millimeter and Terahertz Waves (IRMMW-THz)*, August, 2022

[1]ACST GmbH, Hanau, 63457, Germany;

NOTES:

Ultra-broadband integrated room-temperature Schottky diode based local oscillators for line surveys in the 400-2070 GHz range

Jose V. Siles[1], Jonathan Kawamura[1], Choonsup Lee[1], Robert Lin[1], Alain Maestrini[1], Paul Goldsmith[1] and Jorge Pineda[1]

***Abstract*—We report here on a novel approach to produce room-temperature Schottky diode base local oscillator sources with RF bandwidth potentially up to 40% without compromising conversion efficiency with regards to state-of-the-art traditional designs. The approach consists of using the non-linear capabilities of the device to diplex two or more multiplying cells in a single channel, without requiring external waveguide diplexers. Initial results show coverage from 1350-1550 GHz and 1860-2070 GHz with 20-150 µW output power.**

Keywords—terahertz, local oscillators, Schottky diodes, frequency multipliers.

I. Introduction

Stars form in cores, which condense from interstellar clouds, and thereby incorporate material from their placental core. The process is regulated by stellar feedback, e.g. winds from massive stars and supernova explosions that reshape these clouds. Without this feedback, all the available gas and dust in galaxies like our own would have coalesced into stars long ago. The dynamics of the different gas components of the ISM have to be measured with exquisite precision. Wind-resolved observations require high-spectral resolution receivers tunable to each of the key tracers the govern these processes. For instance: (i) emission and absorption lines of water and its isotopologues from 557 GHz to 1670 GHz; (ii) the [CII] line at 1901 GHz, valuable tracer of the rate of star formation and of the evolution of interstellar atomic diffuse to dense molecular clouds in which new stars form; (iii) the fine structure lines of [NII] at 1461 and 2459 GHz, required to probe ionized regions.

Herschel/HIFI has been an enormously successful space instrument, contributing to our understanding of the ISM and star formation. The GREAT instrument on SOFIA and the GUSTO and ASTHROS balloons are continuing this legacy with modest array receivers. However, future missions will require focal plane arrays covering the entire submillimeter-wave spectrum, which are impossible with state-of-the-art heterodyne receivers. A typical high-spectral resolution heterodyne receiver consists of an antenna, a frequency mixer (SIS or HEB, based on frequency), a Schottky diode-based frequency multiplied LO source, and the IF processor Since the HEB mixers are inherently broadband, as long as the proper antenna structure is used, the bottleneck is the RF bandwidth of the LO (~15% only). Because of this limitation, the HIFI approach required fourteen terahertz local oscillator (LO) sources to cover its RF range of 480-1908 GHz.

II. Approach and Preliminary Results

Using the non-linear characteristics of Schottky multipliers, and with a thorough joint design of the multiplier diodes and matching circuitry, two or more different multiplier "cells" can be diplexed to work together as one, multiplying the bandwidth by the same factor, and potentially achieving full waveguide band coverage. The approach does not require waveguide diplexers and does not compromise the performance (output power and conversion efficiency) with respects to traditional narrow-band multipliers. In the conference, we will present the progress towards a single tunable multiplier chain aiming to cover the 400-2060 GHz with a single channel. Initial results with two multiplier cells combined can be seen in Fig. 1. The chain is driven by a single W-band power amplifier producing ~150 mW and delivers 20-150 µW with coverage between 1350-1550 GHz and 1860-2070 GHz'.

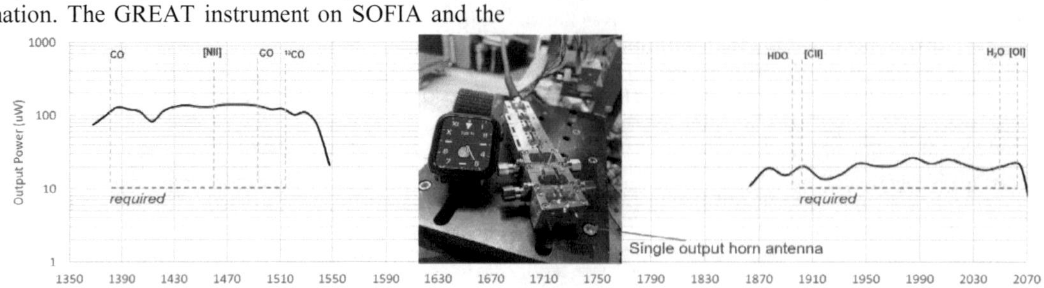

Fig. 1. ASTHROS payload under I&T, showing the 2.5-m primary mirror and the ~4.5-m long sunshade

[1]NASA Jet Propulsion Laboratory, California Institute of Technology, Pasadena CA, 91109, USA.

NOTES:

Terahertz Heterodyne Receiver Frontend Based on SiGe BiCMOS Technology for Space Applications

Nick Rothbart *[1], Alexandra Glück[1,2], Klaus Schmalz[3], and Heinz-Wilhelm Hübers[1,2]

Abstract— **We present a heterodyne receiver based on SiGe BiCMOS technology covering frequencies from 225 GHz to 255 GHz. We have performed emission and absorption spectroscopy of gaseous acetonitrile and have characterized the receiver. The double sideband noise temperature increases from 9 000 to 20 000 K in the frequency band of the receiver.**

Keywords—millimeter-waves, heterodyne receiver, gas spectroscopy, SiGe technology

I. Introduction

The terahertz (THz) spectral range is of great interest in astronomy and in atmospheric research because they cover rotational transitions of many relevant molecules. For instance, carbon monoxide (230.5 GHz) or ozone (235.7 GHz) provide strong transitions in this frequency range. Instruments working at millimeter-wave (mmW)/THz frequencies have been employed successfully to study the Earth's atmosphere, planetary atmospheres, comets and other astronomical objects [1-5].

For missions on small satellites such as CubeSats, compact systems without the need for cryogenic cooling are desired. Silicon-based technology is well suited as it fulfils these requirements and furthermore offers high integration capabilities and relatively low costs. Earlier, we have demonstrated sensitive active gas spectroscopy using a tunable transmitter and receiver based on SiGe BiCMOS technology [6].

In this work, we demonstrate passive absorption and emission spectroscopy around 240 GHz with a SiGe-based receiver frontend and characterize it with respect to space applications.

II. Results

The receiver frontend is implemented in IHP's 0.13 μm SiGe BiCMOS technology [7]. It is mounted on a compact baseband board which provides the supply voltages and the PLL for the local oscillator (LO). The LO covers frequencies between 225 GHz and 255 GHz. We demonstrate passive absorption and emission spectroscopy with a digital fast Fourier transform backend spectrometer and characterize important characteristics of the receiver, e.g. Allan time and noise temperature. Fig. 1 shows emission spectra of acetonitrile measured at different pressures. The noise temperature of the receiver was measured by the Y-factor method and was determined to 9 000 K – 20 000 K depending on the LO frequency [8].

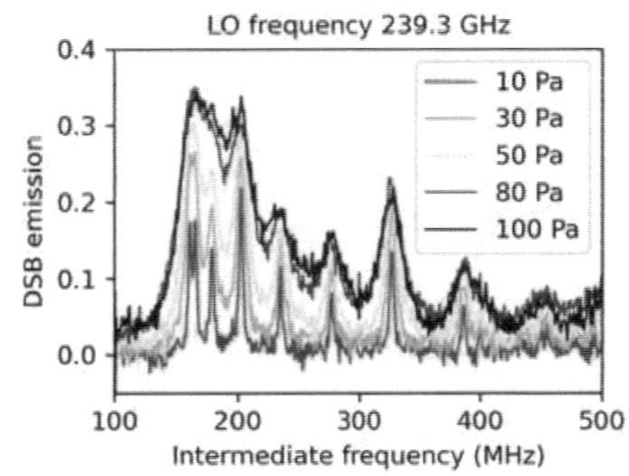

Fig. 1. Emission spectrum of Acetonitrile measured at different total pressures (10 Pa – 100 Pa) with an LO frequency of 239.3 GHz.

References

[1] J. Waters et al., "The Earth observing system microwave limb sounder (EOS MLS) on the Aura Satellite," *IEEE Trans. Geosci. Remote Sens.*, 44 (5), pp. 1075–1092, 2006.

[2] K. Kikuchi et al., "Overview and early results of the Superconducting Submillimeter-Wave Limb-Emission Sounder (SMILES)," *J Geophys Res*, 115 (D23), p. D23306, 2010.

[3] Y. J. Orsolini et al., "Descent from the polar mesosphere and anomalously high stratopause observed in 8 years of water vapor and temperature satellite observations by the Odin Sub-Millimeter Radiometer," *J Geophys Res.*, 115 (D12), p. D12305, 2010.

[4] P. Hartogh et al., "*Herschel* / HIFI observations of Mars: First detection of O_2 at submillimetre wavelengths and upper limits on HCl and H_2O_2," *A & A*, 521, p. L49, 2010.

[5] N. Biver et al., "Molecular composition of comet 46P/Wirtanen from millimetre-wave spectroscopy," *A & A*, 648, p. A49, 2021.

[6] N. Rothbart et al., "Analysis of Human Breath by Millimeter-Wave/Terahertz Spectroscopy," *Sensors*, 19 (12).

[7] K. Schmalz et al., "Dual-Band Transmitter and Receiver with Bowtie-Antenna in 0.13 μm SiGe BiCMOS for Gas Spectroscopy at 222 - 270 GHz," *IEEE Access*, 9, pp. 124805–124816, 2021.

[8] A. Glück et al. "SiGe BiCMOS Heterodyne Receiver Frontend for Remote Sensing with Small Satellites", submitted to IEEE Trans. Terahertz Sci. Technol., 2022.

[1]German Aerospace Center (DLR), Institute of Optical Sensor Systems, Berlin, 12489, Germany [2]Humboldt-Universität zu Berlin, Berlin, 12489, Germany, [3]IHP, Frankfurt/Oder, 15236, Germany

NOTES:

A Solid-State, Non-Cryogenic Receiver Operating at 2.5THz

Theodore Reck[1], Steven Durant[1] and Jeffrey Hesler[1]

***Abstract*—A receiver operating at 2.5THz without cryogenics is presented that utilizes an all-solid state local oscillator (LO). A schottky-diode fourth-harmonic mixer is developed to reduce LO power requirements. This mixer has a system temperature of 65,000K at 2.5THz at an optimum LO power of 1.9mW. A highly DC power-efficient 625 GHz source is presented which leverages a G-band power amplifier. This results in a highly compact receiver system that only consumes 7 W.**

***Keywords*— Remote sensing, radiometers, mixers, amplifiers**

I. Introduction

Space-born radiometer and spectrometers systems have long driven towards low-mass and high-power efficiency, but with the rise of the CubeSat platform, there is even greater pressure to deliver high performance instruments with smaller mass, power and size. One such instrument was ICECube, which used a Schottky-diode 2nd harmonic mixer pumped by a highly integrated 400 GHz LO source [1]. This was a 3U CubeSat demonstration that acquired cloud ice data at 833GHz over 15 months.

Schottky-diode based systems have the advantage of not requiring cryogenics and being fully solid-state. Such a receiver cannot complete with an HEB or SIS on sensitivity, but for certain science goals sensitivity can be traded to lower power-consumption and mass. Solar occultation measurements, which uses the sun as the background for the spectral absorption measurement of an atmosphere, do not require the sensitivities of cryogenic detectors to achieve their science goals. One such instrument is SSOLVE which is currently being developed to better understand the water content of the lunar atmosphere [2].

With a relatively high noise temperature being capable of achieving the science goals of the mission, a fourth-harmonic mixer is chosen to reduce the LO power requirements of the receiver system. This work builds upon Bulcha et.al., which used a single-diode fourth-harmonic mixer to achieve a conversion loss of 30 at 2.7 THz [3]. To improve the conversion loss and sensitivity, an anti-parallel diode pair is used.

II. System Design

The receiver is designed in three separate E-plane split housings to allow for testing of each stage of the system, as shown in Fig. 1. Efficiency of the LO amplifier-multiplier chain is radically improved by including an amplifier after 156 GHz tripler. This amplifier, once packaged, produces over 160mW of power at 10% power added efficiency [4]. This reduces the amount of signal power at the lower stages compared to a system that only used multipliers above 50GHz, thus reducing DC-power consumption. Two varactor doublers follow the amplifier to reach 625GHz. These varactors are put into a separate block because the typically require rework to tune the anode size to target the correct frequency. To keep the system compact, a custom flange is used between the low-frequency block and the varactor block. Finally, a fourth-harmonic mixer is attached in another block with a standard waveguide flange. Details of the fourth-harmonic mixer design can be found in [5].

[1]Virginia Diodes, Charlottesville, VA, USA

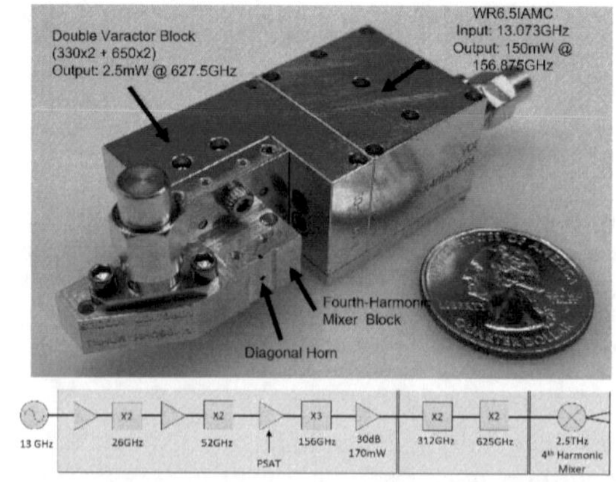

Fig. 1. (Top) The 2.5THz receiver. The size is 6.9x2.5x2cm and it weighs 70g. (Bottom) The system diagram of the receiver.

LO power is adjusted with a voltage applied to the PSAT input of the module, which sets the saturated power of the 52GHz amplifier.

III. Measurements

The power is measured after the 156GHz amplifier and the 625GHz doubler. Fig. 2 (a) shows the powers measured at the amplifier and Fig. 2 (b) is the power at the 625GHz doubler. Bandwidth is limited by the cascaded varactor doublers, which have a fixed bias to simplify the biasing electronics. Over 3 mW is produced at 625 GHz, which is more than sufficient to drive the 2.5THz fourth-harmonic mixer.

Y-factor of the system is measured between room temperature and a liquid nitrogen cooled absorber. Fig. 3 shows the system noise temperature as the LO-driver power is adjusted through the PSAT voltage input. 65,000 K noise temperature is measured across a wide range to PSAT settings, with the mixer turning on at 2.2V, which corresponds to 1.9 mW of LO power.

IV. Conclusion

A compact, low-power 2.5 THz receiver is developed that utilizes a fourth-harmonic Schottky-based mixer. By using a high-frequency, high-efficiency amplifier the amplifier-multiplier chain only consumes 6.8W of DC power. This receiver is currently being integrated at NASA with a 530GHz receiver as part of the SSOLVE instrument.

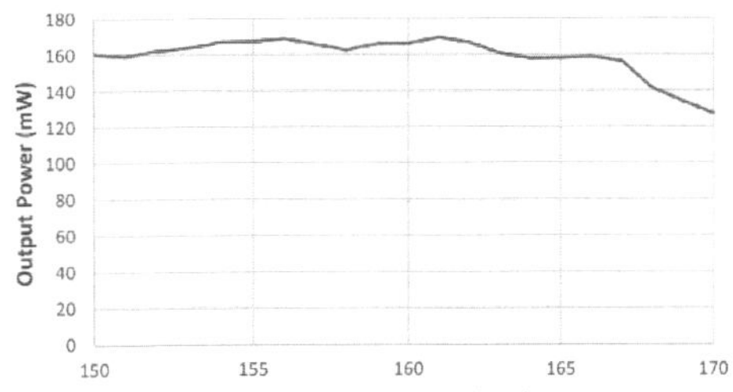

(a)

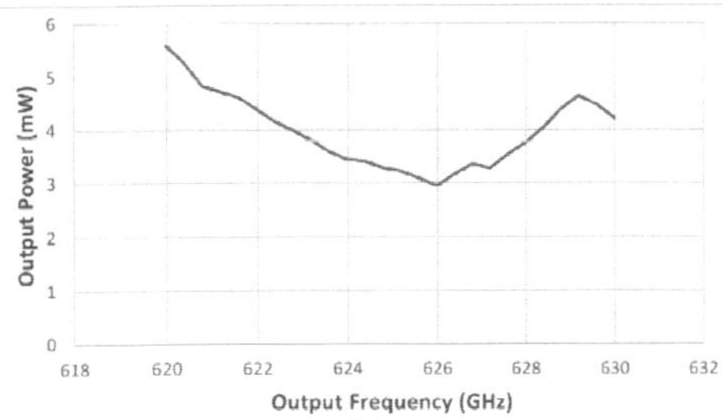

(b)

Fig. 2. (a) Output power for the final amplifier in the LO drive chain. (b) The output power after the two varactor multipliers.

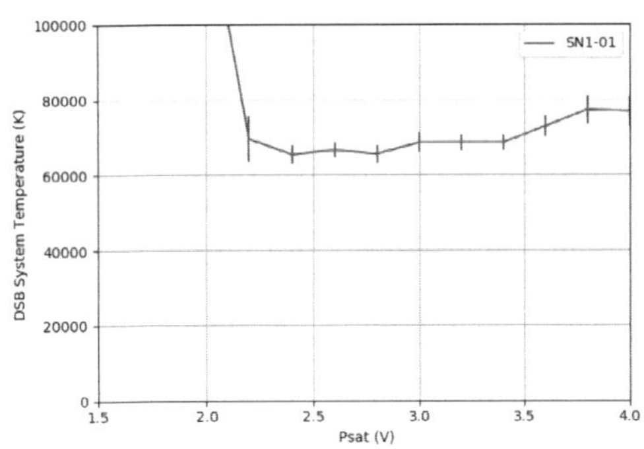

Fig. 3. The system noise temperature of the receiver at 2.5THz as the PSAT voltage is adjusted.

References

[1] Dong L. Wu, Jeffrey R. Piepmeier, Jaime Esper, Negar Ehsan, Paul E. Racette, Thomas E. Johnson, Brian S. Abresch, Eric Bryerton, "IceCube: spaceflight demonstration of 883-GHz cloud radiometer for future science," Proc. SPIE 11131, CubeSats and SmallSats for Remote Sensing III, 1113103 (30 August 2019).

[2] B. Bulcha et al., "Submillimeter Solar Observation Lunar Volatiles Experiment (SSOLVE) Optics and Front-end Spectrometer," 2021 46th International Conference on Infrared, Millimeter and Terahertz Waves (IRMMW-THz), 2021.

[3] B. T. Bulcha et al., "Design and Characterization of 1.8–3.2 THz Schottky-Based Harmonic Mixers," in IEEE Transactions on Terahertz Science and Technology, vol. 6, no. 5, pp. 737-746, Sept. 2016.

[4] Z. Griffith, M. Urteaga, P. Rowell, and L. Tran, "A 150–175 GHz 30-dB S21 G-band power amplifier with 0.25-W Pout and 15.7% PAE in a 250-nm InP HBT technology," in 2022 IEEE/MTT-S International Microwave Symposium - IMS 2022, 2022, pp. 444–447.

[5] T. Reck, S. Durant and J. Hesler, "Design of a 2.5 THz Schottky-Diode Fourth-Harmonic Mixer" in IEEE Transactions on Terahertz Science and Technology, TBD.

THz Receivers for Thermospheric Science

A. Maestrini, J. Siles, C. Lee, R. Lin, D. Hayton, and I. Mehdi
Jet Propulsion Laboratory, California Institute of Technology, Pasadena, CA, 91125 USA

Abstract—**Driven by the JUpiter ICy moon Explorer mission, considerable progress has recently been demonstrated for room temperature Schottky based heterodyne receivers to 1200 GHz. However, pushing operating frequencies of high-spectral resolution receivers to above 2 THz is still considerably challenging due to parasitics associated with traditional planar Schottky diodes and the difficulty to generate the local oscillator signal with sufficient power for optimum operations. This paper will present the on-going work at JPL to demonstrate room-temperature receivers in the 2-5THz band for thermospheric science.**

CRITICAL global measurements of lower thermospheric neutral winds, temperature and density measurements from a low-earth orbit platform are needed to better understand underlying mechanisms of the upper atmospheric composition/dynamics/temperature variability and the role of neutral dynamics on the ionospheric variability. Currently, no reliable satellite remote sensing technique can provide neutral wind measurements in the critical 100-150 km altitude region with complete local time coverage and desired spatial resolution, precision and accuracy. To address this short coming we propose to develop Schottky based receivers to measure the ionized oxygen line at 2.06 and 4.75 THz.

The configuration adopted for the 2.06 THz receiver is shown in Fig. 1. A bias-able subharmonic mixer architecture is used with a local oscillator source at 1.03 THz. The mixer, IF LNA and 1.03THz frequency tripler can be cooled to 120K for improved sensitivity, while the rest of the receiver will be operated at room temperature. The LO chain is predicated on being able to use a high-power amplifier at 114GHz, which are now available. In this configuration, 1.5mW at 1027GHz was recently measured, which is expected to be sufficient to pump the bias-able mixer. An alternate approach relies on a high power G-band PA. The mixer chips have been designed using both the configurations proposed in [2] and in [3]. Respective advantages of these two approaches will be discussed along with recent receiver results.

For the 4.7 THz channel a number of mixer topologies were investigated. The first approach would be to build a fundamental mixer, which would require around 0.5 mW of LO power at 4.7 THz. This is currently not possible with multiplied sources. An alternative to multiplied sources would be to use a QCL. However, QCLs would need cooling to 77K making system design and instrument build more complicated. A second traditional approach investigated was a sub-harmonically pumped mixer such as the one for the 2.06 THz channel. Such a mixer would require LO at 2.35 THz with an output power around 1 mW, which is more than an order of

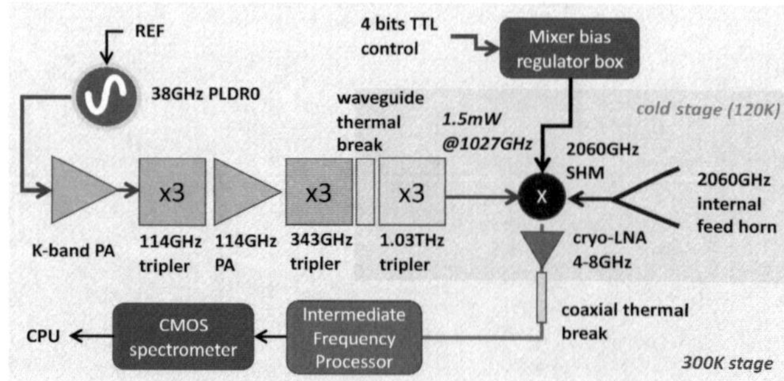

Fig. 1. Top: Receiver architecture designed for the 2.06 THz receiver. The measured LO power at 1027GHz is 1.5mW. This power is sufficient to pump the subharmonic mixer. Bottom: Measured output power from an alternate LO chain featuring two power-combined 114GHz triplers at room temperature [1].

magnitude higher than the state-of-the-art. After detailed consideration we opted to design and build a novel 3rd harmonic mixer for this task. We have designed a novel 3rd harmonic mixer as a practical solution for the 4.7 THz channel.

This novel third-harmonic Schottky mixer which will require around 1 mW at 1.58THz. Given the amount of LO power at 1.58THz that was demonstrated in [1], a single-anode unbalanced third harmonic mixer at 4.75THz is preferred rather than a balanced third-harmonic mixer that would require two diodes on the same circuit.

References

[1] J. Siles, et. al., "A New Generation of Room-Temperature Frequency Multiplied Sources with up to 10x Higher Output Power in the 160 GHz-1.6 THz Range", Proc. of the IEEE Transactions of Space Terahertz Communications, Nov. 2018

[2] E. Schlecht *et al.*, "Schottky Diode Based 1.2 THz Receivers Operating at Room-Temperature and Below for Planetary Atmospheric Sounding," in *IEEE Transactions on Terahertz Science and Technology*, vol. 4, no. 6, pp. 661-669, Nov. 2014, doi: 10.1109/TTHZ.2014.2361621.

[3] A. Maestrini *et al.*, "The 1200GHz Receiver Frontend of the Submillimetre Wave Instrument of ESA JUpiter ICy moons Explorer," *2018 43rd International Conference on Infrared, Millimeter, and Terahertz Waves (IRMMW-THz)*, 2018, pp. 1-2, doi: 10.1109/IRMMW-THz.2018.8509935.

Measurement of Transmission Losses of Superconducting Coplanar Waveguide and Microstrip Lines with On-chip Resonators at 2 mm Wavelength

Wenlei Shan*[1], and Shohei Ezaki[1]

***Abstract*—Transmission losses of superconducting coplanar waveguide and microstrip lines are measured for the purpose of noise breakdown of MMIC SIS mixers operating at 2 mm wavelength. The measurement is done by measuring the resonance curves of on-chip resonators, with SIS junctions as direct detectors and planar OMT as power divider, which enables a calibration reference.**

***Keywords*—SIS mixer, superconducting transmission line, loss, MMIC**

I. Introduction

We have been developing an innovative approach to enable compact focal plane heterodyne detector arrays with SIS mixers for wide field-of-view astronomical observation at mm and sub-mm wavelengths [1]-[3]. The new scheme is characterized by the adoption of superconducting monolithic microwave integrated circuits (MMICs), in which coplanar waveguide (CPW) and microstrip (MS) lines are adopted for guiding signals and LOs. The sensitivity of the SIS mixers is highly dependent on the transmission loss of these transmission lines, which remains not fully understood owing to the prohibitive difficulties in the measurement. We used a new method to measure the transmission loss by making use of our dual-polarization MMIC chips. In one of the polarization routes, half-wavelength resonators are coupled to the signal path, while the other polarization route is used as a calibration channel. By using a sweeping CW source and SIS junctions as direct detectors, the resonance curves of the resonators can be recorded. Finally, the transmission loss is retrieved from the Q-factor of these on-chip resonators. In this presentation, the results about this experiment will be presented.

II. Results

The intrinsic Q factors of CPW resonators and MS resonators were measured to be about 300 and 150 respectively at 4 K. According to these results, the transmission loss of superconducting thin film transmission lines, although not negligible, is not yet a limiting factor for the application of MMIC technology at mm wavelengths. In addition to our specific application, this measurement method can be applied to a wider range of applications for transmission loss investigation.

[1]National Astronomical Observatory of Japan, Tokyo, 181-8588, Japan

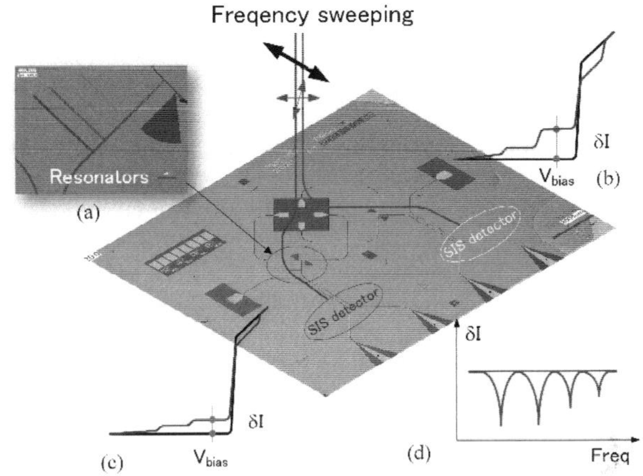

Fig. 1. The meaurement scheme. The central figure shows a dual polarization, balanced SIS MMIC mixer chip. The incident signal is polarized 45 degree with respect to the polarization directions of the planar OMT, so that each of the polarizatin channels received half of the incident power. In one of the two polarization channels, half-wavelength CPW and microstrip resonators are coupled to the signal path as shown in (a), while the other polarization channel is used as reference. The SIS junctions terminated in each polarization channel are used as direct detectors to measure the delivered signal by measuring the photon-asisted current δI as illustrated in (b) and (c). By sweeping the frequency of the signal source, which is composed of a signal generator followed by an amplifer and an x9 multiplexer, the resoance curves of the resonators can be measured after calibration for the freqeuncy dependence of the signal source by using the detection from the reference channel.

References

[1] Wenlei Shan, Shohei Ezaki, et al., "A Compact Superconducting Heterodyne Focal-Plane Array Implemented with HPI (Hybrid Planar Integration) Scheme," IEEE Transactions on Terahertz Science and Technology, vol. 10, no. 6, pp. 677-689, Nov. 2020.
[2] Wenlei Shan, Shohei Ezaki, et al., "Experimental Study of a Planar-Integrated Dual-Polarization Balanced SIS Mixer," IEEE Transactions on Terahertz Science and Technology, vol. 9, no. 6, pp. 549-556, Nov. 2019.
[3] Wenlei Shan, Shohei Ezaki, et al., "A New Concept for Quasi-Planar Integration of Superconductor-Insulator-Superconductor Array Receiver Front Ends," IEEE Trans. on Terahertz Sci. and Tech., vol. 8, no. 4, 472-474, Jul. 2018.

NOTES:

A Micromachined 1.37 THz Waveguide-based 2×2 Beam Divider for HEB Detectors

Haotian Zhu*[1,2], Jerome Valentin[1], Thibaut Vacelet[1], Sylvain Caroopen[1], Samuel Queste[3], Etienne Herth[4], Yan Delorme[1], and Martina Wiedner[1]

***Abstract*— A 2 × 2 local oscillator (LO) beam divider for pumping a hot electron bolometer (HEB) based heterodyne array is designed at 1.37 THz. The WR0.65 rectangular waveguide-based H-plane and E-plane junctions are used to construct the feeding network. The divider comprises two 400-μm thick silicon wafers with multilayer deep reactive-ion etching (DRIE) micromachined circuits, and a Pickett-Potter horn array manufactured in copper by computer numerical control (CNC) machining. The two silicon plates and the copper plate were assembled with a special process using multilayer stacking technology to ensure an assembling misalignment lower than 10 μm. The radiation patterns of the beam divider are measured 1.37 THz.**

***Index Terms*—Beam divider, deep reactive-ion etching (DRIE), micromachining technology, multi-pixel heterodyne receiver, radio astronomy, terahertz (THz).**

I. INTRODUCTION

TO achieve more scientific outputs, the next generation of sub-millimeter heterodyne receivers for both space and ground-based telescopes requires higher frequency bands, higher wide-field mapping speed and wider RF and IF bandwidth. To further improve the observing speed, multi-pixel radio astronomy receivers are favored. The observation speed will be increased by a factor of the pixel number of the receiver (as long as there is no degradation in the sensitivity of the individual pixel.). Therefore, high science throughput can be produced. For example, the observing speed of a 16-pixel focal plane array (FPA) is 16 times faster than a single pixel receiver.

II. RESULTS

This work is focused on the design, fabrication, assembly, and measurement of a 2 × 2 beam divider to couple a LO source to a 1.37 THz heterodyne array receiver using HEB, an extremely sensitive THz mixer. The beam divider consists of a waveguide-based feeding network and four Pickett-Potter horns [1]. The feeding network contains two silicon plates and was fabricated by deep silicon etching [2]. The details of the fabrication process of one of the plates are shown in Fig. 1. The SEM images of the fabricated plates (plate1 and 2) are shown in Fig. 5(b)-(d) and Fig. 5 (e), respectively. Fig. 5 (f) is the fabricated 3D vernier sturcture of plate 2. The vernier will be used in the assembly process.

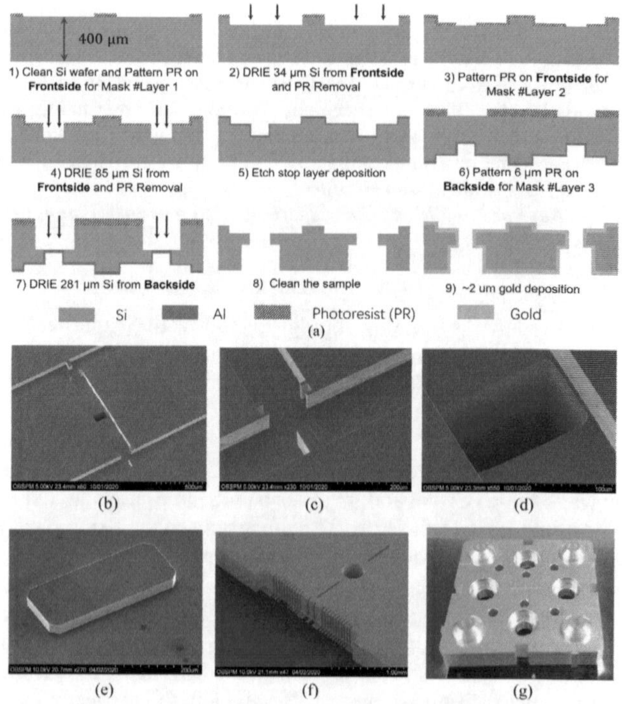

Fig. 1 (a) Fabrication process flow of plate 2. (b)-(d) Detailed SEM images of plate 1, (e)-(f) Detailed SEM images of plate 2 [note the SEM were captured before step (9)]. (g) Fabricated plate 3.

The assembly process of the beam divider was carefully studied and implemented to obtain an accuracy of a few micrometers. The fabricated beam divider is shown in Fig. 2(a). Final, the beam divider was characterized by a LERMA ABmm Vector Network Analyzer in the band of 1.3-1.4 THz. The measured normalized radiation pattern of the beam divider at 1.37 THz is shown in Fig. 2(b). A HDPE lens was set between the beam divider and the measurement plane.

[1] LERMA, Observatoire de Paris, Sorbonne Université, Université PSL, CNRS, Paris 75014, France.

[2] Key Laboratory of Microwave Remote Sensing, National Space Science Center, Chinese Academy of Sciences, Beijing 100190, China. (e-mail: zhuhaotian_2007@126.com, zhuhaotian@nssc.ac.cn)

[3] Institut FEMTO-ST, Université Bourgogne Franche-Comté UBFC, CNRS UMR 6174, 25030 Besançon, France.

[4] C2N, CNRS, Université Paris-Sud, Université Paris-Saclay, Palaiseau, France.

NOTES:

The distance between the beam divider and the HDPE lens is 32 mm, while the distance between the HDPE lens and the measurement plane is 190 mm. For more information about this work, please refer [3].

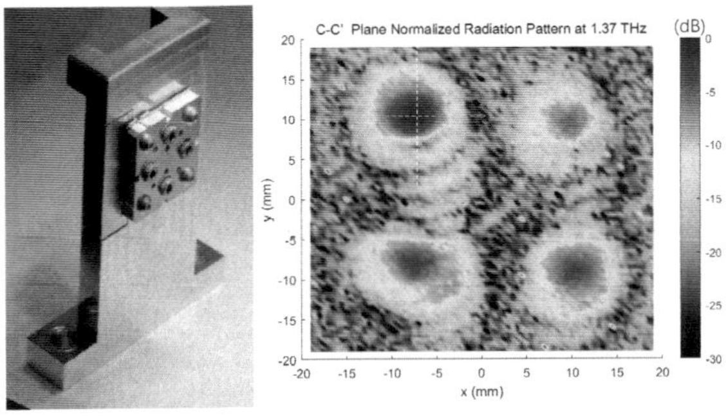

(a) (b)

Fig. 2 (a) Assembled beam divider, (b) Measured normalized radiation patterns of the beam divider at 1.37 THz.

CONCLUSION

This work provides a solution to couple a single LO source to a multi-pixel heterodyne array receiver for the next generation of terahertz instruments for astronomy. This paper presents a relatively new approach in which the complex E- and H-plane junction-based feeding network is engraved on two pieces of silicon by DRIE technology. The silicon stacked architecture is compatible with the standard micromachining fabrication. A third plate was realized using CNC machining. This work presents the first prototype of a multilayer stacked waveguide-based beam divider at 1.37 THz. The future work is to improve the transmission of the beam divider with the aim of pumping efficiently the 2×2 HEB heterodyne array.

ACKNOWLEDGMENT

The authors would like to thank Jean-Louis Leclercq from INL for the gold deposition. The authors would also like to thank Grégory Gay from LERMA, Observatoire de Paris, for programing the radiation pattern measurement program.

REFERENCES

[1] H.-T. Zhu *et al.*, "A 2 X 2 beam divider for an array local oscillator at 1.37 THz," *Proc. 44th Int. Conf. Infrared Millimeter THz Wave*, Paris, France, pp. 1-2, Sep. 2019.

[2] H.-T. Zhu *et al.*, "A 1.37 THz waveguide-based 2x2 beam divider fabricated by two microfabrication technologies," *Proc. 31st International Symposium on Space Terahertz Technology*, Tempe, AZ, pp. 1-1, Apr. 2020.

[3] H.-T. Zhu *et al.*, "A micromachined 1.37 THz waveguide-based 2x2 beam divider," *IEEE Trans. THz Sci. Technol.*, vol. 12, no. 6, pp. 565-573, Nov. 2022.

NOTES:

A Turnstile Quad-Ridge OMT for Full Octave-Bandwidth Receivers

Doug Henke*[1], Neal Kelly[1], Kyle Marshall[1], Ivan Wevers[1], and Lewis B. G. Knee[1]

***Abstract*—We report on measured results of a turnstile quad-ridge OMT suitable for millimeter-wavelength receivers. The quad-ridge OMT is directly compatible with a quad-ridge feed horn to facilitate dual-polarization low-noise receivers that exceed the single-mode bandwidth of rectangular waveguide. We have chosen a bandwidth of 24–51 GHz to demonstrate machinability and implementation.**

Keywords—octave-bandwidth receivers, OMT, orthomode transducer, quad-ridge OMT, ridge waveguide

I. THE TURNSTILE QUAD-RIDGE OMT

In order to exceed the usable single-mode bandwidth of a standard waveguide (i.e., ~1.7:1), the waveguide cross-section must be modified; most often ridges or fins are used to modify the circular and rectangular waveguide sections. There are existing quad-ridge OMTs that exceed an octave bandwidth and make use of coaxial probes/pins, in combination with a backshort, to couple out the polarized signals (see [1] as an excellent recent example). As frequencies exceed 10 GHz, however, the tight clearances required for the ridges and probes become increasingly difficult to manufacture and assemble. Planar OMTs may also be paired with quad-ridge feed horns for wide bandwidth operation [2], [3].

In [4] and [5], a quad-ridge turnstile OMT was demonstrated with performance exceeding an octave bandwidth across impressive frequency ranges of 6–19 GHz and 18–45 GHz. The OMT was also paired with a separate quad-ridge feed horn. In order to manufacture the OMT, the base and upper blocks were each sectioned into 4 quarters, and each ridge section was clamped within the joins. The turnstile tuning stub was replaceable to allow for experimentation with various stub geometries.

In [6], a conceptual design of the turnstile quad ridge OMT was shown following a layer approach where the ridges were directly machined into the adjoining layer, instead of as separate assemblies. Here, we present measured results with details on implementation and machinability. A frequency range spanning 24–51 GHz was used to demonstrate the design.

As shown in Fig. 1 (a), the turnstile quad-ridge OMT is a network of ridge waveguide components: a quad-ridge circular input, a turnstile with single-ridge rectangular outputs, E-bends with double-ridge waveguide, and a T-junction single-ridge combiner.

The six-layered design is compatible with a 3-axis milling machining center and Fig. 1 (c) illustrates how the ridges are milled into the adjacent layer.

To provide compatibility with standard rectangular waveguide for measurement, custom adapters spanning WR-34 and WR-22 were used, as shown in Fig. 2. Tellurium copper was used throughout.

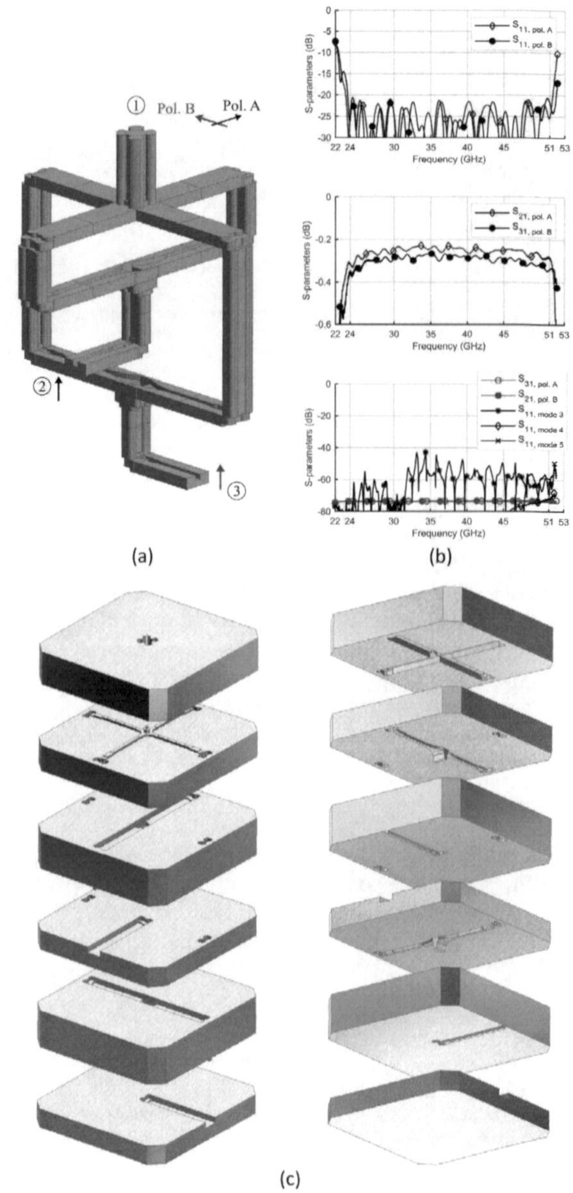

Fig. 1. (a) CAD model of the quad-ridge turnstile OMT showing the waveguide channels and corresponding polarization outputs. Encircled numbers represent the ports for simulation. (b) Simulated results of the OMT. (c) Iso views showing the waveguide channel and ridge details.

The measured OMT demonstrated 15 dB return loss across 24–51 GHz, as shown in Fig. 3 (a) and (b). Cross-polarization leakage [Fig. 3 (c)] and port isolation were both measured < -40 dB, with mean values of ~-50 dB. Insertion loss was found to be 0.25–0.4 dB across the band. Full details can be found in [7].

[1]NRC Herzberg Astronomy and Astrophysics, Victoria, B.C., Canada.

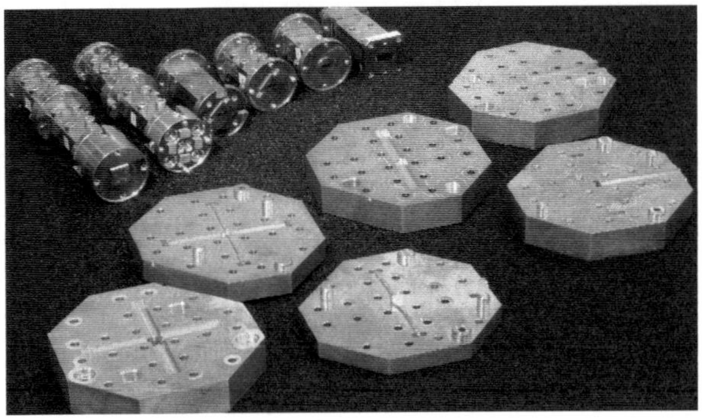

Fig. 2. Machined adapters and quad-ridge OMT. Since the operating bandwidth spans two standard waveguide bands (WR-34 and WR-22), custom adapters were used to transform to single-ridge waveguide. A cloverleaf shaped transition was used to excite the OMT quad-ridge input. 6 layers were used to complete the OMT assembly.

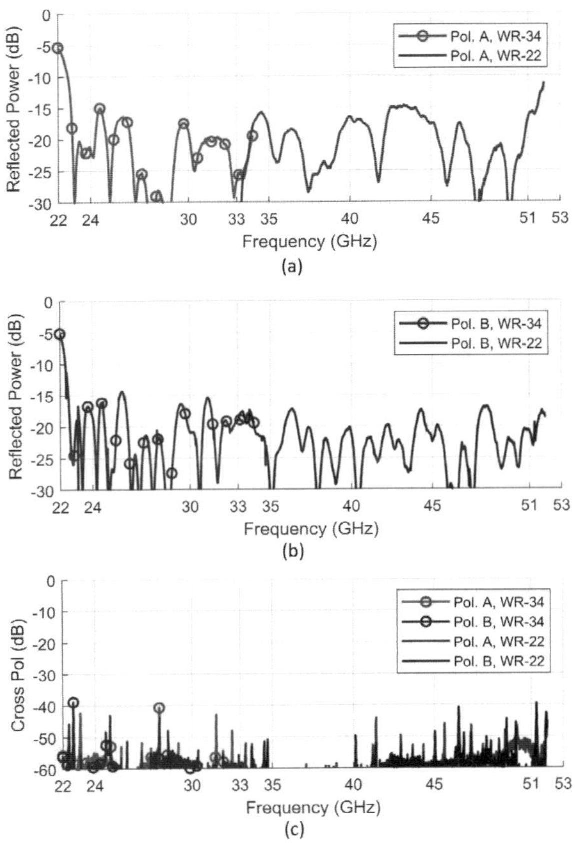

Fig. 3. Measure response of the fabricated quad-ridge turnstile OMT. The reflected power response, i.e., S_{11}, is shows in (a) and (b) for polarization A and B, respectively. Cross-polarization leakage is shown in (c).

REFERENCES

[1] A. W. Pollak and M. E. Jones, "A Compact Quad-Ridge Orthogonal Mode Transducer with Wide Operational Bandwidth," IEEE Antennas and Wireless Propagation Letters, vol. 17, no. 3, pp. 422-425, March 2018.

[2] CMB-S4 Technology Book, First Edition, arXiv:1706.02464 [astro-ph.IM].

[3] S. Shu et al., "Development of octave-band planar ortho-mode transducer with kinetic inductance detector for LiteBIRD," in Proc. SPIE, 2016, vol. 9914, pp. 99142C-1–99142C-6.

[4] N. Jastram, "Passive front-ends for wideband millimeter wave electronic warfare," Ph.D. dissertation, Dept. Elect. Comput. Energy Eng., Univ. Colorado, Boulder, CO, USA, 2014.

[5] P. V. Prasannakumar, M. A. Elmansouri and D. S. Filipovic, "Wideband Decoupling Techniques for Dual-Polarized Bi-Static Simultaneous Transmit and Receive Antenna Subsystem," IEEE Trans. Antennas Propag., vol. 65, no. 10, pp. 4991-5001, Oct. 2017.

[6] D. Henke, "A Full Octave-Band OMT for Millimetre-Wave Receivers," in Proc. 31st Int. Symp. Space Terahertz Technol., Tempe, Arizona, Mar. 2020.

[7] D. Henke, N. Kelly, K. Marshall, I. Wevers, and L. B. G. Knee, "A Turnstile Quad-Ridge Orthomode Transducer (OMT) for Octave-Bandwidth Receiver Front-Ends (24–51 GHz)," submitted for publication, Dec. 22, 2022.

Recent research on (sub)mm-wave OMTs at NAOJ

A. Gonzalez*[1], K. Kaneko, R. Sakai, S. Masui, and T. Kojima

Abstract—**Dual-ridged waveguide orthomode transducers (OMTs) are a well-proven type of high-performance waveguide components used for the separation of orthogonal linear polarizations, and wideband designs have been recently demonstrated at frequencies as high as 500 GHz. In this paper, we review recent work at NAOJ on this type of OMTs and investigate the limits of performance in a narrower bandwidth.**

Keywords*— Orthomode transducers, millimeter-wave circuits, polarization splitter, fabrication techniques, radio astronomy.*

I. Introduction

Orthomode transducers (OMTs) are key waveguide components to separate orthogonal linear polarizations. They are thus very important in applications which require polarization discrimination such as radio astronomy or communications. In radio astronomy, polarization is useful to extract additional scientific information, whereas in communications, it is used to reutilize the available spectrum. OMTs receive two linear orthogonal polarization signals, vertical (V) and horizontal (H) for convenience, within the same frequency band coming from a common square or circular waveguide, and separate them into two separate rectangular waveguides.

Several of the waveguide OMT architectures common at microwave frequencies have been applied to frequencies as high as 300 GHz [1-7], but the number of OMTs at sub-millimeter wavelengths is still limited [8-9], mainly due to fabrication challenges. Among existing OMT types at mm and sub-mm wavelengths, dual-ridged OMTs represent a good balance between performance and ease of fabrication. They can achieve good impedance matching in all ports, good polarization discrimination and isolation between output ports, and low insertion loss, at the same time they can be fabricated in two split blocks by direct machining in combination with wire electro-discharged machining (EDM). At the National Astronomical Observatory of Japan, we developed this type of OMTs towards the production of 73 OMTs for the 125-163 GHz [5] and 385-500 GHz [8] heterodyne receivers for the ALMA (Atacama Large Millimeter/sub-millimeter Array) telescope. In recent years, we have successfully designed and fabricated wideband components covering the 67-116 GHz [10] and 275-500 GHz [11] frequency ranges. In this contribution, we summarize the results in [10] and [11], and apply our know-how to maximize performance of dual-ridged waveguide OMTs at a narrower bandwidth considering practical aspects related to electromagnetic design. In particular, we have designed an OMT with the highest possible performance in the 195-290 GHz band. The chosen bandwidth corresponds to an extended version of the ALMA band 6 (211-275 GHz).

II. Previous Designs and EM Simulations

We have recently developed state-of-the-art wideband OMTs at mm and sub-mm wavelengths [10]-[11] using the dual-ridged waveguide topology shown in Fig. 1. Successful designs have been achieved thanks to the use of modern electromagnetic (EM) software, which allows high parametrization of the design through a model, the combination of different EM analysis methods, and advanced optimization algorithms. In those previous designs, the goal was to design OMTs as broadband as possible (55-60% fractional bandwidth) with the highest possible performance. In this contribution, we target a more moderate but still wide bandwidth, 195-290 GHz (~40% fractional bandwidth), and investigate the limits of performance we can achieve based on our previous models. This OMT was firstly designed using a hybrid Mode-Matching (MM) + Finite Elements (FE) method in the commercial software WaspNET. Then, the same design was evaluated with Finite Elements Boundary Integral (FEBI) in WaspNET and with the FE software HFSS.

The comparison of the results of return loss using the three EM simulation methods showed good agreement for the H polarization, in spite of some degradation of performance to a level around 25 dB, and some discrepancies for the V polarization, for which the two FE methods showed performance between 20 and 25 dB. These results are shown in Fig. 2. In order to understand the discrepancy in the V polarization better, two adjacent FE elements in our model were combined into a single FE element, as shown in Fig. 3, which yielded results closer to FEBI and HFSS. This result highlights the importance to carefully choose the appropriate simulation method for each part of the hybrid model when trying to push the limits of performance. The new MM+FE model produced results consistent with FEBI and HFSS in subsequent designs.

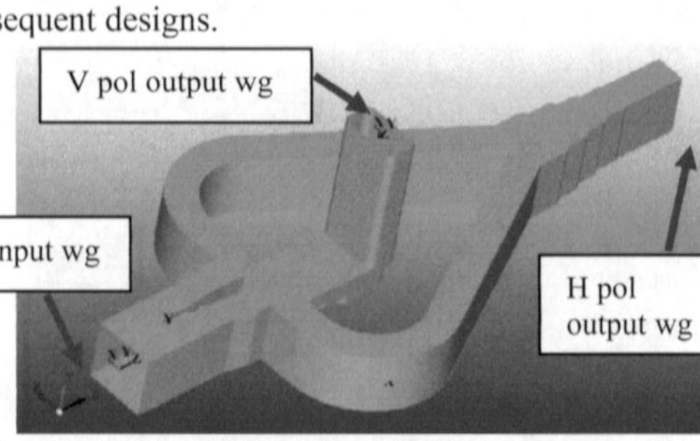

Fig. 1. 3D schematic of the simulation model of a dual-ridged waveguide (wg) OMT. This design targets the 195-290 GHz bandwidth.

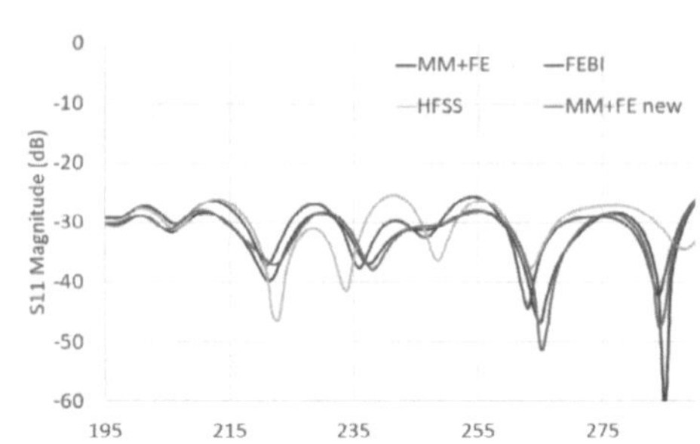

[1]National Astronomical Observatory of Japan, Mitaka, 181-8588, Japan

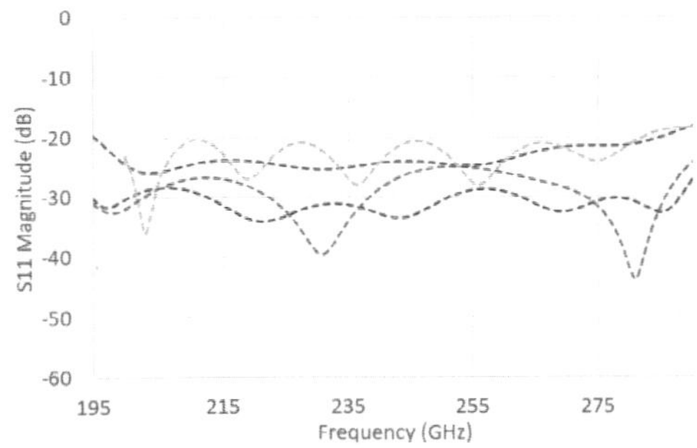

Fig. 2. Return loss simulation results (as S11) for the designed OMT using MM+FE, FEBI and HFSS methods. The top figure corresponds to the H polarization, whereas the bottom figure shows the results for the V polarization. HFSS results for the V polarization include ripples due to impedance mismatch between the rounded-corner output waveguide and the rectangular output port. Results for the improved MM+FE model in Fig. 3 are shown in green color.

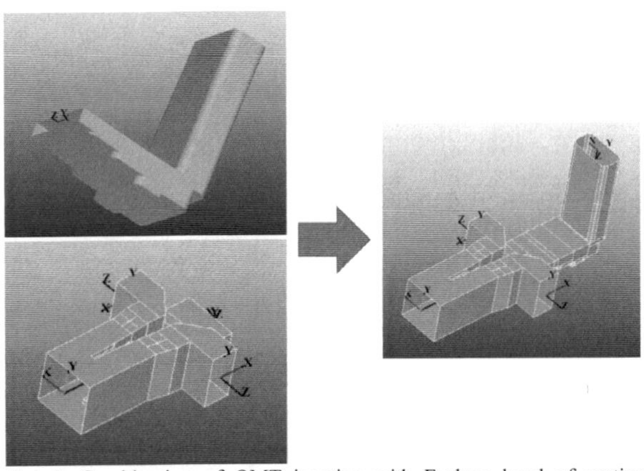

Fig. 3. Combination of OMT junction with E-plane bend of vertical polarization output into a single FE element.

III. Results

Two prototypes of the designed OMT for the 195-290 GHz band were fabricated and tested. In order to speed up testing, the prototype units were measured with standard WR-4.3 (864 μm x 432 μm) waveguide extenders and without transitions to the 940 μm x 470 μm output waveguides. A square to WR-4.3 waveguide transition which could be rotated 90 degrees was used for the common input waveguide. Results are shown in Fig. 4. Reflection loss is better than 20 dB with all degradations introduced by the measurement setup uncompensated, cross-polarization is better than -30 dB and isolation is better than 40 dB. The measurement results for both prototype units were similar.

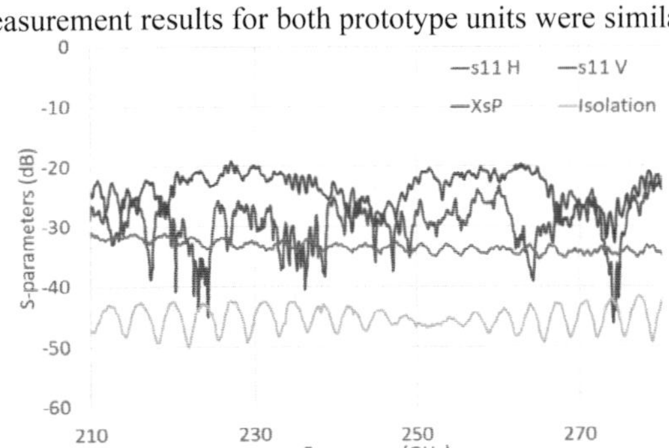

Fig. 4. Measurement results of prototype OMT unit without using output transitions and without compensating for the effect of the rectangular-to-square input transition.

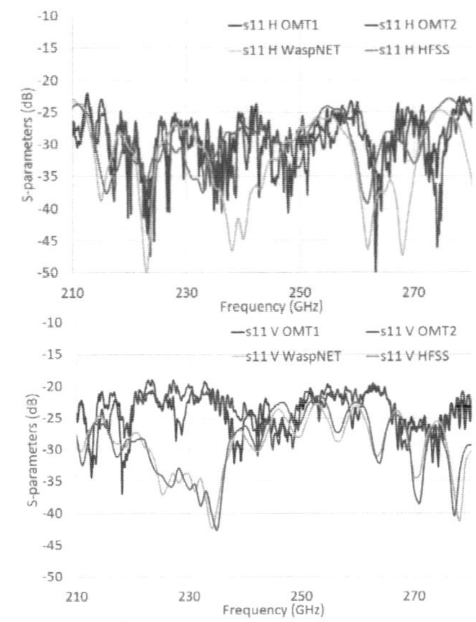

Fig. 5. Comparison of measurement results for two prototype OMT units and WaspNET and HFSS simulations including all transitions and steps in the actual measurement. (Top) H polarization, (Bottom) V polarization

Measurement results were compared with simulations modeling the measurement setup. Results showed that H polarization measurements were similar to simulations, which hints the actual performance of the OMT in terms of insertion loss is in the order of 28 dB. Results in V polarization were not bad but different from simulations. This was tracked down to fabrication errors.

References

[1] A. Navarrini, and R.L. Plambeck, "A turnstile junction waveguide orthomode transducer", IEEE Trans. Microwave Theory and Techniques, vol. 54, no. 1, pp.272-277, Jan. 2006

[2] A.M. Boifot, E. Lier, T. Schaug-Pettersen, "Simple and broadband orthomode transducer", IEEE Proceedings, Vol. 137, No. 6, pp.396-400, Dec. 1990

[3] E.J. Wollack, W. Grammer, and J. Kingsley, "The Boifot Orthomode Junction", ALMA memo 425, May 2002

[4] A. Dunning, "Double ridged orthogonal mode transducer for the 16-26 GHz microwave band", in Workshop on the Applications of Radio Science, February 20-22, 2002

[5] S. Asayama, and M. Kamikura, "Development of Doubled-Ridged Waveguide Orthomode Transducer for the 2mm Band", J Infrared Millimeter and Terahertz Waves, Vol. 30, pp. 573-579, June 2009

[6] A. Navarrini, and R. Nesti, "Symmetric Reverse-Coupling Waveguide Orthomode Transducer for the 3-mm Band", IEEE Trans. Microwave Theory and Techniques, Vol. 57, No. 1, pp.80-88, Jan. 2009

[7] G. Chattopadhyay, et al., "A 96-GHz Ortho-Mode Transducer for the Polatron", IEEE Microwave and Guided Wave Letters, Vol. 8, No. 12, pp. 421-423, Dec. 1998

[8] M. Kamikura, et al., "Development of a Submillimeter Double-Ridged Waveguide Ortho-Mode Transducer (OMT) for the 385-500 GHz Band", J Infrared Millimeter and Terahertz Waves, Vol. 31, pp. 697-707, June 2010

[9] Alessandro Navarrini, Christopher Groppi, Robert Lin and Goutam Chattopadhyay, "Test of a Waveguide Orthomode Trasducer for the 385-500 GHz Band," 22nd International Symposium on Space Terahertz Technology, Tucson, 26-28 April, 2011

[10] A. Gonzalez, and S. Asayama, "Double-Ridged Waveguide Orthomode Transducer (OMT) for the 67-116-GHz Band", Journal of Infrared, Millimeter, and Terahertz Waves, Vol. 39, Issue 8, pp.723-737, Aug. 2018

[11] A. Gonzalez, and K. Kaneko, "High-Performance Wideband Double-Ridged Waveguide OMT for the 275–500 GHz Band", IEEE Transactions on Terahertz Science and Technology, vol. 11, no. 3, pp. 345-350, May 2021.

Results from ALMA Band 2 cryogenic LNA pre-production run

S. Türk*[1], F. Schäfer[1], C. Leinz[1], P. Yagoubov[2], F. Thome[3], A. Leuther[3], S. Heyminck[1], P. Pütz[1], and G. Wieching[1]

***Abstract*—We present the results of a pre-production run of cryogenic low noise amplifiers developed for the upcoming ALMA Band 2 receiver project. We will characterise 14 units against a full set of technical specifications set by ALMA. The ongoing measurements indicate a state-of-the-art noise performance achieved over an extended W-band being fully compliant to specifications. For the first six units we measure Tnoise < 22 K (67 GHz – 90 GHz), with a minimum of 15 K at 76 GHz, and < 30 K (90 GHz – 116 GHz) at 15 K operation temperature.**

Keywords—Cryogenic, high-electron-mobility transistors (HEMTs), low-noise amplifiers (LNAs), metamorphic HEMTs (mHEMTs), millimeter wave (mmW), monolithic microwave integrated circuits (MMICs), noise, radio astronomy, W-band

I. Introduction

The application of radio astronomy sets challenging requirements for noise performance of receiver components and for higher frequencies mandates cryogenic operation. In particular this hold true for the first components of the receiver signal chain, and therewith for a first-stage low noise amplifier. For the Atacama Large Millimeter/submillimeter Array (ALMA), the European Southern Observatory (ESO) has decided to move forward with Band 2 that will cover the entire 67 GHz – 116 GHz atmospheric window. A single signal chain in a receiver cartridge developed by a consortium of international partners is to be used. Similar to the pre-production approach on cartridge level, ESO had awarded the MPIfR/IAF consortium as a potential component supplier a small pre-production batch of LNA units. Performance is evaluated against an extensive specification sheet and, additionally, series production readiness is verified.

A cryogenic LNA covering extended W-band operation using standard WR10 waveguides has been developed. At the heart an ultra-low noise performance MMIC using 50-nm gate length metamorphic HEMT (mHEMT) technology developed at Fraunhofer IAF is used. Here the active device layers are grown on GaAs substrates by means of a metamorphic buffer. This technology has matured over the recent years to provide state-of-the-art cryogenic noise performance, concluding in a benchmark W-band LNA design that is basis of our ALMA Band 2 LNA contribution [1]. The LNA package is a MPIfR WR10 split-block design with quartz substrate waveguide probes at input and output.

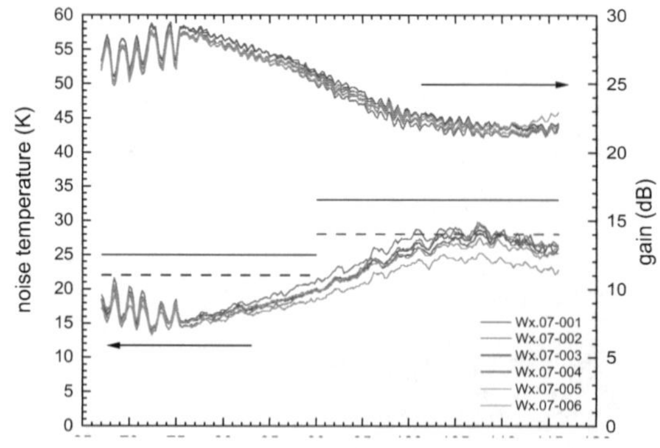

Fig. 1. Noise temperature and gain measured at 15 K for the first six pre-production units. The solid and dashed lines indicate the 100% and 80% requirement on maximum noise temperature, respectively. The standing wave features at low band edge are assumed to be an artefact from the measurement setup.

II. Results

The pre-production units are evaluated against requirements set by technical (e.g. cryogenic and RT S-parameters, noise, gain, gain/phase stability), environmental, mechanical and RAMS specifications. We will present these results for the full 14-unit pre-production run and detail on LNA and package design, our methodology to ensure a high enough production yield, and functionality of the cryogenic test setups involved. As an example, in Fig. 1 noise temperature and gain performance for the first six pre-production units measured at 15 K operation temperature using the cryogenic test-setup at ESO, Munich [1] is plotted. The results closely match those of early prototype units using MMICs from a different wafer. This emphasises uniform performance of MMIC production and packaging process and confirms reproducibility of production.

References

[1] F. Thome, et al., "A 67–116-GHz Cryogenic Low-Noise Amplifier in a 50-nm InGaAs Metamorphic HEMT Technology", *IEEE Microwave and Wireless Components Letters*, vol. 32, no. 5, pp. 430-433, May 2022, doi: 10.1109/LMWC.2021.3134462.

[1]Max Planck Institute for Radio Astronomy, Bonn, 53121, Germany; [2]European Southern Observatory (ESO), 85748 Garching, Germany; [3]Fraunhofer IAF, Fraunhofer Institute for Applied Solid State Physics, 79108 Freiburg im Breisgau, Germany

NOTES:

ALMA Band 2 Cold Cartridge Assembly Design

Belitsky V.[1], Lapkin I.[1], Fredrixon M.[1], Lopez C.[1], Ferm S.-E.[1], Pavolotsky A.[1], Strandberg M.[1], Sundin E.[1], Desmaris V.[1], Hesper R.[2], Adema J.[2], Barkhof J.[2], Baryshev A.[2], Bekema M.[2], Realini S.[2], Koops A.[2], de Haan, R. [2], Rodenhuis M.[2], Cuttaia F.[3], Nesti R.[4], Ricciardi S.[3], Terenzi L.[3], Villa F.[3], Gonzalez A.[5], Kaneko K.[5], Sakai R.[5], Imada H.[5], Kojima T.[5], Phillips N.[6],Yagoubov P.[6]

***Abstract*— As part of the ALMA development, we present the design of the ALMA Band 2 Cold Cartridge Assembly (CCA). The Band 2 is the last band that completes the suit of the 10 receiver channels of ALMA. The originally planned ALMA Band 2 receiver cartridge should cover the RF band of 67 - -90 GHz. The recent progress in technology, optics, OMT design and mm-wave amplifiers, however allowed to implement receiver that has an extended RF band up to 116 GHz. Furthermore, the Band 2 receiver pursues 2SB layout and provides 4-18 GHz IF band for two sidebands in a dual-polarization configuration. Here, we describe the design of the Band 2 CCA that includes optics, amplifier assembly, internal RF transport, mechanics and cryogenics. The downconverter part and performances are described elsewhere.**

***Index Terms*— ALMA receiver, cold cartridge assembly, amplifiers, RF transport, vacuum, cryogenics.**

I. INTRODUCTION

ALMA Observatory is the most advanced and modern radio astronomy observatory. Yet, the development and expanding ALMA capacity observing different frequencies is not completed and the last remaining receiver band, ALMA Band 2, covering 67-90 GHz is under progress. Recent advances in the low-noise amplifier (LNA) technology [1, 2, 3, 4] and optics components [5, 6] allow building ultra-wideband low-noise systems, e.g. 67-116 GHz RF band, with state-of-the-art performance. In such receiver that follows ALMA receiver layout, the CCA provides low-noise gain and is operated inside ALMA Front End (FE) cryostat at cryogenic temperatures while in the Warm Cartridge Assembly (WCA), the 2SB mixer and IF amplifiers provide signal conditioning and down-conversion to deliver it to ALMA Correlator. ALMA Band 2 receiver with it outstanding IF bandwidth 4-18 GHz belongs to a new generation of receivers demanded for new prospective science in the ALMA Roadmap 2030 [7].

II. ALMA BAND 2 CCA DESIGN

A very significant effort in developing technology and making ALMA Band 2 pre-prototype was done during several preceding years [8] within the study led by the European Southern Observatory (ESO) with participation from GARD[1], INAF[3], NAOJ[5], University of Chile, University of Manchester, RPG-Radiometer Physics GmbH, LNF-Low Noise Factory, MPIfR in collaboration with IAF, and other institutions. Building up on these results, we have developed all-new design for ALMA Band 2 CCA with extended RF band and covering 67 - 116 GHz and with interfaces fully in compliance with the ALMA FE cryostat and ALMA antenna.

The CCA employs optics [9] that consists of a short-focus UHMWPE lens placed at room temperature, also acting as a vacuum window, and a corrugated horn. The optics provides frequency independent illumination of the secondary. The design of the lens is optimized based on available ALMA FE cryostat window opening for Band 2, spillover loss, dielectric loss in the lens (the shorter the focal length, the thicker the lens) and position of the corrugated horn inside the cryostat. In the ALMA Band 2, the corrugated horn is placed in the space between the 110K and 15K thermal shields, Fig. 1.

Fig. 1. 3D model. ALMA Band 2 CCA inside the ALMA FE cryostat. The lens and its support are not shown. The simulated antenna

This manuscript was submitted on April 20, 2023. This work was supported by European Sothern Observatory (ESO) under the Agreement #87603/ESO/18/88584/ASP. NAOJ contribution was supported by the national funds of Japan. *(Corresponding author: Victor Belitsky, victor.belitsky@chalmers.se).*

The authors' current affiliations according to the footnotes: [1]GARD, Chalmers University, Gothenburg, 41296, Sweden; [2]Kapteyn Astronomical Institute, University of Groningen, 9747 AV Groningen, the Netherlands; [3]INAF-OAS Via Gobetti, 101, 40129, Bologna, [4]INAF-OAA, Largo Enrico Fermi 5, I - 50125 Firenze, Italy; [5]National Astronomical Observatory of Japan (NAOJ), 181-8588, Mitaka, Japan; [6]ESO, Karl-Schwarzschild-Strasse 2, D-85748, Garching, Germany.

beam clearly shows the lens position and the horn placed in space between 110K and 15K thermal shields.

The original ALMA FE IR filter at 15K thermal shield has to be removed to allow the horn taking the desired position between the 110K and 15K thermal shields of the ALMA FE cryostat. To preclude the leak of the IR radiation into the 15K environment, we have introduced a baffle that blocks the IR radiation yet provides the necessary clearance for the horn during CCA-FE integration procedure. In turn, the clearance space is sealed against IR radiation leaks by introducing an additional reflector placed around the horn but below the 15K shield level and thermally terminated at 15K via the support pillar, see Fig. 2 for details.

The thermal design of the ALMA Band 2 CCA is largely defined by the optics. Because of the optics layout, the horn, OMT and amplifiers are placed at substantial distance from 15K thermal plate. The cooling of the passive components (the horn, OMT, waveguides) and active components (CLNAs) is extremely important for achieving ultimate noise performance. In order to cool down all the components, we used a massive pillar made of Aluminum AW-6082 alloy. The pillar also provides a mechanical reference for the optics and support for all receiver components.

The polarization split occurs in the OMT [8, 9] that is directly attached to the corrugated horn. In each polarization, two wideband low-noise cryogenic amplifiers (CLNA) are used connected in series with an isolator in between to control the ripples. With the total gain of the amplifiers over ~45 dB and the loss in the internal RF transport chain less than 6 dB, the contribution of the warm downconverter electronics of the WCA to the noise temperature is less than 2%.

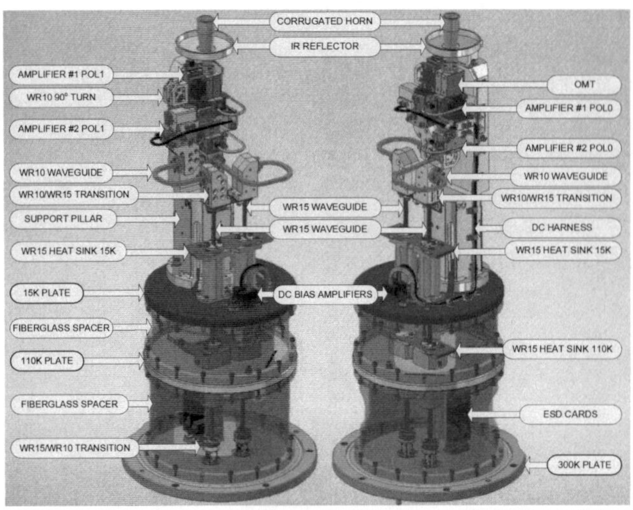

Fig. 2. Annotated picture shows ALMA Band 2 CCA 3D model.

ALMA Band 2 CCA allows to accommodate different types of the available CLNAs, [2, 3, 4]. Together with the amplifier vendors, all the amplifiers were harmonized in their WR10 interfaces, lengths and DC connectors. However, the material used for the amplifier blocks were Aluminum alloy and Tellurium copper. In order to accommodate the difference in materials that causes significant mechanical stress during the cooling because of the thermal contraction, the design uses "adapter approach". Depending on the particular amplifier and material used, the cooling bracket was used as an adapter, which shape, material and design were adjusted to provide sufficient cooling via mechanical contact to the pillar playing the role of the cooling buss while the adapters' material with theirs shapes were chosen to compensate for the difference in the thermal contraction. When possible, with CLNAs with Aluminum bodies, we used all-Aluminum alloy solid thermal adapters and/or semi-rigid links when CLNAs with Tellurium copper bodies were used.

The internal RF transport system employs combination of the WR10 and the oversized WR15 waveguides and is designed to allow controlling the RF insertion loss, ensuring minimum mechanical stress while cooling to cryogenic temperatures. The shape of the RF transport system waveguides was chosen to fit in the tight space within the CCA body while controlling the RF signal insertion loss. The oversized WR15 waveguides ensure low RF insertion loss and were made straight to avoid generation of high-order modes and trapped modes at the turns. All necessary turns were made either by bent copper WR10 waveguides or using milled split-blocks of an Aluminum alloy, Fig. 2. The copper WR10 waveguides are bended in one plane; this facilitates manufacturing and provides required flexibility to absorb/relax the stress caused by thermal contraction of the cartridge body and the pillar supporting the optics and amplifiers. The WR10-to-WR15 transitions and split-block turns were simulated in Ansys HFSS to achieve a better than -20 dB return loss at any frequency within the RF band 67-116 GHz. As example here, Fig. 3 demonstrates the design and performance of the 90 – degree corner WR10 adapter used in the Pol1 RF transport chain right after the CLNA. The option to have the 90-deg corner before the CLNA was studied but abandoned. In the implemented design of the CCA, the corner is just after the CLNA. In this respect the important parameter is the return loss and not the insertion loss.

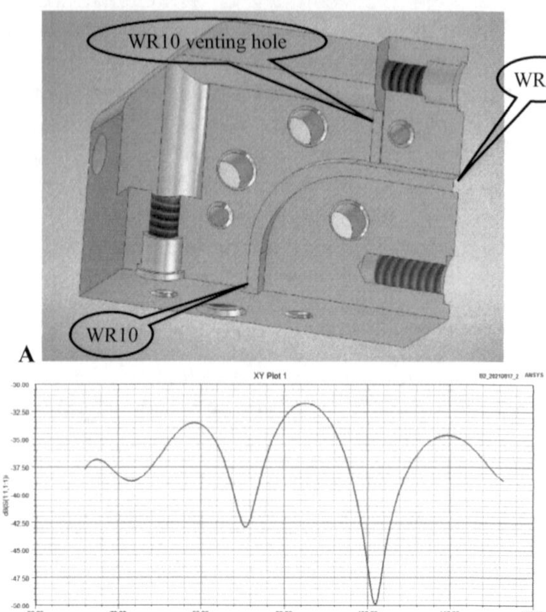

Fig. 3. A) 90 – degree corner WR10 adapter made in split block and

used in the Pol1 RF transport chain. **B**) The HFSS simulated S_{11}/S_{22} of the adapter.

The vacuum WR10 RF feedthroughs were designed and optimized for the entire RF band to provide less than 0.5 dB insertion loss and S_{11}/S_{22} better than -20dB, see reference [10] for the details.

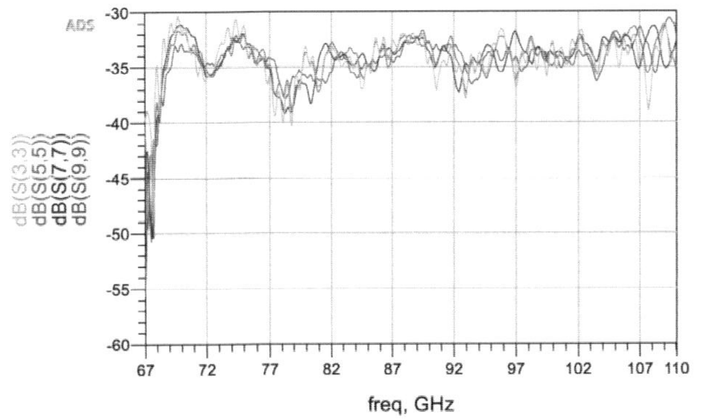

Fig. 3. C) Measured magnitude of the return loss of 4 fabricated 90 – degree WR10 corner adapters; better than -30dB in the RF band of 67-116 GHz.

The stainless steel WR15 thin-wall waveguides ensure necessary thermal isolation between the different temperature stages inside the CCA because of the low thermal conductivity of the material thus providing minimum thermal load to the 15K cryogenic stage with the amplifiers and the intermediate 110 K stage of the ALMA cryostat. The oversized waveguide dimensions ensure reduced insertion RF loss for the signal as compared to WR10 waveguide. Flexible thermal links, Fig. 2, connect the WR15 waveguides to the cartridge thermal plates via additional "chimneys" that allow us to choose the required length of the waveguide separating different temperature levels. By choosing the appropriate length, we could control the heat influx at the specified levels for the CCA 110K and 15K temperature plates.

Fig. 4. ALMA Band 2 CCA cooling simulated with ANSYS™ Mechanical® software. For better scale resolution, only components mounted at the ALMA Band 2 CCA cartridge 15K plate are shown. In the simulation, all LNAs have 22.5mW power dissipation and 100mW of the IR radiation power were added at the corrugated horn to simulate IR radiation thermal influx from 110K environment.

The thermal simulations using Ansys have shown that the chosen cooling structure fulfills its purpose by keeping the passive components and the amplifiers close to the 15K temperature provided by the ALMA FE cryostat. The laboratory measurement results in the ALMA Cartridge Test Cryostat for a single cartridge have shown the temperature at the CLNAs are very close to the simulated levels of the temperature. The simulations results are presented in Fig. 4.

Table 1 below shows *calculated and specified heat loads* at different temperature stages of the CCA. The differences between the passive and total thermal loads give the margin for the heat produced by active components, e.g., 107 mW is available for CLNAs. During the simulations, we assumed that the horn that is exposed to the IR radiation of the 110K environment picks up 100 mW of power.

Table 1. Passive Heat Loading

FE Cryostat Stage	WR15 Heat Load	Harness Heat Load	Total Heat Load (spec)
15 K Stage	64 mW	14 mW	78 (185) mW
110 K Stage	243 mW	18 mW	261 (600) mW

At this stage of the ALMA Band 2 project, the two prototype ALMA B2 CCAs were built and tested, in the extended 67-116 GHz bandwidth, covering both the ALMA Band 2 and Band 3 frequency range, Fig 5. The measurements of the physical temperature of the critical components confirmed the simulations (Fig.4) with inaccuracy of the measured temperature ±0.5 K at the CLNAs bodies. The prototype CCA simulated Eigen frequency for the lowest mode was 147,75 Hz. The CCA was run through the shaking test to confirm the Eigen frequencies are above the specified 75 Hz, the observed lowest resonance was at about 125 Hz.

Fig. 5. Assembled pre-production ALMA B2 CCA.

The CCA was complemented with the Warm Cartridge Assembly [9] and fully characterized including optics. The details on the test setup could be found in [12].

Fig. 6 demonstrate measured noise performance of one of the pre-production ALMA Band 2 receiver cartridges.

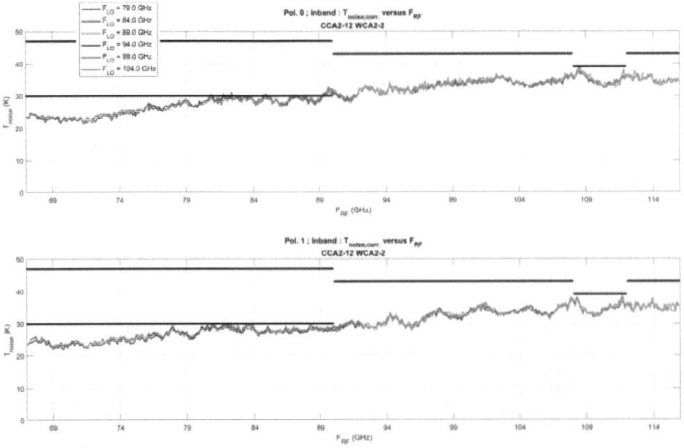

Fig. 5. ALMA Band 2 CCA 2SB-corrected receiver noise temperature vs. RF frequency at 100 MHz resolution, including specification markers (solid black lines).

III. Conclusion

A new generation receiver for ALMA, Band 2 covering 67-116 GHz, dual polarization, 2SB with the IF bandwidth 4-18 GHz has been designed, built and tested. The receiver complies with the ALMA specifications for Band 2 and Band 3 and produces total 56 GHz wide IF band from the two polarization channels and upper and low sidebands of each polarization. The receiver has an excellent noise temperature and fulfill expectations for next-generation ALMA receivers [7] with the extended IF bandwidth. At the time of writing this manuscript, the first ALMA Band 2 receiver was undergoing commissioning at the ALMA Observatory.

References

[1] Chau-Ching Chiong, Yunshan Wang, Kai-Chun Chang, and Huei Wang, “Low-Noise Amplifier for Next-Generation Radio Astronomy Telescopes”, IEEE Microwave magazine, Volume 23, Number 1, January 2022, ISSN 1527-3342.

[2] Y. Tang et al., “Cryogenic W-band LNA for ALMA band 2+3 with average noise temperature of 24 K,” in Proc. IEEE MTT-S Int. Microw. Symp. (IMS), 2017, pp. 176–179. doi: 10.1109/MWSYM. 2017.8058981. Custom model, https://lownoisefactory.com/product/lnf-lnc65_115wb-2/

[3] F. Thome, F. Schäfer, S. Türk, P. Yagoubov and A. Leuther, "A 67–116-GHz Cryogenic Low-Noise Amplifier in a 50-nm InGaAs Metamorphic HEMT Technology," in IEEE Microwave and Wireless Components Letters, vol. 32, no. 5, pp. 430-433, May 2022, doi: 10.1109/LMWC.2021.3134462.

[4] David Cuadrado-Calle, Danielle George, Gary A. Fuller, Kieran Cleary, Lorene Samoska, Pekka Kangaslahti, Jacob W. Kooi, Mary Soria, Mikko Varonen, Richard Lai, and Xiaobing Mei, “Broadband MMIC LNAs for ALMA band 2+3 with noise temperature below 28 K,” IEEE Trans. Microw. Theory Techn., vol. 65, no. 5, pp. 1589–1597, May 2017. doi: 10.1109/ TMTT.2016.2639018.

[5] P. Yagoubov et al, “67-116 GHz optics development for ALMA band 2-3 receivers”, 2016 41st International Conference on Infrared, Millimeter, and Terahertz waves (IRMMW-THz), https://doi.org/10.1109/IRMMW-THz.2016.7758971 .

[6] Gonzalez, A. et al., ALMA band 2+3 (67-116 GHz) optics: Design and first measurements, IEEE International Symposium on Antennas and Propagation 2016, https://doi.org/10.1109/APS.2016.7696305 .

[7] The ALMA Development Roadmap, AKA ALMA 2030, http://www.eso.org/sci/facilities/alma/announcements/20180712-alma-development-roadmap.pdf

[8] Yagoubov P., et.al, “Wideband 67-116 GHz receiver development for ALMA Band 2,” A&A, vol. 634, p. A46, 2020. https://doi.org/10.1051/0004-6361/201936777

[9] Álvaro González, et.al., "Development of receiver optics for ALMA bands 1 and 2, and possible synergies with ngVLA," Proc. SPIE 11453, Mm, Submm, and Far-Infrared Detectors and Instrumentation for Astronomy X, 114533S (December 2020), https://doi.org/10.1117/12.2561595

[10] Igor Lapkin, Cristian López, Mathias Fredrixon, Alexey Pavolotsky, Sven-Erik Ferm, Vincent Desmaris and Victor Belitsky, “Vacuum-seal Waveguide Feedthrough for Extended W band 67-116 GHz”, submitted to IEEE Journal of Microwaves, March 2023.

[11] Bertrand Thomas, “A 67-116 GHz MMIC-based dual-polarized 2SB down-converter for the ALMA Band 2 Warm Cartridge Assembly prototype development”, ISSTT2022.

[12] J. Barkhof et.al., “ALMA Band 2 Receiver Automated Test System”, ISSTT2022.

ALMA Band 9 Sideband Separating Upgrade

R. Hesper*[1], J. Barkhof[1], S. Realini[1], A.M. Baryshev[1], J. Adema[1], P. Yagoubov[2], N. Phillips[2] and C. de Breuck[2]

Abstract—**We present a brief summary of a study, performed over the last few years, on the feasibility of upgrading the existing ALMA Band 9 receivers (602–720 GHz) along the lines of the ALMA-2030 roadmap. The goals: upgrading the mixer architecture from Double-Sideband (DSB) to Sideband-Separating (2SB); extending the IF bandwidth beyond the original 4–12 GHz; extending the RF bandwidth beyond the original Band 9 window; investigating whether the current stock of Band 9 SIS junctions is sufficient for the 2SB upgrade; investigating the possibilities of improving the polarimetric performance of Band 9; investigating upgrade roll-out strategies; and finally estimating the cost of such an upgrade. Most of the performance upgrades (IF bandwidth, RF bandwidth and polarimetry) are either demonstrated or are argued to be feasible. The number of available mixers may just be sufficient for an upgrade with similar noise performance.**

Keywords—Sideband-separating receivers, Sub-mm receivers

I. Introduction

The technical feasibility of an ALMA Band 9 mixer upgrade from the existing double-sideband (DSB) configuration to sideband-separating (2SB) was demonstrated in an ESO study by the NOVA Sub-mm Instrumentation Group several years ago [1]. Since then, the design of the sideband-separating mixer has been developed further with significant improvement of key parameters, namely sensitivity (noise temperature) and image rejection ratio (IRR) [2]–[4]. Subsequently, two left-over DSB Band 9 receiver cartridges were converted to 2SB operation [5], one for the SEPIA facility instrument [6] on the APEX telescope in Chile ("SEPIA-660") and one for the future LLAMA observatory in Argentina. The former has successfully passed its commissioning phase, with several key performance parameters far exceeding the specification [7], and is in full science operation. These receivers both offer a total IF bandwidth of 4×8 GHz (configured as 4×(4–12) GHz), which is double the total bandwidth of the current ALMA DSB and 2SB receivers (with the exception of Band 6, which already exports 4×5.5 GHz).

In this paper, we summarize the main results of a recently concluded follow-up study that targetted both practical issues of such an upgrade (i.e., whether new SIS devices must be produced or how to perform an upgrade without disrupting observatory operations), as well as a boost in bandwidth, sensitivity and polarimetric performance to new levels, in line with upcoming demands to keep ALMA up-to-date for the coming decade. The full study report, with many more aspects than can even be touched upon here, is publicly available [8].

A. Goals of the study

The goals (or tasks) in the study, slightly paraphrased from their original definitions:

1) Investigate the feasibility of extending the IF bandwidth to at least 12 GHz (e.g., to 4–16 GHz or more) with the goal to achieve as broad a bandwidth as possible without compromising the other performance parameters;
2) Investigate the extension the RF bandwidth beyond the current edges of Band 9 (602–720 GHz) without compromising the other performance parameters (at least not in the original core RF range);
3) Verify the availability of a sufficient number of SIS mixer devices at NOVA to enable a 2SB upgrade of all 73 ALMA Band 9 receivers;
4) Investigate possibilities to improve the polarimetric performance beyond that of the currently installed Band 9 receivers;
5) Investigate the possibilities of performing the upgrade while keeping the majority of the old DSB and new 2SB Band 9 receivers in operation;
6) Determine the expected cost, both in new hardware and labour, to upgrade all existing ALMA Band 9 receivers, including the options for increased IF/RF bandwidth and optical performance mentioned above;

In the following sections we report on the results obtained within goals 1, 2 and 4. Goals 3, 5 and 6 will only be briefly summarized, since these are more programmatic than technical. The reader is referred to the full report [8] in case of further interest in these matters, as well as for the full technical details of the project.

B. Overall architecture

The architecture used to obtain sideband separation is the classical I-Q mixer setup shown in fig. 1. Also drawn are the cryogenic components after the IF hybrid. The LNAs amplifying the IF signals are critical for the noise performance, since they are the first elements in the signal chain providing gain, all preceding components being lossy to some extent. Critical for the IRR performance is the suppression of reflections in all

*r.hesper@astro.rug.nl
[1]NOVA Sub-mm Instrumentation Group, Kapteyn astronomical Institute, University of Groningen, Groningen, The Netherlands
[2]ESO Headquarters, 85748 Garching bei München, Germany
This study was performed in the framework of the ESO "Advanced Study for Upgrades of the Atacama Large Millimeter/submillimeter Array (ALMA)" (CFP/ESO/16/11115/OSZ), and carried out under ESO Collaboration Agreement No. 94866/19/86535/ASP.

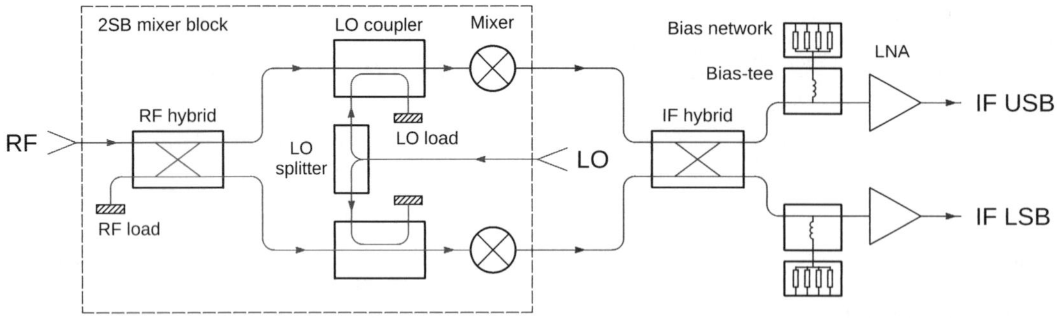

Fig. 1: Schematic diagram of the classical 2SB mixer architecture used in the study (as well as in the SEPIA/LLAMA receivers). Waveguides in green, IF coaxial cables/strip lines in red, DC lines in blue. Each "Mixer" denotes a single-ended (DSB) SIS device. A full receiver typically contains two such assemblies, for orthogonal polarizations. Further details in [8].

parts of the structure: not only the absorption of the waveguide loads which terminate the uncoupled LO power and any LO power reflected off the mixers ("RF load") is crucial, but also the impedance matching of the hybrids (RF and IF) and the LNA inputs [3], [4].

The SIS mixer devices require a DC bias voltage, which must be injected into the IF system somewhere by a bias-tee. In our case, this happens between the IF hybrid and the IF LNAs, which is possible because the IF hybrid is (crosswise) DC-transparent. In the case of SEPIA/LLAMA, the bias-tees (with their associated bias networks) were conveniently integrated into the IF LNAs; for the wide-band experiments presented in the current paper a discrete in-house produced bias-tee was used.

II. Extending the IF bandwidth

A. Scientific driver

Even without any further widening of the total IF bandwidth beyond the demonstrated factor of 2 in SEPIA-660, an upgrade like this would increase the ALMA sensitivity in Band 9 by 20–30% on average for spectral line observations [3]. This ties in with the Recommended Development Path number 2 ("Larger bandwidths and better receiver sensitivity: enabling gain speed") in the ASAC recommendations for ALMA 2030 [9]. It also fits in Pathway No. 04 ("2SB B9/B10") and Pathway No. 05 ("Sensitivity: Lower noise Rxs") in the ALMA Development Working Group Report "Pathways to Developing ALMA" [10] and in the ALMA Development Roadmap [11].

B. Technical aspects

The main technical questions and challenges are the following:

1) What is the actual maximum usable IF bandwidth of the existing Band 9 SIS devices?
2) What are the key performance parameters of the IF components, especially in the view of obtaining high ultra-wide band image rejection?
3) Can IF components be found with the needed bandwith, and if not, how can architectural workarounds be devised?

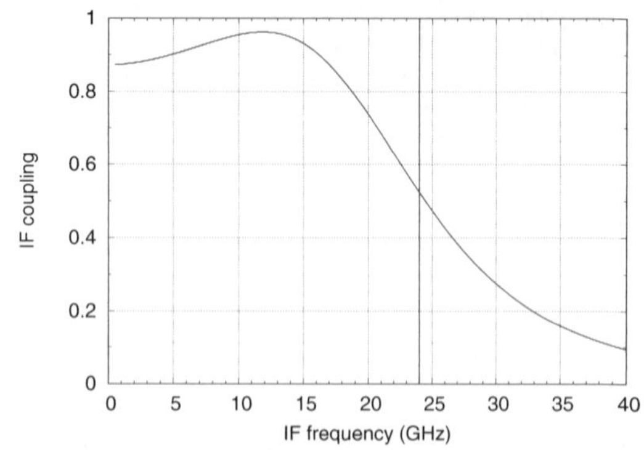

Fig. 2: Simulated IF power coupling of the Band 9 SIS junction at the operating point.

To answer the first question, the employed Band 9 SIS devices were resimulated, using tools that were unavailable at the time the devices were developed almost twenty years ago. We will not go further into the simulation here (details can be found in [12] and of course [8]), except to show one of the important results in this context, namely the mixer's IF power coupling to a $50\,\Omega$ coaxial port. This is shown in figure 2, demonstrating a good match up to at least 20 GHz. The -3 dB point in the power coupling is actually located at about 24 GHz. One of the two main aspects of the study was the experimental verification of these simulations.

The second and third questions above are of course closely related. For instance, circulators, traditionally used to suppress standing waves between poorly matched mixers and LNAs, become impractical (or nonexistant) when an ultra-wide IF bandwidth (say 2–20 GHz) is pursued. This means in turn that either the mixer output or the LNA input (or both) should be well-matched. Since the SIS matching is hard to improve (and in our case, we intend to re-use the existing Band 9 mixers in any case), the focus must be on the input matching of the LNAs. For application in a 2SB receiver, the input matching becomes even more important [3], [4]. Apart from the matching, the noise temperature (as mentioned above) and

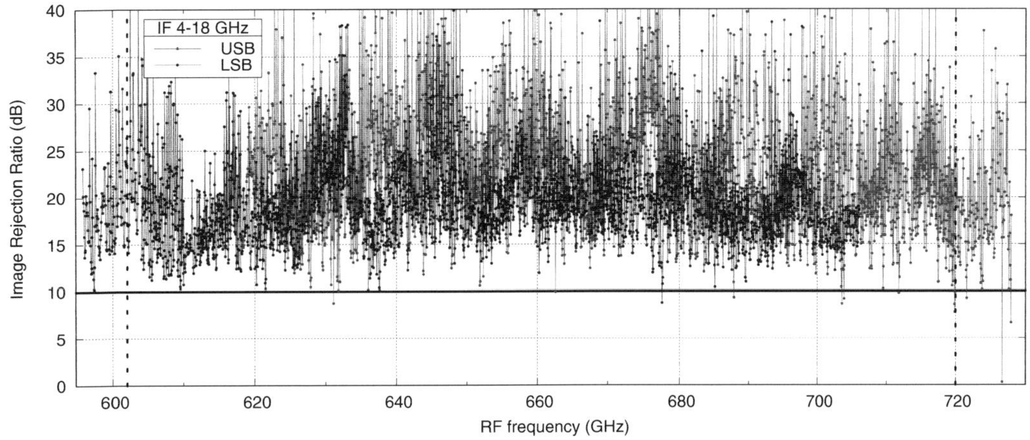

Fig. 3: The combined IRR data for the tested 2SB mixer with ultra-wide band IF infrastructure and 8 GHz-spaced LO settings, as function of the RF frequency. The IF range of each sub-trace is limited to 4–18 GHz. The horizontal black line at 10 dB indicates a typical specification of similar receivers.

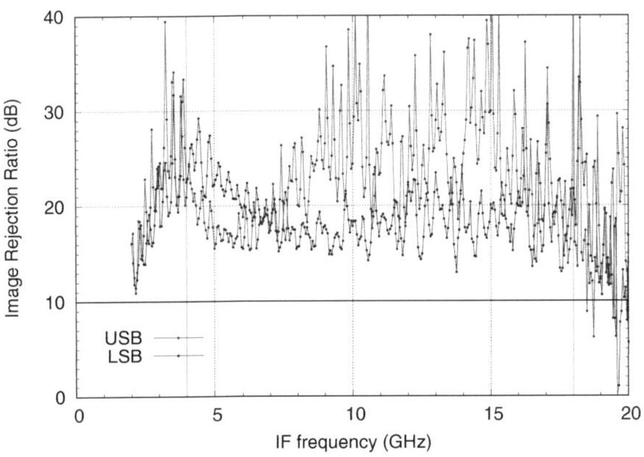

Fig. 4: The IRR singled out for one LO frequency near the middle of the band (662 GHz) over the full 2–20 GHz IF band. The vertical yellow lines indicate a usable band of 4–18 GHz.

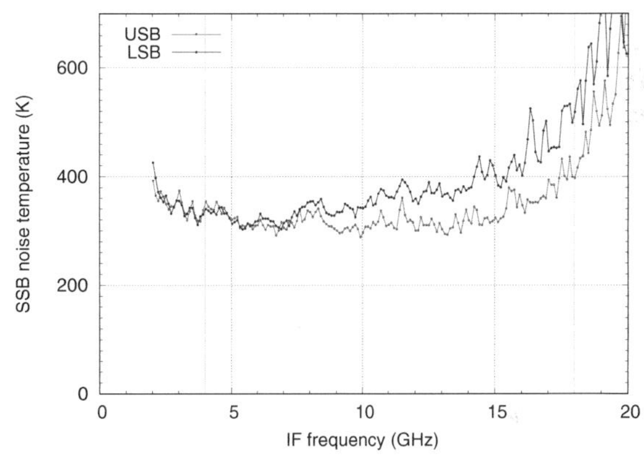

Fig. 5: The noise temperature for both sidebands at the same LO frequency as fig. 4. Note that this is the single-sideband (SSB), which is intrinsically double that of the individual DSB mixers, even any added components are lossless).

the gain flatness are critical.

Several cryogenic wide-band LNAs were used in the course of the study. These were supplied by the Observatorio Astronómico de Yebes (Spain), who have their own program to develop amplifiers along these performance lines. Yebes also supplied us with a cryogenic 4–20 GHz quadrature hybrid for the 2SB demonstration mixer. The LNA type (Y420G) used in he final phases of this study, namely for construction of the single-polarization 2SB receiver has a full gain bandwidth of 4–20 GHz with an input return loss better than 15 dB in at most frequencies, and better than 10 dB overall. The noise temperature is in the range of 5–6.5 K. In the mean time, the development process at Yebes progressed [13], with further improvements in noise performance and matching characteristics.

C. Results

We can only present here a small selection of the experimental results, to show that a wide-band 2SB receiver is indeed feasible at this frequency, using the existing Band 9 mixer devices. It should be noted that these results are a proof of concept, and were not extensively optimized.

Figure 3 shows the obtained image rejection ration (IRR), synthesized out of traces 8 GHz in LO frequency apart. The IF band here is limited to a practical 4–18 GHz. Even with the non-optimized system, the IRR is above 10 dB (a typical specification for the existing ALMA 2SB bands), while in many places 15 dB is obtained. Because the contributions of the different LO settings overlap in fig. 3, for clarity one trace (LO 662 GHz) is singled out in fig. 4, where the IRR can clearly be observed as better than 14–15 dB up to about 18 GHz IF.

The noise temperature at the same LO setting is plotted in fig. 5. As mentioned above, this system was not optimized yet. The best part of the USB is just above 300 K (SSB), while with SEPIA660 values around 150-200 K were obtained [5]. Since the main losses are located in components before the mixer (which did not fundamentally change with respect to SEPIA660), we expect that with careful optimization similar noise performance can be obtained.

III. Extending the RF bandwidth

A. Scientific driver

The reason that ALMA Band 9 originally was chosen to be 602–720 GHz is the presence of strong pressure-broadened water vapour absorption lines at 557 and 750 GHz. In practice, however, very dry conditions still allow useful observations slightly outside of this range, especially if the source is sufficiently bright. A good example, obtained with the SEPIA660 2SB receiver at APEX, is a spectral line survey of the Orion-KL star-forming region [14], with the focus specifically on the frequency ranges around and just outside of the ALMA Band 9 band edges (viz. 581–607 GHz and 701–727 GHz). Even with non-ideal atmospheric conditions, over 100 lines of various molecular species were observed in just minutes of integration time. Although many of these molecules can also be observed in the current ALMA Band 9 frequency range (or in other bands) [15], this clearly demonstrates the feasibility of such observations, especially when combined with a high image-rejection ratio (better than 20 dB at most frequencies for SEPIA660).

Another scientific driver is the expanded access to atomic cooling lines of, e.g., [CII] and [OIII] in distant galaxies over a larger range of redshifts. The mentioned lines appear in Band 9 around $z = 2$ and $z = 4$, respectively, just about bracketing the peak of star formation, making this aspect relevant in the realization of one of the three ALMA 2030 science goals (Origins of Galaxies) [11].

B. Technical aspects

As demonstrated in SEPIA660, the Band 9 SIS junctions typically have an RF bandwidth extending significantly beyond the 602–720 GHz range of the Band 9 specification. Also the reflective optics in the receiver as well as the corrugated feedhorns have much wider relative bandwidths than the specification ($\sim$25–30%, rather than 12%). In the scope of this study, apart from the mixers qualified for SEPIA and LLAMA, about 28 extra SIS mixers have been tested successfully in the extended RF band. We therefore assume that this is representative for all Band 9 mixers.

Since the junctions and surrounding components already have sufficient RF bandwidth, the only modification needed is the extension of the local oscillator (LO) tuning range. Most of the LO resides in the Warm Cartridge Assembly (WCA) just outside the vacuum flange of the front-end cryostat, consisting of a YIG-tuned oscillator (YTO), an active multiplier chain (AMC) and a power amplifier (PA). The final part, the $\times 9$ multiplier, is located in the cryogenic part. It was demonstrated before [5], during the construction and technical commissioning of the SEPIA and LLAMA receivers, that AMC, PA and multiplier are indeed usable beyond the ALMA band edges. For these two receivers, a new LO tuning range of 586–730 GHz was defined. With an IF band of 4–12 GHz, this yields an RF band of 574–742 GHz. Since the edges of this band lie deep within the pressure-broadened tails of the water vapour lines, we can state that the receiver RF bandwidth is then totally atmosphere-limited. In fact, the SEPIA660 receiver has recently been used over this extended range to perform a high-resolution study of the atmospheric transmission itself [16].

Concludingly, we can state that the only technical intervention needed is the modification of the YTO, which is a relatively minor operation.

IV. Improving the polarization purity

A. Scientific driver

In discussions with members from the astronomical community, two science cases were identified that would benefit greatly from improved polarimetric performance of Band 9:

- The study of magnetic fields in very dense environments of circumstellar envelopes around evolved stars and high mass star-forming regions through the vibrationally excited water maser line at 658 GHz (see e.g., [17]).
- The study of dust polarization at high frequency, in combination with similar measurements at low frequency (ultimately down to Band 1), provides a powerful tool to constrain the sizes of dust grains, and thus to study processes such as dust settling and grain growth in protoplanetary disks around young stellar objects.

To a certain degree there may be a chicken-and-egg situation here, as there are hardly any facilities in the world, including ALMA, that *can* do high-quality polarimetry on extended sources at these frequencies. This also means that compelling science cases will rarely be considered by the community. It is very well conceivable that an offer of this capability may bootstrap further interest in it.

B. Technical issues

An internal study at ESO pointed out that the polarimetric performance for extended sources is typically not dominated by the cross-polar level of the receiver, but by station-to-station variations in the the misalignment (the "beam squint") of the two orthogonal polarizations on sky. For meaningful polarimetry, the spread in beam squint for Band 9 (although formally within specification) would have to be reduced by an order of magnitude.

Based on on-sky observation of point sources, it turns out that there is a clear division between the beam squint of the bands employing a grid for polarization splitting (Bands 7, 9 and 10), and the ones using a single feedhorn coupled to a waveguide orthomode transducer (OMT), the latter being better by an order of magnitude or more; exactly the improvement that is needed, as just mentioned.

The question is if such an horn-OMT combination is feasible for Band 9. Not only the required tolerances (of the order of 1.5 μm) are challenging, but especially the extra losses incurred must be minimized at all cost. The double-ridged septum

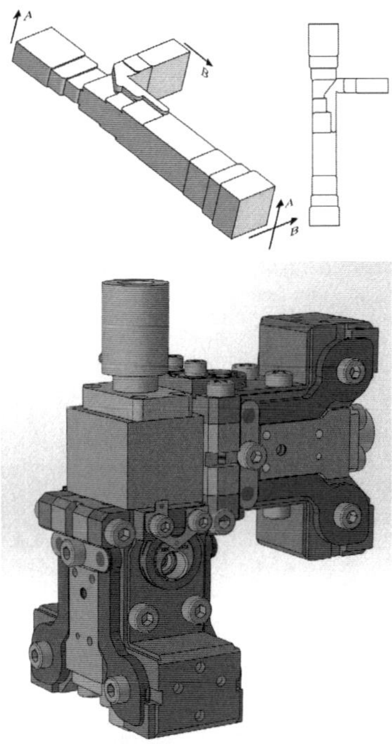

Fig. 6: Top: structure of the OMT by Dunning et al. (image from [18]). Bottom: the basic geometry of two SEPIA660-type Band 9 2SB mixers around a block (green) that could contain a Dunning-type OMT. The OMT structure itself will be about 2 mm long in the Band 9 frequency range and will certainly fit in a block like this.

OMT used in Band 5 (based on NAOJ's Band 4 design [19]) gives excellent cross-polar performance (26-27 dB typically). However, its long internal waveguide runs are estimated to cause an unacceptible increase in noise temperature. As an alternative, we propose the use of another OMT structure, invented by Dunning, Srikanth and Kerr [18], represented in the top panel of fig. 6. This OMT looks pretty much like a minimum-length T-splitter, and a more compact design can hardly be imagined. Since waveguide losses are the main argument against using an OMT at our frequencies, a design like this seems a good candidate for a Band 9 single-feedhorn architecture. In the bottom panel of fig. 6, a straw-man design is shown of how two Band 9 2SB mixer could be fitted around such an OMT.

A simpler (and definitely cheaper) solution is to try and improve the beam squint in the existing optics assemblies. For this, we simulated the Band 9 optics again, with wider fields of tolerance than before. The most likely cause of the large spread in beam squint was pinpointed to the polarization grid itself, rather than machining tolerances of the mirror surfaces or their alignment structures. Perhaps the deviations are stable enough to enable shimming based on beam pattern measurements in the lab (possibly with some iterations). Whether this is doable is unknown at the moment, but could be established in a few test cases. A definite disadvantage is that this method would preclude any improvement of the cross-polar level, so that the total improvement in polarimetric purity is limited.

V. Programmatic Issues

Here we present briefly the outcome of the other study goals, listed in section I-A. As usual, the reader is referred to [8] for the full account.

A. Availability of SIS mixer devices

An extensive statistical data-mining operation was undertaken, together with a small remeasurement campaign, to determine whether a sufficient number of SIS juncions are still available.

For a full Band 9 2SB upgrade a total of about 320 pair-matched SIS devices are needed. For the matching, in order to optimize the image rejection ratio, we expect to need up to 50% more, i.e., a worst-case maximum of 480. The actual number needed for the upgrade will most likely be in between these two extremes. The intention is to re-use as many of the mixers from the existing DSB Band 9 receivers as possible, as well as the delivered spares. Besides that, we will have to produce new mixers from bare SIS chips. We found that our current stock of SIS junctions is probably just enough to achieve this with an average noise temperature very close to the current average in the Band 9 array.

It should be noted that, although efforts are ongoing, e.g., at GARD in Sweden to re-establish Band 9 SIS production, any remanufacturing of these will be quite cosltly. For that reason, our baseline goal is to work with the existing SIS devices. On the other hand, for a significant *improvement* in the overall noise temperature, the complete remanufacturing of the SIS junctions remains the only option.

B. Upgrade strategies

A discussion was held between stakeholders from NOVA, ESO and JAO/OSF concerning the possible on-site strategies to upgrade the existing Band 9 DSB receivers to a 2SB configuration. The main issue here is the availability of Band 9 during the upgrade, which is likely to take at least 2–3 years. Three global options can be envisioned:

- Take Band 9 completely off-line for the duration of the upgrade;
- Recombine the LSB and USB of the upgraded 2SB Band 9 receivers before the IF switch for (part of) the duration of the upgrade, remove upon completion;
- Let the 2SB and DSB CCAs coexist and only use either the USB or LSB of the 2SB CCAs (switchable by bias-inversion) for the duration of the upgrade.

The first one is obviously highly undesirable, and will not be considered further. Also, during the discussion it became clear that the temporary addition of combiners in already-upgraded receivers (to be removed later) would pose significant obstacles for operations and maintenance of the array.

From the front-end point of view, the simplest solution therefore is to let a growing population of 2SB receivers co-exist with a shrinking number of DSB ones. In this case, the already upgraded receivers would always export both USB and LSB, but when correlated with the DSB receivers, only one of the sidebands would be used. Two operation modes could be offered: full 2SB with the growing number of upgraded receivers, or the full array with reduced sensitivity due to the missing sideband in part of the receivers. An advantage of the latter mode is that, although the sensitivity is reduced, all baselines contribute to filling the U-V plane. Apart from a mechanism to flag the data as coming from either DSB or 2SB receivers, the correlator and data pipeline should have no fundamental problems handling the mixed array.

Logistically, since significant re-use of components is taken into account, for a "rolling" upgrade scheme, an upgrade pool of the order of 10 receivers is expected to be required to enable a smooth flow. This could be accomodated by a temporary reduction of the number of operational receivers in the array.

C. Upgrade costing

Any estimate for the total cost of an upgrade is by nature a snapshot. Especially with the current economical developents, a time-stable estimate is hard to give. Nevertheless, the costing exercise performed during the study indicates that a full upgrade of Band 9 (or any other band for that matter) will be dominated by labour costs. Furthermore, the components that have to re-purchased are typically the more expensive in the receiver (e.g., mixers, cryogenic amplifiers). Even while we can re-using several expensive key components (LOs, including cryogenic multipliers, corrugated horns, etc.), an upgrade will likely to come out at 60–80% of the price of a totally new band, depending on chosen upgrade options.

VI. Conclusions

We demonstrated a proof-of-concept side-band separating ALMA Band 9 mixer with an IF band extending up to at least 18 GHz. The RF bandwidth can also be extended beyond the Band 9 limits, up to the very edges of the blocking water vapour lines. With a single-feed architecture, the polarimetric performance (mainly limited by the beam squint) could be improved significantly, probably to a lesser extent by simply shimming the existing grids. The number of remaining SIS devices may just be sufficient to upgrade the array while maintaining the noise performance. For a real improvement of noise temperature, new mixer devices are most likely needed. Finally, upgrade logistics and costing were explored.

It is our intention to build and characterize a full side-band separating ultra-wide IF Band 9 demonstration receiver with improved polarimetric performance in the coming few years.

References

[1] A. Baryshev, A. Khudchenko, R. Hesper, and W. Boland, "Design verification and test results," NOVA Sub-mm Instrumentation Group, ALMA document FEND-40.02.09.00-1874-A-REP, Mar. 2012.

[2] R. Hesper, A. Khudchenko, A. M. Baryshev, J. Barkhof, and F. P. Mena, "A new high-performance sideband-separating mixer for 650 GHz," in *SPIE Conference Series*, vol. 9914, 2016, pp. 99 140G–1–99 140G–11. [Online]. Available: http://dx.doi.org/10.1117/12.2233065

[3] A. Khudchenko, R. Hesper, A. M. Baryshev, J. Barkhof, and F. P. Mena, "Modular 2SB SIS receiver for 600–720 GHz: Performance and characterization methods," *IEEE Transactions on Terahertz Science and Technology*, vol. 7, no. 1, pp. 2–9, Jan. 2017.

[4] R. Hesper, A. Khudchenko, A. M. Baryshev, J. Barkhof, and F. P. Mena, "A high-performance 650-GHz sideband-separating mixer — design and results," *IEEE Transactions on Terahertz Science and Technology*, vol. 7, no. 6, pp. 686–693, Nov. 2017.

[5] R. Hesper, A. Khudchenko, M. Lindemulder, M. Bekema, R. de Haan-Stijkel, J. Barkhof, J. Adema, and A. Baryshev, "A deployable 600–720 GHz ALMA-type sideband-separating receiver cartridge," in *29th International Symposium on Space Terahertz Technology, Pasadena*, 2018. [Online]. Available: https://www.nrao.edu/meetings/isstt/papers/2018/2018098103.pdf

[6] Belitsky, V., Lapkin, I., Fredrixon, M., Meledin, D., Sundin, E., Billade, B., Ferm, S.-E., Pavolotsky, A., Rashid, H., Strandberg, M., Desmaris, V., Ermakov, A., Krause, S., Olberg, M., Aghdam, P., Shafiee, S., Bergman, P., Beck, E. De, Olofsson, H., Conway, J., Breuck, C. De, Immer, K., Yagoubov, P., Montenegro-Montes, F. M., Torstensson, K., Pérez-Beaupuits, J.-P., Klein, T., Boland, W., Baryshev, A. M., Hesper, R., Barkhof, J., Adema, J., Bekema, M. E., and Koops, A., "SEPIA — a new single pixel receiver at the APEX telescope," *A&A*, vol. 612, p. A23, 2018. [Online]. Available: https://doi.org/10.1051/0004-6361/201731458

[7] P. Bergman, M. Olberg, D. Meledin, M. Fredrixon, E. Sundin, M. Strandberg, R. Hesper, J. Barkhof, K. Torstensson, J.-P. Perez, F. Montenegro, R. Parra, M. Ciechanowicz, C. de Breuck, and F. Wyrowski, "SEPIA660 (Band 9) commissioning report," Atacama Pathfinder EXperiment, APEX Doc. APEX-ESO-TRE-0023, Jan. 2019.

[8] R. Hesper, J. Barkhof, S. Realini, A. Baryshev, and J. Adema, "ALMA band 9 sideband separating upgrade - study report," ESO, Study Report FEND-40.02.09.00-1974-C-REP, Feb. 2023. [Online]. Available: https://www.eso.org/sci/facilities/alma/developmentstudies/Band9UpgradeStudyReport.pdf

[9] A. D. Bolatto, J. Carpenter, S. Casassus, D. Iono, R. Ivison, K. Johnson, H. van Langevelde, J. Martín-Pintado, M. Momose, R. Moreno, K. Motohara, R. Neri, N. Ohashi, T. Oka, R. Osten, R. Plambeck, E. Schinnerer, D. Scott, L. Testi, and A. Wootten, "A road map for developing ALMA — ASAC recommendations for ALMA 2030," 2015.

[10] A. Bolatto, S. Corder, D. Iono, L. Testi, and A. Wootten, "Pathways to developing ALMA — ALMA development working group report," 2015.

[11] J. Carpenter, D. Iono, L. Testi, N. W. amd A. Wootten, and N. Evans, "The ALMA development roadmap," 2018.

[12] K. Rudakov, "Development of advanced superconductor-insulator-superconductor mixers for terahertz radio astronomy," Ph.D. dissertation, University of Groningen, Jul. 2021.

[13] I. López-Fernández, J. Gallego, C. Diez, I. Malo, R. Amils, R. Flückiger, D. Marti, and R. Hesper, "A 16 GHz wide band cryogenic amplifier with 4 K noise temperature for the future IF of sub-mm radio astronomy receivers," Apr. 2023, submitted.

[14] F. M. Montenegro-Montes, K. Torstensson, R. Parra, J. P. Pérez-Beaupuits, L. Å. Nyman, C. Agurto, F. Azagra, M. Cárdenas, E. González, F. MacAuliffe, P. Venegas, C. De Breuck, P. Bergman, D. S. Gunawan, F. Wyrowski, T. Stanke, V. Belitsky, M. Fredrixon, D. Meledin, M. Olberg, M. Strandberg, E. Sundin, J. Adema, J. Barkhof, A. Baryshev, R. Hesper, and A. Khudchenko, "Orion-kl observations with the extended tuning range of the new SEPIA660 APEX facility instrument," *The Messenger*, vol. 176, pp. 20–24, Jun. 2019.

[15] F. Kemper, E-mail, 2019-11-08.

[16] J. R. Pardo, C. De Breuck, D. Muders, J. González, F. M. Montenegro-Montes, J. P. Pérez-Beaupuits, J. Cernicharo, C. Prigent, E. Serabyn, T. Mroczkowski, and N. Phillips, "Extremely high spectral resolution measurements of the 450 μm atmospheric window at chajnantor with APEX," *A&A*, vol. 664, p. A153, 2022. [Online]. Available: https://doi.org/10.1051/0004-6361/202243409

[17] A. Baudry, E. M. L. Humphreys, F. Herpin, K. Torstensson, W. H. T. Vlemmings, A. M. S. Richards, M. D. Gray, C. De Breuck, and M. Olberg, "Vibrationally excited water emission at 658 GHz from evolved stars," *A&A*, vol. 609, p. A25, 2018.

[18] A. Dunning, S. Srikanth, and A. R. Kerr, "A simple orthomode transducer for centimeter to submillimeter wavelengths," in *Proc. 20th ISSTT*, 2009, pp. 191–194.

[19] S. Asayama and M. Kamikura, "Development of double-ridged wavegide orthomode transducer for the 2 mm band," *J. of Infrared, Millimeter, and Terahertz Waves*, vol. 30, no. 6, pp. 573–579, Jun. 2009.

Status of ALMA Band 6v2 Receiver Development

J. G. Lambert*[1], A. Navarrini[1], A.R. Kerr[1], J. Effland[1], P. Dindo[1], K. Saini[1], S. Srikanth[1], D. Vaselaar[1], A.W. Lichtenberger[2], M. Cyberey[2], P. Mena[1], and C. Jarufe[1]

Abstract— **We are developing an upgraded version of the existing 211-275 GHz ALMA Band 6 receiver, referred to as "Band 6v2", which complies with the strategies defined in the ALMA Development Roadmap to 2030 and conforms to the recommendations of the ALMA Front-end & Digitizer Requirements Upgrade Working Group. Our goal is to deliver an improved production-level receiver based on (i) a new Cold Cartridge Assembly (CCA) with increased sensitivity over an expanded 4-16 GHz IF band (4-20 GHz goal) and a modestly expanded 209-281 GHz RF band, and (ii) a new low-noise Warm Cartridge Assembly (WCA). The new Band 6v2 receiver will be backward-compatible with the current Band 6 receiver and will directly replace the existing units installed into the ALMA FE (Front-End) cryostat.**

Keywords—SIS, mixer, sideband separation, LNA, receiver, optics, OMT.

I. Introduction

In the 20 years since the current ALMA Band 6 receiver was designed there have been significant advances in technology which will allow a fundamentally re-designed receiver to cover a substantially wider IF (Intermediate Frequency) band with lower and flatter receiver noise temperature over an expanded RF (Radio Frequency) band:

- Cryogenic IF amplifiers are now available with lower noise temperature over a significantly expanded band.

- Cryogenic edge-mode IF isolators can be made with low insertion loss spanning more than two octaves, enabling flatter receiver noise temperatures over a much expanded IF band.

- It is now possible to make low leakage-current SIS junctions with higher critical current density, thereby allowing low noise operation over a wider RF band.

- LO (Local Oscillator) sources can now be made with reduced sideband noise, resulting in lower receiver noise temperature.

According to the board-approved ALMA 2030 Development Roadmap [1], the current development priorities are to broaden the receiver IF bandwidth by a factor of two or more and to upgrade the associated electronics and correlator to handle the increased bandwidth, as described by the Wideband Sensitivity Upgrade (WSU) initiative [2].

Driven by the science needs of ALMA, and following a successful outcome of our ALMA project proposal, we are currently developing an upgrade for the existing 211-275 GHz ALMA Band 6 receiver [3] cartridge with a substantially improved second generation version, referred to here as "Band 6v2" [4], which complies with the ALMA Development Roadmap strategy and with the recommendations of the ALMA Front-End & Digitizer Requirements Upgrade Working Group.

Despite its high sensitivity and scientific productivity, Band 6 has some deficiencies that we plan to remedy. Currently, ALMA can handle only 4 GHz of bandwidth per sideband per polarization. The present ALMA Band 6v1 receivers have low noise, of order 4 to 6 times the photon temperature, over ~6-10 GHz IF, but above and below that range they degrade substantially (Fig. 1). As the LO and IF are changed, the SSB receiver noise T_{rx} varies by a factor of 2 or more, typically from a low of ~40 K to ~80 K when LO sideband noise and IF amplifier noise are at their highest. In Band 6, it is particularly desirable to simultaneously observe the $^{12}C^{16}O$, $^{13}C^{16}O$, and $^{12}C^{18}O$ astronomical lines which fall in the frequency range 219-231 GHz. This requires $f_{LO} \approx 225$ GHz, for which the lines of interest fall in the LSB and USB

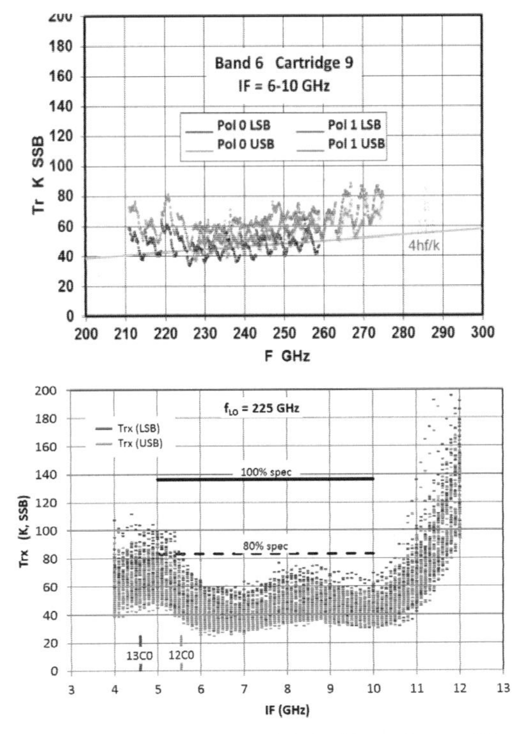

Fig. 1. SSB noise temperatures of Band 6 Cartridge no. 9 versus RF for IF in the range 6-10 GHz (top), and of 80 production Band 6 receivers versus IF for f_{LO}= 225 GHz (bottom).

[1]NRAO Central Development Laboratory, Charlottesville, 22903 VA, USA (e-mail: jlambert@nrao.edu); [2]Innovation in Fabrication (IFAB) Facility, University of Virginia, Charlottesville, 22904 VA, USA.

and produce IF signals in the range 4.3-5.5 GHz. The bottom panel of Fig. 1 shows the SSB receiver noise temperatures versus IF of 80 production Band 6 receivers with $f_{LO} = 225$ GHz. It is clear that astronomical lines appearing in the 4.3-5.5 GHz range will be observed with reduced sensitivity.

The interaction between the Band-6 SIS mixer and the IF preamp, which is connected directly to the mixer without an isolator, can cause the cartridge gain to vary across the IF band to an undesirable degree and resulting in many Band 6 mixer-preamplifiers missing gain flatness specifications.

Further, some Band 6 LO modules, which are part of the WCA, have substantial noise in the 4-12 GHz sidebands, at offsets ~4-5 GHz from the carrier frequency. This LO sideband noise contributes to the overall receiver noise temperature for low IF.

Finally, the original polarization efficiency specification of 99.5% could not be met on most of the Band 6 cartridges. A concession was made which relaxed the required polarization efficiency to different percentages over different blocks of frequencies in the 211-275 GHz band.

The upgraded second-generation Band 6 receiver (Band 6v2), will result in several benefits including a reduction in integration time by a factor of ~1.5 to as much as 3 in the worst parts of the current IF (assuming a typical sky temperature), an increase in the IF bandwidth from the present 5.5 GHz per sideband per polarization to at least 12 GHz (4-16 GHz) and potentially to 16 GHz (4-20 GHz), and an increase in RF coverage by 8 GHz, to 209-281 GHz from the present 211-275 GHz.

The four primary goals of the Band 6v2 upgrade are the following:

1) To increase the IF bandwidth from the present 4 GHz per sideband per polarization (across ~6-10 GHz) to 12 GHz per sideband per polarization (minimum 4-16 GHz) or 16 GHz per polarization per sideband (corresponding to our goal of 4-20 GHz IF). While the current ALMA correlator and IF transmission system can only accommodate 4 GHz per sideband per polarization, the new receiver will take advantage of future upgrades of those sub-systems in the framework of the WSU.
2) To expand the usable RF band from the current 211-275 GHz to 209-281 GHz, thereby adding the ability to observe new diagnostic transitions and other lines in Band 6.
3) To improve and flatten the receiver noise performance and IF power density slope across the full IF range.
4) To improve the OMT and optics to achieve lower sidelobes, lower cross-polarization, and easier fabrication.

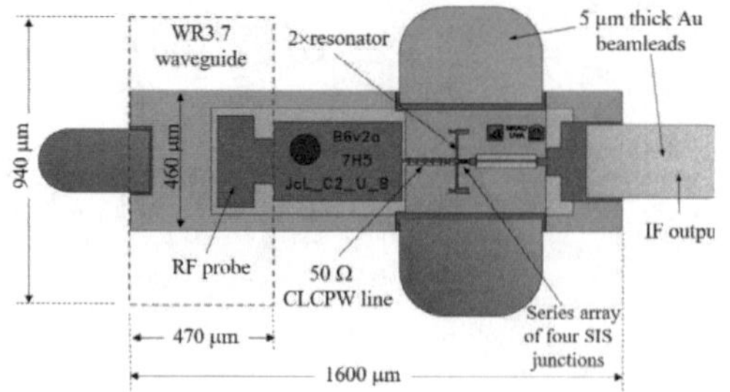

Fig. 3. The layout of the Band 6v2 mixer chip.

The project is exploring several different receiver configurations for the ALMA Band 6v2 sideband separating (2SB) SIS receiver, with improvements to all the major receiver components, including optics, OMT, mixers, IF section and local oscillator. The baseline configuration for one of the two polarizations is shown in Fig. 2. It includes two DSB SIS mixers followed by 4-20 GHz low-loss edge-mode IF isolators from SAO (Smithsonian Astrophysical Observatory) [5], 4-20 GHz IF low noise amplifiers (LNA) from NRAO, and 4-20 GHz IF hybrid from the Yebes Observatory.

In the Band 6v1 SIS mixers, the junctions and associated RF tuning circuits are on an 80-µm-thick quartz substrate. The new Band 6v2 mixer chips (Fig. 3) are fabricated on thin silicon membranes with gold beam leads to provide electrical connections and mechanical support. The drop-in SIS mixer chips, fabricated at UVA IFAB, will utilize 5 µm thick Si with 5 µm thick Au beam leads. The mixer utilizes an array of four Nb/Al-AlOx/Nb SIS junctions in series with target $J_C = 6.9$ kA/cm^2 (higher than for Band 6v1, which targeted $J_C = 5.2$ kA/cm^2). Simulations predict flat noise across the RF band. The mixer chips will be easier to package than for Band 6v1.

The individual Band 6v2 mixer chips will be permanently mounted in DSB (Double Side Band) mixer modules which

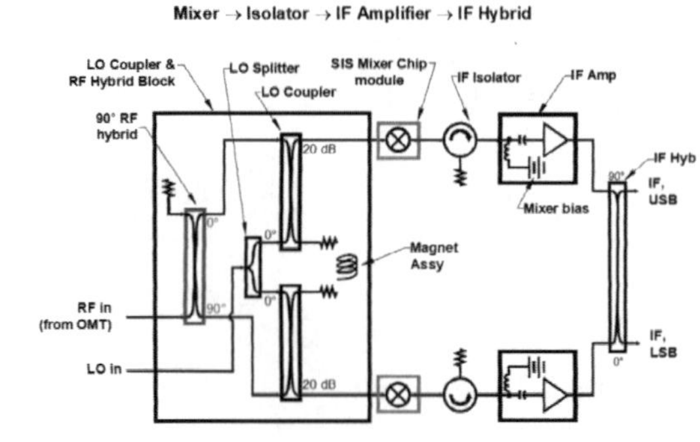

Fig. 2. Schematic of the Sideband Separating (2SB) SIS receiver for Band 6v2 (only one polarization channel is shown). The receiver is based on two DSB SIS mixers cascaded with IF amplifiers preceded by broadband IF isolators.

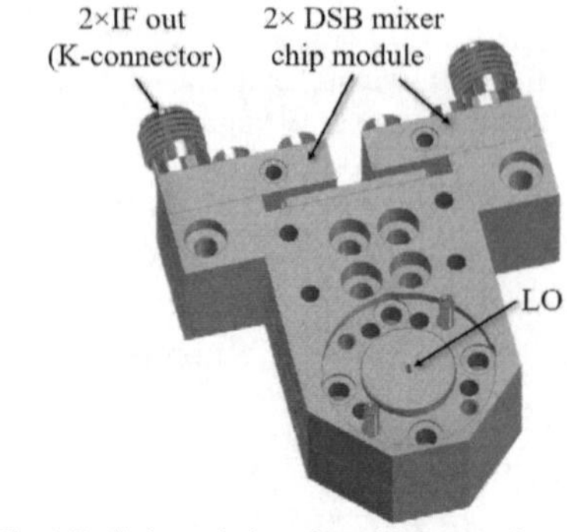

Fig. 4. Preliminary design of Band 6v2 2SB mixer assembly with two DSB mixer chip modules mounted on a passive assembly comprising the RF and LO waveguide circuitry. The RF waveguide input is parallel to the LO waveguide input and is located on the opposite side of the 2SB assembly (not visible).

can be rapidly cooled to determine their I(V) characteristics and DSB mixer properties. This will allow the selection of pairs of well-matched modules for use in sideband-separating mixers. The complete Band 6v2 2SB mixer consists of an E-plane split-block waveguide assembly containing an RF branch-line waveguide hybrid, the LO coupler, and the 4-K image and LO terminations, attached to the two elemental DSB mixer chip modules (Fig. 4). At a later phase of development, a more integrated sideband separating mixer will be considered, after downselection of the receiver architecture.

In Band 6v1, a corrugated feedhorn and a Bøifot junction OMT [6] were adopted, both having a square waveguide (0.94×0.94 mm^2) at their interface flanges. The Band 6v1 OMT incorporates wire pins in the waveguide side-arm, and a thin septum across the square input waveguide whose positioning in the split block is critical to the performance, particularly the polarization properties. A misalignment between the square waveguides of feed and OMT at the interface between the two units can cause the excitation of higher-order spurious modes and generate unwanted cross-polarization. Three alternative waveguide OMT designs that do not use pins or a septum are being considered as replacement for the Bøifot Band 6v1design: a dual-ridge and a reverse-coupler OMT, both designed by NRAO, and an OMT being designed by NAOJ. The NRAO OMT designs are scaled from the W-band models given in [7, 8]. All OMTs adopt circular waveguide inputs (1.29 mm diameter) to match that of the Band 6v2 feedhorn. Unlike the Band 6v1 OMT, the waveguide outputs of the Band 6v2 OMT have E-planes parallel to each other to facilitate integration with the two 2SB mixer assemblies.

By the end of Phase 1 (the current project phase), our goal is to build and demonstrate an improved prototype receiver based on: *a*) a new Cold Cartridge Assembly (CCA) with increased sensitivity over a greatly expanded IF band and modestly expanded RF band; *b*) a new low-noise Warm Cartridge Assembly (WCA) which contains a new LO source with lower AM sideband noise that incorporates a fundamental frequency YIG oscillator operating at twice the frequency of the current one. The receiver optics will also be redesigned. The horn and dual mirror assembly are modified to improve access to the cartridge components for repairs, and an improved low-loss orthomode transducer (OMT) with simpler construction and better reproducibility is being developed.

The improvements from Band 6v2 will unlock discovery space in all three fundamental science areas of ALMA 2030: The Origins of Planets, The Origins of Chemical Complexity, and The Origins of Galaxies.

References

[1] J. Carpenter, D. Iono, L. Testi, N. Whyborn, A. Wotten, N. Evans, "The ALMA Development Roadmap".

[2] J. Carpenter, C. Brogan, D. Iono, T. Mroczkowski, "The ALMA2030 Wideband Sensitivity Upgrade", *ALMA Memo 621*, Nov. 2022.

[3] A. R. Kerr, S.-K. Pan, S. M. X. Claude, P. Dindo, A. W. Lichtenberger, J. E. Effland and E. F. Lauria, "Development of the ALMA-North America Sideband-Separating SIS Mixers," IEEE Trans. Terahertz Science and Technology, v. 4, no. 2, pp. 201-212, Mar 2014.

[4] A. Navarrini, A. R. Kerr, P. Dindo, J. Effland, J. Lambert, A. W. Lichtenberger, K. Saini, S. Srikanth, D. Vaselaar, B. Hawkins, ALMA Band6v2 Receiver Upgrade, Presented at the ALMA FED-2021 Conference, Sep. 30, 2021. DOI: 10.5281/zenodo.5642784.

[5] L. Zeng, E. Tong, R. Blundell, P. Grimes, S. Paine, "A Low-Loss Edge-Mode Isolator With Improved Bandwidth for Cryogenic Operation," *IEEE Trans. MTT-66*, no. **5**, May 2018.

[6] E. Wollack, "A Full Waveguide Band Orthomode Junction," *NRAO Electronics Division Internal Report* 303, May 1996.

[7] G. Moorey, R. Bolton, A. Dunning, R. Gough, H. Kanoniuk and L. Reilly, "A 77-117 GHz cryogenically cooled receiver for radioastronomy," *Proc. Workshop on the Applications of Radio Science (WARS2006)*, Leura, NSW, Australia, 15-17 Feb. 2006

[8] A. Navarrini and R. Nesti, "Symmetric Reverse-Coupling Waveguide Orthomode Transducer for the 3-mm Band," *IEEE Trans. Microwave Theory Tech.* vol. **57**, no. 1, pp. 80-88, Jan. 2009.

16 pixel, 3 mm band, Cryogenic Array Receiver for Users of the Sardinia Observatory (CARUSO)

Nart Daghestani[*1], Hui Wang[1], Nick Brewster[1], Peter Hunyor[1], Eimear Gallagher[1], Ali Obeed[1], Ben Lane[1], Kai Parow-Souchon[1], Simon Rea[1], Peter G. Huggard[1], Mat Beardsley[1], Diego Pardo[1], Mark Merritt[1], Brian N. Ellison[1], Josef Boehm[2], William McGenn[3], Gary Fuller[3], Danielle George[4], Claudio Jarufe[4**], Alessandro Navarrini[5], Luca Olmi[5], Renzo Nesti[5], Pasqualino Marongiu[5], Pierluigi Ortu[5], Andrea Orlati[5], Luca Cresci[5], Alessandro Scalambra[5], Davide Fierro[5], Federica Govoni[5], Tonino Pisanu[5], Catherine Hambly-Symons[1] and Robert Gallafent[1]

Abstract - **The CARUSO instrument provides 16 independent dual polarization heterodyne receivers in 4×4 configuration operating at W-band frequencies and configured as a focal plane imaging array. CARUSO is designed for use on the 64 m diameter Sardinia Radio Telescope (SRT, www.srt.inaf.it) and detects over the frequency range 70 to 116 GHz. Individual cryogenic pixels use HEMT low noise amplifier technology combined with sideband separating subharmonic heterodyne mixing. Polarisation duplexing is achieved by means of a compact, low loss, orthomode transducer. An image of a single pixel is shown in Fig. 1. The receiver array is mechanically cooled to 20 K in a novel conical shaped cryostat. Integrated receiver front-end assemblies are located at the SRT's Gregorian focal plane and provide high quality beam patterns with high antenna efficiency. The field of view covered by the 4×4 array is 2:15 × 2:15 arcmin^2, unfilled, with separation between contiguous elements of 43 arcsec.**

When installed, CARUSO will provide a state-of-the art W-band array receiver and will detect both continuum and molecular spectral lines from astronomical sources and radio emission from the Sun in the 3 mm atmospheric window. The overall system concept and latest receiver performance results will be presented.

The Enhancement of the Sardinia Radio Telescope (SRT) for the study of the Universe at high radio frequencies is financially supported by the National Operative Program (Programma Operativo Nazionale-PON) of the Italian Ministry of University and Research, Notice D.D. 424 of 28/02/2018 for the granting of funding aimed at strengthening research infrastructures, in implementation of the Action II.1 – Project Proposal PIR01_00010. The Italian National Institute for Astrophysics has contracted UK Research and Innovation for the supply of CARUSO, the Cryogenic Multi-Beam Receiver in W-Band (3 mm wavelength) for SRT, Project Code C87E19000000007, Tender Identification Code 8018611F1D, under the PON OR1grant.

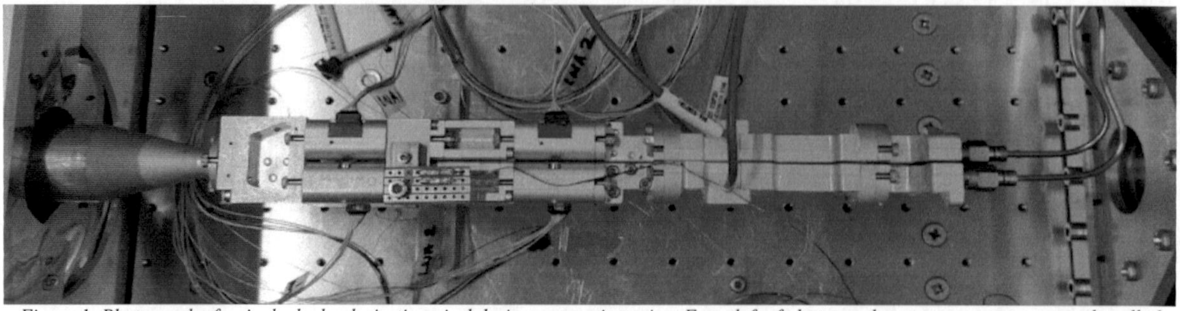

Figure 1: Photograph of a single dual polarization pixel during cryogenic testing. From left of photograph, components are a smooth walled feedhorn, orthomode transducer, low noise amplifiers #1, waveguide isolators, low noise amplifiers #3, subharmonic sideband separating sub-harmonic mixers and frequency triplers

[1]MMT/Chilbolton Group, STFC Rutherford Appleton Laboratory, Harwell, OX11 0QX, UK; [2]Technology department, STFC Rutherford Appleton Laboratory, Harwell, OX11 0QX, UK; [3]Jodrell Bank Centre for Astrophysics, The University of Manchester, Manchester, M13 9PL, UK; [4]Department of Electrical and Electronic Engineering, The University of Manchester, M1 3BB, UK; [5]INAF (National Institute for Astrophysics), Italy; **Now at Central Development Laboratory, National Radio Astronomy Observatory, Charlottesville, VA, USA.

Algorithmic Design of Terahertz Silicon Metaoptics

Conner Ballew*, Sven van Berkel, Cecile Jung-Kubiak, Subash Khanal, and Goutam Chattopadhyay

Abstract—**Patterning dielectric materials at the near- or sub-wavelength scale unlocks a versatile design-space for optical components. These components, often termed "metaoptics", have been widely explored in two-dimensional architectures such as high-contrast gratings (HCGs) and metasurfaces. More recently, volumetric (i.e., 3D) metaoptics designed through multi-objective inverse design have been explored that are capable broadband multifunctional behavior. This talk discusses the design of these 3D metaoptics, their fabrication using silicon microfabrication, and their integration into THz systems. Furthermore, we discuss the shortcomings of the various metaoptics, and how the integration of 3D metaoptics, HCGs, and metasurfaces together can address these shortcomings.**

Keywords—Metaoptics, inverse-design, metasurfaces, high-contrast gratings

I. Introduction

A principle of optical design that has long been used by designers is that optical systems must be built from a modular arrangement of elements. This principle is of course not a physical law, but rather a tool to break large problems down into reasonably sized sub-parts (lens, gratings, etc.) which that can optimized. However, modular arrangements inevitably must be comprised of many discrete components to reach high-performance for complex tasks. Particularly with regards to space technology, this convenient tool becomes a limitation to realizing compact systems.

Approaching optical design with more holistic procedures can enable the synthetization of multiple discrete optical elements into single elements. In fact, the number of degrees of freedom available in even a single sub-wavelength layer is sufficient to enable high-efficiency multi-functional elements such as polarizing beam splitters with simultaneous focusing capabilities [1], [2], albeit over a relatively small bandwidth [3], [4]. More degrees of freedom can be introduced by increasing the thickness of the device, thus allowing greater bandwidth and functionality. To access the degrees of freedom, the volume should be patterned with subwavelength precision in 3D. The primary design questions are: what is the optimal shape of a device, and how can it be fabricated?

This talk focuses on answering these two questions, specifically for THz frequencies. A relatively new class of 3D metaoptics comprised of stacked, patterned Si wafers is introduced that is designed with adjoint-based inverse-design techniques to allow for extremely versatile functionality, allowing the control of all fundamental properties of light over broad spectral bands [5]. Practically, these devices must be small or periodically tiled with small periodicity due to computational limitations. In recent years, these devices have been studied in isolation, rather than integrated into systems. This talk will therefore delve into the unexplored area of coupling these devices with algorithmically designed metasurfaces, high-contrast gratings, and other optical components to enable extremely compact optical systems for space technology.

References

[1] N. Yu and F. Capasso, "Flat optics with designer metasurfaces," *Nature Materials*, vol. 13, no. 2. Nature Publishing Group, pp. 139–150, Jan. 23, 2014, doi: 10.1038/nmat3839.

[2] A. Arbabi, Y. Horie, M. Bagheri, and A. Faraon, "Dielectric metasurfaces for complete control of phase and polarization with subwavelength spatial resolution and high transmission," *Nat. Nanotechnol.*, vol. 10, no. 11, pp. 937–943, Nov. 2015, doi: 10.1038/nnano.2015.186.

[3] D. A. B. Miller, "Fundamental limit for optical components," *J. Opt. Soc. Am. B*, vol. 24, no. 10, p. A1, Oct. 2007, doi: 10.1364/josab.24.0000a1.

[4] A. Arbabi and A. Faraon, "Fundamental limits of ultrathin metasurfaces," *Sci. Rep.*, vol. 7, no. 1, pp. 1–9, Mar. 2017, doi: 10.1038/srep43722.

[5] P. Camayd-Muñoz, C. Ballew, G. Roberts, and A. Faraon, "Multifunctional volumetric meta-optics for color and polarization image sensors," *Optica*, vol. 7, no. 4, p. 280, Apr. 2020, doi: 10.1364/optica.384228.

[1]Jet Propulsion Laboratory, Pasadena, CA 91109, USA

NOTES:

Optics for the TEMPERA-C polarimetric middle atmosphere temperature sounder

Roland Albers*[1], Witali Krochin[1], Gunter Stober[1], Axel Murk[1]

***Abstract*— This paper presents the optical design, calibration, and receiver of the new fully polarimetric temperature radiometer operating around 53 GHz (TEMPERA-C) from the University of Bern for ground-based measurement campaigns. The optical setup will allow for elevation scanning, minimises cross-polar characteristics, and is more compact than the previous iteration of the instrument. The optics have been optimised using the GRASP software and the results are presented here.**

Keywords—radiometry, polarimetric, optics, atmospheric temperature, calibration

I. Introduction

The current iteration [1] of the TEMPERA instrument is a linearly polarised radiometer at 50 GHz in a stationary configuration at the University of Bern which has been used operationally to retrieve atmospheric temperatures up to 50 km altitude. However, there is interest in extending the altitude of these retrievals and enabling campaign measurements at other locations such as the Jungfraujoch (3500 m) which is ideal for reaching higher retrieval altitudes but requires an environmentally resistant instrument. Hence, an updated version of the instrument has been developed which is fully polarimetric and is combined with a more optimised retrieval that considers the Zeeman effect [1] which is currently limiting the vertical range of the retrieval. Resolving the difference between the left- and right-hand polarised emission will make it possible to push the temperature sounding higher into the lower mesosphere. Along with redesigning the optics for polarimetry, emphasis was placed on reducing the size to create a compact and portable instrument. This optimisation has been carried out using the GRASP software by Ticra.

II. Optical Setup

The optical setup (see Fig. 1) consists of a custom designed corrugated, highly gaussian feedhorn (yellow) positioned in the horizontal with a gain of 20.13 dB illuminating a 22.5° off-axis parabolic reflector (orange) with a diameter of 190mm and an edge taper >30dB. The off-axis angle is minimised to avoid induced asymmetries in the farfield beam. The beam is reflected upwards to a circular plane reflector (green). A path length modulation mechanism is used for the mounting of this reflector to account for standing waves. The reflector alternates between two fixed positions by rotating 180°, either directing the beam to the internal calibration target (black) or to the skyward plane reflector (blue). The skyward reflector is mounted in a periscope structure (grey) which can be rotated in the elevation plane to allow for off-zenith measurements. The purpose of the periscope structure (which is not shown in its full extend in Fig. 2) is to protect the instrument during rain- and snowfall, by pointing it towards ground. Some microwave transparent material could be used to seal the periscope but would introduce unnecessary reflections or standing waves especially if the window is painted with a water-resistant coating.

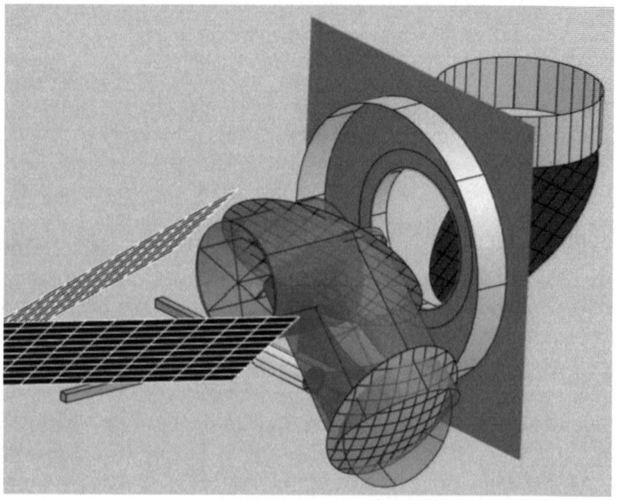

Fig. 1. GRASP model of TEMPERA-C optics in OBCT scan mode

III. Receiver

The receiver of Tempera-C is designed to simultaneously observe all four Stokes parameters. The frontend of Tempera-C splits the signal into vertical and horizontal polarisation, using an ortho-mode transducer (OMT). Both polarisation chains are identical, eliminating phase differences between the two when they are combined in the correlator. The following sentences describe a single chain for simplicity. A coupler right after the OMT facilitates adding a noise diode signal for gain calibration. After a bandpass filter, the signal is downconverted using a heterodyne mixer and then split into two. Both signals are supplied to the

[1]University of Bern, Bern, 3012, Switzerland

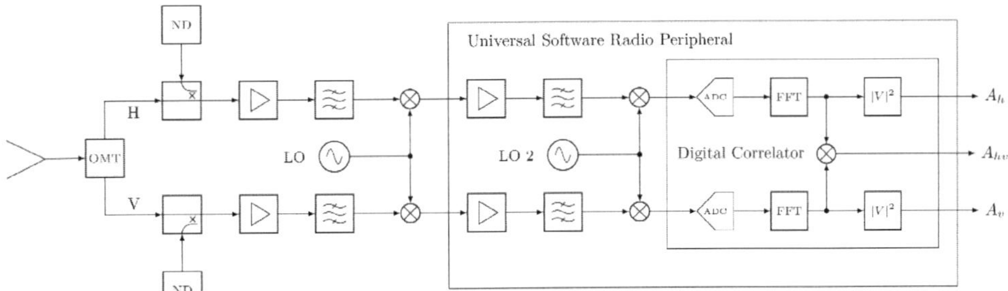

Fig. 2. Schematic of Tempera-C receiver

spectrometer (so four inputs – two per polarisation). In the spectrometer the signals are downconverted further to two different frequencies and digitised with a sampling rate of 200MHz. This allows the spectrometer to process two lines simultaneously, in this case corresponding to 53.0669 GHz and 53.596 GHz. The digital correlator inside the spectrometer then combines both polarisations to extract the third and fourth Stokes parameter. Since the correlation is done purely digitally, it removes a potential error source and is the key advantage of this setup for polarimetry. The spectrometer used for Tempera-C is based on a commercially available Universal Software Radio Peripheral and a FPGA firmware developed in-house for the real-time FFT processing and complex cross-correlation.

IV. Simulated Instrument Performance

The optics of Tempera-C were designed using an ideal gaussian beam approximation and consequently optimised and simulated with the simulated pattern of the actual feedhorn using the GRASP package from Ticra. The horn was selected based on its very low crosspolar levels. A comparison of the simulated pattern against measurements can be found below in Fig. 3. Key performance criteria for the instrument are a full width half maximum of <4°, minimised spillover and

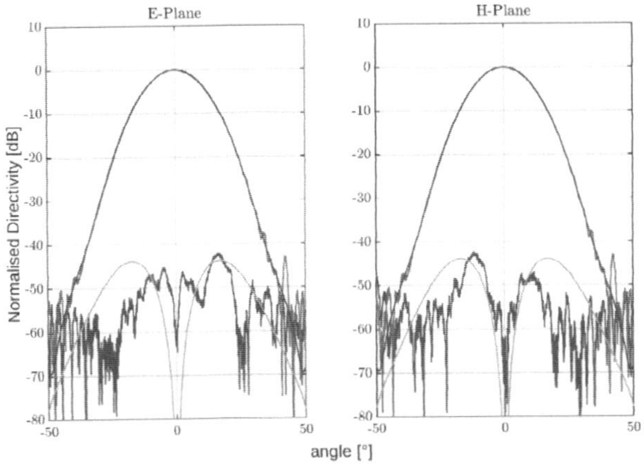

Fig. 3. Tempera-C horn measurements in comparison to simulation. Co- and Cross-polar simulation in violet and yellow. Measurements in red and blue.

low cross-polar component. Fig. 4 shows the farfield beams in zenith for two orthogonal polarisations as simulated in GRASP on a Phi, Theta grid. At low power levels (< -20dB), some small aberrations can be seen, due to the off-axis reflector. The total spillover of the instrument is less than 0.2%, including consideration of the periscope tube.

To check the bias introduced by the optics the product of the co-polar pattern of one plane and the cross-polar pattern of the orthogonal plane is calculated. In Fig. 4 it is the product of the two fields forming a column. The sum of these two products is the bias seen in the third stokes by the digital correlator. Assuming the horn pattern is symmetrical, the bias of the horn farfield using this methodology would be zero, as any bias in the individual chain would be identical to the other chain, but with a flipped sign and cancel out. This was verified in GRASP with the real pattern and a gaussian pattern to verify the approach. In the farfield of the instrument (including the optics) at zenith pointing the bias is the equivalent of –2.6mK, for a homogenous scene of 300K.

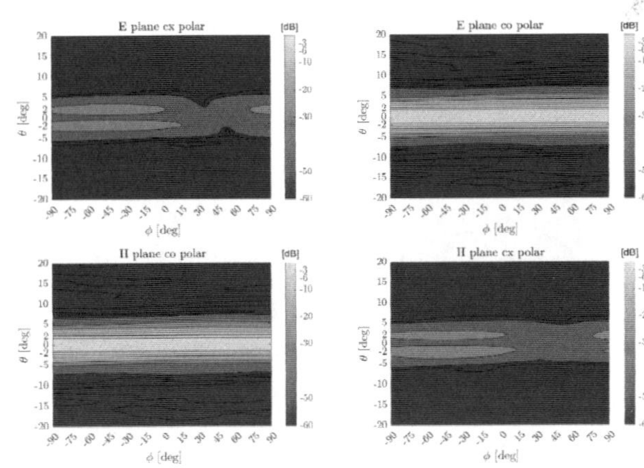

Fig. 4. Farfield patterns of Tempera-C for two orthogonal polarisations

V. CALIBRATION

The instrument includes a wedge-shaped ambient load for calibration. In addition to the ambient load, the noise diodes in both frontend chains allow calibration without the constant use of a cold load. Periodically, this calibration methodology is validated using a liquid nitrogen cold load positioned below the skyward reflector, which can be rotated to point to ground. The ambient wedge is rotated 45° to both the horizontal and vertical polarisation for consistent performance across both polarisations. The ambient load is based on an existing design developed by University of Bern for the Arctic Weather Satellite [3], including an absorber mixture which was developed inhouse. Test data of the Arctic Weather Satellite wedge load show a return loss of 55 to 65 dB across the Tempera-C band. A cone would be a preferable shape since it is polarisation independent geometry but is harder to manufacture.

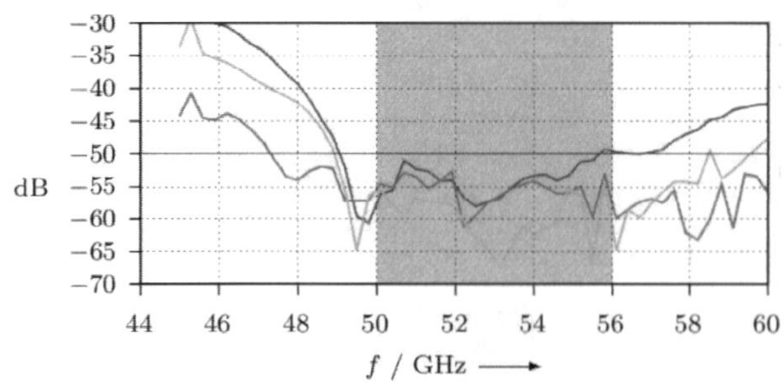

Fig. 5. Comparison of TE, TM and 45° rotation (blue, red and yellow respectively) return loss of OBCT.

VI. CONCLUSION

The Institute of Applied Physics is building a new fully polarimetric radiometer called TEMPERA-C, which allows to measure the for Stokes parameters simultaneously and to resolve the circular polarization which is caused by the Zeeman splitting of the oxygen emission lines. The new design raises the retrieval altitude into the lower mesosphere and the instrument will also be capable of measurement campaigns. As part of the optics redesign, an ultra-gaussian feedhorn was optimised along with more compact optics. The radiometer includes two identical receiver chains and a digital correlator which will eliminate error sources for the correlation. However, the optics could introduce a bias, which was investigated and found 2.6mK for a 300K homogenous scene in zenith. Further work is required to investigate this bias for other scan angles and inhomogeneous scenes.

ACKNOWLEDGEMENTS

Witali Krochin has been supported by the Schweizerischer Nationalfonds zur Förderung der Wissenschaftlichen Forschung (grant no. 200021-200517/1).

REFERENCES

[1] Francisco Navas-Guzman, et al, "Intercomparison of stratospheric temperature profiles from a ground-based microwave radiometer with other techniques" *Atmos. Chem. Phys*, Nov 2017

[2] Witali Krochin, et al, "Development of a Polarimetric 50 GHz Spectrometer for Temperature Sounding in the Middle Atmosphere", *"Journal of Selected Topics in Applied Earth Observations and Remote Sensing"*, Jun 2022

[3] Tobias Plüss, et al, "An Onboard Calibration Target for the Arctic Weather Satellite", *"Space Microwave Week"*, May 2023

Monolithic focal plane array concept for space and ground applications

Andrey M. Baryshev*[1], Martina Wiedner[2]

Abstract—**Single pixel heterodyne receivers has been a common for ground based and especially for space based applications in mm/submm wavelengths. Arranging receiver pixels in a Focal Plane Arrays (FPA) is the next step in improving observing efficiencies in many observing tasks. While development of FPA's has been successful for balloon and airborne projects, the main problem for achieving large pixel count has been the cost per pixel. Typically FPA is constructed out single array pixels which are typically assembled and tested separately. This leads to relatively large handling and testing costs in addition to cost of an optical element which is either lens or horn. We propose an array concept which combines a silicon lenslet array technology, that has been demonstrated with an antenna coupled microwave kinetic inductance detector arrays with a planar array of antennas, fabricated on the same chip. With F/D ratios in the range of 1.7..2.5 the array size limited by 2 F lambda/D will be of order of several millimeters which allows for dramatic decrease size, weight and of the cost per pixel of the RF part of an FPA.**

Keywords—Focal Plane Array, Silicon Lens, Aplanatic Lens, Heterodyne Mixers

I. Introduction

Single pixel heterodyne receivers has been a common for ground based and especially for space based applications in mm/submm wavelengths. Arranging receiver pixels in a Focal Plane Arrays (FPA) is the next step in improving observing efficiencies in many observing tasks. In fact, practically all space mission proposals dealing with high resolution spectrometers discuss exclusively FPAs. While development of FPA's has been successful for balloon and airborne projects like GUSSTO and SOFIA, the main problem for achieving large pixel count has been cost per pixel which is driven by adopting a combination single pixels into a FPA configuration. These single array pixels are typically assembled and tested separately prior to integrating to an FPA structure. This leads to relatively large handling and testing costs in addition to cost of an optical element which is either lens or horn.

We propose to revisit a monolythic array concept which combines a silicon lenslet array technology with a planar array of antennas, fabricated on the same chip. The silicon lenslet array technology has been recently developed and demonstrated with an antenna coupled microwave kinetic inductance detector arrays. Large lenslet arrays can be manufactured in one technological cycle by a laser evaporation technique. Lenselet arrays of several thousand pixels has been successfully produced with very small cost per pixel. One limitation of this technology is that manufacturing time and cost per pixel is proportional to a volume of material which needs to be removed to form the lenslets. This leads to a relatively small lens diameter and thus to a large beam divergence. Lenslets of only several wavelengths size has been demonstrated to work efficiently.

With F/D ratios in the range of 1.7..2.5 the array size limited by 2 F λ/D will be of order of several millimeters which allows for a dramatic decrease of the cost per pixel of the RF part of an FPA. An example of array/chip profile calculated for F/D=1.72 is shown in fig. 1. The array raw size calculated for 650 GHz central frequency is only 4 mm. This means a rectangular array of 9 pixels will use only 4x4 mm chip size which is pretty common for a single pixel planar heterodyne mixer. Since cost of detector wafer fabrication is fixed, cost per detector is inversely proportional to detector chip area, so the proposed approach allows to decrease production costs 9 times for a given example. This number does not take into account yield considerations

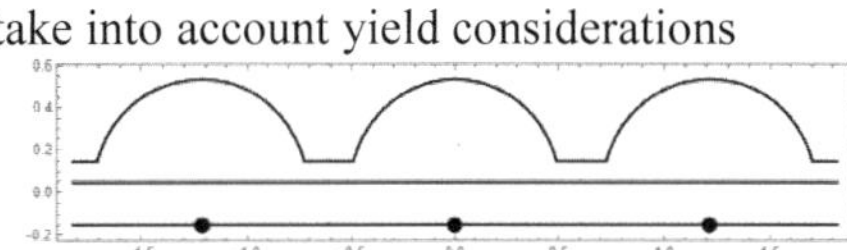

Fig. 1. Cross section of one array raw, showing silicon lens profile (top), array chip (bottom) and planar antenna positions (large dots).

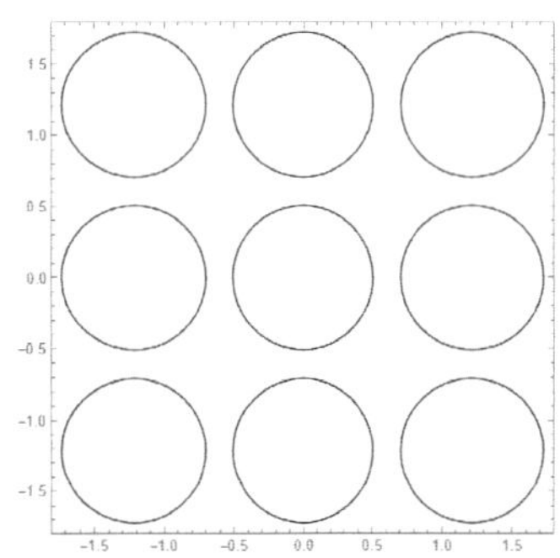

Fig. 2. Cross section of one array raw, showing silicon lens profile (top), array chip (bottom) and planar antenna positions (large dots).

We will discuss optical configuration and realizable chip layouts, including IF distribution, RF cross talk, aperture efficiency and application for both SIS and HEB technology.

[1]Kapteyn Astronomical Institute (NOVA) University of Groningen, Groningen, 9747 AD, The Netherlands; [2] LERMA Observatoire de Paris, Paris, France;

NOTES:

3D Printed Submillimeter Reflectors: a New Design and Manufacturing Methodology

Ryan T. Watkins *[1], Paul F. Goldsmith[1], Youngmin Seo[1], and Samad A. Firdosy[1]

***Abstract*—Submillimeter wavelength (1mm > λ > 0.1mm) reflectors are being utilized to an increasing degree in NASA missions due to their ability to observe phenomena in disciplines ranging from climatology to astrophysics. Due to the inherent form factor of reflectors and the need for stiffening backing structures, production is often limited by manufacturing restrictions that can lead to mass, cost, and/or schedule issues. To address these issues, we have developed a new manufacturing technique based on additive manufacturing and the use of light weight structures to enable a new class of low cost, high performance submillimeter reflectors.**

Keywords*—*submillimeter antenna, additive manufacturing, 3D printing, lattice, topology optimization

I. Introduction

Submillimeter wavelength (1mm>λ>0.1mm) reflectors are being utilized to an increasing degree in NASA missions. For example, in astrophysics, the ASTHROS mission will study how new stars are formed by observing fine structure lines of ionized nitrogen at 0.21 and 0.12 mm wavelengths [1]. In climatology, the Earth Observing System Microwave Limb Sounder (EOS MLS) instrument on the Aura Satellite has been studying the Earth's atmosphere through observation of 0.47 mm and 0.12 mm spectral lines [2].

Due to the inherent form factor of reflectors and the need for stiffening backing structures, production is often limited by manufacturing restrictions that can lead to mass, cost, and/or schedule issues [3], [4]. Small reflectors (cm scale) are now generally made by direct machining, but are relatively heavy due to limitations on minimal wall thicknesses. For larger (meter scale) reflectors, the reflecting surface is often electroformed in sections on a glass mold and then bonded to a stiffening backing structure (such as an aluminum honeycomb). The glass molds used in these processes are usually expensive (>$100k) and long-lead (6 months to a year), not only impacting project budgets, but also limiting design flexibility and agility to incorporate changes. In the case of off-axis antennas, as are increasingly desired, the cost further increases as a single mold cannot be used for as many panels as for a symmetric reflector. From a performance standpoint, the mismatched Coefficients of Thermal Expansion (CTE) found in these bonded structures can lead to thermal related performance issues. A new and fast manufacturing technique capable of producing light, monolithic reflectors is therefore highly desirable.

To address these manufacturing issues and enable a new class of low cost, high performance submillimeter reflectors, we developed a new methodology based on Additive Manufacturing (AM). AM is a relatively new manufacturing technique that builds a structure through a layer-by-layer process, allowing for the production of highly complex parts at a low cost. In the context of reflectors, this allows for rapid development and production of symmetric or off-axis parabolas with a fully integrated backing structure without the need for complicated molding/bonding procedures or CTE mismatches within the structure.

II. Manufacturing

Although Additive Manufacturing (AM) has the potential to overcome many of the issues associated with conventional manufacturing methodologies, features inherent to the process could cause performance issues. Of particular concern are thermal distortions due to the complex thermal history of the part and subsurface defects that result from the melt pool dynamics inherent to the process (such as keyholing and voids), both of which could negatively impact achievable surface profile accuracy of the reflector [5]. Furthermore, the limited length-scale resolution of modern printers could limit the areal density (projected area of the reflector divided by the mass) [6].

Due to the complexity associated with predicting manufacturing outcomes via modeling, reflectors were manufactured to verify the applicability of AM. A total of three reflector concepts were investigated, including 4x 8cm on-axis reflectors with an integrated lattice backing structure, a 20 cm off-axis reflector with an integrated lattice backing structure, and a 20 cm off-axis reflector with a topology optimized backing structure (see Figure 1). The focal lengths were half the nominal reflector diameter (4 cm and 10 cm) and the targeted wavelength range was 0.6 mm (500 GHz). All reflectors were manufactured from Aluminum A6061-RAM2 -- a printable version of the commonly used aerospace material Al 6061 -- using Laser Powder Bed Fusion (LPFB) due to its overall process maturity and feature resolution; all parts were printed on an EOS M290. Due to surface finish limitations in the LPFB process, and to help compensate for potential thermal distortions, the reflector surfaces were overbuilt in thickness by 3 mm, which was subsequently removed in a post machining step. To minimize the potential for thermal distortions and subsurface defects, the parts were Hot Isostatic Pressed (HIPed) at 510°C for 120 min at 101.7 MPa per ASTM B998. To improve strength, they were then solution heat treated at 529°C for 120 minutes, quenched in a 20% Glycol solution and subsequently aged at 160 °C for 18 hrs (which is equivalent to the commonly used T6 heat treatment as per Ams2770 for wrought Al 6061). Lastly, the reflector surfaces were post machined on a 3-axis mill to achieve the desired profile accuracy.

The surface profiles for 2x 8 cm and 1x 20 cm reflectors were measured and found to meet common specifications for operation at wavelengths (λ) of 0.6 mm and higher. The influence of profile accuracy on antenna gain is approximated by

$$\frac{G}{G_0} = e^{-\left(4\pi\frac{e}{\lambda}\right)},$$

*Email: ryan.t.watkins@jpl.nasa.gov

[1]Jet Propulsion Laboratory, California Institute of Technology, Pasadena, CA, 91109, USA

where e is RMS surface profile error, and G/G_0 is the normalized gain loss [7]. In line with common requirements for spacecraft reflectors, the target gain efficiency here is at least 67% (-1.7 dB). The required surface profile error must therefore be λ/20 RMS or less, which corresponds to 30 μm RMS when operating at 500 GHz. A Mitutoyo AVANT Formtrace profilometer was used to measure each reflector, measuring multiple linear traces at a fixed spacing across the geometry (see Figure 2 a-b). To measure the pointwise surface accuracy, the nominal parabolic surface was fit to the measured point cloud and the difference between expected and measured position was calculated. In the case of the 8 cm diameter reflectors, the surface profile tolerances were measured to be 10.3 μm RMS (SN002) and 12.7 μm RMS (SN001), while the larger 20 cm reflector was measured at 16.2 μm RMS, all of which are lower than the 30 μm RMS requirement.

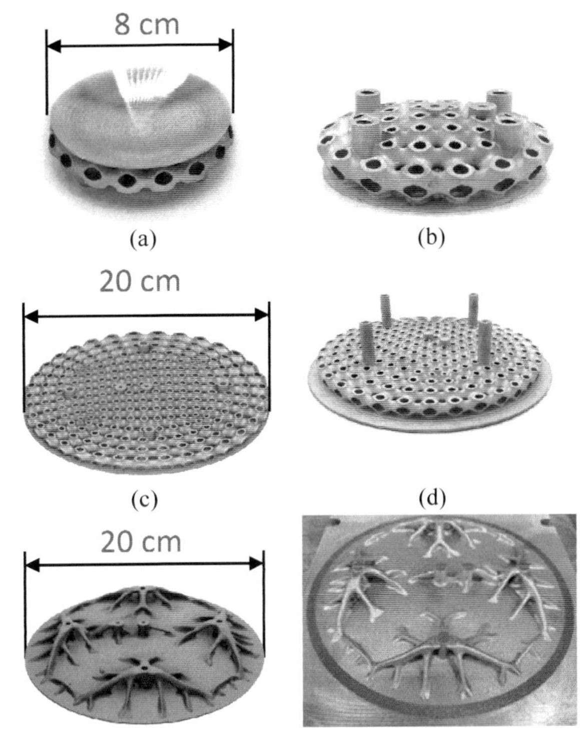

Figure 1: To demonstrate the manufacturing precision requirements for reflectors, three AM reflectors geometries were manufactured: (a, b) an on-axis 8 cm diameter subscale geometry used for process development (areal density of 13.4 kg/m²), (c, d) an off-axis 20 cm diameter geometry with a lattice backing structure (areal density of 8.1 kg/m²), and (e, f) an off-axis 20 cm diameter geometry with a topology optimized backing structure (areal density of 6.3 kg/m²).

To overcome the feature resolution limitations associated with LPBF (predominantly minimum thickness) and the corresponding need for light weight aerospace structures, a chemical etching process was developed to refine the reflector backing structure (see Figure 3). For the reflector size range examined here (< 20 cm in diameter), state of the art reflector areal density is on the order of 14 kg/m² [8] In the case of the 8 cm on-axis geometries, when running the printer at its minimum printable thickness of approximately 1 mm, the as-printed areal density was 19.2 kg/m²; however, removing 165 μm off the backing structure surfaces via chemical etching reduced the areal density down to 13.4 kg/m². It should be noted that this 8 cm geometry was not originally optimized for mass or performance, but rather as a unit to aid in process development. These results therefore indicate the ability to refine feature size rather than the ability to achieve a specific areal density.

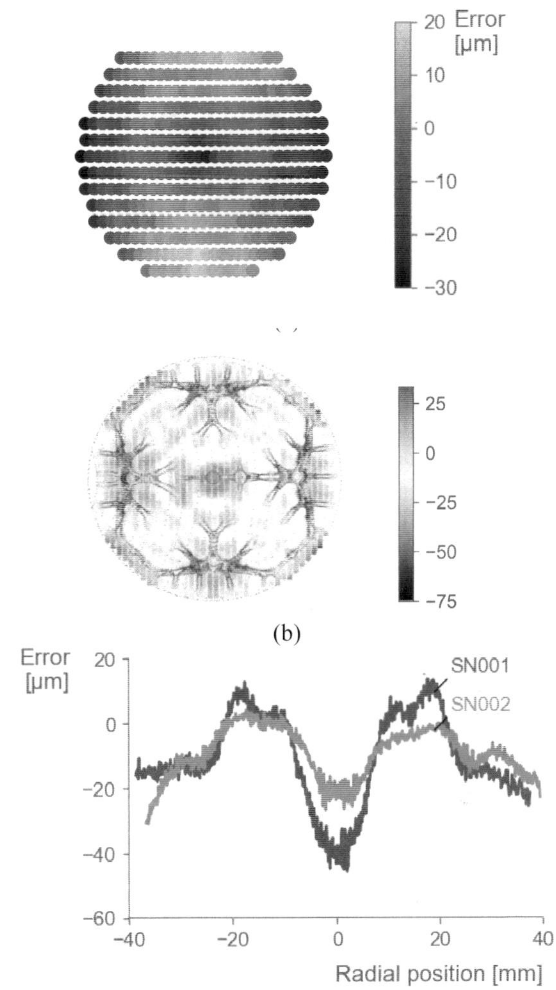

Figure 2: The reflector profile tolerance of 30 μm RMS was verified through profilometry, taking multiple line scans over the reflector surface and comparing the data to the nominal geometry: (a) 10.3 μm RMS for 8 cm reflector SN002 and (b) 16.2 μm RMS for the 20 cm topology optimized reflector (noting that the overlay illustrates the backing structure layout). (c) The consistency of the manufacturing process was verified through the measurement of two 8 cm reflectors, with a single radial trace shown to illustrate surface roughness and part to part variation.

III. Design

The added design flexibility of AM allows for the possibility of reconceptualizing the design of reflectors; in particular, the stiffening backing structure. A reflector backing structure provides two primary functions: sufficient stiffness to survive the spacecraft launch environment and sufficient stiffness to alleviate the need for gravity compensation during ground handling and calibration (i.e., negligible gravity sag). This is conventionally achieved through an isogrid (triangular webbing pattern) that is machined into the reflector or hexagonal honeycomb that is bonded on to the back of a reflector [3], [4], [9]. In the context of AM reflectors, these stiffening concepts can be generalized into a broader range of lattice structures or directly optimized using topology optimization.

(a)

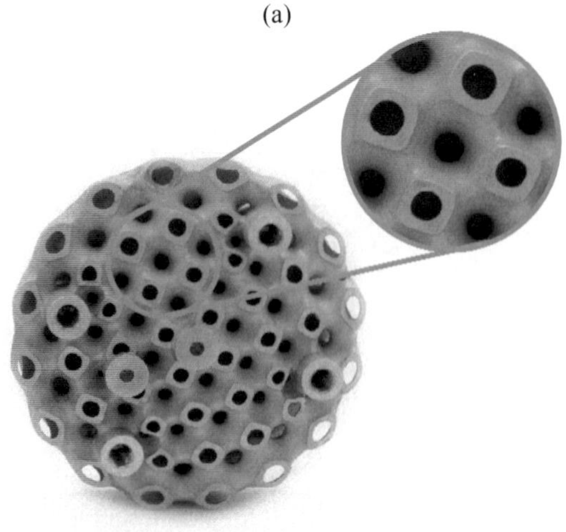

(b)

Figure 3: Chemical etching was demonstrated to be an effective geometry refinement technique, reducing the areal density of the as-printed 8 cm reflector from (a) 19.2 kg/m^2 down to (b) 13.4 kg/m^2.

The use of a lattice structure (such as an isogrid or honeycomb) as the backing of a reflector leverages a similar concept as an I-beam: it acts as webbing to support shear forces while pushing the majority of the material away from the bending neutral axis to increase area moment of inertia and thus bending stiffness [4]. The primary design driver is to therefore identify a lightweight lattice geometry with a high shear stiffness. Of the many possible lattice candidates, a Schwarz Triply Periodic Minimal Surface (TPMS) structure was identified as an ideal lattice geometry for not only its high shear stiffness, but its self-supporting, printable geometry and the easy flowability of etchant throughout the structure. This design methodology was implemented for both the on-axis 8 cm reflectors (see Figure 1 a-b) and the off-axis 20 cm diameter reflector with the lattice backing structure (see Figure 1 c-d), resulting in areal densities of 13.6 kg/m^2 and 8.1 kg/m^2, respectively. When compared to a conventionally isogrided reflector with equivalent mass and geometry (cell size and height), the corresponding stiffness and fundamental frequency are similar. Further study into alternative lattice unit cells may yield better performance, but were not investigated here.

Designing a reflector via Topology Optimization (TO) moves beyond the engineering intuition of lattice structures and directly optimizes for the desired end behavior: high bending stiffness. TO is a computational design framework that relies on physics-based simulations (typically via the Finite Element Method) to inform the layout of a structure based on governing design objectives and requirements. In the case of aerospace reflector design, the ideal optimization problem is to minimize structural mass given requirements on fundamental frequency (which relates to stiffness), gravity sag, and material strength; however, in practice, this problem formulation is challenging to solve with existing commercial tools and therefore requires some reformulation. A better behaving optimization is to maximize stiffness given mass and stress constraints. Using the same off-axis reflector profile as the 20 cm reflector with the lattice backing structure, the stiffness was optimized with respect to pressure loading on the surface of the reflector with a mass constraint of 0.2 kg (which corresponds to an areal density of 6.3 kg/m^2). Additionally, the design is restricted to have one plane of symmetry and overhangs limited to 45° (which is required for AM). The final optimized design is shown in Figure 1 (e-f) and was generated using Autodesk's Fusion 360 software tool.

For the design case examined here, the TO backing structure resulted in a lighter and stiffer design than the lattice backing structure. Under pressure loading across the reflector surface, the TO design has 1.75x higher specific stiffness (which is defined by the inverse of total strain energy -- a measure of stiffness -- divided by mass); additionally, the TO design has 1.83x higher specific fundamental frequency (which is defined by the fundamental frequency divided by mass).

IV. Conclusions

Additive Manufacturing (AM) of aerospace reflector structures was demonstrated as a proof-of-concept for submillimeter applications. In the case of reflector structures, the primary limitation of AM is its minimum feature resolution, affecting the potential mass efficiency and reflector profile accuracy. To overcome this limitation, a thick version of the reflector was printed and then refined to the desired final geometry through chemical etching of the backing structure and post machining of the reflector profile. This process methodology was implemented on on-axis 8 cm diameter and off-axis 20 cm diameter reflector geometries, demonstrating areal densities ranging from 6.3 kg/m^2 to 13.4 kg/m^2 and surface profile accuracies ranging from 10.3 μm RMS to 16.2 μm RMS (which corresponds to diffraction limited observations on the order of 500 GHz). AM therefore has the potential to reduce cost and lead time of reflectors while maintaining the mass and geometric tolerances required for aerospace operations.

The added design flexibility of AM allows for the possibility of reconceptualizing the design of reflectors. Through the use of generalized lattice structures and computational design tools such as Topology Optimization (TO), lightweight, yet stiff geometries are possible. A Schwarz Triply Periodic Minimal Surface (TPMS) lattice was found to have comparable stiffness to conventional isogridding approaches; however, topology optimization improved stiffness by approximately 1.75x. AM therefore has the potential to significantly increase the specific stiffness (stiffness-to-mass ratio) or reflectors, resulting in lighter weight structures.

V. Acknowledgements

The research was carried out at the Jet Propulsion Laboratory, California Institute of Technology, under a contract with the National Aeronautics and Space Administration (80NM0018D0004).

VI. References

[1] J. V Siles, J. Pineda, J. Kawamura, and P. Goldsmith, "ASTHROS-Astrophysics Stratospheric Telescope For High-Spectral Resolution Observations at Submillimeter-waves Mission & Payload Overview," in *Proc. SPIE 11445, Ground-based and Airborne Telescopes VIII*, 2020. [Online]. Available: www.nasa.gov

[2] J. W. Waters *et al.*, "The Earth Observing System Microwave Limb Sounder (EOS MLS) on the aura satellite," *IEEE Transactions on Geoscience and Remote Sensing*, vol. 44, no. 5, pp. 1075–1092, May 2006, doi: 10.1109/TGRS.2006.873771.

[3] S. K. Sharma, L. Shafai, and S. Rao, *Handbook of Reflector Antennas and Feed Systems: Theory and Design of Reflectors*, vol. Volume I. 2013.

[4] W. A. (William A.) Imbriale, Steven. Gao, and Luigi. Boccia, *Space antenna handbook*. John Wiley & Sons, 2012.

[5] S. Chowdhury *et al.*, "Laser powder bed fusion: a state-of-the-art review of the technology, materials, properties & defects, and numerical modelling," *Journal of Materials Research and Technology*, vol. 20, pp. 2109–2172, Sep. 2022, doi: 10.1016/J.JMRT.2022.07.121.

[6] Z. Wu, S. P. Narra, and A. Rollett, "Exploring the fabrication limits of thin-wall structures in a laser powder bed fusion process," *The International Journal of Advanced Manufacturing Technology* , vol. 110, pp. 191–207, 2020, doi: 10.1007/s00170-020-05827-4/Published.

[7] J. Ruze, "Antenna Tolerance Theory—A Review," *Proceedings of the IEEE*, vol. 54, no. 4, pp. 633–640, 1966, doi: 10.1109/PROC.1966.4784.

[8] D. Mihai *et al.*, "Design, analysis and evaluation of titanium antenna reflector for deep space missions," *Acta Astronaut*, vol. 184, pp. 101–118, Jul. 2021, doi: 10.1016/J.ACTAASTRO.2021.04.006.

[9] S. Kendrick, D. Chaney, R. E. Brown Stephen Kendrick, R. J. Brown, and S. E. Kendrick, "Optical characterization of the beryllium semi-rigid AMSD mirror assembly," *https://doi.org/10.1117/12.506386*, vol. 5180, no. 22, pp. 180–187, Dec. 2003, doi: 10.1117/12.506386.

Towards the heterodyne receiver for balloons and SmallSats above 2 THz

Chris A. Curwen[1], Daniel P. Cunnane[1], Jonathan H. Kawamura[1], Darren J. Hayton[1], Weibing Yang[2], Ke Chen[2], Xiaoxing Xi[2], Anthony D. Kim[3], Benjamin S. Williams[3], and Boris S. Karasik[*1]

***Abstract*— We demonstrate a broadband THz heterodyne receiver for astrophysics using Hot-Electron Bolometer (HEB) mixers based on high-T_C superconducting MgB_2 and Quantum Cascade Laser (QCL) as local oscillator (LO). This detector will advance the SOA by allowing an operation at elevated temperature (~20 K) and providing the intermediate frequency (IF) bandwidth of about 10 GHz. Such a bandwidth is required for high-resolution spectroscopy of interstellar molecular clouds at frequencies 3-5 THz and is not currently available in NbN based HEB mixers. A lab receiver system is being developed to prototype the payload for application on balloons and SmallSats.**

Keywords—hot-electron bolometer, quantum cascade laser, MgB_2, THz heterodyne receiver

I. Introduction

Terahertz (THz) heterodyne receivers play a critical role in astrophysical studies of star- and planet-formation, and the overall structure and lifecycle of the interstellar medium. Along with the most studied [CII] line at 1.9 THz, [OI] (4.74 THz), HD (2.68 THz), [NII] (1.46 THz, 2.46 THz), [NIII] (5.23 THz), and [OIII] (3.39 THz , 5.79 THz) lines are of great importance. Observations in the THz remain challenging due to the lack of well-developed technologies in the THz frequency range, the need for cryogenically cooled superconducting detectors, and the need for doing measurements from space or suborbital platforms.

II. MgB_2 HEB Mixer

Quasioptical mixer devices were made from 10-nm thin MgB_2 films grown on C-terminated face of 6H-SiC using HPCVD process. The HEB device is a 2μm (W) × 5μm (L) bridge made using ion milling and integrated with log-spiral planar antenna for coupling to the THz radiation. The critical temperature of the HEB devices was $T_C \approx$ 34-36 K.

The high critical current density $j_C > 10^7$ A/cm^2 @ 4.2 K indicated the film quality and uniformity. This value is higher than in our previous films prepared by a combination of the HPCVD and ion mill processes [1]. We also found a clear correlation between the mixer performance and the degree of granularity of the MgB_2 films traceable via the bias dependence of the output mixer noise.

III. THz Quantum Cascade VECSEL

THz quantum-cascade vertical-external-cavity surface-emitting-lasers (QC-VECSEL) were developed recently with the goal of achieving high-power, tunable sources with nearly Gaussian beam. A laterally extended emitting metasurface allows the beam to be minimally distorted by the diffraction. An external cavity allows continuous tuning over the fractional bandwidth ~ 20% around 3.4 THz. Such a device operated in a LN_2 cooled cryostat (77K) was used as local oscillator. Similar devices at 2.7 THz and 4.7 THz have been fabricated and will be employed in the THz breadboard receiver system being constructed at JPL.

IV. THz Heterodyne Receiver

Currently, we have characterized a number of MgB_2 HEB mixers using the 3.4 THz QCL, the 600 GHz Schottky-diode frequency multiplier source, and several discrete frequency lines in the 1.9-4.3 THz range of the far-IR molecular gas laser. The best data are shown in Fig. 1 along with the data for other JPL THz receivers.

The near temperature independent (up to ≈ 20K) sensitivity of the MgB_2 HEB mixer is an important factor enabling applications on small spaceborne platforms where SWaP limits are tight. The MgB_2 mixer and the QC-VECSEL make a pair of novel components, very promising for future THz instruments.

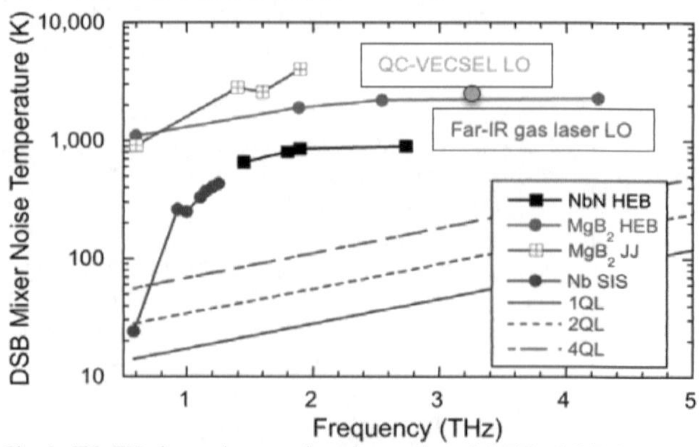

Fig. 1. JPL THz heterodyne receiver data to date. 0.6 THz, 1.9 THz, and 4.3 THz data points for MgB_2 HEBs are from the devices described in [1].

References

[1]. D. Cunnane, et al., "Optimization of Parameters of MgB_2 Hot-Electron Bolometers," *IEEE Trans. Appl. Supercond.*, vol. 27, no.4, pp. 2300405, Jun. 2017.

[1] Jet Propulsion Laboratory, California Institute of Technology, Pasadena, CA, 91109, USA; [2] Temple University, Philadelphia, PA, 19122, USA; [3] University of California, Los Angeles, Los Angeles, CA, 90095 USA.

NOTES:

Performance analysis of superconductor insulator superconductor mixer for 260 GHz atmospheric window.

Andrey M. Baryshev*[1], D. A. Ronso da Costa Lima [1], Ronald Hesper [1], Rob de Haan Stijkel [1], Jan Barkhof [1], Kirill Rudakov[3], Valery P. Koshelets[2]

***Abstract*—Heterodyne receivers in 260 GHz atmospheric transmission window (ALMA band 6) play key role in several scientific facilities and initiatives such as Atacama Large Millimeter Array (ALMA), Event Horizon Telescope (EHT), Latin American Millimetre Array (LLAMA) and Millimetron project. It is crucial to develop an superconductor insulator superconductor (SIS) mixer with large RF and extremely large IF bandwidth up to 20 GHz with noise performance close to quantum noise limit. We have designed and manufactured a waveguide SIS mixer for 200-280 GHz atmospheric window. We report detailed laboratory measurement of such a mixer which operates close to a quantum limit and provide analysis based on mixer theory of all gain/noise contributions. We will discuss feasibility of actually reaching quantum limit with such a system.**

***Keywords*—Supercnductor Insulator Superconductor mixer, cryogenic heterodyne receiver, quantum limit,**

I. Extended Rationale

Heterodyne receivers in 260 GHz atmospheric transmission window (ALMA band 6) play key role in several scientific facilities and initiatives such as Atacama Large Millimeter Array (ALMA), Event Horizon Telescope (EHT), Latin American Millimetre Array (LLAMA) and Millimetron project. Due to presence of quantum limit, performance of heterodyne receiver can be improved in two major ways: increase of instantaneous bandwidth and reduction of the receiver noise up to the quantum limit. It is crucial to develop an superconductor insulator superconductor (SIS) mixer with large RF and extremely large IF bandwidth up to 20 GHz with noise performance close to quantum noise limit. While many existing designs demonstrate performance which is several times the quantum limit of h f/(2 k) non has actually touched this limit. It must be recognized that in any low heterodyne noise system the receiver noise is made out of contribution of several receiver elements, not only SIS mixer. To further approach to a quantum limit a careful analysis of these contribution needs to be performed based on the mixer theory and precise measurements.

We have designed and manufactured a waveguide SIS mixer for 200-280 GHz atmospheric window. The mechanical layout and mixer chip layout is shown in fig. 1. The mixer consist of a backport cavity, a substrate channel with quartz substrate mixer chip. The mixer contains a Nb superconducting low pass filter, waveguide probe, SIS junction and integrated tuning circuit. It has been designed to cover 4-12 GHz IF band.

We report detailed laboratory measurement of such a mixer which operates close to a quantum limit and provide analysis based on mixer theory of all gain/noise contributions. As result of this analysis we will discuss feasibility of actually reaching quantum limit with such a system and possible ways to further improve the performance.

The developed mixer will be a basic building block of a sideband separating (2SB) mixer which is described in the contribution of S. Realini et al at this conference. Two of such 2SB mixers will be used in a LLAMA telescope receiving system.

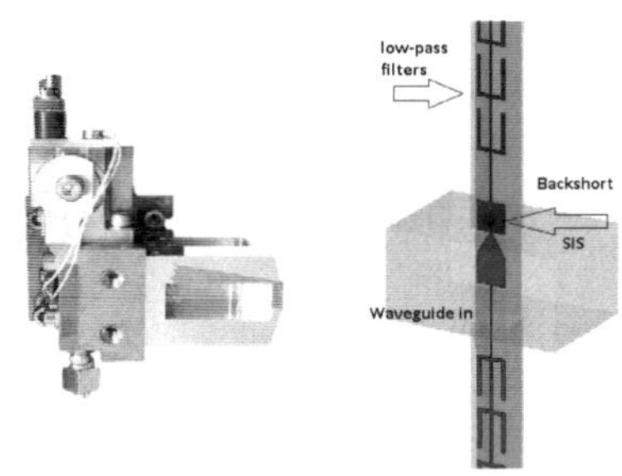

Fig. 1. SIS mixer mechanical construction (left) and mixer chip layout as calculated in 3D

[1]Kapteyn Astronomical Institute (NOVA) University of Groningen, Groningen, 9747 AD, The Netherlands; [2]Kotel'nikov Institute of Radio Engineering and Electronics, Russian Academy of Science, 11/7 Mokhovaya St., 125009, Moscow, [3] SRON Netherlands Institute for Space Research, P.O. Box 800, 9700 AV Groningen, the Netherlands;

NOTES:

Low noise, wide band MgB_2 hot electron bolometer mixer at 5.3 THz and 20 K

B. Mirzaei[*1], Y. Gan[2,3], J. R. G. Silva[2,3], S. Cherednichenko[4], and J. R. Gao[1,2]

***Abstract*—We present a low noise and wide band superconducting MgB2 hot electron bolometer (HEB) mixer characterized at 5.3 THz and 20 K, where a DSB receiver noise temperature of 2920 K is obtained and an IF bandwidth is 9.5 GHz. The frequency of 5.3 THz represents the highest frequency, where an MgB2 HEB mixer has ever been characterized at so far, providing an insight to their applicability for probing important higher frequency molecular/atomic emission lines. The operating temperature of 20 K enables an HEB mixer to operate in compact, low power consumption coolers, which can significantly reduce the cost and complexity of any relevant instruments specially in the case of space, where tight supply conditions and limitations apply.**

***Keywords*— MgB2, HEB mixer, 20 K cooler**

I. Introduction

Superconducting HEB mixers have been the most applicable heterodyne detectors for frequencies above 1 THz. Thanks to their superior sensitivity they have been used for very high spectral resolution spectroscopy in different THz observatories, including space born. Great sensitivities are acquired based on the optimal properties of the employed superconducting material i.e. NbN, although it by nature can only provide a limited intermediate frequency (IF) bandwidth (~3.5 GHz) and requires an operating temperature around 4 K, which are both influential parameters to determination of the instrument science targets, the complexity, and the cost etc. Therefore, any considerable improvement of these two figures of merit requires the change of the superconducting material.

MgB_2 has been under study [1-3] as a suitable candidate mainly for enhancing the IF bandwidth of HEBs significantly to above 13 GHz, which is a good compromise for their reduced sensitivity comparing to NbN HEBs. However, so far no published results of their performance above 2 THz can be found in literature although some data exist in other labs, such as Chalmers University and NASA JPL . Here we report the first experimental study of an MgB_2 HEB mixer [4] at 5.3 THz, when it is operated at 20 K as a projected temperature for compact space born cryo coolers.

II. Measurement Setup and Results

We focus on an HEB with dimensions of 285 x 850 nm^2 and a T_c of 38.4 K, fabricated at TU Delft. It is based on a 7 nm thick MgB_2 film on SiC substrate, developed at Chalmers University. An elliptical silicon lens was used to focus the radiation to a spiral antenna coupled HEB, which was mounted on a mixer block with an integrated heater and temperature sensor. An FIR gas laser was used to provide a 5.25 THz local oscillator. The intermediate coupling optics were set up in vacuum to minimize THz absorption in air.

A DSB receiver noise temperature (T_{rec}^{DSB}) of 3960 K was measured at 20 K operating temperature, which could be reduced to 2920 K with an optimally coated silicon lens and a thinner beam splitter. This value is about 2.5 times T_{rec}^{DSB} of a NbN HEB operated at 4 K in our lab. Fig. 1 shows a relatively weak increase of noise by 12 % when temperature rises from 6.5 to 20 K.

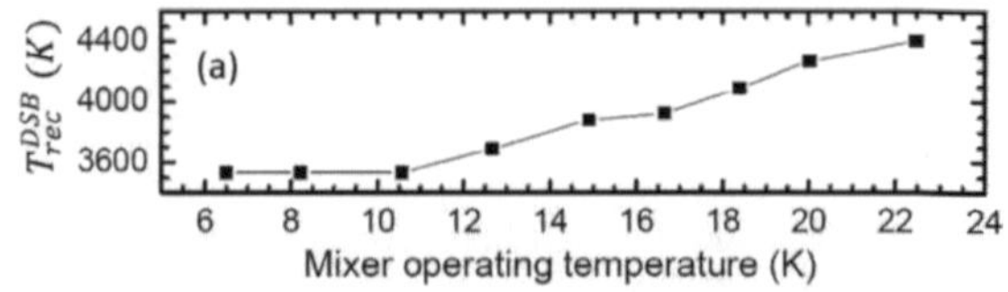

Fig. 1. Receiver noise temperature versus operating temperature

An IF bandwidth of 9.5 GHz was measured at 20 K, which is around 3 times wider than what NbN HEBs normally provide.

We also observed an interesting change in the current voltage (IV) curves of this device when it was pumped at 1.6 and 5.3 THz, which can be attributed to the existence of two superconducting gaps uniquely in MgB_2.

References

[1] D. Cunnane, et al., "Optimization of Parameters of MgB_2 Hot-Electron Bolometers", *IEEE Transactions on Applied Superconductivity,* Jan. 2017.

[2] E. Novoselo, et al., "Low Noise Terahertz MgB_2 Hot-Electron Bolometer Mixers with an 11 GHz Bandwidth", *Applied Physics Letters*, Jan. 2017.

[3] N. Acharya, et al., "Analysis of the Broad IF-Band Performance of MgB_2 HEB Mixers", *IEEE transactions on Terahertz Science and Technology*, Oct. 2019.

[4] Y. Gan, et al., "Low noise MgB2 Hot Electron Bolometer Mixer Operated at 5.3 THz and at 20 K", *Applied Physics Letters*, Jan. 2017.

[1]Optics Group, Department of Imaging Physics, Delft University of Technology, The Netherlands; [2]SRON Netherlands Institute for Space Research, The Netherlands; [3]Kapteyn Astronomical Institute, University of Groningen, Groningen, the Netherlands, [4]Terahertz and Millimeter Wave Laboratory, Department of Microtechnology and Nanoscience, Chalmers University of Technology, Sweden.

NOTES:

On embedding of an HEB mixer into a THz photonic integrated circuit

Alexander Shurakov*[1,2], Ivan Belikov[1,2], Anatoliy Prikhodko[1,2], and Gregory Goltsman[1,2]

***Abstract*—Photonic integrated circuits are worth using at frequencies beyond 1 THz, where they greatly outperform conventional hollow metallic waveguides. In this paper, we propose design of a THz HEB mixer that comprises Si photonic crystal waveguide integrated with two silicon dielectric rod antenna arrays via multimode interference power dividers. These arrays are to couple an HEB element with local oscillator and signal at 2.5 THz. The design is justified by EM modeling and performance tests of the mixer key elements.**

Keywords—terahertz, hot electron bolometer, photonic crystal waveguide, dielectric rod antenna

I. Introduction

Photonic integrated circuits (PICs) are worth using for terahertz applications. Above 300 GHz, they mostly rely on Si platform and make use of either an effective media or photonic crystal waveguides (PCWG) [1]. Si PICs enable creation of RF circuits with numerous passive and active components interacting through low-loss dielectric waveguides. They are especially in demand at frequencies beyond 1 THz, where performance of conventional CNC-machined hollow metallic waveguides is compromised.

In this paper, we propose design of a terahertz (THz) hot electron bolometer (HEB) mixer that comprises Si PCWG integrated with 2 Si dielectric rod antenna (DRA) arrays through multimode interference power dividers. These arrays are to couple an HEB element with local oscillator (LO) and signal at 2.5 THz. Details on the design considerations are provided further in the text.

II. Results

The proposed design relies on a linear 4-element [2] Si DRA array with a taper angle $\alpha = 12°$ and input cross-section of $0.25\lambda_0 \times 0.275\lambda_0$ (height × width) of each element, where λ_0 is the free-space wavelength. It provides a half power beamwidth $HPBW$ = 24.6° and 65.9° in H- and E-planes, respectively. The values are justified by both EM modeling at 2.5 THz and beam profile measurements of the array 1:50 scale model at 50 GHz. EM modeling also suggests that acceptable beam properties can be achieved even in case of a single DRA. The best performance is observed for $\alpha \leq 2°$, when $HPBW \approx 50°$ for both H- and E-planes, gain equals 13.7 dB or more and a side lobe level is well below -30 dB. Since fabrication of the narrow-angle DRA is challenging, we first focus on fabricating the structure with $\alpha = 10°$.

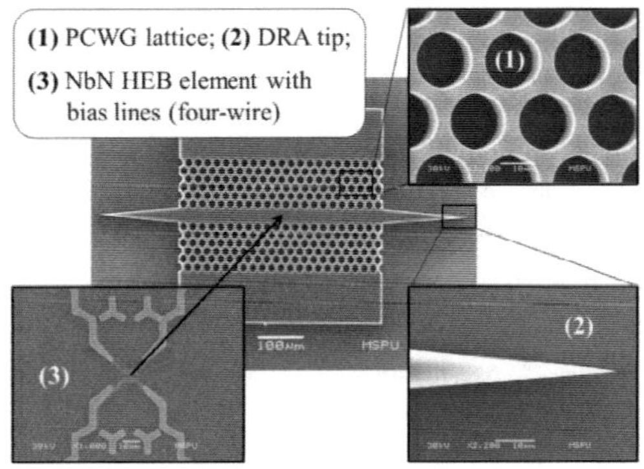

Fig. 1. Key elements of an HEB mixer intended for integration with a THz photonic circuit.

Fig. 1 provides SEM image of a PCWG integrated with 2 DRAs fabricated from SOI via Bosch process. No violation of the PCWG lattice (1) or the DRA tip (2) geometries is observed. The fabricated structure is intended for integration with an NbN HEB element (3) operated at 2.5 THz. The PCWG footprint implemented on SOI with a 30 μm thick topmost layer provides insertion loss due to the energy leakage into a free space $L \leq 0.1$ dB/mm. This is justified by EM modeling and insertion loss measurement of the PCWG 1:16.7 scale model at 150 GHz. Coupling efficiency between the HEB element and the PCWG is to be enhanced by a resonator [3]. Additionally, to increase the NbN/Si interface area and not to drastically enlarge the required LO power, NbN bridge has a size of 1.5×10 μm^2 (length × width) and 2 symmetric 0.25×4.25 μm^2 slots in its central part. We preliminary observe reasonable sensitivity for conventional waveguide HEB mixer with a similar NbN bridge (4 nm thick film) at the LO frequencies below 1 THz. Details on the performance tests of the proposed HEB mixer design at 2.5 THz are to be presented at the conference site.

References

[1] Withawat Withayachumnankul, et al., "Integrated Silicon Photonic Crystals Toward Terahertz Communications", *Advanced Optical Materials*, 2018

[2] Henning Tesmer, et al., "Fully Dielectric Rod Antenna Arrays with Integrated Power Divider", *Frequenz*, 2019

[3] Andreas Vetter, et al., "Cavity-enhanced and ultrafast superconducting single-photon detectors", *Nano letters*, 2016

[1]Moscow Pedagogical State University, Moscow, 119991, Russia; [2]National Research University Higher School of Economics, Moscow, 101000, Russia; *alexander@rplab.ru

NOTES:

2-18 GHz Ultra-wideband Cryogenic Amplifier with 4 K Noise Temperature

Isaac López-Fernández*[1], Juan D. Gallego[1], Carmen Diez[1], Inmaculada Malo[1], Ricardo I. Amils[1], Ralf Flückiger[2] and Diego Marti[2]

***Abstract*—A cryogenic low noise amplifier with 4 K average noise temperature in the 2-18 GHz band has been developed with new high performance InP devices. This amplifier meets the demands of ALMA next generation receivers, doubling the present IF bandwidth with state-of-the-art noise performance. A balanced configuration is feasible to improve the input return loss.**

***Keywords*—Broadband amplifiers, cryogenic amplifiers, InP, high electron mobility transistors (HEMTs), low-noise amplifiers, submillimeter wave receivers, radio astronomy**

I. Introduction

One of the priorities of modern radio astronomical receivers is to achieve ever larger instantaneous bandwidths. The goal for ALMA (Atacama Large Millimeter/Sub-millimeter Array) 2030 development roadmap [1] is to implement an IF of 16 GHz which would provide 32 GHz per polarization in 2SB receivers. The Event Horizon Telescope (EHT) consortium aims for the next generation sub-millimeter receivers are similar. The front-end critical component to fulfill these planned upgrades is the cryogenic LNA, given that SIS mixers have already demonstrated IF bandwidths spanning up to 20 GHz [2]. These demands are not exclusive of high frequency receivers, but also of centimeter wave radio telescopes with the LNA before down-conversion, although the feed fractional bandwidth also plays a limiting role here.

Yebes Observatory amplifier group designed some years ago a 2-14 GHz LNA for the front-end of wideband VLBI (Very Long Baseline Interferometry) receivers which is being widely used by VGOS (VLBI Geodetic Observing System) qualified antennas, generally combined with high performance custom made cryogenic 3 dB 90° hybrids in a balanced configuration [3]. More recently, and fruit of a long-term and extensive collaboration with Diramics, a new generation of InP HEMT transistors in 100 nm technology has been developed that incorporate improvements which translate in a very significant leap in performance.

The availability of these active devices and the need for an extension in IF bandwidth motivated the design of a new amplifier based on the existing 2-14 GHz LNA, optimizing its noise performance, and widening the bandwidth to 2-18 GHz. This represents a fractional bandwidth of 9:1 (almost a decade). A prototype unit was built and tested.

II. Design

The key modifications that were introduced in the previously mentioned experimental HEMT devices are the improvement of stability and cryogenic noise performance. Better stability is achieved by pad metallization reduction and the use of a two-finger layout in combination with source air bridges. Noise temperature was improved mainly by the optimization of the gate to channel distance, through extensive on-wafer cryogenic DC measurements and cryogenic noise characterization of devices on test amplifiers.

The design requires three common-source stages in order to achieve the desired levels of gain and noise optimization and was built in hybrid microstrip technology (chip & wire) with independent bias for each of the stages. This fabrication methodology grants access to high number degrees of freedom allowing fine performance optimization. Assembly and manufacturing procedures were followed similar to those of the IF amplifiers for ALMA bands 5, 7 and 9 [4].

Only the first stage transistor comes from the enhanced batch, and the gate periphery selected to maximize wide band noise matching was 200 μm. The other stages have HRL transistors with 150 μm total gate widths. Figs. 1(a) and 1(b) show a picture of the LNA and of the first stage optimized Diramics device, respectively. Another significant feature of the design was the use of a small quartz chip spiral inductor to couple the first stage gate bias, improving the noise contribution and gain equalization of other approaches (like big resistors or wideband conical inductors). All this, together with precise cryogenic modeling of devices and components allowed obtaining the state-of-the-art results and similarity to the simulated performance that is presented in Fig. 2 and Fig. 3.

III. Results

The cryogenic noise and gain of the amplifier was characterized using two different measurement methods: a variable temperature cryogenic load [5] and the cold attenuator method [6] at 6 K and 15 K ambient temperature, with better than 0.5 K average noise temperature discrepancy between both methods (a 1 K noise difference was measured between 6 and 15 K ambient with the variable temperature cryogenic load).

Average noise temperature of 4.1 K was obtained at 6 K, not exceeding 5 K in the 2-18 GHz band. The amplifier also exhibits a flat gain around 31.8 dB (less than 1.6 dBpp

(a) (b)

Fig. 1. (a) Photograph of the three-stage 2-18 GHz CLNA without cover. Dimensions excluding connectors are 20×22×9 mm. (b) 2×100 μm gate finger InP HEMT in 100 nm technology used in the 1st stage of the amplifier.

*i.lopez@oan.es
[1]Observatorio de Yebes, CDT (IGN), Guadalajara, 19141, Spain; [2]Diramics AG, Steinhausen, 6312, Switzerland.

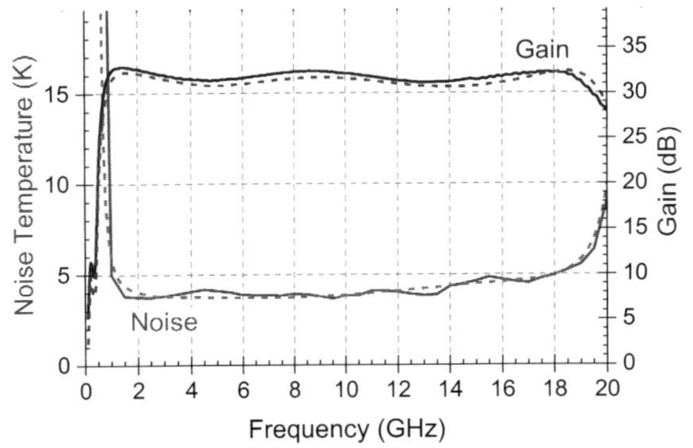

Fig. 2. Measured (solid lines) and simulated (dashed lines) gain and noise temperature of the LNA at 6 K ambient.

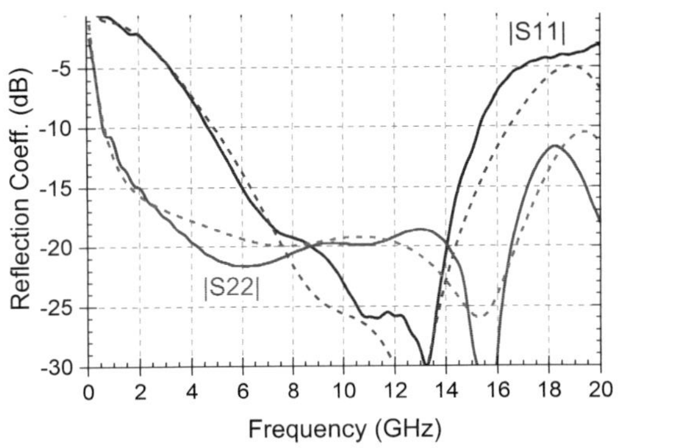

Fig. 3. Measured (solid lines) and simulated (dashed lines) input and output reflection of the LNA at 15 K ambient.

ripple). As shown in Fig. 2, the design easily covers 1-19 GHz with similar performance (4.2 K noise, gain flatness below 1.8 dBpp). To the best of our knowledge, this is the lowest noise temperature reached in an 18 GHz wide band.

Cryogenic S parameters were also measured at 15 K as in [6], obtaining better than -15 dB output reflection (Fig. 3). Input matching of InP HEMT based LNAs with very large fractional band is challenging at the low frequency end and requires trade-offs in noise temperature [7]. Applications demanding minimum input return loss can address this problem at the cost of complexity and power dissipation by employing balanced architectures (the alternative of using a cryogenic isolator is not practical for such large fractional bandwidths). The needed 3 dB 90° 2-18 GHz hybrid has not been designed yet but could be produced by scaling the existing one developed by our group for the 1.5-15.5 GHz band [8].

The input 1 dB compression point at 3 GHz is -40.5 dBm. The amplifier is unconditionally stable in the band and the absence of high frequency oscillations was verified by DC curves measuring. Power dissipation is 17.7 mW and although it was not a design goal it can be reduced by redesigning the bias networks and including optimized transistors in the second and third stages.

REFERENCES

[1] J. Carpenter, D. Iono, L. Testi, N. Whyborn, A. Wootten and N. Evans, "The ALMA Development Roadmap," 2018. [Online]. Available: http://www.almaobservatory.org/wp-content/uploads/2018/07/20180712-alma-development-roadmap.pdf

[2] E. C.-Y. Tong, R. Blundell, K. G. Megerian, J. A. Stern, Shing-Kuo Pan and M. Pospieszalski, "A distributed lumped-element SIS mixer with very wide instantaneous bandwidth," IEEE Trans. Appl. Supercond., vol. 15, no. 2, pp. 490-494, June 2005, doi: 10.1109/TASC.2005.849885.

[3] J. A. Lopez-Pérez *et al,* "Description of RAEGE Yebes VGOS Receiver Upgrades," *Proc. 12th General Meeting International VLBI Service for Geodesy and Astrometry*, Helsinki, Finland, Jan. 2023, vol. NASA/CP–20220018789, pp. 47–51.

[4] I. Lopez-Fernandez, J. D. Gallego Puyol, C. Diez and A. Barcia, "Development of Cryogenic IF Low-Noise 4-12 GHz Amplifiers for ALMA Radio Astronomy Receivers," *IEEE/MTT-S Int. Microw. Symp.*, San Francisco, CA, USA, 2006, pp. 1907-1910, doi: 10.1109/MWSYM.2006.249788.

[5] D. Bruch *et al.*, "A Noise Source Module for In-Situ Noise Figure Measurements From DC to 50 GHz at Cryogenic Temperatures," *IEEE Microw. Wireless Compon. Lett.*, vol. 22, no. 12, pp. 657-659, Dec. 2012, doi: 10.1109/LMWC.2012.2228176.

[6] J. D. Gallego, I. López-Fernández and C. Diez, "A Measurement Test Set for ALMA Band 9 Amplifiers," *Radionet FP7 1st Eng. Forum Workshop*, Gothenburg, Sweden, June 2009. [Online]. Available: https://www.radionet-eu.org/fp7wiki/lib/exe/lopez-fernandez_final092b.pdf_%3b?media=na:engineering:ew:lopez-fernandez_final.pdf

[7] R. I. Amils, J. D. Gallego, C. Diez, I. López-Fernández, A. Barcia, S. Muñoz, J. L. Sebastián and I. Malo, "Low input reflection cryogenic low noise amplifier for Radio Astronomy multipixel receivers," *J. Inst.* v.11, Oct. 2016, doi: 10.1088/1748-0221/11/10/P10018

[8] J. D. Gallego, I. López-Fernández, I. Malo-Gómez, A. García, C. Diez and R. I. Amils, "Ultra-wide Band LNAs for BRAND Front-ends: Single-ended and Balanced Approaches," Yebes Observatory, Spain, BRAND/Yeb-TR-2018-001, doi.org/10.5281/zenodo.3980987.

Versatile Radiometric Testbed for the Submillimeter Wave Instrument

Mikko Kotiranta[*1], Karl Jacob[1], Michael Olberg[2], Paul Hartogh[3], and Axel Murk[1]

***Abstract*— This paper presents a radiometric testbed for the characterization of the 600-GHz and 1200-GHz double side band heterodyne receivers of the Submillimeter Wave Instrument on the Juice mission of ESA. The testbed enables measurements of receiver noise temperature, gain stability, side band ratio, and beam coalignment in a thermal-vacuum chamber. Receiver spectral response and spurious signal investigations are possible with an optional transmitter. Measurement results obtained for engineering and protoflight models of the instrument are presented.**

***Keywords*— Beam coalignment, Juice, microwave radiometry, side band ratio, Submillimeter Wave Instrument, testbed**

I. Introduction

The Submillimeter Wave Instrument (SWI) [1] on the Jupiter Icy Moons Explorer (Juice) spacecraft of ESA consists of two double side band heterodyne receivers that are independently tunable in the frequency bands 530 GHz - 625 GHz and 1080 GHz - 1275 GHz, respectively. The Schottky diode mixers and IF low-noise amplifiers of the receivers are cooled passively to a temperature of 150 K. The instrument backend consists of broadband autocorrelation spectrometers (ACS), high-resolution chirp transform spectrometers (CSH), and continuum channels.

In addition to the receiver noise temperature and gain variations that limit the receiver sensitivity and the accuracy of absolute brightness temperature determination, knowledge of the receiver side band ratio (SBR) is necessary for obtaining exact spectral line intensities present in the upper and lower side bands. The SBR of Submillimeter Wave Satellite (SWAS) receiver system [2], the Odin radiometer [3], and the Heterodyne Instrument for the Far Infrared (HIFI) on board the Herschel Space Observatory [4],[5] were determined before the launch using a gas cell that enables observations of known molecular spectral lines in a laboratory. The SBR of submillimeter wave receivers has also been determined with a setup based on a Martin-Puplett interferometer [6].

The 600- and 1200-GHz beams of SWI must be well coaligned to allow simultaneous observations without excessive signal loss in either of the channels. A facility such as the Low-temperature Near-field Terahertz Chamber (Lorentz) [7] allows beams to be characterized by a two-dimensional RF near-field measurement in space-quality vacuum in a wide range of temperatures. The beam positions inside the instrument, for example during the alignment process of the receivers and the relay optics, may be determined with a compact setup described in [8].

A radiometric testbed allowing the characterization of the SWI receivers in a realistic environment prior to the launch is essential for maximizing the science output during the mission. This paper presents the design and operation of such a testbed.

To understand how the testbed can be used for characterizing the SWI receivers, the instrument optics is briefly reviewed in Section II. The design of the testbed is then described in Section III, the employed test methods are presented in Section IV, and the measurement results demonstrating correct operation of the testbed are given in Section V. Finally, conclusions based on the results are drawn in Section VI.

II. SWI Optics

SWI receives thermal radiation from Jupiter's atmosphere with a 29-cm off-axis paraboloid antenna that is steerable in the angular range ±72°. The primary (M1) and secondary (M2) mirrors are mounted on a rocker that is attached to the receiver unit box with two bearings that allow the rocker to be tilted up to ±4.3°. There is a cut-out in the rocker to allow the signal from M2 to enter the receiver unit (RU) box, in which relay optics consisting of planar and elliptical mirrors (M3-M5) and a free-standing wire grid (WG) guide the signal to the 600- and 1200-GHz receivers. The signal of the 600-GHz channel is collected by a corrugated feedhorn with a nonlinear profile whereas the 1200-GHz receiver is equipped with a smooth-walled, spline-profile horn. Further, a planar flip-mirror (MF) allows the receivers to view a conical calibration hot load (CHL). The beam path is illustrated schematically in Fig. 1.

III. Testbed Design

The testbed is built around a Martin-Puplett interferometer (MPI), and its main parts are depicted schematically in Fig. 2a. The testbed is attached to the receiver unit (RU) box using an interface part that fits one of the bearing holes. The rocker and the mirror M3 are not present during the tests, so instead of reflecting from the mirror M3, the 600-GHz and 1200-GHz receiver beams exit the RU box through the bearing hole. After entering the testbed, they first meet a parabolic mirror from which they are reflected to a photolithographic polarizer that splits the signals to a fixed

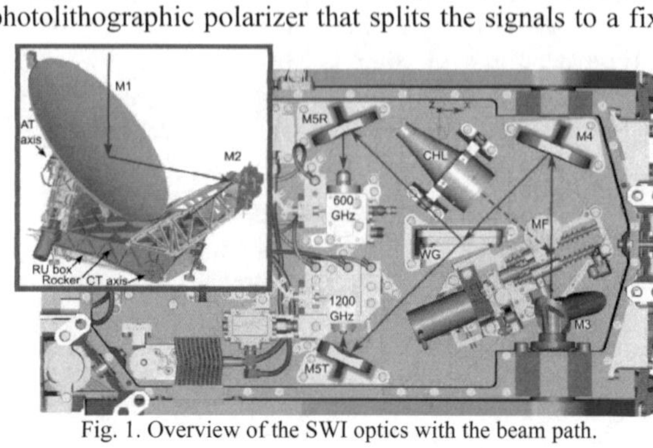

Fig. 1. Overview of the SWI optics with the beam path.

[1]University of Bern, Bern, 3012, Switzerland; [2]Chalmers University of Technology, Onsala, 43992, Sweden; [3]Max Planck Institute for Solar System Research, Göttingen, 37077, Germany

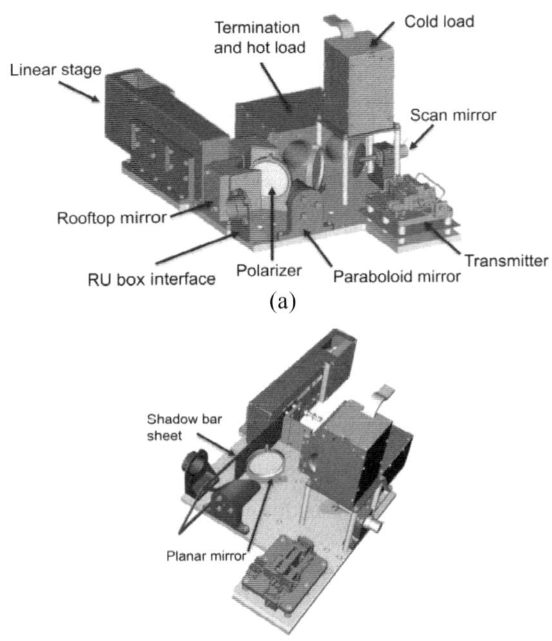

Fig. 2. CAD model of the SWI testbed: configuration for measuring a) side band ratio, b) beam coalignment.

rooftop mirror and to a movable rooftop mirror that is mounted on a vacuum-compatible linear stage. After reflecting at the roof top mirrors and recombining at the polarizer, the signals then meet a second photolithographic polarizer at the output of the MPI. Only the vertical components are transmitted to a scan mirror that reflects the signals either towards a hot or cold conical cavity blackbody calibration target. The targets have a nonlinear profile and utilize an epoxy-based composite material for efficient microwave absorption. The cold target stands on thermally insulating supports, and its metal housing is connected to a cryo-cooler. A third scan mirror position allows a signal from a transmitter to be injected towards the receivers to investigate the receiver spectral response and generated spurious signals.

While a receiver side band ratio measurement calls for an interferometric setup, a more direct beam path to the calibration targets is sufficient for noise temperature and stability measurements: instead of the polarizers, a single planar mirror is mounted to the testbed, as shown in Fig. 2b. The planar mirror configuration is employed also during beam coalignment measurements. The movable roof mirror is replaced with a heated shadow bar sheet. The receivers

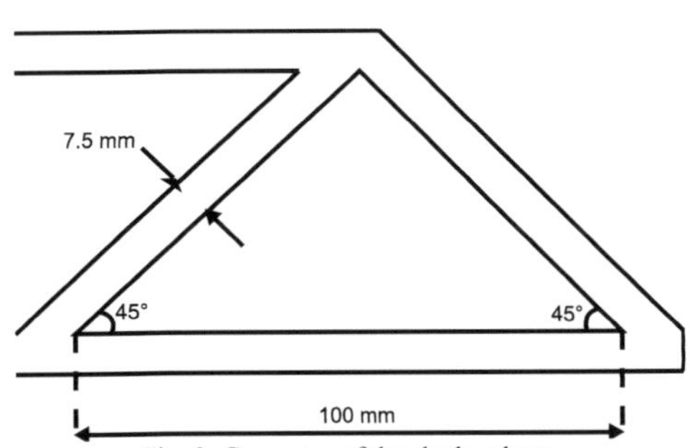

Fig. 3. Geometry of the shadow bars.

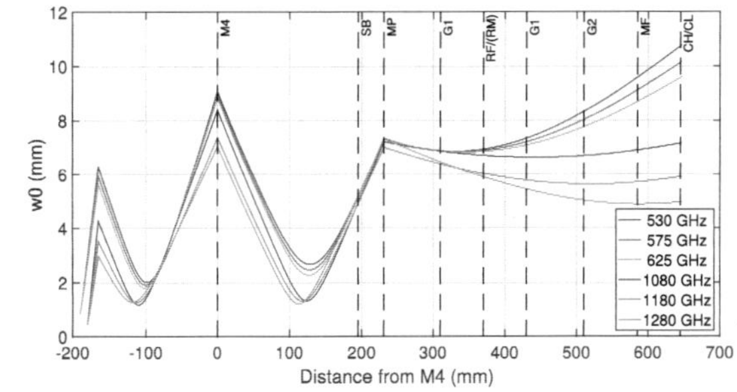

Fig. 4. Beam radius in the testbed computed using Gaussian beam mode propagation. SB: shadow bar sheet; MP: paraboloid mirror (reflected focal length 101.6 mm); G1/G2: polarizer; RF/RM: fixed/movable rooftop mirror; MF: scan mirror; CH/CL: hot/cold load.

view the cold load, and the sheet with two shadow bars that are oriented in an angle of ±45° with respect to the vertical are then moved across the 600-GHz and 1200-GHz beams. The geometry of the shadow bars is illustrated in Fig. 3. Beam offsets at least up to 10 mm with respect to optical axis can be determined.

The propagation of the receiver beams in the testbed was determined using Gaussian beam mode analysis (Fig. 4). The radii of both beams are about 5 mm at the plane of the shadow bar. The shadow bars have a non-projected width of 7.5 mm, so at the linear stage position, where the greatest beam blockage is obtained, more than 98 % of the power should hit the shadow bar if ideal Gaussian beams are assumed.

IV. Measurement Methods

A. Beam Coalignment Measurements

The horizontal and vertical offsets of the beams with respect to the optical axis are determined with the relations given in [8]:

$$\Delta h = \frac{(\Delta x_1 + \Delta x_2)}{2}, \tag{1}$$

$$\Delta v = \frac{(\Delta x_1 - \Delta x_2)}{2}, \tag{2}$$

where $\Delta x_1 = x_1' - x_1$, $\Delta x_2 = x_2' - x_2$, x_1' and x_2' are the linear stage coordinates at which the peak IF power is measured, and x_1 and x_2 the reference coordinates defined by the testbed mechanical design. The peak IF power is measured at the position where the blockage of the cold load view by the warm shadow bar is the greatest.

The beam coalignment, or the positions of the beams with respect to each other, can be determined also from a relative measurement: the measured coordinates of one beam are used as the reference coordinates during the measurement of the second beam.

B. Side Band Measurements

In a double side band receiver both the upper and lower side bands contribute to the IF spectrum. In an ideal DSB receiver one half of the signal originates from the upper side band and the other half from the lower side band. In case of SWI, an equal contribution of both side bands is not important but rather that the fractions are known so that the spectral line intensity calibration may be performed correctly.

Signal transmission through the MPI is dependent on the electrical path length and follows a cosine curve. When the rooftop mirror position is swept and the transmission curves

Fig. 6. Beam coalignment measurement with the testbed in a thermal-vacuum chamber. 600-GHz receiver beam path is indicated with a red line.

at USB and LSB frequencies are superimposed, a beat pattern is formed. Following the procedure presented in [6], the USB and LSB conversion gains are computed by fitting two cosine functions to the beat pattern that is obtained by measuring the Y-factor for each ACS channel at different rooftop mirror positions. The Y-factor is obtained by looking at the hot and cold loads through the MPI and by computing the power ratio. The USB and LSB gains g_{USB} and g_{LSB} are determined from the amplitudes of the cosine-functions. The side band ratio is then

$$SBR = \frac{g_{USB}}{g_{LSB}}. \quad (3)$$

V. Results

Fig. 6 shows the RU box that is attached to the testbed in a TVAC during a beam coalignment measurement. It was first performed on the 600-GHz receiver of the SWI engineering model (EM) at room temperature. The 1200-GHz receiver was not switched on due to concerns related to the degradation of mixer lifetime. The measurement was then repeated with both receivers when the RU box was

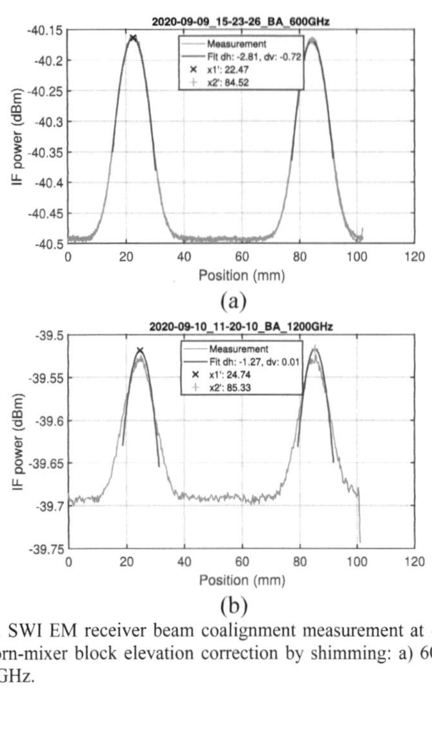

Fig. 7. SWI EM receiver beam coalignment measurement at -80°C after feedhorn-mixer block elevation correction by shimming: a) 600-GHz, b) 1200-GHz.

TABLE I
HORIZONTAL AND VERTICAL BEAM OFFSET WITH RESPECT TO THE NOMINAL BEAM AXIS AND THE DISTANCE D BETWEEN THE BEAM CENTERS.

	600 GHz		1200 GHz		D (mm)
	Δh (mm)	Δv (mm)	Δh (mm)	Δv (mm)	
Room temp.	-2.9	-3.3	n/a	n/a	n/a
Cold	-2.7	-3.3	-0.4	-3.5	2.3
Cold, shimmed	-2.8	-0.7	-1.3	0.0	1.7

cooled down to a temperature of -80 °C. Coordinate measurement machine measurements had shown that the feedhorn-mixer blocks of the receivers had a misalignment of about 400 μm in the elevation direction with respect to the mounting surface. This misalignment was corrected by shimming, and the shadow bar measurement was repeated. Fig. 7 shows as an example the measured IF power as a function of the linear stage position for both receivers at a RU box temperature of -80 °C. The measured beam offsets as well as the distance between the beam centers D are summarized in Table I.

The side band ratios of the SWI protoflight model (PFM) receivers are presented in Fig. 5 at three LO-frequencies near the center and edges of the operating frequency bands. The ratios vary between 0.68 and 1.23 in the 600-GHz band and between 0.74 and 1.36 in the 1200-GHz band. They exhibit a ripple, the magnitude of which is dependent on the LO-frequency.

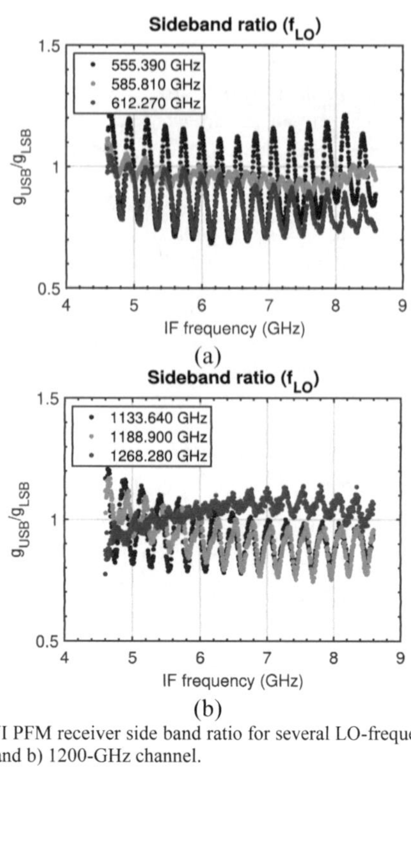

Fig. 5. SWI PFM receiver side band ratio for several LO-frequencies in a) 600-GHz and b) 1200-GHz channel.

VI. Conclusions

A versatile testbed for radiometric characterization of SWI has been built and successfully employed in the testing of SWI EM and PFM receivers in a thermal-vacuum chamber (TVAC).

The shadow bar measurements show that the cooling of the receiver unit box has only a minor effect, -0.2 mm in horizontal direction, on the beam position at 600 GHz. Further, the results confirm the reduction of vertical beam offset as the feed horns of both channels were shimmed in the elevation direction. The final measured distance between the beams centers, 1.7 mm, applies only in the shadow bar plane if the beams are not propagating parallel to each other.

The side band ratios of both SWI receivers have been determined successfully. A ripple in the results with respect to the IF frequency has been observed and further investigations of its origin are on-going.

Acknowledgments

The work at the University of Bern has been funded by the Swiss National Science Foundation under the Grant No. 200020-165744 and the ESA PRODEX program PEA 4000111892.

References

[1] M. Kotiranta, et al., "Optical Design and Analysis of the Submillimeter-Wave Instrument on JUICE," in IEEE Trans Terahertz Sci, vol. 8, no. 6, pp. 588-595, Nov. 2018.

[2] V. Tolls et al., "Submillimeter Wave Astronomy Satellite Performance on the ground and in orbit," ApJS, vol. 152, no. 1, pp. 137-162, May 2004.

[3] U. Frisk et al., "The Odin satellite - I. Radiometer design and test," A&A, vol. 402, no. 3, pp. L27-L34, May 2003.

[4] Teyssier, D., Dartois, E., Deboffle, D., *et al.*, "A Multi-path Far-Infrared and Sub-millimetre Gas Cell for Spectral Tests of HerscheL/HIFI," in ISSTT 2004, Northampton, Massachusetts, USA, 2004.

[5] Higgins, R., Teyssier, D., Borys, C. et al., "The effect of sideband ratio on line intensity for Herschel/HIFI," Exp Astron no. 37, pp. 433-452, May 2014.

[6] M. Renker, et al., "Sideband characterization and atmospheric observations with various 340 GHz heterodyne receivers", Rev. Sci. Instrum. 85, 084706, 2014.

[7] P. Moseley et al., "Introducing LORENTZ: A Novel Low-temperature Near-field Terahertz Chamber for Instrument Characterisation," in ISSTT 2022, Baeza, Spain, 2022.

[8] U. U Graf, "A Compact Beam Measurement Setup", J Infrared Milli Terahz Waves, no. 37, pp. 770–775, Mar. 2016

Harmonic phase and amplitude beam characterization of a wideband on-chip spectrometer

S. J. C. Yates*[1], K. Karatsu[2,3], A. Moerman[3], S.O. Dabironezare[3], A. Endo[3], J. J. A. Baselmans[2,3]

***Abstract*—We present a phase and amplitude beam pattern measurement technique using harmonic mixers. This allows a simultaneous multi-frequency phase sensitive characterization of a low resolution and wideband (220-420GHz) on-chip spectrometer using microwave kinetic inductance detectors. We investigate the beam quality, in particular the beam pointing and inferred telescope coupling and hence aperture efficiency. The measurements match the goal requirements for the DESHIMA-2 instrument for the ASTE telescope. The technique would be of interest for any (direct detector) spectrometer with a wide instantaneous bandwidth, particularly ones with dispersive components.**

Keywords—direct detectors, Microwave kinetic inductance detector, complex field mapping, optical characterization.

I. Introduction

Low resolution wideband direct detector on-chip spectrometers have a number of advantages in terms of compactness, bandwidth, sensitivity (for single dish applications) and scalability. As such they are a current topic of active research [1 and refs. there-in], with the recent first light instrument DESHIMA-1 [1] demonstrating the proof of principle on sky. DESHIMA-2 plans [2] to go beyond this to a full science grade instrument for the ASTE telescope. One area of improvement required is in the beam characterization, coupling and in particular pointing, which needs to be verified over the entire band prior to deployment.

Phase sensitive beam characterization has long been an important tool for sub-mm heterodyne instruments [3]. The complex beam pattern allows near to far field transformation, beam propagation and hence coupling and pointing in a single measurement prior to (on-sky) deployment. Recently the authors showed how this could be extended to direct detector (phase insensitive) arrays to allow such systems to be similarly characterized [4]. These techniques rely on two phase locked signal sources, and as such to do multi-spectral measurements require multiple measurements. We here extend this technique by using wide bandwidth harmonic mixers which produce sub-mm signals at multiple frequencies, harmonics of a base ~10 GHz signal. This allows a sparse frequency sampling over the whole band of interest in a single measurement. For dispersive spectrometers, like filter bank spectrometer used for DESHIMA 2, these signal frequencies are measured by individual filter bank detectors. Since there is only a single signal per detector, the signal to noise is then be optimized over the entire band: lower frequency harmonics with higher power do not mask the higher frequencies. Studying different harmonics enables study of the crosstalk, while wideband detectors on-chip enable the system response to be (quasi-) normalized.

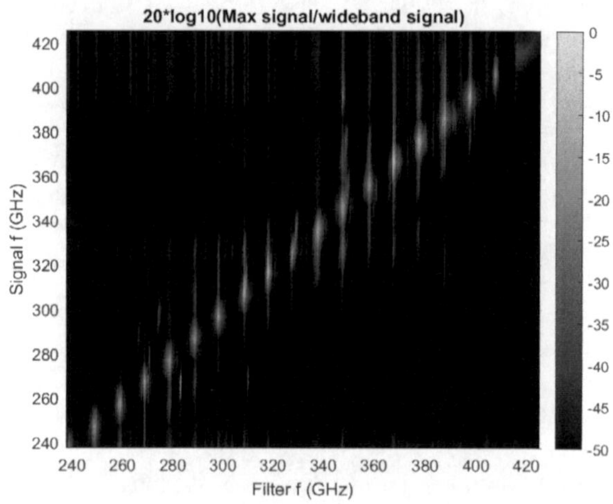

Fig. 1. Signal response matrix, the response in dB at each detector. Vertical aixs is the applied signal frequency, horizontal axis the detector frequency (from [2]).

II. Results

Some results are shown in fig.1, showing the peak normalized beam pattern response for the array from [2], measured in the far-field of the DESHIMA 2 cold optics. The stray light control and aperture efficiency, including pointing towards the telescope is matching simulations and requirements and will be discussed further.

References

[1] A. Endo et al., Nature Astronomy pp. 1–8 (2019). DOI 10.1038/s41550-019-0850-8

[2] A. Taniguchi et al, preprint arXiv:2110.14656, (2021)

[3] C. Y. E. Tong et al., IEEE Microw. Wireless Compon. Lett., vol. 13, no. 6, pp. 235–237, (2003)

[4] K. K. Davis et al., IEEE Trans. THz Sci. Tech. vol. 9, no. 1, pp. 67-77 (2019).

[1]SRON, Netherlands Institute for Space Research, Landleven 12, Groningen, The Netherlands; [2]SRON, Netherlands Institute for Space Research, Niels Bohrweg 4, 2333 CA Leiden, The Netherlands; [3]THz Sensing group, Microelectronics Department, Delft University of Technology, 2628 CD Delft, The Netherlands.

NOTES:

Characterization of widefield THz optics using phase shifting interferometry

Nicolás Reyes[1], Ivan Cámara-Mayorga[1], Gerrit Grutzeck[1], Stephen Yates[2], Andrey Baryshev[1], Bernd Klein[1]

***Abstract*—Characterization of wide-field optics in the Terahertz regime imposes new and demanding requirements for testing systems. From basic characterization using scalar sources or thermal sources located in the instrument focal plane, basic beam measurements can be determined, whereas important features, such as coupling to the antenna system or the effect of curved focal planes, cannot be determined by such approaches. These limitations can be overcome by performing phase and amplitude measurements of the beam. From its complete optical information of the system is obtained, allowing the estimation of all relevant optic parameters. In this work we present and demonstrate a technique to perform such measurement based on the use of photonic sources. The system was implemented to fully characterize the A-MKID instrument, a wide-field camera based on kinetic inductance detector (KID) technology to be deployed at the APEX Telescope in Chile.**

I. Introduction

The characterization of wide-field optics operating in the terahertz regime is a major challenge as it involves very demanding requirements many of which are hard to meet with the available testing systems technology. Whilst scalar or thermal sources placed in the focal plane allow basic characterization, important features such as spillover efficiency, wave front error, or aperture efficiency cannot be determined by such approaches. In addition, when instruments have a curved focal plane, designed to match the Petzval surface of the hosting telescope, even basic parameters are difficult to extract from scalar planar measurements. By contrast, from complex planar (amplitude and phase) measurements, the complete information of the optical system can be obtained, allowing the estimation of all relevant optical parameters [1][2]. The present paper reports on our work to implement a broad frequency tunable testing system capable of performing amplitude and phase measurements based on two interfering photonic sources of coherent terahertz radiation [3][4]. The described testing system allowed to characterize the wide-field optics of the A-MKID instrument, which will be shortly installed in the Atacama Pathfinder Experiment telescope (APEX) in Chile [5]. The A-MKID instrument is a wide-field, dual-color, bolometer camera. It is based on kinetic inductance detectors [6][7] and operates at 350 GHz and 850 GHz. The instrument optics consists of six reflective mirrors, possessing complex extended polynomial shapes optimized to achieve a diffraction limited performance and Strehl ratio better than 0.8 over the complete field of view [8].

This work presents a characterization system which consists of tunable terahertz photonic sources to perform amplitude and phase measurements by phase shifting interferometry. The experimental setup involved two terahertz sources located at the instrument focal plane, radiating at the same frequency but with the relative phase controlled using optical fiber stretchers. The use of fiber stretchers allows for electronical control of the phase shift, eliminating the need for mechanical delay lines, traditionally required by the phase shifting interferometry technique.

The interference between both sources -reference and signal source- is used to measure the complex beam pattern for every pixel in the detector. The signal source is mounted on an XY translation stage, whereas the reference source is static and couples to the detectors via reflection on a polarization grid.

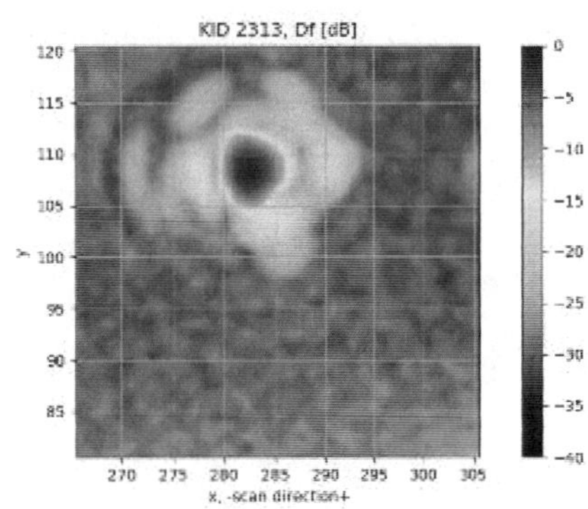

Figure 1: AMKID measured amplitude beam at 850GHz band. Sidelobe level is around -13 dB, in agreement with simulations, no pedestal response is observed down to -40 dB.

Nicolas Reyes, Ivan Camara Mayorga, Gerrit Grutzeck and Bernd Klein are with the Max Planck Institute for Radio-Astronomy, Auf dem Hugel 69, 53121, Bonn.
Stephen Yates is with SRON, Netherland Institute for Space Research, Groningen, The Netherlands.
Andrey Baryshev is with the Kapteyn Astronomical Institute, University of Groningen, The Netherlands and with the Max Planck Institute for Radio-Astronomy, Bonn.

NOTES:

Both terahertz signals propagate through the AMKID optical system and reach the detector array, where they interfere. The detected signal by the KID readout is proportional to the squared electric field at the detector, corresponding to the addition of the fields generated by the two photonic sources. As the phase of the sources, reference, and signal, was swept by a saw-tooth signal, the extraction of the amplitude and phase of the beam pattern is straightforward.

II. Conclusion

The setup probed to show an excellent terahertz signal phase stability of few degrees over several hours and allowed to characterize the optics of the instrument in two frequency bands: 350GHz and 850 GHz. The measured phase and amplitude beam maps yielded typically a signal to noise ratio of more than 40 dB, Figure 1, and allowed the extraction of aperture efficiency, spill-over efficiency, wavefront distorsion, beam ellipticity etc.

The system achieved a high level of accuracy and demonstrated a reliable method for measuring amplitude and phase in the terahertz regime.

References

[1] K.K. Davis, et al., "Complex Field Mapping of Large Direct Detector Focal Plane Arrays", IEEE Transactions on Terahertz science and technology, Vol 9, No 1, January 2019.

[2] S.J.C. Yates et al., "Complex beam mapping and Fourier optics analysis of a wide-field Microwave Kinetic Inductance detector camera" Journal of Low Temperature Physics. 199:156-163, 2020.

[3] I. Cámara Mayorga, et al. "Terahertz photonic mixers as local oscillators for hot electron bolometer and superconductor-insulator-superconductor astronomical receivers," J. Appl. Phys. 100 (4), 043116 (2006).

[4] I. Cámara Mayorga, et al., "First in-field application of a full photonic local oscillator to terahertz astronomy", *IEEE Trans. THz Sci. Technol.*, vol. 2, no. 4, pp. 393-399, Jul. 2012.

[5] R. Güsten, et al., "The Atacama pathfinder experiment (APEX), a new submillimeter facility for southern skies" A&A, 454(2): L13–L16, 2006.

[6] P. Day, et al., "A broadband superconducting detector suitable for use in large arrays," *Nature*, vol. 425, no. 6960, pp. 817–821, Oct. 23, 2003.

[7] R. M. J. Janssen *et al.*, "High optical efficiency and photon noise limited sensitivity of microwave kinetic inductance detectors using phase readout," *Appl. Phys. Lett.*, vol. 103, no. 20, 2013

[8] L. Esteras, "The optical system and the astronomical potential of A-MKID, a new camera using microwave kinetic inductance detector technology" Ph.D. dissertation. University of Bonn, 2014

NOTES:

CCAT-prime: 'Multi-map' Holographic Measurement for the FYST Testbed – Near-field beam measurement and Data analysis

Xiaodong Ren[1], Urs U. Graf[1], Pablo Astudillo[2], Sebastian Jorquera[2], Richard E. Hills[3], Bojan Nikolic[3], Stephen C. Parshley[4], Nicolás Reyes[5], Lars Weikert[1] and Jürgen Stutzki[1]

***Abstract*—A novel method based on microwave holography has been developed to align the optics of the Fred Young Sub-millimeter Telescope (FYST). The telescope uses a crossed-Dragone configuration with a 6 meters primary dish and a similar size sub-reflector. To reach high efficiency in the sub-millimeter band, it requires a surface precision better than 10.7 µm. The proposed method requires measuring four or more beam maps by putting the receiver at different points in the focal plane to break the phase degeneracy between the reflectors. Therefore, we refer to this method as 'Multi-map' holography. The system is designed to operate at around 300GHz to achieve a measurement accuracy better than 2 µm. In this report, a small model of the FYST has been constructed to demonstrate the feasibility of this new holography metrology in the laboratory. Conventional one-beam holography and the new 'Multi-map' measuring were carried out to measure the artificial surface errors on the reflectors of the small telescope. The results prove that the errors on the two reflectors can be discriminated and measured with a statistic error lower than 1 µm. The measurement also indicates that the large spatial errors existing on the two reflectors also can be observed.**

***Index Terms*—microwave, holography, alignment, sub-millimeter telescope, crossed-Dragone, CCAT-prime.**

I. Introduction

MEASURING and correcting the surface deformation of large radio telescopes are critical to preserving the electromagnetic performance of the telescope at the highest operating frequency. One conventional surface diagnosis method is microwave holography [1] [2], which directly measures the telescope's beam in both amplitude and phase and converts the beam to the complex fields on the reflector surface. Then the offset of the reflector from its original position can be computed by using the phase term of the solved surface fields. This technique is regularly used at almost all sub-millimeter telescopes around the world, and offers an efficient and accurate way for adjusting the reflectors. But It is always used in conventional telescopes where only one large reflector needs to be adjusted. For systems having two reflectors, e.g., the crossed-Dragone optics [3] [4] used in the FYST [5] [6], employing this method only indicates that reflector deformations exist, but cannot identify which of the two mirrors is deformed, i.e., it cannot resolve the surface error degeneracy in the two-mirror system. A new method, which measures several beams with the receiver placed in different well-separated points in the focal plane, was proposed for the FYST telescope in paper [7]. In this report, a 1/15th-scale model of the FYST was constructed as a testbed to demonstrate the feasibility of the new proposed method.

[1]I. Physikalisches Institut, Universität zu Köln,Germany; [2]Departamento de Astronomía, Universidad de Chile,Chile; [3]Cavendish Laboratory, University of Cambridge, UK; [4]Department of Astronomy, Cornell University, USA; [5]Max Planck Institute for Radio Astronomy, Germany.

II. Laboratory Holography Design

The FYST reflectors are built using 146 rectangular panels, 77 on the primary mirror and 69 on the secondary, with panel sizes of 670 × 750 mm and 700 × 710 mm respectively. We would use a spatial resolution of 10×10 cm^2 to properly solve the panel deformations on the measured surface error maps. Translating this requirement to the laboratory model indicates that a spatial resolution of 10×10 mm^2 is reasonable for the 400mm diameter laboratory antenna. This means a 5.7×5.7 deg^2 beam map extension needs to be measured at 300GHz. The technical details of the holographic design of the scale model and the full FYST system are compared and summarized in Table I. The laboratory setup is illustrated in Fig. 1. The Antenna and transmitter are separated by ~5 meters, and the beams are measured by scanning the transmitter that is moved by an XY-scanner. The reference receiver is mounted beside the antenna, 275mm from the primary reflector (M1) center. The pathlength from the transmitter to the reference receiver is not fixed. The phase modifications on the observed beam by the pathlength changes have to be corrected by geometrical optics.

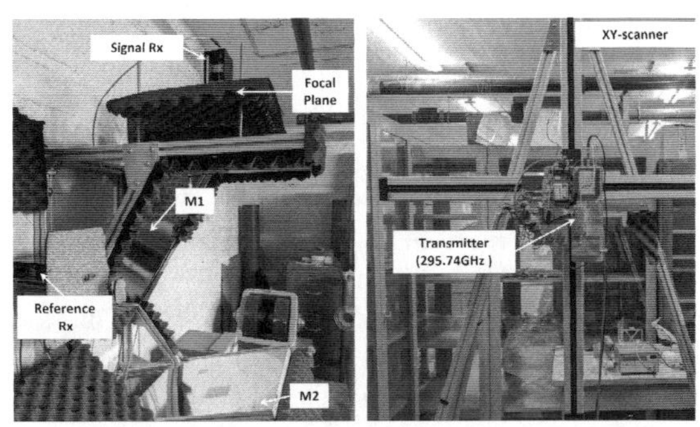

Fig. 1. Laboratory holographic testbed. Left: the 1/15th-scale FYST telescope, 400mm in diameter, and the location of the signal and reference receiver; right: the transmitter mounted on an XY-scanner is located at ~5 meters away from the antenna to illuminate the telescope. The scanner can move in the range of ±550mm along x and y directions.

TABLE I
CRITICAL HOLOGRAPHIC DESIGN

	Lab	FYST
Operating Wavelength (λ):	1.0137mm	1.0137mm
Telescope Aperture:	400mm	6000mm
Distance from source to antenna:	5m	300m
New focus (behind nominal focus):	200mm	705mm
Accuracy:	$< 2\,\mu m$	$< 2\,\mu m$
Spatial Resolution:	10mm	100mm
Required angular range:	5.7°	0.57°
Source scan range:	$\pm 250mm$	$\pm 1500mm$
S/N:	>64dB	>67dB
Number of field points	51×51	(71×71)
Receiver position spacing:	100mm	800mm

The two receivers and transmitter adopt the same diagonal horn [8] whose half-power beamwidth is about 16 degrees. The extra optics designed for reference receiver and transmitter modules in the original FYST holography [7] is removed because of the laboratory setup's shorter distance and broader beam map size requirement.

The beam maps of the Lab antenna are measured by placing the signal receiver to 5 points in the focal plane, at the center and the four corners of a square with side length of 100mm, see Fig. 2. Since the source is around 5 meters away from the telescope, to refocus the optics the receiver is moved back 200mm from the astronomical focus.

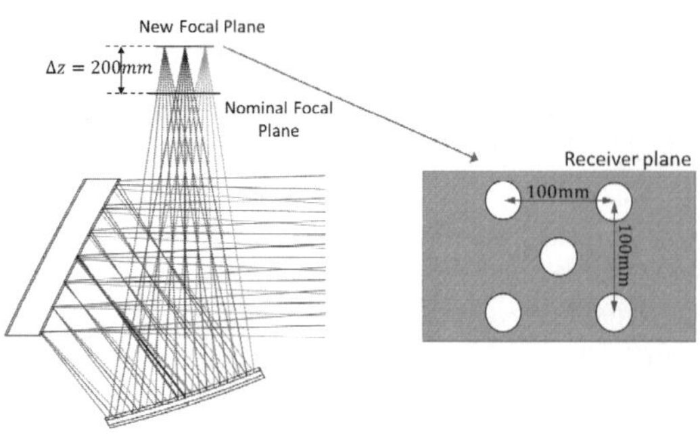

Fig. 2. Optical layout of the laboratory antenna (left). 5 receiver positions in the receiver plane (right).

III. BEAM MEASUREMENT AND ANALYSIS

The beams are measured by scanning the transmitter column by column with a constant speed. The field points are recorded using a digital cross-correlation spectrometer with ~100Hz sampling rate (~10ms integration time). The recorded data and position of field points are synchronized by time stamp. To calibrate the effect of systematic drift, the transmitter is moved to the beam center every 20s. After the beam scan, the impact of the phase and amplitude structure of the reference receiver beam needs to be removed from the recorded data.

To check the feasibility and accuracy of the holographic system, a copper foil with a thickness of $50\,\mu m$ is used to create artificial piston errors on the mirror surfaces, and plastic tapes are applied to produce equivalent piston error in the opposite direction. Fig. 3 shows the artificial error patches on the mirrors. Then the one-beam holography and multi-map analysis are both measured to diagnose the antenna optics.

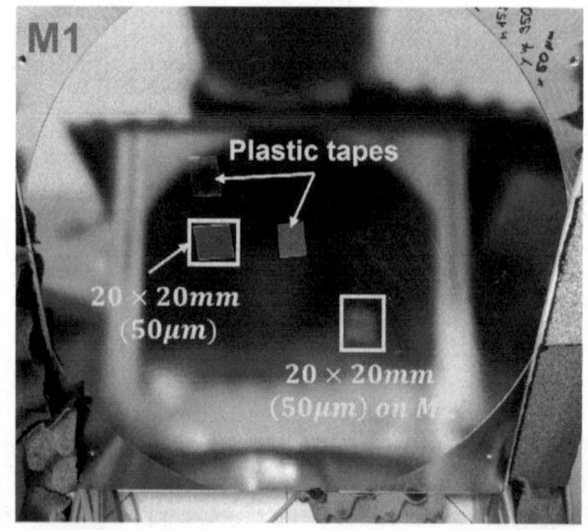

Fig. 3. Artificial piston errors on M1 and M2 made by copper foils and plastic tapes.

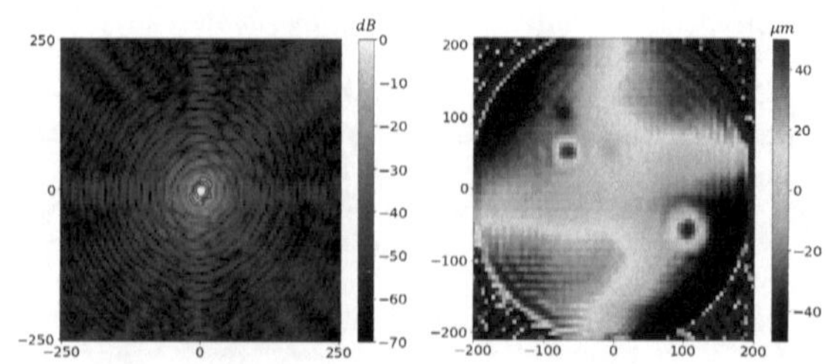

Fig. 4. Conventional one-beam holographic analysis. Left: the measured focused central beam; Right: equivalent surface deformations at the surface of M1.

A. Conventional One-beam Holography

The conventional holography just measures the focused beam with the receiver mounted in the center of the focal plane, see Fig. 4 (left). Since the beam is measured in the source plane that is parallel to the aperture plane of the antenna, we can convert the beam map labelled by $f_A(x, y)$ to the aperture fields expressed by $f_B(x, y)$ using technique of physical optics propagation [10]. data analysis processes are summarized below:

1) Calculate angular spectrum $F_A(u, v)$ of the observed fields using fast Fourier transform, where u and v are expressed by $\sin\theta \cdot \cos\phi$ and $\sin\theta \cdot \sin\phi$ respectively. θ and ϕ represent the elevation and azimuth angles.
2) Compute the angular spectrum of the field in aperture $F_B(u, v)$ using physical optics propagation technique expressed by the formula of $F_B = F_A \cdot exp(j\frac{2\pi}{\lambda}\Delta z \cdot \sqrt{1 - u^2 - v^2})$, where Δz is the distance between field planes.
3) Make inverse Fourier transform on the new angular spectrum $F_B(u, v)$ to get the aperture fields $f_B(x, y)$.
4) Based on the phase distribution of the solved aperture fields, the antenna geometry, and operating wavelength, the equivalent surface errors on M1, which is the sum

of errors on M1 and M2, can be computed, see Fig. 4 (right).

The main beam is obviously distorted. The inferred surface deformation in Fig. 4 right indicates all error patches, also points out the large spatial twist error in the antenna optics. But only using this error map cannot discriminate where the errors come from. These will be analyzed by following the 'Multi-map' holography.

B. Multi-map Holographic Analysis

The 'Multi-map' approach uses all five complex beam maps measured with the receiver at the five positions shown in Fig. 2. Fig. 5 displays the observed on-axis and four off-axis beam maps. The off-axis beams shift the projection of mirror errors in the aperture plane, allowing to separate the error contribution by their respective parallax. We can convert the five complex beams to two mirror surface maps using the numerical fitting algorithm described in paper [7], which expresses the surface errors by a set of parameters and find their values that can make the simulated beam best fit to the observed beam maps. We first research the error source of the twist-like errors in Fig. 4. The mirror surfaces of the small antenna are represented by Zernike polynomials [9] with maximum order of 7th (36 orthogonal polynomials). The fitting results shown in Fig. 6 top indicate that the twist-like errors come from M1.

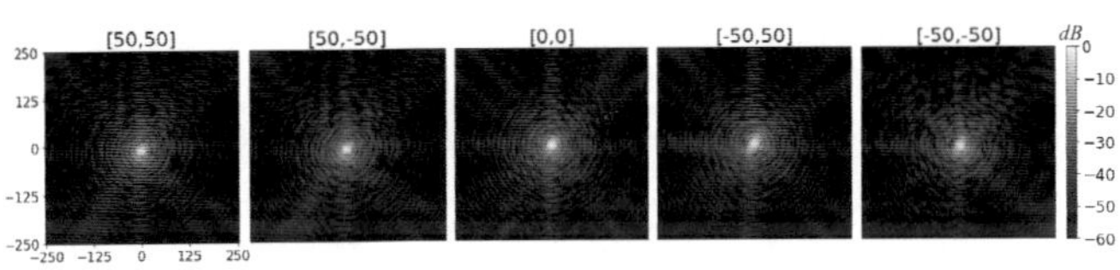

Fig. 5. The observed 5 distorted focused beams with the receiver located at positions [50mm,50mm], [50mm,-50mm], [0,0], [-50mm,50mm], and [-50mm,-50mm] relative to the center of the focal plane.

To find out the small artificial errors, more polynomials are required to achieve high spatial resolution, for example, employing a set of polynomials with maximum order of 30th (496 parameters on each mirror needed to be fitting), more details of the surface quality are resolved, see Fig. 6 bottom. All four artificial piston errors (3 patches on M1 and 1 copper foil on M2) are clearly detected in their correct location. The two plastic tape patches introduce a fraction of phase delay that is equivalent to negative piston panel errors. Their precise phase delay is not well known because of the multi-reflection between the two surfaces and the difficulty of attaching them tightly to the mirror surface.

The statistical error of the measurement can be studied by repeating the measurements, which shows a random variations of the surface with a RMS scatter of < 1 μm.

The resolved surface errors in Fig. 6 show a twist-like error on M1. To verify this, the deformation of M1 was checked mechanically, and was found to be caused by misalignment of the mounting frame. Correcting this misalignment fixed the beam distortion. Fig. 7 shows the comparison of the beam affected by the twist error on M1 and the beam after error

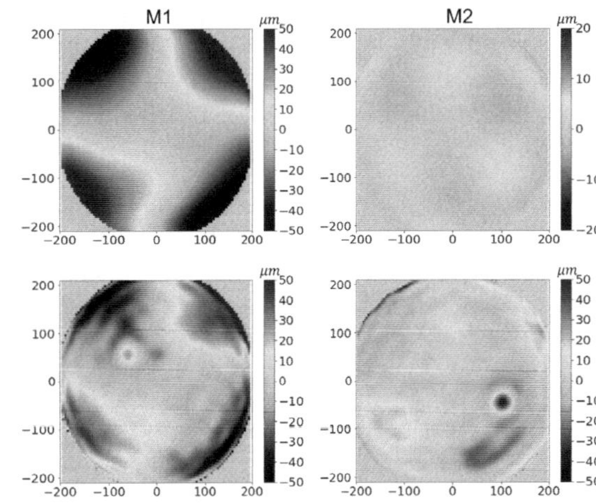

Fig. 6. The deduced surface errors analyzed by the 'Multi-map' holography algorithm. Top is the fitted surface errors only in large spatial scale using 36 Zernike polynomials with maximum order of 7th for each mirror surface. Bottom is the more detailed surfaces that are resolved by fitting 496 parameters per mirror (992 in total).

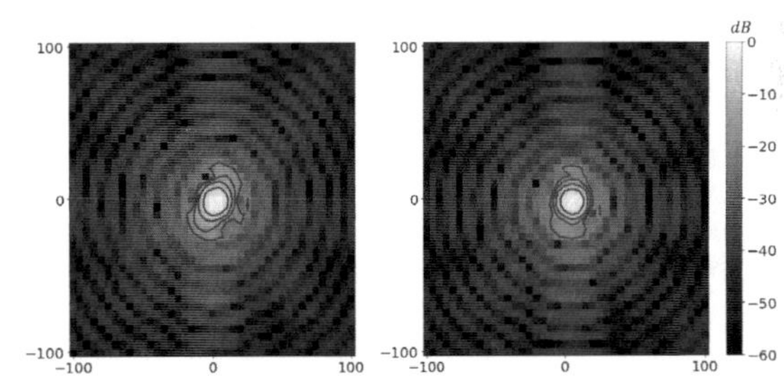

Fig. 7. Observed focused beam of the laboratory antenna. Left is the measured beam distorted by the twist error on M1; Right is the one after the large error was corrected. Blue curves are the contour map of the observed beam.

correction. This shows that the novel 'multi-map' holography technique can also correctly identify and discriminate the large-scale errors of the mirror surfaces.

IV. Conclusion

A near-field beam measurement system has been built to measure the surface profile of the two reflectors of the small version FYST antenna. The feasibility of the new 'Mutli-map' holographic technique is proven. Employing this technique, the small-scale surface deviations on the two reflectors can be measured and discriminated with a repeatable error less than $< 1\mu m$. The deformations in large spatial scale of the reflectors, such as errors caused by gravity and thermal expansions, can also be diagnosed. The experiments cannot check the accuracy of the measured large spatial errors because the laboratory antenna cannot be adjusted with predictable behavior. But we believe that repeating a couple of times of the holographic measurements and panel adjustments for the full-size FYST telescope can correct the large spatial errors.

Acknowledgments

The work is carried out at UniversitUniversität zu Köln and supported by the Colaborative Research Centre 956, sub-projects D2 and S, and funded by the Deutsche Forschungsgemeinschaft (DFG)-project ID 184018867. The team at Universidad de Chile acknowledge the support from ANID through project QUIMAL 180004. We would like to dedicate this paper to the memory of Richards Hills.

REFERENCES

[1] J. Bennett, et al., "Microwave holographic metrology of large reflector antennas," in *IEEE Transactions on Antennas and Propagation*, vol. 24, no. 3, pp. 295-303, May 1976.

[2] Scott, P. F. and Ryle, M., "A rapid method for measuring the figure of a radio telescope reflector.", *Monthly Notices of the Royal Astronomical Society*, vol. 178, pp. 539–545, 1977.

[3] C. Dragone, "Offset multireflector antennas with perfect pattern symmetry and polarization discrimination," in *The Bell System Technical Journal*, vol. 57, no. 7, pp. 2663-2684, Sept. 1978.

[4] Niemack, M. D., "Designs for a large-aperture telescope to map the CMB 10X faster", Applied Optics, Vol. 56, No. 7, pp.1688-1696 (2016).

[5] Parshley, S. C., et al., "CCAT-prime: a novel telescope for sub-millimeter astronomy," Proc. SPIE 10700, Ground-based and Airborne Telescopes VII, 107005X (6 July 2018).

[6] Parshley, S. C., et al., "The optical design of the six-meter CCAT-prime and Simons Observatory telescopes," Proc. SPIE 10700, Ground-based and Airborne Telescopes VII, 1070041 (6 July 2018);

[7] Ren, X., et al., "Holographic surface measurement system for the Fred Young Submillimeter Telescope," Proc. SPIE 11445, Ground-based and Airborne Telescopes VIII, 114456D (13 December 2020)

[8] H. J. Gibson, et al., "A Novel Spline-Profile Diagonal Horn Suitable for Integration Into THz Split-Block Components," in IEEE Transactions on Terahertz Science and Technology, vol. 7, no. 6, pp. 657-663, Nov. 2017, doi: 10.1109/TTHZ.2017.2752423.

[9] Born, M. and Wolf, E., "Principles of Optics", Elsevier Science (1980).

[10] Goodman, J. W. 2005, Introduction to Fourier optics, 3rd ed., by J.W. Goodman. Englewood, CO: Roberts & Co. Publishers, 2005

Introducing LORENTZ: A Novel Low-temperature Near-field Terahertz Chamber for Instrument Characterisation.

P. Moseley*[1], L. Rolo[1], A. Baryshev[2,3], T. Vos[4], A. Belitskaya[3], D. A. Ronso da Costa Lima[2],P. de Maagt[1] and P. Hartogh[5]

***Abstract*— The Low-temperature Near-field Terahertz Chamber (LORENTZ) is a novel facility recently installed and commissioned at ESTEC, ESA. This facility has the unique ability to characterize the antenna performance off full sub-millimeter instruments in operational environments down to 80k. We provide an overview on the various design and commissioning steps that were required to ensure all parts of the test facility would operate reliability in such challenging conditions. We also present how the facility performed during the first full measurement run of flight hardware and a roadmap for future developments.**

Keywords—Antenna, RF Testing, Cryogenic, THz, SWI

I. Introduction

The Low-temperature Near-field Terahertz Chamber (LORENTZ) is a novel facility recently installed and commissioned at ESTEC, ESA. It offers the unique ability to measure the RF antenna performance of full instruments in their representative environmental conditions they will experience during mission operation.

Currently antenna patterns of sub-millimeter instruments are performed at room temperature and rely on thermal modeling to predict if any deformation in the optical elements will occur. Such predications increase in complexity and uncertainty as the operational frequency rises. Due to complexity and cost of future instruments, relying on such simulations can be high risk. Therefore, LORENTZ was conceived to mitigate this.

II. Test Facility Architecture

The unique capability of the facility is that it can accommodate a full instrument up to 1900mm cubed with a weight up to 500kg and cool it down to 80k inside a RF anechoic chamber. Within the same chamber, a 1x1m planar scanner offers micron positioning accuracy. This combination allows testing to be performed between 50GHz and 1500GHz. The RF signal is sampled at various positions with the planar scanner, which is transformed to retrieve key performance indicators of the instrument such as gain and pointing accuracy.

The most challenging aspect in designing this test facility was creating a way to thermally isolate the scanner from the anechoic chamber. Due to the large scan plane and accuracy requirements, any large temperature gradients placed on the scanner would heavily degrade its performance. Therefore, a moveable RF and thermal shield was designed. This allows the RF probe an unobstructed view to inside the anechoic chamber, while blocking all thermal radiation. A schematic of this setup is shown in Fig.1.

Fig. 1. Cross-section of the test facility. Left side shows the cryogenic anechoic chamber, the right side shows the scanner held under vacuum but at 280K. Between them is a movable shield that contains the RF probe.

The facility was successful in performing its first full measurement run of flight hardware, by verifying the RF performance of the SWI instrument for JUICE[1], Fig. 2. A more detailed view will be given on the various design and commissioning steps that were required to ensure all parts of the test facility would operate reliability in such challenging conditions.

III. Future capabilities and opportunities

This successfully test is just the beginning for the facility. In the initial design stages of the facility, it was already foreseen that it should be upgradable to allowing cooling down to 4k, to increase the suitability for future THz instruments, including full end-to-end testing of cryogenic receivers.

The facility is open to all industry and institutions that could benefit from its unique capability. It is hoped that this conference will be the right platform to make the wider community aware.

References

[1] https://www.mps.mpg.de/planetary-science/juice-swi

Fig. 2 Testing of SWI inside the LORENTZ chamber

[1]ESA ESTEC, Netherlands; [2] Kapteyn Astronomical Institute, University of Groningen, Netherlands; [3]Dutch Terahertz Inspection Services B.V. Netherlands; [4]SRON, Leiden, Netherlands; [5]MPS (Max Planck Institute for Solar System Research), Germany

NOTES:

GUSTO Payload Design and Performance

Gal/Xgal U/LDB Spectroscopic/ Stratospheric THz Observatory (*GUSTO*) is a NASA Explorers Mission of Opportunity that will make large scale maps of the Milky Way and Large Magellanic Cloud in three important interstellar lines: [CII], [OI], and [NII] at 158, 63, and 205 μm, respectively. During its ~50+ day stratospheric (~120,000 ft) flight, GUSTO's 0.9-meter balloon-borne telescope and THz heterodyne array receivers will provide the spectral and spatial resolution needed to untangle the complexities of the interstellar medium by probing all phases of its Life Cycle. The GUSTO payload consists of (1) a telescope; (2) three 8-pixel heterodyne array receivers; (3) autocorrelator spectrometers; (4) instrument control electronics; and (5) a cryostat. The GUSTO gondola is derived from successful APL designs. Much of the GUSTO instrument architecture and hardware is based on the experience gained in developing and flying the Stratospheric Terahertz Observatory (STO). GUSTO is currently undergoing integration and test and will launch from the NASA Long Duration Balloon (LDB) Facility near McMurdo, Antarctica in December 2023. During our presentation we will provide an overview of the GUSTO payload design and performance characteristics.

The Terahertz Intensity Mapper (TIM): design and development status of a far-infrared balloon for spectroscopic galaxy evolution studies

Christopher E. Groppi*[1], Reinier M.J. Janssen[2,3], James E. Aguirre[4], Justin S. Bracks[4], Charles M. Bradford[2,3], Brockton Brendal[5], Anthony J. Corso[4], Jeffrey P. Filippini[5], Jianyang Fu[6], Steve Hailey-Dunsheath[3], Dylan Joralmon[7], Ryan P. Keenan[8], Lun-Jun Liu[3], Ian Lowe[8], Daniel P. Marrone[8], Philip D. Mauskopf[1], Evan C. Mayer[8], Rong Nie[5], Vesal Razavimaleki[5], Talia Saeid[1], Isaac Trumper[9], and Joaquin D. Vieira[6]

Abstract—**The Terahertz Intensity Mapper (TIM) is a NASA far-infrared long-duration balloon mission, which aims to study the 3D structure of our star forming universe using far-infrared spectral lines. We will present the scientific goal of TIM, the architecture of the balloon payload and spectrometer, and reports upon the current status of hardware components.**

Keywords—far-infrared, balloon, grating spectrometer, intensity mapping, galaxy evolution

I. Science Goal

Understanding the formation and evolution of galaxies over cosmic time is one of the foremost goals of astrophysics and cosmology today. The cosmic star formation rate has undergone a dramatic evolution over the course of the last 14 billion years, and dust obscured star forming galaxies (DSFGs) are a crucial component of this evolution. A variety of important, bright, and unextincted diagnostic lines are present in the far-infrared (FIR) which can provide crucial insight into the physical conditions of galaxy evolution, including the instantaneous star formation rate, the effect of AGN feedback on star formation, the mass function of the stars, metallicities, and the spectrum of their ionizing radiation.

TIM, the Terahertz Intensity Mapper [1], is a NASA balloon mission that will observe the universe in the crucial gap between the spectroscopic coverage of the Atacama Large Millimeter/submillimeter Array (ALMA) in the sub/mm, and the James Webb Space Telescope (JWST) in the mid-IR; something which is impossible to do from the ground. TIM will survey two fields: ~0.1 deg^2 centered on GOODS-S and ~1 deg^2 on the South Pole Telescope (SPT) Deep Field. TIM will produce deep maps of the 3D structure of the Universe by redshift tomography ("intensity mapping") with [CI], and [CII] X [NII] cross-spectra, to constrain the cosmic star formation history at cosmic noon. In addition, it is expected to achieve spectroscopic line detections of ~100 galaxies in the atomic fine structure lines of C, N, and O, as well as establish mean galaxy properties such as star formation rate, metallicity and AGN content, using a stacking analysis of known sources and the wealth of ancillary data available in the GOODS-S and SPT Deep Field.

II. Payload & Instrument Architecture

To achieve these science goals, TIM will fly two long-slit (1 degree slit length) cryogenic grating spectrometers, which cover the 240-317 um and 317-420 um wavelength bands at R~250 simultaneously. Each of these spectrometers is serviced by a ~3600 pixel array of horn-coupled kinetic inductance detectors (KIDs), which will be read out using a Xilinx RFSoC based readout system.

The TIM gondola and cryogenic system are loosely based upon proven BLAST hardware. However, a new 2-meter low-emissivity high-throughput primary mirror will be used to achieve maximum mapping speed. This smaller mirror (compared to BLAST) has allowed the gondola design to become more compact. In contrast, the 4 Kelvin volume of the cryogenics system has grown to accommodate all optics of the spectrometers.

III. Current Status

The TIM consortium is currently receiving hardware for many of its subsystems. In particular, the primary mirror is complete and currently being shipped to the US. The detectors for the long wavelength spectrometer have demonstrated the required single pixel optical performance [2,3] and kilopixel arrays are currently under development [4]. All other major subsystems (gondola, readout, cold optics, flight software) are in their final stages of design, or have already begun fabrication. An initial full assembly of the payload is envisioned for late 2022.

* cgroppi@asu.edu ; [1]School of Earth and Space Exploration, Arizona State University, Tempe, AZ 85287, USA; [2]Jet Propulsion Laboratory, California Institute of Technology, Pasadena, CA 91109, USA; [3]Department of Astronomy, California Institute of Technology, Pasadena, CA 91125, USA; [4]Department of Physics and Astronomy, University of Pennsylvania, Philadelphia, PA 19104, USA; [5]Department of Physics, University of Illinois at Urbana-Champaign, Urbana, IL 61801, USA; [6]Department of Astronomy, University of Illinois at Urbana-Champaign, Urbana, IL 61801, USA; [7]School of Engineering of Matter, Transport and Energy, Arizona State University, Tempe, AZ 85287, USA; [8]Department of Astronomy, University of Arizona, Tucson, AZ 85721, USA; [9]Intuitive Optical Design Lab LLC, Tucson, AZ 85701, USA

NOTES:

References

[1] J. Vieira, et al., "The Terahertz Intensity Mapper (TIM): an Imaging Spectrometer for Galaxy Evolution Studies at High-Redshift", *Proceedings of the 30th International Symposium on Space Terahertz Technology*, Apr. 2019.

[2] R.M.J. Janssen, et al., "Single Pixel Performance of the kinetic inductance detectors for the Terahertz Intensity Mapper", *Journal of Low Temperature Physics*, submitted 2022

[3] R. Nie, et al., "Absorber design and optimization of Kinetic Inductance Detectors for the Terahertz Intensity Mapper", *Journal of Low Temperature Physics*, accepted 2022

[4] L.-J. Liu, et al., "Design of the Kinetic Inductance Detector based focal plane assembly for the Terahertz Intensity Mapper", *Journal of Low Temperature Physics*, submitted 2022

NOTES:

The OSAS-B instrument: a balloon-borne heterodyne spectrometer for atomic oxygen in the mesosphere and lower thermosphere

M. Wienold*[1], A. Semenov[1], H. Richter[1], E. Dietz[1], S. Frohmann[1], P. Dern[1], X. Lü[2], L. Schrottke[2], B. Klein[3], H.-W. Hübers[1]

***Abstract* — We present the design and evaluation of the OSAS-B instrument which is dedicated to probing the 4.7 THz emission of atomic oxygen in the atmosphere of Earth.**

The Oxygen Spectrometer for Atmospheric Science on a Balloon (OSAS-B) has been developed for probing the emission of atomic oxygen in the mesosphere and lower thermosphere (MLT), a region of Earth's atmosphere, where atomic oxygen is the dominant species and plays an important role for the chemistry and energy balance [1, 2]. OSAS-B is a heterodyne receiver for the ground state fine-structure transition of atomic oxygen at 4.745 THz. Due to water absorption, this line cannot be observed from ground. The only instrument, which is currently capable of observing and resolving this transition, is the upGREAT heterodyne receiver (German REceiver for Astronomy at THz Frequencies) on board of SOFIA [3]. However, observations are limited to night time.

Figure 1 shows a CAD visualization of the instrument. The central unit is a dewar with a liquid helium stage for the hot-electron-bolometer (HEB) mixer as the heterodyne detector and a solid nitrogen stage for cooling the quantum-cascade laser used as local oscillator. The Si lens of the quasi-optical HEB mixer allows for an angular resolution of 30'. A pointing mirror is used to measure the atomic oxygen emission for different elevation angles. The backend of the receiver is a digital Fourier transform spectrometer. The instrument weights approximately 60 kg and consumes 100 W in operation.

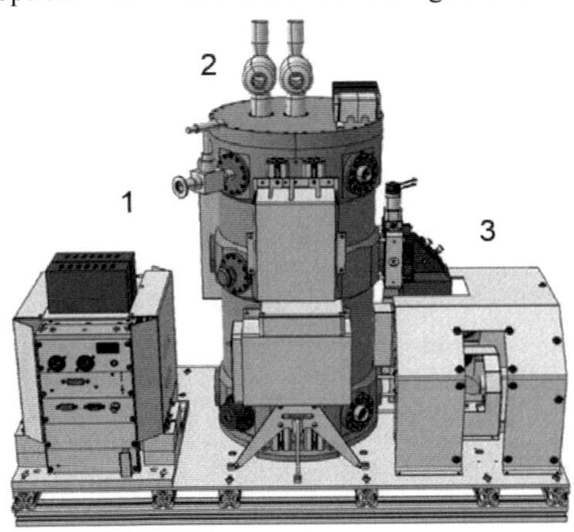

Fig. 1. CAD visualization of the OSAS-B instrument. 1: central control unit with digital Fourier transform spectrometer. 2: cryostat with pressure and periphery control. 3: optics compartment.

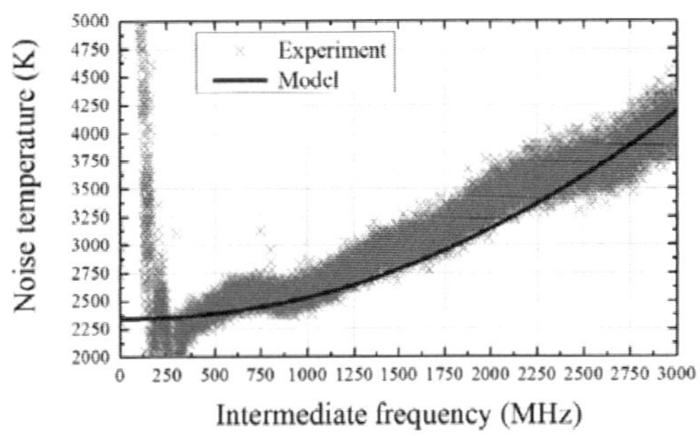

Fig. 2. Measured system noise temperature as a function of the frequency. The solid line represents a simulation of the HEB.

In order to evaluate the performance of the instrument, we studied the stability and system noise temperature T_n in a thermal vacuum chamber using two blackbodies at 90 K and room temperature. T_n is shown in Fig. 2 as a function of the intermediate frequency together with a noise model of the HEB mixer. A noise temperature of 2500 K is achieved up to an intermediate frequency of 1 GHz, similar to the typical performance of the upGREAT receiver in SOFIA. The OSAS-B flight will take place at Esrange, Sweden, in autumn 2022 in the frame of the European HEMERA program. A total flight duration of 12 h is anticipated.

References

[1] M. G. Mlynczak and S. Solomon, "A detailed evaluation of the heating efficiency in the middle atmosphere," *J. Geophys. Res.*, vol. 98, 10517, 1993.

[2] H. Richter, C. Buchbender, R. Güsten, R. Higgins, B. Klein, J. Stutzki, H. Wiesemeyer, H.-W. Hübers, "Direct measurements of atomic oxygen in the mesosphere and lower thermosphere using terahertz heterodyne spectroscopy," *Commun. Earth Environ.*, vol. 2, 19, 2021.

[3] C. Risacher, *et al.*, "The upGREAT Dual Frequency Heterodyne Arrays for SOFIA," *J. Astron. Instrum.*, vol. 7, 1840014, 2018.

[1]DLR Institute of Optical Sensor Systems, Rutherfordstr. 2, 12489 Berlin, Germany; [2]Paul-Drude-Institut für Festkörperelektronik, Hausvogteiplatz 5-7, 10117 Berlin, Germany; [3]Max-Planck-Institut für Radioastronomie, Auf dem Hügel 69, 53121 Bonn, Germany

NOTES:

ASTHROS - Astrophysics Stratospheric Telescope for High-Spectral Resolution Observations at Submillimeter-Wavelengths: Architecture Design and Subsystem Level Integration & Test Status

Jose V. Siles*[1], Jonathan Kawamura[1], Pietro Bernasconi[2], Christopher Groppi[3], Erich Schulze[2], Darren Hayton[1], Michael Carpenter[2], Harry Eaton[2], Robert Coker[2], Daniel Young[2], Heather Borowski[2], Giuseppe Valsecchi[4], Claudio Franchini[4], Mauro Maggioni[4], Josef Eder[4], Marco Pedrali[4], Robert Lin[1], Joshua Gundersen[5], Paul Horton[3], Jeffrey Olson[3], Richard Cofield[1], Bruce Bumble[1], Choonsup Lee[1], Alejandro Peralta[1], Alain Maestrini[1], Paul Goldsmith[1] and Jorge Pineda[1]

***Abstract*—We report on the overall architecture design and subsystem level integration & test status of the ASTHROS mission. ASTHROS is a 2.5-m long duration balloon-borne observatory for high-spectral resolution observations in the 1.4-2.7 THz range, the largest balloon telescope ever built for terahertz science. It is planned to belaunched from Antarctica in December 2023 and fly for at least 3-4 weeks, studying the impact of stellar feedback on star formation in the Milky Way and other galaxies.**

Keywords—suborbital, balloon missions, far-infrared, terahertz, Antarctica, HEB, Schottky diodes.

I. Introduction

ASTHROS builds on the success of heterodyne instruments and technology on Herschel and SOFIA, STO-2, ALMA, etc., providing a low-cost stepping stone for future heterodyne missions. The first Antarctic flight will focus on determining the key mechanisms by which stellar feedback (e.g. supernovae explosions) affects the interstellar medium and regulates star formation, which is a key parameter in galactic evolution. We will map two template Galactic star forming regions and the entire disk of the M83 barred spiral galaxy at high angular resolution, complementing existing datasets from SOFIA, WISE, Herschel, Spitzer and HST. We will simultaneously observe and use the ratio of the [NII] 205um and [NII] 122um to determine the electron density distribution in these regions.

ASTHROS will also be capable of tuning to nearby spectral lines (OH, HDO, HF, HD, CO) for Target of Opportunity observations. One compelling target is the HD 112μm (2.674 THz) line that traces the gas mass in protoplanetary disks. ASTHROS' angular resolution of 12” at 122 μm and 20” at 205 μm correspond to 0.2 pc and 0.35 pc at 122μm and 205μm, respectively, for a source 4 kpc from the Sun. This high angular resolution will enable us to resolve structures ~750 times smaller than the typical size of star forming regions (~150 pc). Combined with large-scale mapping, we will begin to understand how different stellar feedback mechanisms affect ionized gas over a wide range of spatial scales in the Milky Way and other galaxies.

Fig. 1. ASTHROS payload under I&T, showing the 2.5-m primary mirror and the ~4.5-m long sunshade frame,

II. Architecture Overview

The overall architecture of ASTHROS has been designed with the intention of providing NASA with a robust far-infrared platform that allows multiple flights and instrument upgrades. This posed big challenges to the design of both the gondola and the telescope itself. The ASTHROS payload consists of two 8-pixel heterodyne receiver channel, covering the 1.4 THz to 2.7 THz range, and featuring waveguide HEB mixers pumped by Schottky diode based local oscillators. The HEB mixers are cryogenically cooled using a low-power 4K-class cryocooler which allows ASTHROS to perform long duration flights without being limited by the cryogenics on board. The 2.5-m antenna unit features a 9-panel segmented telescope based on ALMA light-weight panel technology. The backing structure and cradle are based on high-modulus carbon fiber, not only to reduce weight, but also to decouple the science instrument from the mirror and to overcome the gravity pull in order to maintain the overall surface RMS below 8 μm during flight. The gondola is designed to support science payloads of more than 1000 kg and over 1500 W. It also provides ~2-5” pointing capability. More details of the design and I&T status will be presented at the conference.

[1]NASA Jet Propulsion Laboratory, California Institute of Technology, Pasadena CA, 91109, USA; [2]Johns Hopkins University – Applied Physics Laboratory, Laurel MD, 20723, USA, [3]Arizona State University, Tempe AZ, 85287, USA; [4]Media Lario S.r.l, Bosisio Parini LC, 23842, Italy, [5]University of Miami, Coral Gables FL, 33124, USA, [6]Lockheed Martin Corporstion, USA.

NOTES:

Frequency stabilization of 3.5-THz and 4.7-THz quantum-cascade lasers by a phase-locked loop

Heiko Richter*[1], Nick Rothbart[1], Martin Wienold[1], Xiang Lü[2], Lutz Schrottke[2], Divya Jayasankar[3], Vladimir Drakinskiy[3], Mats Myremark[3], Jan Stake[3], Peter Sobis[3,4], Anders Emrich[5], and Heinz-Wilhelm Hübers[1,6]

Abstract—**We report on a phase-locked loop for quantum-cascade lasers operating at 3.5 and 4.7 THz. These frequencies are particularly relevant in atmospheric science and astronomy. The QCLs are locked to the multiplied output of a frequency synthesizer using a Schottky diode harmonic mixer. The resulting linewidths are in the order of a few Hz.**

Keywords— frequency stabilization, phase-locked loop, terahertz, quantum-cascade laser

I. Introduction

Terahertz (THz) quantum-cascade lasers (QCLs) are of great importance as local oscillators (LOs) in high-resolution heterodyne spectrometers [1, 2], which are commonly used for astronomical and atmospheric observations [2, 3]. The frequency resolution of such spectrometers depends on the frequency stability of the QCL-LO. An optimal stabilization of temperature and driving current allows for a resolution of a few MHz.

Phase locking to a reference source is a well-known approach for inreased stabilization. However, application of this method to THz QCLs faces two major problems: Both, the reference sources and the mixers, are not readily available at THz frequencies. The mixer, which generates the difference frequency of QCL and reference source, is a key part of the phase-locked loop (PLL). We report on a PLL for QCLs operating at 3.5 THz and 4.7 THz. These are two important frequencies, because the hydroxyl radical, OH, and atomic oxygen, O, have transitions at these frequencies. Both species are significant for atmospheric science and astronomy, and several heterodyne spectrometers are currently in operation or under development for airborne and spaceborne missions [2, 4].

II. Results

The QCLs which are frequency-stabilized by the PLL are mounted in a mechanical cryocooler. They are designed for frequencies of 3.5 and 4.7 THz. The active medium has been developed for continuous-wave operation, high output powers, and low electrical pump powers. Both QCLs have a single-plasmon waveguide with a Fabry-Pérot resonator. The resonators are sufficiently short so that single mode emission over most of the driving current range of the laser is achieved. The beam of the QCL is formed with dedicated optics into an almost Gaussian profile. The free running linewidth of the QCLs is a few MHz. The PLL is based on a Schottky diode harmonic mixer and a multiplier source, which pumps the harmonic mixer. The output frequency is tunable from 570 GHz to 610 GHz with more than 3 mW of output power. The Schottky diode harmonic mixer generates the sixth and eighth harmonic for phase locking of a 3.5-THz QCL and a 4.7-THz QCL, respectively. With this setup, phase locking was achieved for both frequencies (Fig. 1).

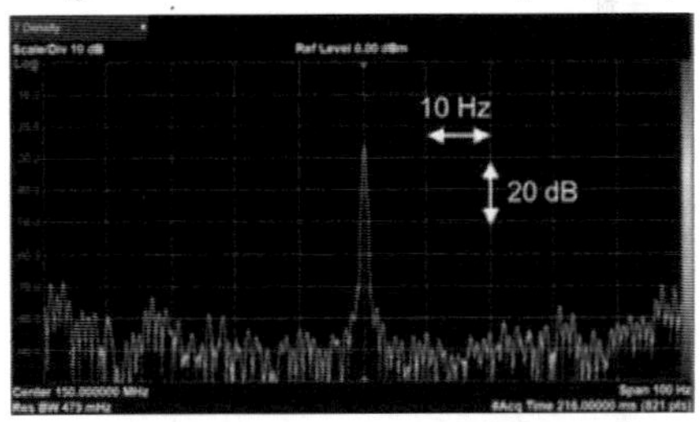

Fig. 1. Locked signal obtained from a 4.7-THz QCL and a multiplier source.

References

[1] H. Richter et al., "4.7-THz Local Oscillator for the GREAT Heterodyne Spectrometer on SOFIA," IEEE Transactions on Terahertz Science and Technology, vol. 5, no. 4, pp. 539-545, 2015.

[2] C. Risacher et al., "The upGREAT Dual Frequency Heterodyne Arrays for SOFIA," Journal of Astronomical Instrumentation, vol. 7, no. 4, 1840014 (16 pages) (2018).

[3] H. Richter et al., "Direct Measurements of Atomic Oxygen in the Mesosphere and Lower Thermosphere Using Terahertz Heterodyne Spectroscopy", Communications Earth and Environment, vol. 2, 19 (2021).

[1]German Aerospace Center (DLR), Institute of Optical Sensor Systems, 12489 Berlin, Germany, [2]Paul-Drude-Institut für Festkörperelektronik, Leibniz-Institut im Forschungsverbund Berlin e.V., 10117 Berlin, Germany, [3]Terahertz and Millimetre Wave Laboratory, Department of Microtechnology and Nanoscience (MC2), Chalmers University of Technology, 41296 Göteborg, Sweden, [4]Low Noise Factory AB, Nellickevägen 24, 412 63 Göteborg, [5]Omnisys Instruments AB, 42132 Vastra Frölunda, Sweden, [6]Humboldt-Universität zu Berlin, 12489 Berlin, Germany

NOTES:

[4] S. P. Rea et al., “The low-cost upper-atmosphere sounder (LOCUS),” Proc. of the 26th Int. Symposium on Space Terahertz Technology (Cambridge, MA, 2015).

NOTES:

The ammonia laser: a possible local oscillator for space applications?

Marie-Hélène Mammez[1], Thomas S. Hearne[2], Dominique Mammez[1], Olivier Pirali[2], Marie-Aline Martin-Drumel[2], Guillaume Ducournau[3], Sophie Eliet[3], Francis Hindle[1], Stefano Barbieri[3], Pascale Roy[4], Gaël Mouret[1], and Jean-François Lampin[*3]

***Abstract*—Local oscillators are essential parts of THz heterodyne receivers for space applications. We report here our progress concerning the ammonia laser which is a molecular laser pumped by a mid-infrared quantum cascade laser. We show the generation of a large number of lines in the 0.7-5.5 THz range. We show also that it can be used to pump a hot electron bolometer for THz heterodyne spectroscopy application.**

***Keywords*—local oscillator, molecular laser, ammonia, quantum cascade laser**

High wall-plug efficiency continuous wave (CW) THz sources are needed for space applications, mainly as local oscillators for heterodyne receivers. Multiplication chains are used for these applications in the 0-3 THz range. Due to the lack of efficient and powerful amplifiers at frequencies >0.3 THz, multiplication chains are not efficient at frequencies >1 THz and the generated power is low close to 3 THz. THz quantum cascade lasers (QCL) have a good efficiency in the range 2-5 THz but they require cooling at cryogenic temperatures which requires additional electrical power. Molecular lasers were historically the first coherent sources in the upper THz range ($f > 1$THz) but the efficiency was small. Since 1970, CO_2 discharge lasers are used to pump optically molecular terahertz lasers but the number of laser lines available for each molecule is limited. The efficiency is also generally low.

With the progress of mid-infrared (MIR) QCL lasers the situation has changed: in 2016 we proved experimentally that it is possible to attain the threshold with pump powers as low as a few mW using ammonia (NH_3) on a 1.07 THz transition [1]. Recently we have studied more systematically the THz laser transitions of NH_3 that can be pumped with a wavelength close to 10 µm where room-temperature CW QCL are available [2]. For this purpose we have defined a molecular gain factor G_m which contains all the parameters related to the molecule itself, excluding the technical parameters. Fig. 1 shows round dots showing the relation between the frequency of the transition and its gain factor. Stars show the THz transitions that were obtained experimentally by pumping $^{14}NH_3$ or $^{15}NH_3$ with an external cavity tunable QCL. The frequencies spans from 0.7 to 5.5 THz, at 1.07 THz a power of 1 mW has been obtained.

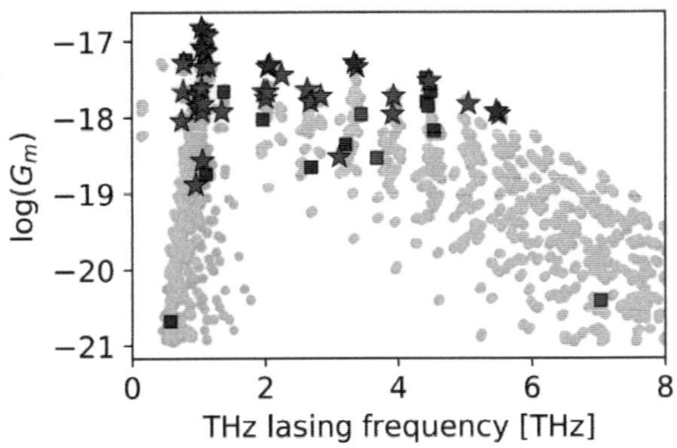

Fig. 1. Logarithm of the molecular gain factor as a function of the THz laser frequencies of $^{14}NH_3$ (red) and $^{15}NH_3$ (blue). CW laser emission obtained by QCL pumping are highlighted by a star symbol and CW laser emission previously reported by CO_2 or N_2O laser pumping are indicated by squares.

We have used the NH_3 THz laser for pumping a hot electron bolometer (HEB) in an high-resolution heterodyne spectroscopy experiment [3]. A broadband THz beam emitted by synchrotron radiation cross a gas cell containing the molecule to be analyzed and the transmitted beam is down converted by HEB pumped by one line of the NH_3 laser. The intermediate frequency is analyzed in a 2.5 GHz bandwidth thanks to a XFFTS. We think that this type of NH_3 laser (or a laser with another molecule [4]) can be used in the future as a laboratory source for testing heterodyne receivers or perhaps in space applications.

REFERENCES

[1] Antoine Pagies, et al., "Low-threshold terahertz molecular laser optically pumped by a quantum cascade laser", *APL Photonics*, 1, 031302 2016.

[2] Marie-Hélène Mammez, et al., "Optically pumped terahertz molecular laser : gain factor and validation up to 5.5 THz", *Advanced Photonics Research*, 2100263 2022.

[3] T. S. Hearne, et al., "Unlocking synchrotron sources for THz spectroscopy at sub-MHz resolution", *Optics Express* 30, 7372 2022.

[4] A. Amirzhan, et al., "A quantum cascade laser-pumped molecular laser tunable over 1 THz", *APL Photonics* 7, 016107 2022.

[1]LPCA, Université du Littoral Côte d'Opale, Dunkerque, F-59140, France; [2]ISMO, CNRS, Orsay, F-91405, France; [3]IEMN, CNRS, Villeneuve d'Ascq, F-59652, France; [4]SOLEIL Synchrotron, Saint-Aubin, F-91190, France

NOTES:

A tunable, high power source for GUSTO's local oscillator at 4.74 THz

Ali Khalatpour*[1], Andrew K. Paulsen[1], Sadhvikas J. Addamane[2], Chris Deimert[3], John. L. Reno[2], Zbig R. Wasilewski[3], Qing Hu[1]

***Abstract*—The Galactic/Extra Ultra long Duration Balloon Spectroscopic-Stratospheric Terahertz Observatory (GUSTO), is a NASA balloon-borne project and is scheduled for launch in late 2022. The balloon will carry a spectroscopic telescope that will detect three brightest emission lines from interstellar medium. GUSTO measurements will shed light on the life-cycle of the gases in the Milky Way and Large Magellanic Cloud (LMC). In this talk, we will discuss the details of a quantum cascade laser used in the local oscillator for detecting the oxygen line at 4.74 THz.**

Keywords*—*THz QCL, GUSTO, OI line

I. Introduction

Terahertz (THz) radiation is loosely defined by the frequency range of 0.5 to 10 THz (1 THz = 10^{12} cycles per second). The THz spectral range has found unique applications in astronomy owing to a vast amount of THz radiation from the interstellar medium (ISM), the matter between stars consisting of low-density gases and dust in planetary atmospheres [1]. The nitrogen line [NII] at 1.4 THz, the carbon line [CII] at 1.9 THz, and the oxygen line [OI] at 4.7 THz are among the brightest emission lines in ISM. [OI] line is rich with information about star formation as it mostly probes warm natural gas heated by massive newborn stars [1]. This spectral line has been inaccessible until recently due to the lack of local oscillators (LOs) in THz frequencies. The invention of THz quantum cascade laser (QCL) [2] made it possible to perform heterodyne observation on this line, and even map it within the Milky Way galaxy. Following up on the STO-2 successful flight, which demonstrated the feasibility of a balloon borne terahertz telescope, GUSTO will be launched in late 2022. GUSTO will survey 124 square degrees of the Milky Way and all of the Large Magellanic Cloud (LMC) in three important interstellar lines with higher angular and velocity resolution compared to previous measurements. The required performance metrics for the local oscillator at 4.7 THz were calculated by researchers at SRON and the University of Arizona to ensure the proper performance of HEB receiver array and phase grating integrated with the optical set-up. By considering the optical loss due to the non-Gaussian beam pattern of THz QCLs, phase grating, and the available cooling power, the overall performance metrics were calculated and listed in Table I. The continuous frequency tunability is essential for detection of large Doppler shift for mapping in the Milky Way. The low cooling power in GUSTO (3.5 W at 60 K) is mainly due to the limits on weight and heat exchange with environment for sterling coolers in balloon borne projects (GUSTO). This higher operating temperature with lower cooling power posts additional challenges in the LO development as compared to previous missions. In this talk, the development of a unidirectional antenna coupled third order distributed feedback grating (UADFB) [3] at 4.744 THz to overcome these limitations is discussed.

TABLE I
GUSTO minimum requirements for the local oscillator at 4.74 THz

performance metric	requirements
nominal frequencies (GHz)	4740.0, 4741.6, 4743.6, and 4749.2
c.w. output power	>1.6 mW at 55 K in single lobe
tunability	1.5 GHz at nominal frequencies
heat dissipation	< 2.7 W at 55 K, < 3.5W at 60 K

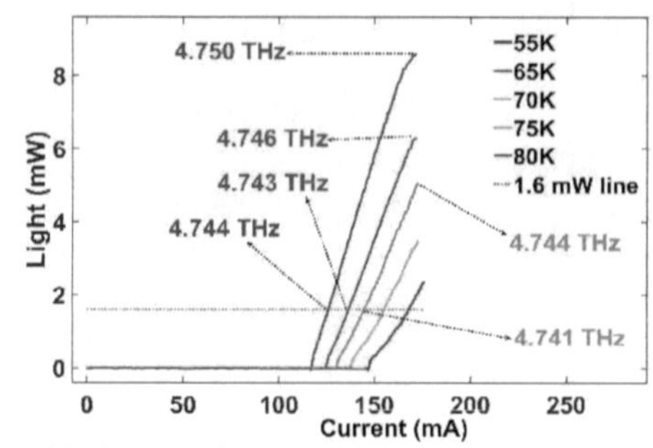

Fig. 1. Light-Current with measured frequencies indicated in the inset.

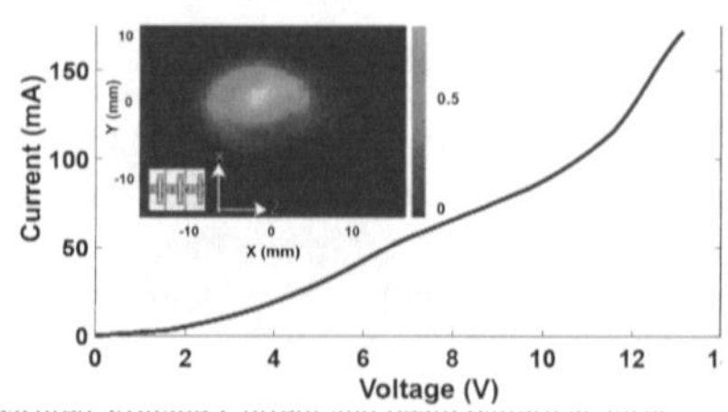

Fig. 2. Measured Voltage-Current and beam pattern as inset.

II. Results

The light-current-spectrum results for a flight ready device are shown in Fig. 1. A combination of red sift in lasing frequency with temperature and blue shift with electrical bias (Stark shift) in in one of the UADFB devices could cover the entire frequency band required by GUSTO. The large tuning with electrical bias is inherent in intersubband lasers However, larger tuning with injection requires high dynamic range, high power, and a robust single mode operation offered in the UADFB platform. Because of near perfect phase matching for this device, a tight single lobe beam is achieved. The measured beam pattern without any collimating optics is shown in Fig 2.

References

[1] A. G. Tielens, The physics and chemistry of the interstellar medium. Cambridge University Press., 2005.
[2] R. Kohler, *et al* ,."Terahertz semiconductor-heterostructure laser," Nature, 2002.
[3] A. Khalatpour, *et al.*, "Unidirectional photonic wire laser," Nature Photonics, 2017

[1]Department of Electrical Engineering, Massachusetts Institute of Technology, Cambridge, MA 02139, USA; [2]Sandia National Laboratories, Center of Integrated Nanotechnologies, Albuquerque, NM 87185-130, USA; [3]Department of Electrical and Computer Engineering, University of Waterloo, 200 University Ave W, Waterloo, Ontario N2L 3G1, Canada.

Kinetic Inductance Detector based focal plane arrays for the Terahertz Intensity Mapper

Reinier M.J. Janssen*[1,2], Lun-Jun Liu[2], Rong Nie[3], Bruce Bumble[1], Jianyang Fu[4], Joseph Redford[2], Jeffrey P. Filippini[3], Steve Hailey-Dunsheath[2], Charles M. Bradford[1,2], James E. Aguirre[5], Justin S. Bracks[5], Brockton Brendal[3], Anthony J. Corso[5], Christopher E. Groppi[6], Dylan Joralmon[7], Ryan P. Keenan[8], Ian Lowe[8], Daniel P. Marrone[8], Philip D. Mauskopf[6], Evan C. Mayer[8], Vesal Razavimaleki[3], Talia Saeid[6], Isaac Trumper[9], and Joaquin D. Vieira[4]

***Abstract*— The Terahertz Intensity Mapper (TIM) is a NASA far-infrared balloon mission, which will fly 2 arrays of ~3600 kinetic inductance detectors (KIDs). The horn-coupled low-volume aluminum lumped-element KIDs are photon noise limited at 80 fW of loading, satisfying the TIM requirements. Kilopixel arrays of these KIDs have been made with >95% yield and we will report on their latest performance evaluation.**

Keywords—kinetic inductance detector, focal plane array, terahertz, kilopixel arrray

I. Terahertz Intensty Mapper

The Terahertz Intensity Mapper (TIM) [1] is a NASA far-infrared balloon mission which aims to study the three-dimensional structure of star formation in the universe. To achieve this, TIM will perform spectroscopic observations of far-infrared fine-structure lines, in particular those of ionized carbon and nitrogen, at the peak of cosmic star formation ($0.5<z<1.5$). These far-infrared emission lines are relatively un-extincted by dust, enabling TIM to measure the star formation activity in dust-obscured galaxies, which account for >70% of the star formation at cosmic noon.

To perform its spectroscopic observations TIM will employ two long-slit (1° slit length) grating spectrometers that cover the 240-317 μm (SW) and 317-420 μm (LW) wavelength bands, respectively, at a spectral resolution R~250. Each of these spectrometer arms is serviced by a focal plane containing ~3600 pixel. In turn, each focal plane will contain four quadrants of ~900 hex-packed horn-coupled kinetic inductance detectors (KIDs).

II. Single Pixel Performance

The fully-aluminum lumped-element KIDs for TIM use a novel "chain-link" absorber design [2] that achieves >90% absorption efficiency in both polarizations across the entire SW or LW band. It achieves this despite the low sheet resistance of 30 nm thick aluminum (Rs~0.7 Ω/sq), while limiting the volume to 100 μm^3 to achieve the required sensitivity, and using a minimum feature size of 500 nm to increase fabrication yield. We have demonstrated that these KIDs are photon noise limited above 80 fW of optical loading with a white noise spectrum down to 1 Hz [3]. A detector limited NEP~1.3×10^{-18} W/$Hz^{0.5}$ is achieved at 215 mK. This easily satisfies the TIM requirements.

III. Array Development

The first quadrant arrays of 864 pixels for the LW module have now been fabricated and tested. They demonstrate over 95% of detector yield, which we attribute to the simple single-layer fabrication and a KID design focused on robustness. It also demonstrates our packaging design, which mounts the array against the hornblock using 50 μm bosses to maintain the air gap between the horns and the array. Our first measurements show the KIDs' quality factor and thermal response are consistent with those of the single pixel test devices, which builds confidence for the optical performance of the entire array. We will present our progress towards a full optical performance evaluation as well as our first results of post-processing aimed at removing resonator collisions.

References

[1] J. Vieira, et al., "The Terahertz Intensity Mapper (TIM): an Imaging Spectrometer for Galaxy Evolution Studies at High-Redshift", *Proceedings of the 30th ISSTT*, Apr. 2019.

[2] R. Nie, et al., "Absorber design and optimization of Kinetic Inductance Detectors for the Terahertz Intensity Mapper", *Journal of Low Temperature Physics*, accepted 2022

[3] R.M.J. Janssen, et al., "Single Pixel Performance of the kinetic inductance detectors for the Terahertz Intensity Mapper", *Journal of Low Temperature Physics*, submitted 2022

[4] L.-J. Liu, et al., "Design of the Kinetic Inductance Detector based focal plane assembly for the Terahertz Intensity Mapper", *Journal of Low Temperature Physics*, submitted 2022

*reinier.m.janssen@jpl.nasa.gov; [1]Jet Propulsion Laboratory, California Institute of Technology, Pasadena, CA 91109, USA; [2]Department of Astronomy, California Institute of Technology, Pasadena, CA 91125, USA; [3]Department of Physics, University of Illinois at Urbana-Champaign, Urbana, IL 61801, USA; [4]Department of Astronomy, University of Illinois at Urbana-Champaign, Urbana, IL 61801, USA; [5]Department of Physics and Astronomy, University of Pennsylvania, Philadelphia, PA 19104, USA; [6]School of Earth and Space Exploration, Arizona State University, Tempe, AZ 85287, USA; [7]School of Engineering of Matter, Transport and Energy, Arizona State University, Tempe, AZ 85287, USA; [8]Department of Astronomy, University of Arizona, Tucson, AZ 85721, USA; [9]Intuitive Optical Design Lab LLC, Tucson, AZ 85701, USA

NOTES:

Reflector-Based Phased Array for High Power G-band radars

A. Maestrini, K. Cooper, S. van Berkel, M. Lebsock, C. Lee, G. Chattopadhyay, I. Mehdi
Jet Propulsion Laboratory, California Institute of Technology, Pasadena, CA, 91125 USA

***Abstract*—We will present the ongoing development of a reflector-based phased array for high power G-band radars and beyond to address the need for global measurements of water vapor content within clouds, and the detection of small-size particles in clouds. The array will use a limited number of medium-gain antennas and state-of-the-art low-power-density solid-state power modules with high DC efficiency. Our concept is applicable to high-power 155-165 GHz Differential Absorption Radar (DAR) for probing the planetary boundary layer (PBL), a targeted observable for incubation in the 2017 Decadal Survey [1], with high vertical resolution. The technology is also applicable to 240 GHz radars which, when combined with observations at more traditional lower frequencies, can better characterize particle size distributions within clouds.**

As emphasized in the NASA PBL Incubation Study Team Report [2], 100 W of emitted power at G-band is the necessary characteristic of any spaceborne DAR system dedicated to PBL remote sensing. As a complementary remote sensing technique to passive radiometer counterparts, a high-power and widely tunable G-band DAR can achieve high resolution and high precision profiling of the water vapor content inside PBL clouds and precipitation, as well as total-column water vapor estimates, both over ocean and terrain [3].

We will present a two-element phased array, as a unit cell of an envisioned 2-meter aperture modular array with 100 W of transmit power, able to steer the beam by +/- 25 beamwidths with sub-second speed. When upgraded with tilt-able secondary reflector, the array architecture will enable fast electromechanical beam-steering to improve global coverage of humidity profiles inside of clouds and precipitation, compared to a fixed-beam radar which lacks the agility to target a cloud-track path within the PBL. The phased array will use closely packed low-profile parabolic antennas of 15-20cm, each featuring fully solid-state, high DC-efficiency electronics that can work under pulsed DC condition. For each array element, the input signal will be up-converted at G-band using JPL-built GaAs Schottky-diode-based frequency multipliers and commercially available W-band and G-band amplifiers, resulting in an emitted power of about 0.5W per element. The relative phase of the emitted signal is electronically tuned to the desired value without any additional phase-shifter, reducing complexity and RF power losses. The transmit and receiver modules are isolated via a custom circular-polarization orthomode transducer. The received signal is down-converted using an LNA and a mixer designed at JPL, before being digitally equalized, phase-shifted, and combined.

The array elements, when tiled over a flat hexagonal surface, will result in a thinner antenna than a classic single solid-dish approach. In addition, the array can be dynamically subdivided for multi-beam forming, pointing simultaneously at different directions. Our concept departs from the focal plane array approach that requires large, high power density MMICs at G-band with both high DC-efficiency for the integrated emitters and low noise for the integrated receivers. Instead, our approach is based on a combination of existing commercial state-of-the-art components, along with some JPL-built strategic ones. It also offers key advantages over the legacy technology of high-power vacuum electronic sources. Among them are active components with long lifespan and wide RF bandwidth, failure mitigation with "graceful degradation" in case of component failure, scalability to large apertures, low power-generation density for easy thermal management, low DC bias voltages, and much less challenging transmit/receive isolation requirements. Our concept takes full advantage of the rapid evolution of analog and digital electronics owing to the continuing consumer electronics revolution, including the breakthroughs in component availability due to the emerging 5G market. These advantages make our approach well suited for space-based missions for future earth science missions.

The paper will present the radar architecture and field test results on a calibration target situated in the far-field, at a distance of about 250m.

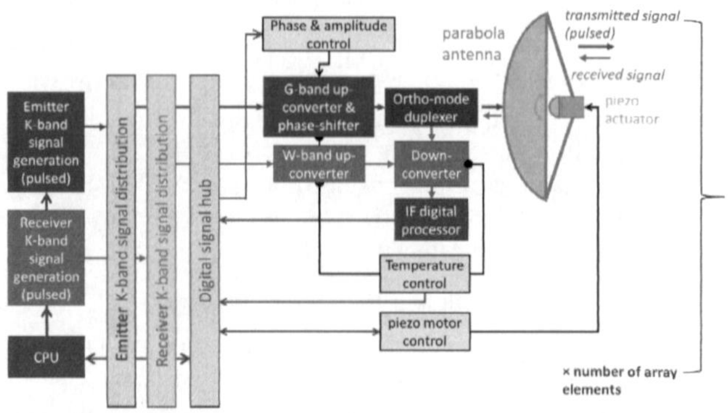

Fig. 1. schematic of our reflector-based phased array concept for G-band DAR operations. For simplicity, the DC power distribution is not represented.

References

[1] National Academies of Sciences, Engineering, and Medicine 2018. Thriving on Our Changing Planet: A Decadal Strategy for Earth Observation from Space. Washington, DC: The National Academies Press. https://doi.org/10.17226/24938 - This is the final product of the 2017-2027 decadal survey for Earth science and applications from space ("ESAE 2017").

[2] João Teixeira et al., The NASA PBL Incubation Study Team Report

[3] K. B. Cooper et al., "G-Band Radar for Humidity and Cloud Remote Sensing," in IEEE Transactions on Geoscience and Remote Sensing, vol. 59, no. 2, pp. 1106-1117, Feb. 2021, doi: 10.1109/TGRS.2020.2995325.

Architecture of ASTHROS' Spectral Data Collection and Onboard Anomaly Detection Pipeline

Paul Horton*[1], Christian Thompson[1], Christopher Groppi[1],
Jose V. Siles[2], and Jonathan H. Kawamura[2]

Abstract— **Balloon borne missions have unique requirements concerning data collection, storage, and analysis. Due to the limited bandwidth between the instrument and ground team, all data collected must be stored on the instrument, requiring specialized storage solutions in cases with high data volume. Additionally, in-flight data assessment proves challenging as data typically cannot be analyzed until after the flight due to low bandwidth.**

ASTHROS (short for Astrophysics Stratospheric Telescope for High Spectral Resolution Observations at Submillimeter-wavelengths) is a high-altitude balloon mission utilizing multiple spectrometers to produce on-the-fly scans of gas and dust clouds in order to measure the density and dynamics of the specific elements in the clouds. For the spectrometers, we will be pairing four spectrometers to a Raspberry Pi Compute Module 4 (CM4) for readout. These readout computers will stream their data to a Network Attached Storage (NAS) device that will synchronize and safely store all the data collected on the flight. Collecting all of the data in a redundant and centralized location makes the data easier to analyze by other systems and makes the system easier to scale with more spectrometers.

Once all of the data is centralized to the NAS, a novelty detection pipeline will be utilized to identify bad calibration spectra. Using a variational autoencoder, we are able to create a system that can identify sudden changes in the science data quality by comparing the latent space representation of a given spectrum with a history of previous spectra. This will allow us to quickly assess changes in the spectrometer's behavior and downlink specific spectra to identify problems with the system. The variational autoencoder used here is a low power solution that can easily be run on-board on either the NAS or an additional Field Programmable Gate Array (FPGA). This solution allows us to identify quality issues with the spectrometer without having to blindly downlink data, thus helping in detecting problems as soon as they occur instead of after the flight.

Keywords—spectroscopy, system architecture, novelty detection, machine learning, operations, balloon-borne, stratospheric

REFERENCES

[1] Růžička, Vít, et al., "Unsupervised Change Detection of Extreme Events Using ML On-Board", *Conference on Nerual Information Processing Systems (NeurIPS)*, Nov. 2021.

[2] Kerner, Hannah R., et al., "Domain-agnostic Outlier Ranking Algorithms (DORA): A Configurable Pipeline for Outlier Detection in Scientific Datasets", *American Geophysical Union (AGU) Fall Meeting,* Dec. 2021.

[3] Siles, Jose V., et al., "ASTHROS-Astrophysics stratospheric telescope for high-spectral resolution observation at submillimeter-waves: Mission overview and development status", *IEEE International Symposium on Space Terahertz Technology (ISSTT),* Mar. 2020

[1]Arizona State University, School of Earth and Space Exploration, Tempe, Arizona, 85281, USA; [2]NASA Jet Propulsion Laboratory, Pasadena, California, 91109, USA

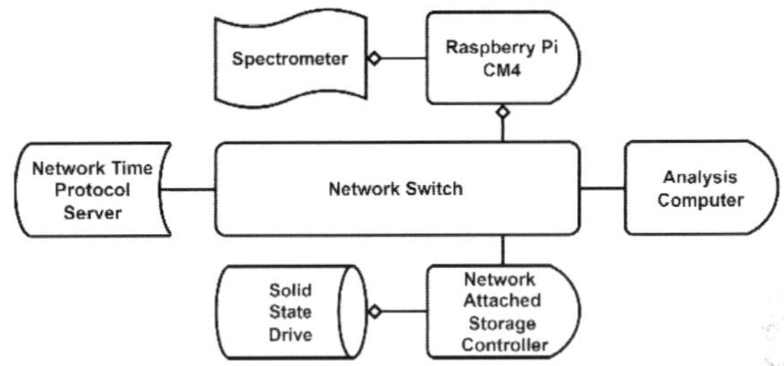

Fig. 1. System diagram for the spectrometer readout on ASTHROS. Up to 4 **Spectrometers** can be paired with a single **CM4** and multiple **CM4s** can be connected to the **Network Switch**. Each **CM4** will write data to **Solid State Drives** through the **Network Attached Storage Controller.** An **Analysis Computer** will be able to access the data and perform anomaly detection on the spectra. Everything will be synced with the time on the gondola using the gondola's **Network Time Protocol server.**

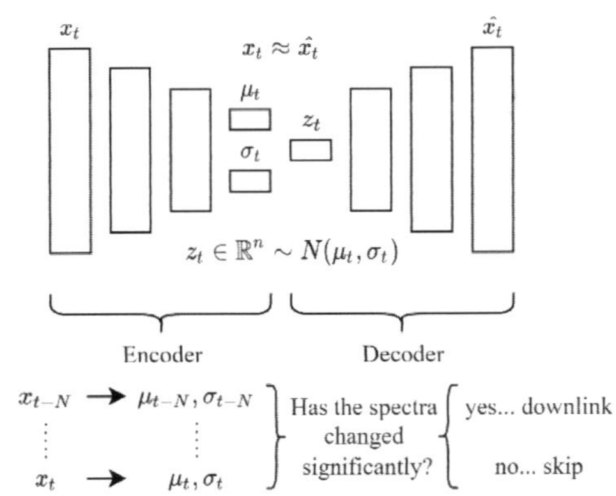

Fig. 2. Variational Autoencoder (VAE) diagram for anomaly detection on ASTHROS. New samples are encoded using the autoencoder and compared with previous samples to identify signifigant changes in calibration spectra.

NOTES:

Atmospheric Phase Monitoring System Evolution for the NOEMA interferometer

S. Mahieu[1], B. Pissard[1], M. Bremer[1], Jan Martin Winters[1], C. Risacher[1], R. Blundell[2], R. Kimberk[2], J. Test[2]

Abstract— **After having adapted the Smithsonian Astrophysical Observatory (SAO) atmospheric phase monitoring system that was developed for the Submillimeter Array (SMA), IRAM is currently refining the concept for an implementation on the NOEMA site.**

I. Introduction

In this paper, we present the development of the atmospheric phase monitoring system and show the latest results obtained on the NOEMA site. The final goal of this project is to provide a permanent monitoring system of the observing conditions on the plateau de Bure. After first onsite observations with simple antennas in 2018, later acquisitions took place during 2019 and 2020 autumns with improved infrastructure. Studies are currently ongoing to make the system compliant with the site harsh weather conditions.

II. System Description

The atmospheric phase monitoring system is based on two off-axis aluminium satellite dishes (Fuba DDA 110: 1090 x 991 mm^2) that receive a broadband white noise-like Ku Band (~11.85 GHz) signal from a geostationary satellite and focus it to the center of the feed. The Low Noise Blocks (LNB) that down convert this signal to the Intermediate Frequency (IF) one (~1.2 GHz) have been modified to be fed with a common Local Oscillator signal source.

The IF signals are then amplified, filtered and transported through optical fibers down to the building where they are processed by a commercial analog correlator (IQ demodulation card) that produces the phase delays between pairs (only one so far) of antennas from the I & Q signals. Then the data are further processed by a LINUX pc that runs software, which calculates a least squares sin fit to the unwrapped phase, subtracts that from the unwrapped phase, calculates the RMS values from the residuals and also provides the correlated power.

III. System Implemetation

After the first autumn 2018 data acquisition session on the plateau de Bure, some infrastructure optimizations were implemented in order to minimize environmental hazards. The second trials on the plateau with antenna sheltered under a radome, in parallel with the NOEMA interferometer confirmed the good agreement between the two systems. A new infrastructure is being developed for future more permanent installation on site.

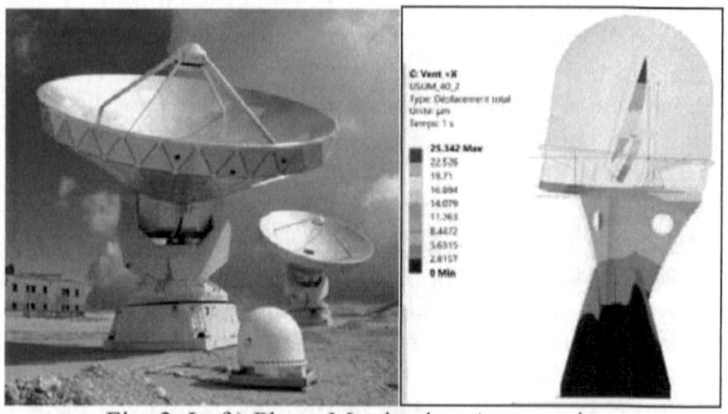

Fig. 2. Left) Phase Monitoring Antenna in a radome in front of two NOEMA antennas, Right) Finite Element Analysis study of mechanical stand under 40km/h of wind.

IV. Conclusion

This abstract has briefly described the phase monitoring system work that has been ongoing for the past three years at IRAM. More detailed description of the system will be provided during the conference, together with up-to-date results and analysis.

References

[1] Robert S. Kimberk & All "A Multi-Baseline 12 GHz Atmospheric Phase Interferometer with One Micron Path Length Sensitivity", May. 2012.

[1] Institut de Radio Astronomie Millimétrique, IRAM, 300 rue de la piscine, 38400 Saint Martin d' Hères, France.

[2] Harvard-Smithsonian Center for Astrophysics, 60 Garden Street, Cambridge, MA 02138, USA.

NOTES:

Heterodyne Array Receiver Study for the FIRSST Space Mission

Martina C. Wiedner[*1], Asantha Cooray[2], Meredith MacGregor[3], Yan Delorme[1]
Gary Melnick[4], the FIRSST team and the heterodyne receiver group

Abstract—**The Far IR Spectroscopy Space Telescope is a study to be submitted for the NASA probe mission. We are carrying out the design of a heterodyne array receiver for FIRSST to be able to observe the fundamental water lines at very high spectral resolution.**

Keywords—heterodyne receiver, far-IR, Terahertz missions, systems and instrumentation

I. Introduction

The Astro2020 decadal review has recommended that a Probe-class mission, with a cost cap of ~$1B, should be launched around 2030 in order to either carry out far-IR or for x-ray observations. The announcement of Opportunity is expected in 2023. Currently there are a number of x-ray studies under way as well as at least four studies for far-IR missions underway: A Single Aperture Large Telescope for Universe Studies (SALTUS) lead by C. Walker, the PRobe far-Infrared Mission for Astrophysics (PRIMA) led by J. Glenn and M. Bradford, the Space Infrared Interferometric Telescope (SPIRIT) by D. Leisawitz and the Far IR Spectroscopy Space Telescope (FIRSST) by A. Cooray.

II. FIRSST

FIRSST has 3 major science goals: It will accurately measure the amount gaseous material in protoplanetary disks in order to determine the maximum mass of potential planets and the timescale for planet formation. FIRSST will trace the trail of water from the interstellar medium to planets and FIRSST will study how galaxies assemble their material. FIRSST will consist of a 1.8m primary mirror actively cooled to ~20K and will have a simple, low-risk architecture. The satellite will be equipped with closed-cycle coolers to guarantee a lifetime of at least 5 years. The observatory will be equipped with two instruments: a direct detection camera with two selectable resolution modes and a heterodyne array receiver. The bolometer camera will be able to observe between 30 to 270 microns at low spectral resolving power, i.e., R~ 200, or in 4 distinct wavelength ~~frequency~~ bands between 50 and 250 microns at high spectral resolving power, i.e., $R\sim10^5$, by employing a Virtually Imaged Phased Array (VIPA) spectrometer.

[1]LERMA, Paris Observatory, Paris, 75014, France; [2]Center for Cosmology, Department of Physics and Astronomy, University of California, Irvine, CA 92697, USA; [3]CASA, University of Colorado, 389 UCB, Boulder, CO 80309, USA; Smithsonian Astrophysical Observatory, Cambridge, MA 02138, United States

III. Heterodyne Array Recevier for FIRSST

The heterodyne array receiver for FIRSST will carry out very high resolution ($R = 10^6$ to 10^7) observations of the fundamental water lines at 557 GHz (ortho) and at 1113 GHz (para) as well as of water isotopes. Therefore, it has two bands: one between 550 to 750 GHz, and a second band between 950 and 1250 GHz. Each band will be covered with a small array receiver. The instrument design is based heavily on HIFI/Herschel, upGREAT/SOFIA and the recent HERO/Origins study.

As we are writing this abstract the design of the heterodyne receiver is rapidly evolving, but the design will need to be well advanced for the first satellite concurrent design lab at the end of July and nearly finished in October for the second design lab. Currently, we are focusing on cost reduction and aim to keep the receiver small, simple and low risk. Each band will see one polarization, allowing us to superimpose the two bands for simultaneous observations. We plan to use high TRL components (from HIFI or other space missions), e.g. SIS mixers for the lower band and HEB or SIS mixers for the higher band. For the water study an IF bandwidth of 6 GHz is sufficient. The IF will be amplified by either InP or SiGe amplifiers, depending on the exact array size and the available cooling power. We would detected the IF signal either by FPGA or CMOS backends. The local oscillator signal will be produced by ambient multiplier-amplifier chains, making use of the recent development for SWI/JUICE (see related papers).

NOTES:

Jupiter Icy Moon Explorer, Submillimeter wave Instrument: Delivery status of the 1200 GHz high spectral resolution receiver front end

J.Treuttel[1], L.Gatilova[1], S. Caroopen[1], A. Feret[1], G. Gay[1], T. Vacelet[1], J. Valentin[1], Y.Jin[2], A. Cavanna[2], K.Jacob[3], and S. Mignoni[1], J-M. Krieg[1], C. Goldstein[3], A. Ravanbakhsh[4], J-P. Garcia[4], A.Maestrini[5], P. Hartogh[4]

***Abstract*—The Jupiter Icy Moons Explorer (JUICE) is a mission chosen in the framework of the Cosmic Vision 2015-2025 program of the Science and Robotic Exploration Directorate of the European Space Agency. The Submillimeter Wave Instrument (SWI) is a spectrometer/radiometer instrument operating in two submillimeter channels between 530 – 625 GHz and 1080 – 1275 GHz to study the dynamics of Jupiter's stratosphere, vertical profiles of wind speed, temperature, composition and structure of exospheres of Ganymede, Europa and Callisto. LERMA is responsible for the delivery of critical sub-systems of the two channel front- ends, including its 1200GHz mixer and last frequency stage local oscillator. In this paper we will describe the SWI radiometer front-end system and address the different procurement steps of the flight hardware. We will present some of the test structures used, the tests conditions as well as some of the failure criterias and allowable drifts during screening of the flight models and include discussion on endurance testing of its qualification models.**

Keywords*—Schottky diode, GaAs, heterodyne detection, submillimeter wave, THz, space qualification, SWI, JUICE, Jupiter.

I. Introduction

Interplanetary space missions such as JUpiter ICy moons Explorer (JUICE) implies particularly severe constraints on devices reliability over a time range that includes approximately 8 years cruising to Jupiter and operations in varied environments. In addition to the requirement for high scientific return, high technological demands are required on the instruments [1][2]. A critical aspect of the system relates to failure of one device within the local oscillator chain or the mixer that represents a single point failure causing non-operationality of the full instrument. The goal of this study is define a safe operating range for the SWI based on III-V Terahertz Schottky junction in particular and to demonstrate operation equivalent to the mission duration.

II. Results

The tuning conditions of the system have to fulfill safe and long term operation as define for each models. In particular, each of the multiplier and mixer diodes require to sustain a certain amount of incoming and dissipated power expressed per junction area (in mW/μm^2) related to DC current density during operation and over time and taking into account the increase of efficiency of frequency multipliers at cryogenic operation discussed in this presentation and more extensively in [3].

References

[1] "Exhaustive Qualification and endurance testing of the 300 GHz Frequency Doubler of the Sub-Millimeter Instrument of the Jupiter Icy Moon Explorer Mission", L.Gatilova, J. Valentin, J.Treuttel, A. Feret, G.Gay, S. Caroopen, T.Vacelet, S. Mignoni, Y. Jin J-L. Roux, In Proceeding of SPIE, Astronomical Telescopes and Instrumentation, Motreal, July 2022.

[2] "Space qualification of MMIC Schottky diodes chips for SWI instrument of JUICE mission", This conference, J.Valentin et al.

[3] "Jupiter Icy Moon Explorer, Submilimeter wave Instrument: 1200 GHz high spectral resolution receiver front end Part I : RF Performance optimization and definition of acceptance levels <by J.Treuttel, L.Gatilova, J.Valentin et al.> and , Part II: Qualification and endurance testing <by J.Valentin, L.Gatilova et al.> - Two papers in preparation

[1]Observatoire de Paris - PSL, LERMA, Paris, 75014, France; [2]Université Paris-Saclay, CNRS, Centre de Nanotechnologies et Nanostructures C2N, Palaiseau, 91120, France; [3]Centre National d'Etudes Spatiales, CNES, Toulouse, 31401, France, [4]Max Planck Institute for Solar System Research, Gottingen, Germany ; [5]was at Observatoire de Paris PSL, LERMA now at JPL California Institute of Technology, 4800 Oak Grove Driva, Pasadena, CA 91109, USA

NOTES:

Cold Optics on the Terahertz Intensity Mapper (TIM)

Talia Saeid*[1], Dylan Joralmon[2], Isaac Trumper[3], Mark Sullivan[4], Nick Emerson[5], Daniel P. Marrone[5], Jianyang Fu[6], Christopher E. Groppi[1], James E. Aguirre[7], Justin S. Bracks[7], Charles M. Bradford[8], Brockton Brendal[6], Anthony J. Corso[7], Jeffrey P. Filippini[6], Steve Hailey-Dunsheath[9], Reinier M. Janssen[8], Ryan P. Keenan[5], Lun-Jun Liu[9], Ian Lowe[5], Philip D. Mauskopf[1], Evan C. Mayer[5], Rong Nie[6], Vesal Razavimaleki[6], and Joaquin D. Vieira[5]

I. Abstract

The Terahertz Intensity Mapper (TIM) will be one of the first instruments to use intensity mapping in the far-infrared (240-420 μm full band) and will do so to produce a 4-dimensional data set probing the cosmic evolution spanning five billion years[1]. Intensity mapping is an observational technique that, instead of resolving individual objects, measures the power spectrum of spatial and spectral intensity fluctuations in a data cube. This power spectrum can reveal statistical information about the underlying galaxy population.

TIM's optical system consists of a Cassegrain telescope, an Offner relay, and two Czerny-Turner spectrometers, each covering a different band– short-wave (240-317 μm) and long-wave (317-420 μm)– and has distinct optical designs[1]. The Offner relay serves to form a Lyot stop (cold stop), which reduces the amount of flare from diffraction.

The spectrometer modules and the Offner relay are housed within a cryostat. The Cassegrain telescope creates a focus within the cryostat, which is then re-imaged by the Offner relay. The focus passes through a slit that splits it in two directions, depending on the wavelength, to either of the spectrometer modules. The optical design will also include a calibration source modeled after the spectrometer calibration source (SCal) on the Herchel-SPIRE[2]. The purpose of this calibration source is to produce a thermal input for detector health checks and calibration measurements.

Here we present the designs of the TIM cold optics as well as our plans for analyzing the system through the design and development phases. We are conducting analyses during the design stage, including finite element analysis (FEA), to ensure the optical elements remain in tolerance for all telescope orientations during flight. These tests allow us to evaluate and modify the optical system optomechanical design until it is ready for construction. The current optomechanical design is shown in Figure 1. After the surfaces are constructed, we will test the optics using a coordinate measurement machine (CMM) and analyze the results to predict the as-built optical performance.

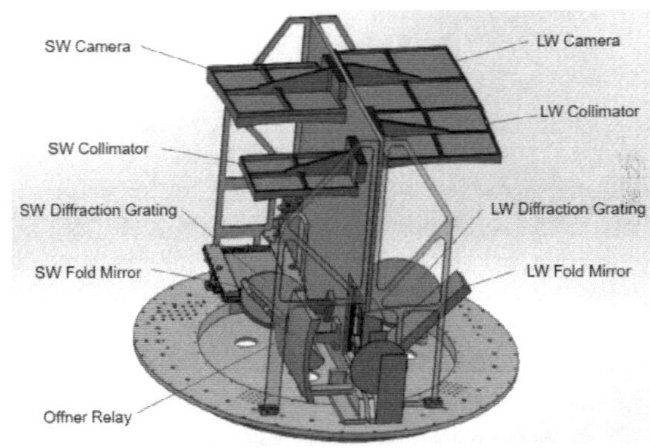

Fig. 1 The optomechanical design contains a short-wavelength (SW) module and long-wavelength (LW) module, corresponding to the two Czerny-Turner spectrometers, components of which are labeled.

References

[1] Vieira, et al., "The Terahertz Intensity Mapper (TIM): an Imaging Spectrometer for Galaxy Evolution Studies at High-Redshift", Proceedings of the 30th International Symposium on Space Terahertz Technology, Apr. 2019.

[2] Hargrave, et al., "Performance of flight-model on-board calibration sources on Herschel-SPIRE", Millimeter and Submillimeter Detectors and Instrumentation for Astronomy III, Jun. 2006.

*tsaeid@asu.edu; [1]School of Earth and Space Exploration, Arizona State University, Tempe, AZ 85287, USA; [2]School for Engineering of Matter, Transport and Energy, Arizona State University, Tempe, AZ 85287, USA; [3]Intuitive Optical Design Lab LLC, Tucson, AZ 85701, USA; [4]Steward Observatory, University of Arizona, Tucson, AZ 85721; [5]Department of Astronomy, University of Arizona, Tucson, AZ 85721, USA; [6]Department of Astronomy, University of Illinois at Urbana-Champaign, Urbana, IL 61801, USA; [7]Department of Physics and Astronomy, University of Pennsylvania, Philadelphia, PA 19104, USA; [8]Jet Propulsion Laboratory, California Institute of Technology, Pasadena, CA 91109, USA; [9]Department of Astronomy, California Institute of Technology, Pasadena, CA 91125, USA

NOTES:

Multipixel 2-color Superconducting Receiver for ASTHROS

Jonathan Kawamura*[1], Bruce Bumble[1], Darren Hayton[1], Richard Cofield[1], Christopher Groppi[2], Jose Siles[1], Paul Goldsmith[1], Robert Lin[1], Choonsup Lee[1]

***Abstract*—We describe the receiver package for the ASHTROS mission: a 2-color 4-pixel THz superconducting receiver that targets a pair of ionized nitrogen fine-structure lines at 1.46 THz and 2.46 THz. The latter mixer will be able to operate at ~2.7 THz in order to observe HD. The design of the receiver package allows observations in both polarizations toward a source to maximize sensitivity. The receiver is operated near 6 K physical temperature using a mechanical cryocooler. A bank of low-power commercially sourced ASIC spectrometers are used to produce astronomy data.**

Keywords—superconducting mixers, array receivers, cryocoolers.

I. THE ASTHROS RECEIVER

The Astrophysics Stratospheric Telescope for High Spectral Resolution Observations at Submillimeter-wavelengths is a high-altitude balloon-borne astronomy project that will map the sky in two key spectroscopic lines of ionized nitrogen at 1.46 THz and 2.46 THz. The key requirements for the receiver package include ability for simultaneous observations at the two frequencies and in two polarizations using 2 spatial pixels for a total of a simultaneous operation of 8 mixers. The number of pixels is largely determined by the constraints of the thermal system: on its projected 21-day flight, the receiver package will be cooled by a low-power mechanical refrigerator.

At the heart of the receiver are waveguide-based NbN hot-electron bolometer mixers, which are coupled to free-space by densely packed cylindrically symmetric drilled horns. The output is coupled to a low-noise first-stage SiGe cryogenic amplifier placed on the third stage of the cryocooler. Following conditioning at room temperature, the microwave signal is processed using an advanced ASIC-based low-power polyphase filter bank with ~4 GHz bandwidth.

[1]Jet Propulsion Laboratory, California Institute of Technology, Pasadena, CA 91109; [2]Arizona State University, Tempe, AZ.

NOTES:

Progress toward Antarctic Terahertz Intensity Interferometry

Hiroshi Matsuo*[1], Hajime Ezawa[1], Ayako Niwa[2], Tomohiro Koseki[2], Rina Enohi[3], Norio Okada[1]
and Antarctic Terahertz Astronomy Group in Japan

***Abstract*— Terahertz Intensity Interferometry can be a technology for future high angular resolution and high sensitivity observations in far-infrared and terahertz frequencies. Optical demonstration with a laboratory interferometer is being prepared using submillimeter-wave SIS photon detectors, cryogenic readout electronics, 0.8-K sorption cooler and a 4-K cryostat with a pulse tube cooler. Some of the design details of the setups are presented. Using the technology, Antarctic intensity interferometer with two 30-cm terahertz telescopes at Dome Fuji station is proposed using 500 GHz SIS mixers and 1.5 THz SIS photon detectors.**

Keywords*— *intensity interferometer, SIS photon detector, cryogenic readout electronics, sorption cooler

I. Introduction

Astronomical observations of atomic fine-structure lines in far-infrared and terahertz frequencies, such as [OIII], [NII] and [CII], have advanced by observing high redshift galaxies with ALMA to identify redshifted emission lines observable in millimeter and submillimeter-waves. However, observing rest-frequencies of these emission lines is difficult owing to atmospheric absorption. From high altitude sites in Antarctica, Dome Fuji and Dome A, some of the far-infrared emission lines can be observed. We are proposing Antarctic terahertz interferometers to observe such emission lines with high angular resolution. Intensity interferometry is a possible technology applicable to the terahertz interferometer for aperture synthesis imaging by utilizing photon bunches for delay time measurements [1-3].

II. Preparation of Lab. Experiments

The key technology for aperture synthesis imaging with intensity interferometry is to measure intensity correlation with large bandwidth and high sensitivity. SIS photon detectors are developed for 500-660 GHz which could be used for photon counting experiments. Details of the detector development are presented in Ezawa et al. [4]. To observe [NII] emission line, SIS photon detectors at 1.5 THz is being developed in Advanced Technology Center of National Astronomical Observatory of Japan, which is presented in Niwa et al. in this symposium.

Fig. 1 shows a setup inside a cryostat. On a 4-K stage (30 cm × 20 cm), a 0.8-K stage supported by CFRP plates is cooled by a compact 0.8-K sorption cooler. An SIS photon detector on hyper-hemispherical silicon lens is attached to the 0.8-K stage and DC magnetic field is applied by a pair of superconducting magnets at 4 K to suppress Josephson currents. The magnets and the detector are covered by a magnetic shield made of cryophy (not shown in Fig. 1) to avoid magnetic traps on SIS junctions. Another SIS photon detector can be attached to the other side of 0.8-K stage for interferometer experiments.

Fig. 1. A setup for cooling SIS photon detectors with 0.8-K sorption cooler on 4-K stage in a cryostat.

Fig. 2. A cryostat with a pulse-tube cooler for demonstration of terahertz intensity interferometry. There are two optical windows, one in front and another at the oposite side.

The cryostat shown in Fig. 2 is cooled by Sumitomo pulse-tube cooler RP-062BS. Thermal link from the cold head to 4-K and 60-K stage is made by 6N copper cables to suppress mechanical vibration. Two optical windows are made of UHMW polyethylene with anti-reflection coating of Zytex. Black polypropylene and Yoshinaga-filters are used on 60-K and 4-K windows, which are anti-reflection coated with Zytex and Teflon, respectively.

III. Cryogenic Readout Electronics

Schematic diagram of a readout electronics is shown in Fig. 3. To realize fast readout of high impedance detector, a first stage source follower with GaAs-JFET is used aside of the SIS photon detector at 0.8 K. Power dissipation of the first stage is less than 100 μW and output impedance of a few kΩ is measured. Then a second stage source follower with GaAs-HEMT at 4-K lowers the output impedance down to 50 Ω, followed by a Si:Ge amplifier (CITLF1). Readout noise is limited by the first stage source follower with transconductance of 0.3 mS with thermal limited noise voltage of about 0.4 $nV/Hz^{0.5}$ and noise temperature of about

[1]National Astronomical Observatory of Japan, Mitaka, Tokyo 181-8588, Japan; [2]University of Tsukuba, Tsukuba, Ibaraki 305-8577, Japan; [3]Toho University, Funabashi, Chiba 274-8510, Japan

60 K. Bandwidth of the readout electronics is less than 1 GHz limited by output impedance of the first-stage source-follower and the second stage gate capacitance. The noise and bandwidth could be improved by using low gate leakage HEMT transistors instead of GaAs-JFET.

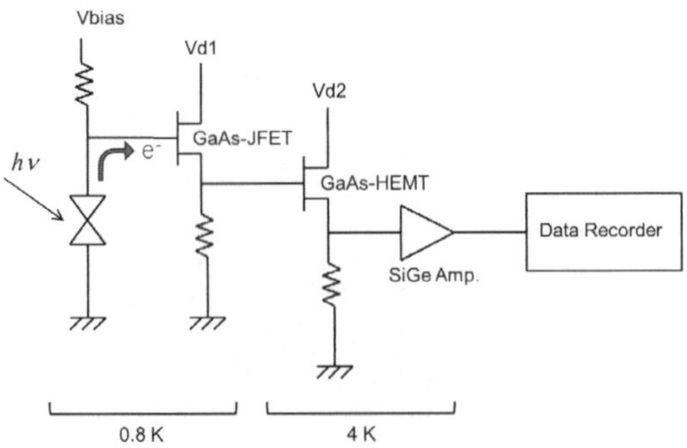

Fig. 3. Schematic diagram of the cryogenic readout electronics for SIS photon detectors, using dual-stage source-followers and a Si:Ge amplifier.

IV. Optical Setup of Interferometer Experiment

Fig.4 shows a picture of optical setups for intensity interferometry. Optical mirror assembly and off-axis parabolic mirror focus input radiation to SIS photon detectors inside the cryostat. Using a high temperature blackbody source and ϕ600-mm spherical mirror, radiation from astronomical source is simulated. Input radiation is then reflected by two input mirrors on movable stages to obtain baseline separation between 10 and 50 cm. Rotation of the baseline can be simulated by rotating a shaped aperture in front of the blackbody. Input radiation is focused by off-axis parabolic mirror in front of each window as in Fig.2. Inside each cryostat window, an SIS photon detector on silicon lens measures photo current which is readout by cryogenic readout circuits. Signals from two SIS photon detectors are recorded simultaneously by a wide bandwidth oscilloscope for further analysis of intensity correlation. Data will be collected for different baseline configuration and shaped aperture rotation to obtain intensity correlation and delay time, which will be used to make aperture synthesis imaging.

Fig. 4. An optical setup for laboratory demonstration of intensity interferometry. A ϕ600 mm spherical mirror (R=4800 mm) out of the picture is used to collimate radiation from the blackbody (on a cryostat) to two input mirrors on moving stages. Separation of the input mirror can be changed between 10 and 50 cm.

V. Toward Antarctic Intensity Interferometry

Antarctic observation from Dome-Fuji is supported by National Institute of Polar Research (NIPR). Dome-Fuji astronomy program with 30-cm terahertz telescopes is starting by a general research observation of the 10th term Japanese Antarctic research program. In Early 2025 by the 66th Japanese Antarctic Research Expedition (JARE-66), the first 30-cm telescope will be installed in Dome-Fuji to start observation of Galactic plane survey of CO and CI emission lines using 500-GHz SIS mixer receivers. Experiment on terahertz intensity interferometry is planned with another 30-cm telescope with similar design.

A schematic presentation of the Antarctic Terahertz Intensity Interferometer is shown in Fig. 5. Two 30-cm telescopes with SIS mixers and/or SIS photon detectors will be used to record intensity fluctuation caused by photon bunches from astronomical sources.

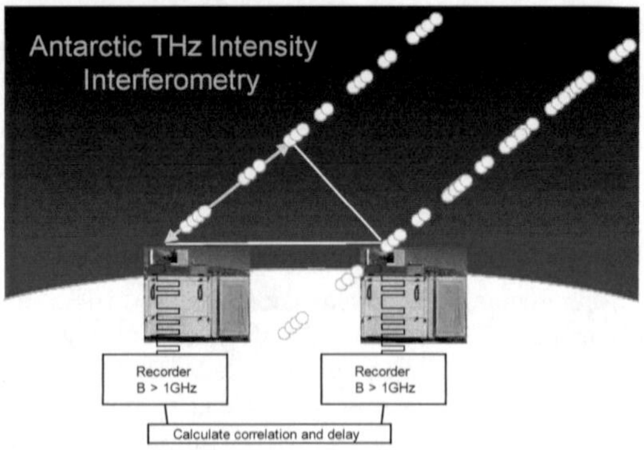

Fig. 5. A concept of Antarctic Terahertz Intensity Interferometer.

The second 30-cm telescope is proposed to be installed in Dome-Fuji station in early 2027 to start test observation of intensity interferometry using 500 GHz SIS mixers. During the 10th term Antarctic program until 2028, baseline separation would be limited to 10's of meters in Dome-Fuji station. We will then install 1.5-THz SIS photon detectors to observe [NII] 1.46 GHz emission lines toward massive star-forming regions in our Galaxy and external galaxies.

Acknowledgments

This development is supported by JSPS KAKENHI Grant Number 20K20346, 20K20345 and JAXA basic R&D for future space science missions.

References

[1] Matsuo, H., Ezawa, H., "Advantages of Photon Counting Detectors for Terahertz Astronomy", Journal of Low Temperature Physics 184, 718, 2016

[2] Matsuo, H., "Far-infrared intensity interferometry for high angular resolution imaging", Proc. SPIE 11443, 114431N, 2020

[3] Matsuo, H., "Development of Terahertz Intensity Interferometry for Very High Angular Resolution Observations", Impact 2020(5), pp. 39-41, 2020

[4] Ezawa, H., Matsuo, H., Ukibe, M., Fujii, G., Shiki, S., "Optical Performance of SIS Photon Detectors at Terahertz Frequencies", JLTP 200, pp. 226-232, 2020

Breaking the 10 mW/pixel Limit for Kinetic Inductance Detector Readout Electronics

Adrian K. Sinclair, James R. Burgoyne, Yaqiong Li, Cody Duell, Scott C. Chapman, Anthony I. Huber, and Ruixuan Xie

***Abstract*—We demonstrate a prototype kinetic inductance detector (KID) readout system that uses less than 10 mW per pixel. The CCAT-prime RFSoC based readout is capable of reading four independent detector networks of up to 1000 KIDs each. The power dissipation was measured to be less than 40 W while running multi-tone combs on all four channels simultaneously. The system was also used for the first time to perform sweeps and resonator identification on a prototype 280 GHz array.**

***Index Terms*—kinetic inductance detectors, RFSoC, CCAT-prime, Prime-Cam, FPGA readout**

I. Introduction

Millimeter and submillimeter direct detector multiplexing methods experienced rapid growth and experimentation in the early 2000's [1]. Some of these early efforts have directly led to a substantial increase in detector counts for existing telescopes and some promise to enable even more for upcoming telescopes [2]–[4].

The Kinetic Inductance Detector (KID) was developed during this period and employs a natural frequency division multiplexing [5]. Each KID is a superconducting thin film shunted resonator in which the absorption of photons above the band gap energy changes the inductance and thus the resonant frequency. These changes in frequency can be monitored as changes in transmission at the KIDs original resonant frequency. By designing an initial frequency spacing for each KID and coupling each to the same microwave feed line, frequency multiplexing factors above 1000 can be achieved with a single pair of coaxial cables. In addition to the high multiplexing factor the fabrication and cryogenic complexity is significantly reduced compared to other methods at the cost of higher complexity for the room temperature readout electronics.

KIDs were chosen as the ideal detector technology to efficiently fill in the wide field-of-view and large optical throughput of the upcoming Prime-Cam instrument on the 6 meter Fred Young Submillimeter Telescope (FYST) [6], previously known as CCAT-prime. Prime-Cam will observe the sky with over 100,000 mm/sub-mm direct detectors at 5600 meters in the Atacama desert of northern Chile. The nearly factor of ten increase in detector count from existing telescopes that Prime-Cam requires has spurred the development of a novel frequency multiplexing readout system.

A.K. Sinclair, J.R. Burgoyne, S.C. Chapman, and R. Xie are with the University of British Columbia, Vancouver, BC Canada (e-mail:adriansinclair@phas.ubc.ca). A.I Huber is with the University of Victoria, Victoria, BC Canada. Y. Li and C. Duell are with Cornell University, Ithaca, NY USA.

We present the first power dissipation measurements of the system and discuss the advantage of low resonant frequency KIDs. We also compare it to a previous generation readout and comment on the implications for planned and future balloon-borne mm/submm telescopes.

II. RFSoC Readout and the Low Frequency Advantage

The Radio Frequency System on Chip (RFSoC) developed by Xilinx has high speed digitizers, reconfigurable logic, and ARM microprocessors integrated all into a single chip. The integrated architecture has lowered the total power dissipation substantially when compared to equivalent implementations separated into multiple chips. All metrics of size, weight, power, cost, and bandwidth make it an ideal platform for the readout of frequency multiplexed detectors. An RFSoC based readout has been recently developed for Prime-Cam [7].

All Prime-Cam KID arrays are being developed to resonate below the Nyquist frequency of the RFSoC digitizers allowing for direct RF sampling and generation. The digital design uses interpolation and decimation filters to limit the instantaneous bandwidth to 512 MHz and will use the numerically controlled local oscillator (NCLO) and digital mixers to up and down convert this band anywhere below Nyquist (2.048 GHz). This eliminates the earlier generation electronics requirement for analog mixers simplifying the warm readout considerably. The image tone suppression on the digital mixers were measured to be greater than 50 dB, a nearly impossible spec to meet with analog mixers.

An additional benefit when choosing lower frequency resonators (< 2 GHz) is that the number that can be multiplexed by the readout electronics increases with decreasing resonant frequency. This is due to the fact that the resonator bandwidth decreases with frequency $\Delta f = f/Q_r$. If the combined detector and readout electronics are viewed as an information carrying system then the low frequency advantage brings it closer to the Shannon channel capacity, although there is still a long way to go (see [8]).

Power dissipation within the RFSoC can be described by a static and dynamic term which scales with operational frequency. Increasing the instantaneous bandwidth of the readout requires either operating at a higher frequency within the fabric or with an increased resource utilization. Both lead to increased power dissipation and thus provide another line of support for the low frequency advantage.

Fig. 1: Power dissipation measurement of four channel design. The RFSoC ZCU111 on the left was powered by the supply in the top right with 12 V and was drawing 3.028 A (36.34 W). The RFSoC was programmed with the four channel firmware and configured to generate frequency combs from each channel. Each frequency comb was digitally mixed with a different numerically controlled local oscillator frequency to separate the combs for visual display on the spectrum analyzer.

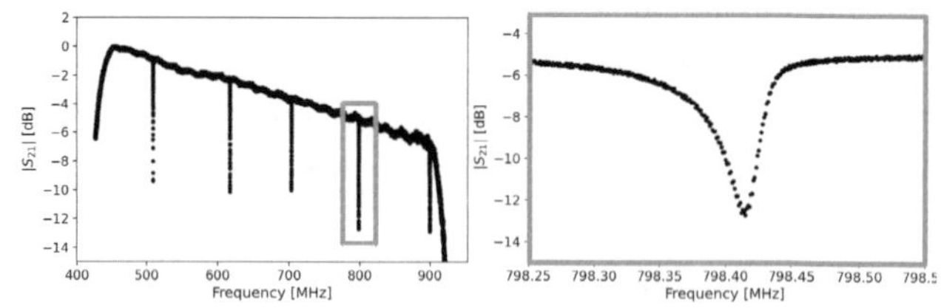

Fig. 2: Plots of the forward transmission as a function of frequency for a 280 GHz prototype array measured by the RFSoC. The plot on the left shows the transmission normalized to the max for the operational bandwidth of 512 MHz and a numerically controlled local oscillator center frequency of 676 MHz. All five resonances corresponding to the five prototype detectors are easily identified. A zoom in of one resonance is shown on the right.

Some of the benefits described above were also found nearly a decade ago with detector performance advantages best described in Swenson et al. 2012 [9].

III. Power Dissipation Measurements

Two power measurements were performed to verify the dissipation, one with a handheld multi-meter and the other with a benchtop power supply.

First the ZCU111 development board power adapter output was measured with a multi-meter to be 12.23 V. The current was measured with a multi-meter in series between the output of the power adapter and the ZCU111 power Molex connector J52. Upon boot the system drew 1.83 A. In the boot state the ARM core loads a lightweight linux operating system and allows communication via Ethernet. The next step was to load the firmware configuration bitstream using the open source PYNQ library. After the bitstream was loaded the current rose to 2.7 A and stabilized. A frequency comb containing 1000 tones spanning 512 MHz of bandwidth were written to all four independent channels. The frequency comb waveform is normalized to maximize the available dynamic range of the D/A. With continuous operation of all four channels the current rose to 3.09 A and stabilized. The total power draw for the multi-meter measurement was 37.8 W.

The second power measurement was performed at the Herzberg Astronomy and Astrophysics Research Centre mm-wave lab with a DC power supply. The system was powered by applying 12 V directly to the Molex connector J52. After the bitstream was loaded and each channel was configured to generate multi-tone combs the current rose to 3.028 A and stabilized. This corresponds to a total power dissipation of 36.34 W. To show all four combs on a spectrum analyzer simultaneously a 4-way power combiner (Mini-Circuits ZFSC-4-1-S+) was found in the lab and connected to all four active D/A's of the ZCU111 development board. Each numerically controlled local oscillator within the RFSoC were set to different center frequencies to spread the combs for visual effect. This measurement is summed up in one photograph show in figure 1.

IV. Comparison with Previous Generation Readout

The author was involved in the development of an earlier generation KID readout [10] which has been deployed in two imaging polarimeter instruments at the Large Millimeter Telescope [11], [12], and soon an on-chip spectrometer [13]. It has also flown on two balloon-borne telescopes [14], [15]. The digital signal processing algorithm used for these projects was the starting point for the RFSoC based readout system.

The earlier generation readout mentioned above was implemented on the reconfigurable open architecture computing hardware version 2 (ROACH2) designed by the collaboration for astronomy and signal processing electronics research (CASPER) [16]. The gateware design implemented on the ROACH2 platform was capable of reading 1000 channels and drew 50 W (Not including intermediate frequency (IF) electronics such as the two analog mixers and external amplifiers), This corresponds to 50 W/1000 pixels = 50 mW/pixel. The new four channel RFSoC design can read out 4000 pixels for less than 40 W meaning $<$ 10 mW/pixel. BLAST-TNG flew five ROACH2s for a total of $\sim$ 250 W (not including IF) with a maximum possible number of detectors of 5000. This could be replaced with two RFSoCs running the four channel design for a significantly relaxed power dissipation, size, and weight. Utilizing all 250 W would correspond to six RFSoCs to read out a total of 24000 detectors. Future balloon-borne telescopes will benefit greatly from the RFSoC platform and a few are already planning to fly them [17], [18].

V. First Measurement with Detector Array

Preliminary measurements were made with the RFSoC on a witness chip for the Aluminum 280 GHz array fabricated at NIST by the quantum sensors group. The array was cooled in a dilution fridge to 100 mK at Cornell University. The system contained 23 dB of cold attenuation to the devices and was followed by a low noise amplifier with 30 dB of gain and input referred noise temperature of 5 K. The RFSoC system was connected via an SMA breakout board to the cryostat with coaxial cables and a room temperature 10 dB fixed attenuator on the drive side. A 1000 tone comb was produced initially and measured on the spectrum analyzer to have approximately –50 dBm per tone. A preliminary version of the readout control software [19] was used to command the channel to generate the comb, perform a frequency sweep, and output a file containing the sweep measurement. The sweep used all 1000 tones and stepped the NCLO such that bandwidth between each tone was covered with a 1 KHz resolution. The result of the sweep is shown in figure 2 where each of the five detector resonances are identified and a zoom in of one resonator is also shown.

VI. Conclusion

We have demonstrated a kinetic inductance detector readout that can achieve less than 10 mW/pixel. This system developed for the Prime-Cam instrument on the Fred Young Submillimeter telescope is ideal for KID-based balloon-borne telescopes. The systems on-board digital mixer and numerically controlled oscillator make transmission and resonator identification measurements with ease and considerably lower systematics. Future improvements in the readout gateware are anticipated and we expect the power per pixel to decrease even further on the same hardware.

VII. Acknowledgment

The author would like to thank the mm-lab at the Herzberg Astronomy and Astrophysics Research Centre for the use of their space and measurement equipment. The CCAT-prime project, FYST and Prime-Cam instrument have been supported by generous contributions from the Fred M. Young, Jr. Charitable Trust, Cornell University, and the Canada Foundation for Innovation and the Provinces of Ontario, Alberta, and British Columbia.

References

[1] M. Dobbs *et al.*, "Multiplexed readout of cmb polarimeters," *Journal of Physics: Conference Series*, vol. 155, no. 1, p. 012004, mar 2009. [Online]. Available: https://dx.doi.org/10.1088/1742-6596/155/1/012004

[2] S. C. Parshley *et al.*, "CCAT-prime: a novel telescope for sub-millimeter astronomy," in *Ground-based and Airborne Telescopes VII*, ser. Society of Photo-Optical Instrumentation Engineers (SPIE) Conference Series, H. K. Marshall and J. Spyromilio, Eds., vol. 10700, Jul. 2018, p. 107005X.

[3] A. Lee *et al.*, "The Simons Observatory," in *Bulletin of the American Astronomical Society*, vol. 51, Sep. 2019, p. 147.

[4] P. Klaassen *et al.*, "The Atacama Large Aperture Submillimeter Telescope (AtLAST)," in *Bulletin of the American Astronomical Society*, vol. 51, Sep. 2019, p. 58.

[5] P. Day *et al.*, "Antenna-coupled microwave kinetic inductance detectors," *Nuclear Instruments and Methods in Physics Research Section A: Accelerators, Spectrometers, Detectors and Associated Equipment*, vol. 559, no. 2, pp. 561–563, 2006, proceedings of the 11th International Workshop on Low Temperature Detectors. [Online]. Available: https://www.sciencedirect.com/science/article/pii/S0168900205024824

[6] E. M. Vavagiakis *et al.*, "Prime-Cam: a first-light instrument for the CCAT-prime telescope," in *Millimeter, Submillimeter, and Far-Infrared Detectors and Instrumentation for Astronomy IX*, ser. Society of Photo-Optical Instrumentation Engineers (SPIE) Conference Series, J. Zmuidzinas and J.-R. Gao, Eds., vol. 10708, Jul. 2018, p. 107081U.

[7] A. K. Sinclair *et al.*, "CCAT-prime: RFSoC based readout for frequency multiplexed kinetic inductance detectors," in *Millimeter, Submillimeter, and Far-Infrared Detectors and Instrumentation for Astronomy XI*, ser. Society of Photo-Optical Instrumentation Engineers (SPIE) Conference Series, J. Zmuidzinas and J.-R. Gao, Eds., vol. 12190, Aug. 2022, p. 121900W.

[8] K. D. Irwin, "Shannon Limits for Low-Temperature Detector Readout," *AIP Conference Proceedings*, vol. 1185, no. 1, pp. 229–236, 12 2009. [Online]. Available: https://doi.org/10.1063/1.3292320

[9] L. J. Swenson *et al.*, "MAKO: a pathfinder instrument for on-sky demonstration of low-cost 350 micron imaging arrays," in *Millimeter, Submillimeter, and Far-Infrared Detectors and Instrumentation for Astronomy VI*, ser. Society of Photo-Optical Instrumentation Engineers (SPIE) Conference Series, W. S. Holland and J. Zmuidzinas, Eds., vol. 8452, Sep. 2012, p. 84520P.

[10] S. Gordon *et al.*, "An open source, fpga-based lekid readout for blast-tng: Pre-flight results," *Journal of Astronomical Instrumentation*, vol. 05, no. 04, p. 1641003, 2016. [Online]. Available: https://doi.org/10.1142/S2251171716410038

[11] S. Rowe *et al.*, "The MUSCAT Readout Electronics Backend: Design and Pre-deployment Performance," *Journal of Low Temperature Physics*, Dec. 2022.

[12] G. W. Wilson *et al.*, "The TolTEC camera: an overview of the instrument and in-lab testing results," in *Millimeter, Submillimeter, and Far-Infrared Detectors and Instrumentation for Astronomy X*, J. Zmuidzinas and J.-R. Gao, Eds., vol. 11453, International Society for Optics and Photonics. SPIE, 2020, p. 1145302. [Online]. Available: https://doi.org/10.1117/12.2562331

[13] J. G. Redford *et al.*, "SuperSpec: On-chip spectrometer design, characterization, and performance," in *Millimeter, Submillimeter, and Far-Infrared Detectors and Instrumentation for Astronomy X*, J. Zmuidzinas and J.-R. Gao, Eds., vol. 11453, International Society for Optics and Photonics. SPIE, 2020, p. 1145336. [Online]. Available: https://doi.org/10.1117/12.2563110

[14] A. Paiella *et al.*, "In-Flight Performance of the LEKIDs of the OLIMPO Experiment," *Journal of Low Temperature Physics*, vol. 199, no. 1-2, pp. 491–501, Feb. 2020.

[15] G. Coppi *et al.*, "In-flight performance of the BLAST-TNG telescope platform," in *Ground-based and Airborne Telescopes VIII*, H. K. Marshall *et al.*, Eds. SPIE, dec 2020.

[16] J. Hickish *et al.*, "A Decade of Developing Radio-Astronomy Instrumentation using CASPER Open-Source Technology," *Journal of Astronomical Instrumentation*, vol. 5, no. 4, pp. 1 641 001–12, Dec. 2016.

[17] G. Cataldo *et al.*, "Overview and status of EXCLAIM, the experiment for cryogenic large-aperture intensity mapping," *arXiv e-prints*, p. arXiv:2101.11734, Jan. 2021.

[18] J. Vieira *et al.*, "The Terahertz Intensity Mapper (TIM): a Next-Generation Experiment for Galaxy Evolution Studies," *arXiv e-prints*, p. arXiv:2009.14340, Sep. 2020.

[19] J. Burgoyne, "Ccatphive," https://github.com/TheJabur/CCATpHive, 2023.

The Space Interferometer for Cosmic Evolution (SPICE):
The Far-Infrared Universe at High Spatial Resolution

Locke D. Spencer*[1], on behalf of the SPICE Science Team[2]

***Abstract*— The Space Interferometer for Cosmic Evolution (SPICE) is a Far-Infrared probe mission concept. SPICE's unprecedented sub-arcsecond resolution and superlative sensitivity in the wavelength range 25 - 400 microns will enable the mission's scientific objectives including observations of large statistical samples of distant galaxies, protoplanetary disks, and debris disks. SPICE will greatly improve understanding the physical processes that drive the evolution of galaxies and their central massive black holes throughout cosmic time, planet formation and water delivery to nascent planets, and planetary system architectures and their evolution.**

Keywords*— *Far-infrared, interferometry, high spatial resolution, Fourier Spectroscopy, Astrophysics, Cosmology

I. SPICE Mission Concept

The SPICE concept design employs a two-input variable baseline interferometer that achieves u-v plane coverage through scanning of the baseline and rotation of the instrument (see figure 1). The two interferometric inputs are directed towards a central instrument employing a Fourier Transform Spectrometer (FTS). The SPICE Far-infrared Interferometry Probe mission offers:

- Wavelength range 25 - 400 μm (approximate),
- Angular resolution 0.3 (λ/100 μm) arcseconds for high quality imaging,
- Dense u-v plane coverage,
- Integral field spectroscopy over a 1 x 1 arcmin Field of View (FoV),
- Spectral resolution λ/Δλ > 3000 in each spatial resolution element, through using a single "double Fourier" beam combiner.

SPICE has the ~same angular resolution as the Webb Telescope (i.e., JWST), at ten times longer wavelengths - more than an order of magnitude better spatial resolution than any other far-infrared observatory (see Fig. 2). Fig. 3 illustrates this enhanced resolution using three examples: an extragalactic deep field (left), a protoplanetary disk (center), and a debris disk (right).

Fig. 1. Illustration of SPICE Probe design concept. Two receivers separated by a variable baseline direct light to the central instrument.

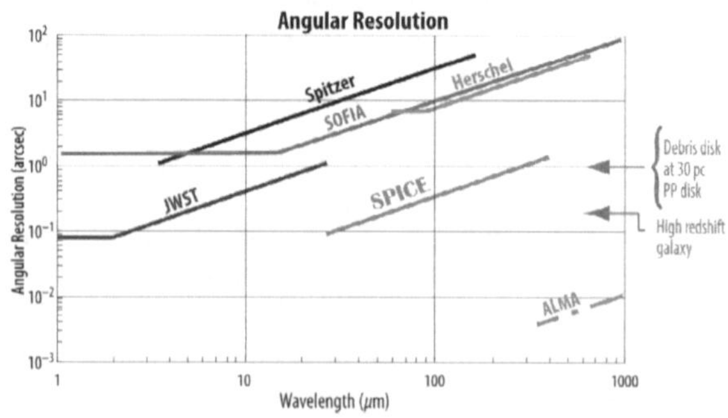

Fig. 2. Angular Resolution of SPICE mission, in context.

II. SPICE Science

The SPICE Science team is comprised of an instrument simulation and modelling working group (WG) and several science working groups. Science WG topics include galaxy evolution, star and planetary formation, and debris disks.

A. Galaxy Evolution

The key goal of galaxy formation studies is to understand the physical processes that drive the evolution of galaxies and their central massive blackholes throughout cosmic time. A crucial component of the baryonic content of galaxies is the gas and dust in between stars, i.e., the interstellar medium (ISM). Indeed, the infrared emission of galaxies makes up about half of the cosmic radiation background from galaxy formation and evolution processes, and the star formation rate density of the Universe is dominated by dust-obscured star formation emitting in infrared, at least out to z~4. Dust attenuation severely biases UV and optical observations.

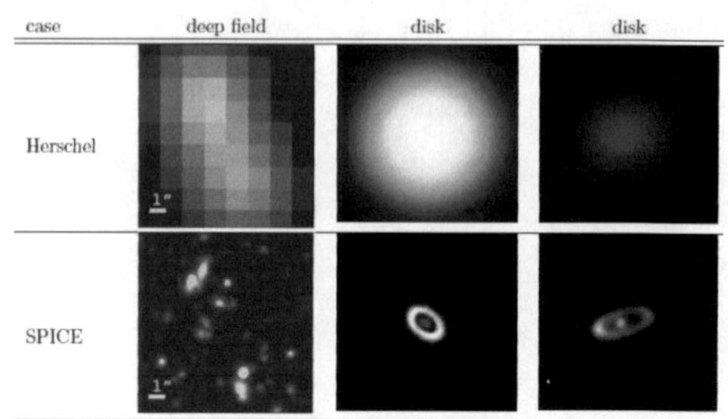

Fig. 3. Comparison of Herschel and SPICE spatial resolution for a variety of observation targets.

[1]Locke.Spencer@uLeth.ca, Institute for Space Imaging Science, University of Lethbridge, Lethbridge, T1K 3M4, Canada; [2] see https://spice.astro.umd.edu

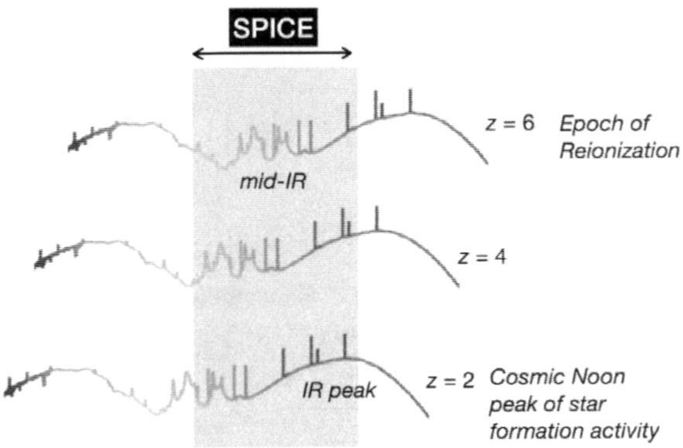

Fig. 4. SPICE spectral coverage demonstrated with redshifted galactic spectra.

Probing the stellar, gaseous, metal and blackhole content of galaxies in infrared wavelengths will provide a complete and unbiased view of the physical and chemical processes inside galaxies.

The high spatial resolution of SPICE (~0.3 arcsec) compared to its predecessors, Spitzer (~6-18 arcsec) and Herschel (~5-40 arcsec), and its >10x larger field of view compared to that of ALMA, will enable us, for the first time, to disambiguate the infrared emission of *individual* (unlensed) galaxies beyond the local Universe in statistically large samples. In addition, the high spatial resolution of SPICE will allow for studies of far-IR cooling lines in *individual* star forming regions and the diffuse ISM of galaxies other than the Milky Way, providing context for the ALMA detections of these lines at high redshifts and connecting the local Universe to the epoch of reionization and first galaxies.

B. *Protostar and Planetary Formation*

Planetary disks, which give rise to planets, comets, and asteroids, are the gas-rich disks that surround young stars. Protostar observations at far-infrared wavelengths, dominated by the emission from gas-rich protoplanetary disks surrounding these young stars, remain spatially unresolved. Detecting and resolving the location of gas species at high-resolution in protoplanetary disks is the essential next step to constrain models of planet formation. Of those species, water is perhaps the most important molecule driving the origins of life, and much of the water is thought to be frozen. With sub-arcsecond spatial resolution, and spectral resolving powers $R>10^3$ from space, SPICE will be the first instrument to detect and resolve the location of water-ice in protoplanetary disks in nearby star forming regions and define how this is delivered to young planets. Moreover, probing for the carbon and oxygen bearing species only accessible at far-infrared wavelengths that are necessary for the complex chemical interactions that occurred in our Solar System, SPICE will constrain the chemical evolution with which planets, comets, and asteroids form.

C. *Debris Disks*

Debris disks are analogues of the Solar system's small-body populations seen to orbit other stars. The origin of a debris disk's radial profile, whether it be structured or not, remains unclear. Current sensitivity can detect disks about ten times dustier than Solar system levels, with disks detected around approximately 20% of stars. We know that the Solar system's debris disk lies somewhere among the faintest 80%, but do not know whether this is unusually faint or not.

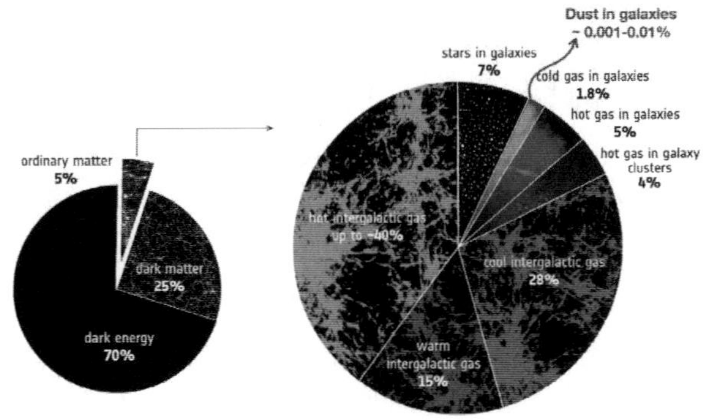

Fig. 5. Proportion of dust in the composition of the Universe.

Although dust is only present in trace amounts compared to the other components of our universe, it plays important roles in many astrophysical processes. Fig. 5 illustrates the proportional dust composition of the universe.

D. *SPICE Instrument Simulation and Modelling*

The Instrument Performance Modelling working group has developed and continues to refine a Python-based Far-Infrared Interferometer Simulator (PyFIInS) [5]. This tool is used in parallel with other instrument modelling and design activities with a goal to understand and predict the abilities and limitations of a SPICE-like system more accurately. This will allow exploration of limits on the scientific questions that can be answered with the high spatial resolution far-infrared observations of SPICE and translate these results into technical specification definitions for a far-infrared space-based interferometer mission such as SPICE. Publicly available versions of these tools, and associated documentation, are in preparation.

III. Conclusions and Future Work

We must now consider the next big leap in far-infrared astronomical observations. Although both Herschel and Spitzer operated successful missions with high spectral resolution, they were both limited in spatial resolution due to the fundamental diffraction limit of their 3.5 m and 0.85 m primary mirrors, respectively. Clearly, high spatial resolution (sub-arcsecond) should be our next step in observing in the Far-Infrared.

References

[1] Rinehart, et al., "The Path to Interferometry in Space," *Proc. SPIE 9907*, Aug. 2016.

[2] Farah, et al., "Review: far-IR instrumentation and technology development for the next decade," *JATIS*, Aug. 2019.

[3] Linz, et al., "Bringing high spatial resolution to the far-infrared," Experimental Astronomy, May 2021.

[4] Helmich and Ivison, "FIRI – A far-infrared interferometer," Experimental Astronomy, 23, 245-276, 2009.

[5] Juanola-Parramon, "A Far-infrared Spectro-Spatial Space Interferometer. Instrument Simulator and Testbed Implementation," PhD Thesis, April 2014.

G-Band Metamaterial-Based Circulator for FCC-Compliant Space-to-Earth Communication

Ali J. Alqaraghuli*, Sergi Aliaga*, Arjun Singh†, and Josep M. Jornet*

*Department of Electrical and Computer Engineering, Northeastern University, Boston, MA, USA
†Department of Engineering, SUNY Polytechnic Institute, Utica, NY, USA
E-mail: {alqaraghuli.a, jmjornet}@northeastern.edu, singha8@sunypoly.edu

***Abstract*—A metamaterial-based circulator operating at G-band is presented for the purpose of high-data-rate space-to-Earth communication. The circulator is designed to be equipped on a CubeSat using a single antenna, integrating a metamaterial within an interface of three WR3 waveguides measuring 3 mm in length per waveguide. The isolation is achieved between transmitting (downlink) signals at 295 GHz and reception (uplink) at 218 GHz. The frequencies are chosen in accordance with FCC regulations and aim to provide a high-data-rate download link for CubeSats deployed in non-terrestrial networks.**

***Index Terms*—Terahertz Space Communication; Metamaterial-based Circulator; Terahertz CubeSat Networks**

I. Introduction

With the inclination towards high-data-rate space-based internet and the increasing demand of efficient communication across scientific instruments, terahertz (THz)-band communication (0.1-10 THz) has been explored as a candidate for space communication leveraging existing technology. The potential for high data rates in the terahertz band is enabled by the high available bandwidth, as well as existing technology that can support such bandwidth [1]. Furthermore, THz CubeSat networks have been explored as part of the vision for sixth-generation (6G) wireless communication through multi Giga-bit-per-second (Gbps) inter-satellite links, with physical layer and medium access control solutions already being explored [2]. In our previous work, THz communication was studied extensively for inter-satellite as well as space-to-Earth links in an end-to-end simulation, demonstrating multi-Gbps in both scenarios when high antenna gain is available [3]. The simulation was performed in compliance with FCC regulations, crowning 217.5 GHz and 295 GHz as ideal central frequencies for uplink and downlink respectively using 8320 small satellites.

In order to deploy a massive number of satellites, reductions in size, weight, and power must be leveraged in the form of CubeSats, which have been considered a staple of 6G communication and a key element of the increasingly-popular non-terrestrial networks (NTN) [2]. Similarly, in order to fully leverage size reductions, CubeSats would ideally use a single antenna for transmission and reception; given that THz antennas will require high gain, using a single antenna would majorly contribute to saving real estate.

The interest in saving real estate by duplexing transceivers and, therefore, use a single antenna for transmission and reception has long been present in the radar community. Recently, with the development of the first active weather radar operating within G-band (110 GHz to 300 GHz), full-duplex capability has been obtained thanks to the use of quasi-optical circular polarization duplexers [4], [5]. These encode transmission and reception signals in the horitzontal and vertical polarizations, respectively, for full duplexing. This approach, however, is not appealing from a communications perspective since multiplexing capabilities through the polarizations of the transmitted wave would be lost. Thus, the objective then is to design a G-band circulator that satisfies the compactness and power criteria of terahertz CubeSat for high-bandwidth communication.

Achieving full duplexing capabilities through the use of circulators has been studied for satellite communications at lower frequencies. Traditional circulator designs targeting modest microwave frequencies are based on semiconductor devices, in intermediate bands (i.e. Ku/Ka band), ferrite based devices are used almost exclusively. These kind of circulators are based on Turnstile designs implementing Okada resonators [6], [7], which mainly consist of two ferrite disks of identical geometry and dimensions separated by a dielectric material. Ferrite based circulators are being designed for wide band space applications at Ku band (25% bandwidth) [8] and at Ka band (13% bandwidth) [9]. However, the implementation of this type of circulator at frequencies close to the THz band is technically challenging; ferrite components suffer from high cost, high weight, and incompatibility with an integrated circuit design, which are critical restrictions when considering Cubesats dimensions. Here, research in the field of metamaterials has enabled the creation of broadband circulators at frequencies of tens of GHz using spoof surface plasmon polaritons (SSPPs) [10], [11]. Near-zero-index metamaterials (NZIMs) have been investigated for the design of magnetless circulators [7] in which controlling the properties of EM waves (propagation, absorption and polarization) is possible without the need of any external magnetic field, although using a restrictive angle-selective methodology.

In this paper, we propose a frequency-selective solution leveraging a frequency-dependant metamaterial as explored in the design methodology.

II. Design Methodology

In order to achieve simultaneous transmission and reception, the feeding network of the dual-band antenna proposed in [12] is realized in Fig. 1, where a circulator serves to isolate between the transmitting and receiving frequencies of a super-heterodyne terahertz system similar to the one in [1].

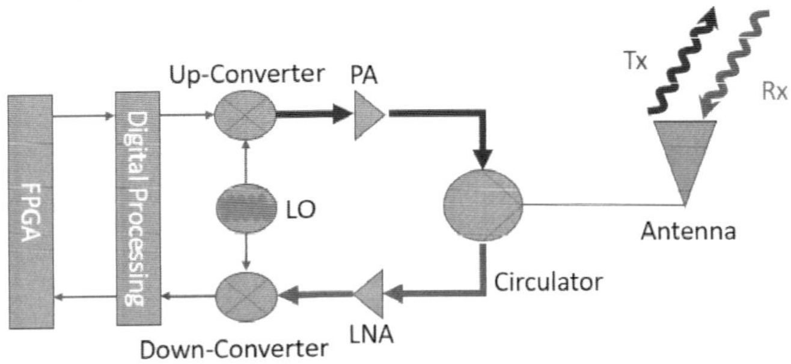

Fig. 1. The working principle of the dual band antenna, enabled by the proposed circulator.

A commercially-available WR3 waveguide is utilized as to allow both 218 GHz and 295 GHz signals to operate in accordance with the rectangular waveguide cutoff frequency given by

$$f_c = \frac{c}{2}\sqrt{\left(\frac{m}{a}\right)^2 + \left(\frac{n}{b}\right)^2}, \tag{1}$$

where a is width and b is thickness of the waveguide, m and n are the mode propagation numbers and c is speed of light inside the waveguide. The metamaterial is then inserted at the interface of the receiving waveguide and the waveguide junction as visualized in Fig. 2. The metamaterial's special behavior, mainly frequency-dependent complex-valued permittivity, is defined by

$$\epsilon(\omega) = \epsilon'(\omega) + j\epsilon''(\omega) = \epsilon_r(\omega)\epsilon_0 + j\frac{\sigma(\omega)}{\omega}, \tag{2}$$

where ω is the wave frequency, ϵ is effective permittivity, ϵ_r is relative permittivity of the material, ϵ_0 is free-space permittivity, and σ is conductivity of the material.

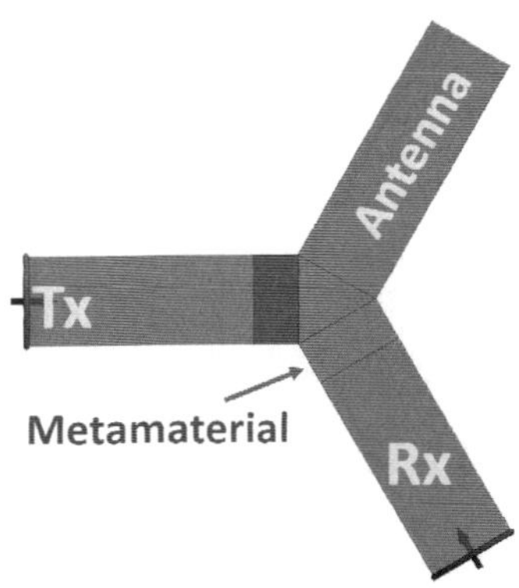

Fig. 2. The working principle of the circulator, interfacing the three WR3 waveguides.

III. Results

The results show the fundamental mode waves propagating from the waveguide source port throughout the circulator, which is the feeding waveguide (Tx) for the transmitter and the antenna for the receiver. Figure 3 shows the surface currents across the circulator for the transmit and receive cases, proving the metamaterial effective at achieving isolation. The reflection coefficient for the transmit case is -21 dB and -20 dB for reception. The simulations are completed using Altair FEKO software utilizing the multilevel fast multipode method (MLFMM). The next step is to develop and simulate an integrated full-duplex communication system involving the antenna deploying from the CubeSat.

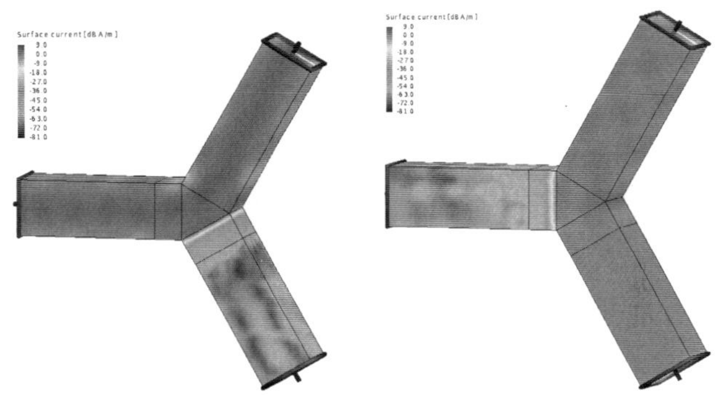

Fig. 3. The surface currents present on the waveguides in transmission mode, showing behavior at 218 GHz (left side) and at 295 GHz (right side).

References

[1] I. Mehdi, J. Siles, C. P. Chen, and J. M. Jornet, "Thz technology for space communications," in *2018 Asia-Pacific Microwave Conference (APMC)*. IEEE, 2018, pp. 76–78.

[2] I. F. Akyildiz, A. Kak, and S. Nie, "6g and beyond: The future of wireless communications systems," *IEEE access*, vol. 8, pp. 133 995–134 030, 2020.

[3] A. J. Alqaraghuli, H. Abdellatif, and J. M. Jornet, "Performance analysis of adual terahertz/ka band communication systemfor satellite mega-constellations," in *Proc. of the 22nd IEEE World of Wireless, Mobile, and Multimedia networks (WoWMOM)*, 2021, pp. 1–7.

[4] K. B. Cooper, R. J. Roy, R. Dengler, R. R. Monje, M. Alonso-Delpino, J. V. Siles, O. Yurduseven, C. Parashare, L. Millán, and M. Lebsock, "G-band radar for humidity and cloud remote sensing," *IEEE Transactions on Geoscience and Remote Sensing*, vol. 59, no. 2, p. 1106–1117, Feb 2021.

[5] K. B. Cooper, N. Llombart, G. Chattopadhyay, B. Dengler, R. E. Cofield, C. Lee, S. Filchenkov, and E. Koposova, "A grating-based circular polarization duplexer for submillimeter-wave transceivers," *IEEE Microwave and Wireless Components Letters*, vol. 22, no. 3, p. 108–110, Mar 2012.

[6] J. Helszajn and J. Sharp, "Adjustment of in-phase mode in circulators using turnstile junctions," *IEEE Transactions on Microwave Theory and Techniques*, vol. 33, no. 4, p. 339–343, Apr 1985.

[7] W. Xue, J. Zhang, J.-W. Ma, Z.-L. Hou, Q.-L. Zhao, Q. Xie, and S. Bi, "A highly directional metamaterial-based terahertz circulator that does not require an external magnetic field," *Journal of Physics D: Applied Physics*, vol. 54, no. 10, p. 105103, 2020.

[8] A. Ghadiya, K. Trivedi, S. Soni, and P. Bhatt, "Wide band stripline circulator at ku band for space applications," in *2018 IEEE MTT-S International Microwave and RF Conference (IMaRC)*, Nov 2018, p. 1–4.

[9] S. K. Jain and P. Pradeepkumar, "Design of broadband waveguide circulator at ka-band," in *2021 IEEE MTT-S International Microwave and RF Conference (IMARC)*, Dec 2021, p. 1–4.

[10] T. Qiu, J. Wang, Y. Li, J. Wang, and S. Qu, "Broadband circulator based on spoof surface plasmon polaritons," *Journal of Physics D: Applied Physics*, vol. 49, no. 35, p. 355002, Sep 2016.

[11] T. Qiu, J. Wang, Y. Li, and S. Qu, "Circulator based on spoof surface plasmon polaritons," *IEEE Antennas and Wireless Propagation Letters*, vol. 16, p. 821–824, 2017.

[12] A. J. Alqaraghuli, A. Singh, and J. M. Jornet, "Compact high-gain dual-band antenna for full-duplex terahertz communication in cubesat mega-constellations," in *2021 IEEE International Symposium on Antennas and Propagation and USNC-URSI Radio Science Meeting (APS/URSI)*, 2021, pp. 1827–1828.

Optimization of a PID controlled TES detector array for use in an FIR double-Fourier interferometer

Chris S. Benson*[1], Jeremy P. Scott, Locke D. Spencer

***Abstract*—As part of the ongoing efforts in the development of a laboratory-based double-Fourier transform interferometer in the far-infrared, we present the characterization of a custom-developed cryogenic transition edge sensor detector array system.**

Keywords—Detectors, Bolometers, Transition-edge Sensors Fourier Interferometry, Spectroscopy, Far-infrared, Instrumentation

I. Introduction

Double Fourier interferometry provides a promising path to achieve broadband observations of the far-infrared (~0.3—10 THz) with sub-arcsecond spatial resolution [1]. We have developed a double-Fourier Transform interferometer (2FTI) instrument coupled to a unique array of 25 multiplexed TES bolometers to provide laboratory-based validation of this technique [2].

Each detector in the array is maintained at a bias temperature on the TES thermometer's superconducting transition by conduction to the surrounding cryogenic environment, optical loading, the bias circuitry of the TES, and a heating element controlled by a proportional-integral-derivative (PID) loop. At the start of each measurement pulse, the heater is switched on for a selected time period before a measurement of the TES voltage is made and fed into the PID feedback calculation. This voltage is a proxy for TES temperature and regulates the heater power used to maintain the temperature setpoint. In this way, the heater power provides a measure for the optical loading on the TES detector [3]. The feedback loop corresponding to a single detector measurement can be seen in Fig. 1. This detector system was developed by QMC Instruments [3].

The optimization of the detector array requires consideration of the adjustable timing of the heater duty cycle, the delay after heater shutoff before measurements of the TES state are used in PID calculation, and the tuning of the PID feedback loop. Also, the performance of the detector array should be considered within the context of the 2FTI as the scanning speed of the interferometer's spectral arm is limited by detector time constants [4].

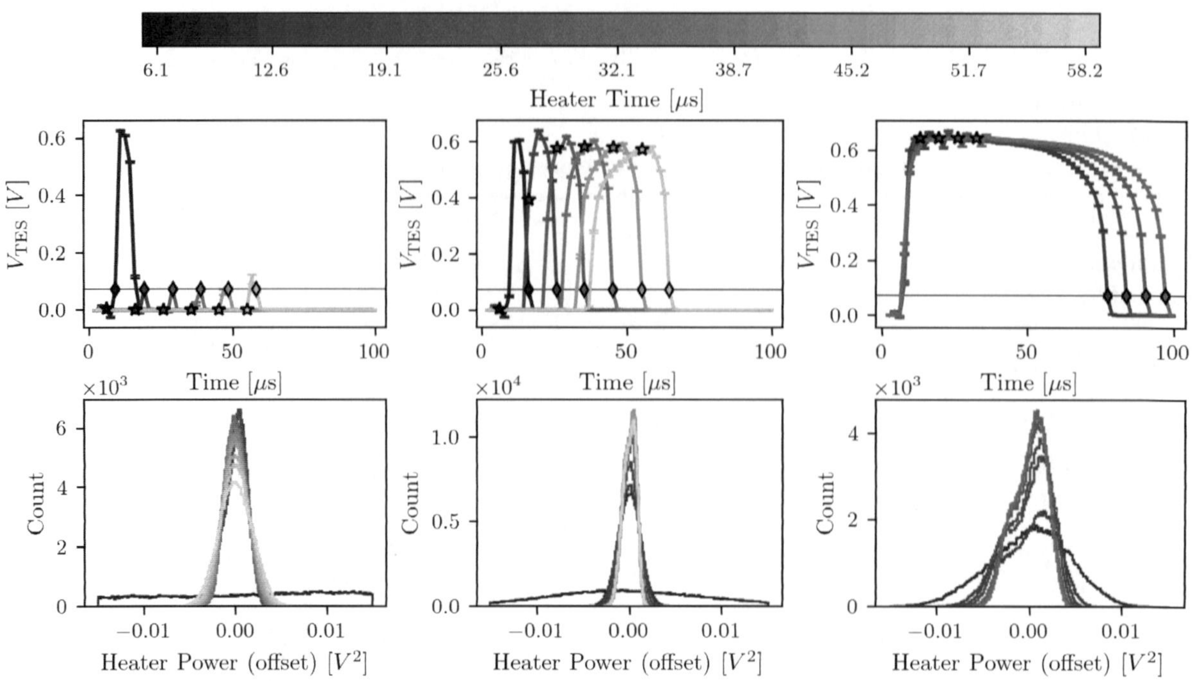

Fig. 1. The timing between when the heater is shut off and when measurements of the TES voltage are used in PID calculation is shown for time delays of 3.25 (left), 9.75 (center), and 65.00 (right)\,μs for a range of heater duty cycles. The top row of plots shows the TES signal averaged over several thousands of detector measurements. TES voltage measurements marked by a star indicate the TES state corresponding to when the heater is switched off while those marked by a diamond show the TES state that is used in the PID calculation. The PID setpoint is marked by the horizontal black line. The bottom row of panels displays the distribution of heater powers, represented by the square of the voltage across the heating element, used to maintain the setpoint for each case. The distribution of heater power values corresponds to the detector's measurements of optical loading and demonstrates the stability of the detector signal. Adjusting the timing of the detector measurement loop results in an offset in detector signal with shorter heater times and longer time delays requiring more heater power.

[1]Institute for Space Imaging Science, University of Lethbridge, Lethbridge, T1K 3M4, Canada

II. Timing Parameters

The timing of a single detector measurement is defined by the time for which the heating element is active and the time that is waited before the TES state is measured and used to calculate the heater value for the subsequent measurement (see Fig. 1). These timing effects have been studied by considering the signal stability of the central detector as its timing parameters were varied while viewing a ~6 K static optical load.

We have found that a time delay of 9.75 μs between when the heater is shut off and when the TES state is read for PID calculation causes the TES to be heated above its transition into the non-linear region before cooling down towards the setpoint and provides the most stable detector signal. This is shown by the central panels of Fig. 1. If the time delay is too short, the measurement of the TES state used in the PID loop is taken on the rising edge and can be subject to electrical pickup noise from the heater switching off resulting in a less stable detector signal (shown by the leftmost panels of Fig. 1). If there is too long of a delay between the heater and TES state measurements for PID calculation, the TES spends most of the time of a single measurement pulse in a non-linear state and a much less stable detector signal. Additionally, we have found that if the heater is on for less than 10 μs, the PID loop is unstable unless the TES voltage is read ~65 μs after the heater is switched off.

III. PID tuning

Tuning the gain values associated with each term of the PID calculation can optimize the performance of the associated detector.

The frequency response of the central detector was measured using the spectral arm of the 2FTI while viewing a quasi-monochromatic source provided by a frequency-tunable photomixer [5]. Through scanning the spectral arm at different speeds, sinusoidal optical modulation on the detectors is achieved up to ~600 Hz. The detector response to this modulation is shown in Fig. 3 at two different gain states associated with the P and I terms of the PID calculation. Increasing the gain associated with the I term while decreasing the gain of the P term results in a faster detector time constant and thereby a higher cutoff frequency.

The faster detector time constant and higher spectral scanning speeds that can be reached as a result can be useful in increasing the frequency of the spectral modulation of the 2FTI to avoid 1/f noise. In spite of this, we have measured decreased signal-to-noise in detector signal with the PID loop is I-term dominated [3]. Optimizing the PID parameters of the detector array for use in the 2FTI will require consideration of both effects.

IV. Conclusion

We have found optimal TES performance with a time delay of ~9.75 μs between heater switch-off and reading the TES state for PID calculation. Additionally, PID tuning of the detector array will require consideration of the SNRs of the detectors as well as their frequency response. The effective time constant of each detector should be selected such that the noise resulting from increasing the gain associated with the I parameter of the PID loop does not outweigh the reduction in 1/f-noise obtained from the scanning speed of the interferometer's spectral arm.

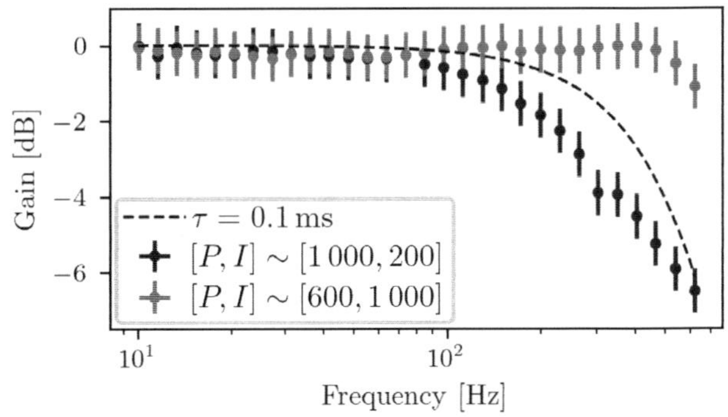

Fig. 3. The frequency response of the central detector under two PID tuning conditions is shown. The modelled frequency response behaviour of a nominal TES detector with an effective time constant of 0.1 ms is shown by the dashed black line [6].

References

[1] D. Farrah, K. E. Smith, D. Ardila, Bradford et al., "Review: far-infrared instrumentation and technology development for the next decade", *Journal of Astronomical Telescopes, Instruments, and Systems*, vol. 5, p. 020901, Apr. 2019.

[2] L. D. Spencer, P. A. Ade, W. F. Grainger, M. J. Griffin, R. Juanola-Parramon, D. A. Naylor, E. Pascale, G. Savini, and B. M. Swinyard, "Development of a far-infrared double Fourier spatial/spectral interferometery laboratory testbed instrument", *Fourier Transform Spectroscopy*. Optical Society of America, 2015, pp. FT4A-6.

[3] L. D. Spencer, R.V. Sudiwala, C. S. Benson, J. P. Scott, M. D. De Silva, T. Jones, J. Cox, K. Wood, and P. A. Ade, "Optimization of a cryogenic transition-edge sensor detector array for far-infrared astrophysics", *Space Telescopes and Instrumentation 2022: Optical, Infrared, and Millimeter Wave,* vol. 12180. SPIE, 2022, pp. 1018-1035.

[4] L. D. Spencer, D. A. Naylor, B. M. Swinyard, A. A. Aramburu, T. R. Fulton, T. L. Lim, S. D. Yonayette, and I. S. Schofield, "A Fourier transform spectrometer for ground testing of the Herschel/SPIRE instrument", *Optical, Infrared, and Millimeter Space Telescopes,* vol. 5487. SPIE, 2004, pp. 501-512.

[5] A. J. Deninger, A. Roggenbuck, S. Schindler, and S. Preu, "2.75 THz tuning with a triple-DFB laser system at 1550 nm and InGaAs photomixers" *Journal of Infrared Millimetre Terahertz Waves,* vol. 36, no. 3, pp. 269-277.

[6] K. D. Irwin and G. C. Hilton, "Transition-edge sensors", *Cryogenic particle detection*, pp. 63-150, 2005.

Particle Swarm Algorithm Applied to Quadrature Hybrid Multi-Branch Directional Coupler Optimization for ALMA Band 3

Jorge Hernán Cárdenas *1, Óscar Restrepo[2], Fausto Patricio Mena[3], David Monasterio[4] and Germán Chaparro[1]

***Abstract*—We introduce a novel optimization method based on the evolutionary algorithm Particle Swarm Optimization (PSO). Simulations were conducted for the design of quadrature hybrids intended to operate in the 85-115 GHz band using a fully tuned and validated version of the algorithm. We present quadrature hybrids designs which are optimized with respect to operational requirements for the scattering parameters and amplitude imbalance. Furthermore, the resulting designs take into account machining constraints related to cost and feasibility requirements. This method can be easily extended to optimize other microwave devices and waveguides for radio astronomy applications, with the benefit of speeding up the design process as well as reducing the computational costs.**

***Keywords*—quadrature hybrid, directional coupler, optimization algorithm, amplitude imbalance, millimeter and submillimeter device.**

I. Introduction

Radio astronomy has highly demanding standards for microwave devices, often requiring ultra-wide band operability, high sensitivity, and noise rejection. For this reason, research into the design of microwave components for band widening within realistic manufacturing constraints is an ongoing effort [1][2]. In this paper we introduce a novel geometry optimization based on an evolutionary algorithm, that has been applied to quadrature hybrids operating in ALMA (Atacama Large Millimeter/submillimeter Array) Band 3 (85-115 GHz). Our reported results are part of an ongoing research project focusing on the design of a quadrature hybrid covering the ultra-wide ALMA Band 2+3 (67-115 GHz). Our approach to the problem of quadrature hybrid optimization is based on the exploration of the geometric parameter space of standard multi-branch couplers using the Particle Swarm algorithm.

II. Classical Quadrature Hybrids

Multi-branch couplers are widely used in heterodyne receivers. Special attention has been paid to standard designs consisting of two main waveguides connected by branches without further modifications to the guides [4]. Even though an eight-branch model has been proven to comply with the requirements for ALMA Band 3 [3], it has been proposed to increase the number of branches or to use overmoded waveguides to extend the operational band [2]. Further significant modifications have been introduced to the hybrid geometry by drilling extra cavities to the main guides, to add ripples to the signal and consequently reducing the amplitude imbalance at the center of the band [5]-[7]. We applied our method to 8-, 12- and 14-branch geometries, while also including extrusions in the main waveguide to assess their impact on the amplitude imbalance.

III. Quadrature Hybrid Optimization

Our optimization method is based on the Particle Swarm (PSO) evolutionary algorithm together with the Surrogate-Based Optimization (SBO) heuristic method. The algorithm seeks to find the waveguide and branches dimensions stored in a vector called a *particle,* that minimizes a fitness function (FF), while complying with any electromagnetic constraints. In this work we have used those set by ALMA, namely: a) scattering parameters S_{31} and S_{21} are expected to be close to -3 $\pm$0.5 dB, which also ensures keeping the amplitude imbalance below 1 dB [5], and b) the S_{11} and S_{41} parameters must stay below -15 dB. To allow the use of standard drilling tools, we set an additional constraint, defined as the ratio between the height of the main waveguide and the width of the branches, with a maximum value of 7.

As for hyperparameter and FF weights tuning, we settled on a scheme that we benchmarked against a well-defined and characterized quadrature hybrid designed for ALMA Band 5 (163–211 GHz) [9]. After we ran several optimization batches with this geometry as a starting point to fine-tune the algorithm and validate our resulting hybrids, we obtained an optimized geometry that improves the result obtained in [9] (see Fig. 1(a)).

[1]Institue of Physics, University of Antioquia, Medellín, Colombia (jorge.cardenas3@udea.edu.co) and (german.chaparro@udea.edu.co); [2]ECCI University and Universidad de Chile, Bogotá, Colombia (orestrepog@ecci.edu.co); [3]NRAO, Charlottesville, VA, USA (pmena@nrao.edu); [4]DAS Universidad de Chile, Santiago 8370451, Chile (David.monasterio@raig.uchile.cl).

NOTES:

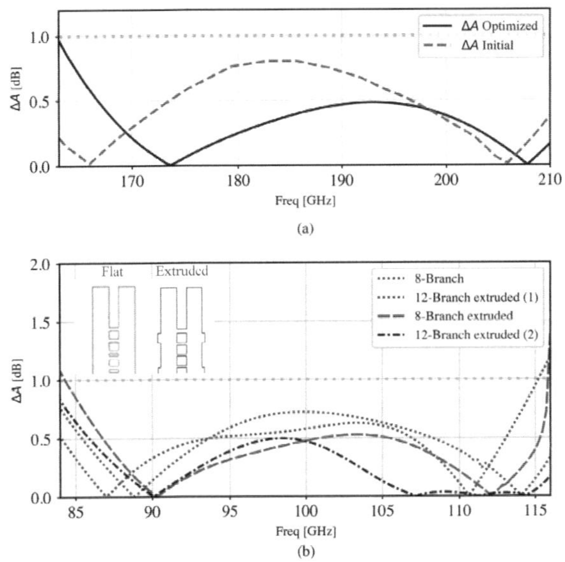

Fig. 1. Optimized quadrature hybrid designs (a) for ALMA Band 5 and (b) ALMA Band 3.

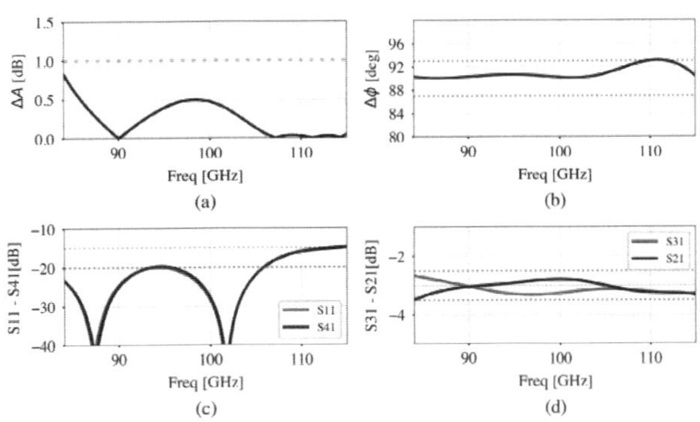

Fig. 2. Resulting (a) Amplitude imabalance, (b) phase imbalance and S-parameters (c,d) for a 12-branch quadrature hybrid incluiding extrusions optimized for ALMA Band 3.

IV. Results in ALMA Band 3

The resulting quadrature hybrid designs obtained for ALMA Band 3 using our optimization method show an excellent simulated performance vis-à-vis the S-parameter and amplitude imbalance requirements. Moreover, resulting geometries are found to be comparable or to improve upon prior results for ALMA Band 3 by fully covering the band with an amplitude imbalance below 1 dB. Fig. 1(b) shows the results for different standard geometries with amplitude imbalance under the expected value, with some hybrids staying below 0.5 dB. The 12-branch hybrid also shows a good S_{11}-S_{41} behavior (Fig. 2) thus an excellent amplitude imbalance contributing to an appropriate IRR [8], all that making it subject of interest for the further analysis and possible fabrication. We performed an additional optimization batch based on our 14-branch hybrid result aimed at widening the current band by an additional 10 GHz, yielding a band spanning from 80 GHz to 116 GHz with full compliance to the established constraints.

V. Conclusions

This novel optimization method provides validated and consistent results for quadrature hybrid designs that comply with electromagnetic constraints that would allow it to operate and improve upon current ALMA Band 3 designs. This method speeds up the design process and it can be extended to the development of other microwave devices for radio telescopes such as feedhorns, orthomode transducers (OMT) and even microstrips. We also found that standard geometries do not provide a viable solution to find a quadrature hybrid to fully cover the entire ALMA band 2+3, as we were only able to reach a solution compatible with (but not wider than) Band W under realistic manufacturing constraints. To achieve the whole coverage, attention must be paid to alternative and novel geometries that can be optimized using the tool herein proposed; this is part of our ongoing work.

References

[1] P. Yagoubov et al., "Wideband 67-116 GHz receiver development for ALMA Band 2," Astronomy and Astrophysics, vol. 634, no. 36777, pp. 1–23, 2020.

[2] A. Gonzalez, T. Kojima, K. Kaneko, and S. Asayama, "275-500 GHz Waveguide Diplexer to Combine Local Oscillators for Different Frequency Bands," IEEE Transactions on Terahertz Science and Technology, vol. 7, no. 6, pp. 669–676, 2017.

[3] S. Srikanth and A. Kerr, "ALMA Memo 343: Waveguide quadrature hybrids for ALMA receivers," ALMA Memo Series, vol. 9, January 2001.

[4] Reed, "The Multiple Branch Waveguide Coupler," IEEE Transactions on Microwave Theory and Techniques, vol. 6, no. 4, pp. 398–403, Oct. 1958.

[5] J. Ding, Y. Zhao, J.-X. Ge, and S. Shi, "A 90° Waveguide Hybrid with Low Amplitude Imbalance in Full W-Band," Journal of Infrared, Millimeter, and Terahertz Waves, vol. 40, no. 4, pp. 429–434, Apr. 2019.

[6] M. A. Hamid and M. M. Yunik, "On the Design of Stepped Transmission-Line Transformers," IEEE Transactions on Microwave Theory and Techniques, vol. MTT-15, no. 9, pp. 528–529, 1967, doi: 10.1109/TMTT.1967.1126525.

[7] H. Rashid, V. Desmaris, V. Belitsky, M. Ruf, T. Bednorz, and A. Henkel, "Design of Wideband Waveguide Hybrid with Ultra-Low Amplitude Imbalance," IEEE Transactions on Terahertz Science and Technology, vol. 6, no. 1, pp. 83–90, 2016.

[8] A. Khudchenko, R. Hesper, A. M. Baryshev, J. Barkhof, and F. P. Mena, "Modular 2SB SIS Receiver for 600-720 GHz: Performance and Characterization Methods," IEEE Transactions on Terahertz Science and Technology, vol. 7, no. 1, pp. 2–9, 2017.

[9] B. Billade et al., "Performance of the first ALMA band 5 production cartridge," 22nd International Symposium on Space Terahertz Technology 2011, ISSTT 2011, vol. 2, no. 2, pp. 56–56, 2011.

NOTES:

The effect of complex dispersion and impedance in the gain of superconducting traveling-wave kinetic inductance parametric amplifiers

Javier Carrasco[1,2,*], Daniel Valenzuela[1], Claudio Falcón[2], and F. Patricio Mena[1,3]

***Abstract*—Superconducting traveling-wave parametric amplifiers are a promising amplification technology suitable for applications in submillimeter astronomy. Their implementation relies on the use of Floquet transmission lines in order to create strong stopbands to suppress undesired harmonics. For design, amplitude equations are used to predict their gain. However, they do not take into account the complex dispersion and impedance that result from the use of Floquet lines, hindering reliable design. In order to overcome this limitation, we have used the multiple-scales method to include those effects. We demonstrate that complex dispersion and impedance have a stark effect on the gain, even suppressing it completely in certain cases. The equations presented here can, thus, guide to a better design and understanding of the properties of this kind of amplifiers.**

Keywords—Superconductors, parametric amplifiers, gain.

I. Introduction

Achieving larger bandwidths at the RF and IF bands, and improving receiver sensitivity are major challenges for future millimeter and submillimeter heterodyne observations. Consequently, extensive work is being performed in order to improve the performance of SIS junctions and HEMT amplifiers, the key components of current state-of-the-art receivers. However, this approach may soon hit fundamental limitations and improving bandwidths will eventually only be obtained at the expense of sacrificing noise temperature [1]. Recently, a promising superconducting technology that could overcome this problem has emerged [2]. It uses the kinetic inductance (KI) manifest in superconductors to produce parametric amplification in a long transmission line (TL). Devices working with this principle are dubbed Traveling-Wave Kinetic-Inductance Parametric Amplifiers (TKIPAs).

The KI modifies the wave-equation of the current through the TL by adding a nonlinear term which allows mixing of amplitudes when more than one monochromatic signal is injected [3]. Hence, it is possible to amplify the input signal if other tones, called pumps, are simultaneously injected. Nonetheless, more signals, including undesired harmonics, are also generated, deterring the amplification process. Eom et al. solved this problem by implementing a Floquet TL, conformed by a periodically repeating unit cell, creating stopbands that avoid the propagation of the main undesired harmonic of the pump signal. The use of such a line, however, translates into a TL with more intricate properties, namely a complex dispersion and impedance with strong frequency dependencies, particularly close to the stopbands.

In order to design TKIPAs, a nonlinear wave equation must be solved. This is usually done by approximating the process of amplitude gain as a dynamical evolution occuring at a much larger length scale than the wavelength of the involved signals. Within this approximation, not taking into account the complex nature of the Floquet TL, a set of nonlinear amplitude equations can be obtained [2]. An attempt to introduce a complex propagation constant in the process has been reported [4], but some unjustified approximations that are not valid near the stopbands were used.

We have tackled this problem by formally solving the nonlinear wave equation using the *multiple-scales method*, widely used in mathematical nonlinear physics, and especially useful in traveling-wave equations [5]. We demonstrate that the properties of the Floquet TL have a profound effect on the attainable gain, in particular when the pump signal is close to a stopband. Depending on the specific properties of the used Floquet TL and the amplitude and frequency of the pump signal, our equations depart notably from the predictions given by the traditional amplitude equations.

II. Amplitude Equations Including Complex Dispersion and Impedance

From the telegrapher's equations and the total inductance in a superconductor, $L(I) = L_0(1 + I^2/I_*^2)$, it is found that the dynamics of the electric current I through a TKIPA is described by a nonlinear wave equation,

$$\left(\frac{\partial^2}{\partial z^2} - CL_0\frac{\partial^2}{\partial t^2} - (CR + GL_0)\frac{\partial}{\partial t} - RG\right)I = \frac{L_0}{3I_*^2}\left(G\frac{\partial}{\partial t} + C\frac{\partial^2}{\partial t^2}\right)I^3, \quad (1)$$

where I_* determines the scale of the nonlinear term. C, R, G, and L_0 are the parameters per unit length of the TL used to implement the TKIPA, i.e., capacitance, resistance, conductance, and inductance at zero current, respectively.

Equation (1) can be solved with the multiple-scales method considering that the nonlinear term will produce a modification of the wave amplitudes at a much smaller rate than the wavelength. In order to correctly apply the method

[1]Electrical Engineering Department, University of Chile, Santiago, Chile; [2]Department of Physics, University of Chile, Santiago, Chile; [3]National Radio Astronomy Observatory, Charlottesville, VA, USA.

NOTES:

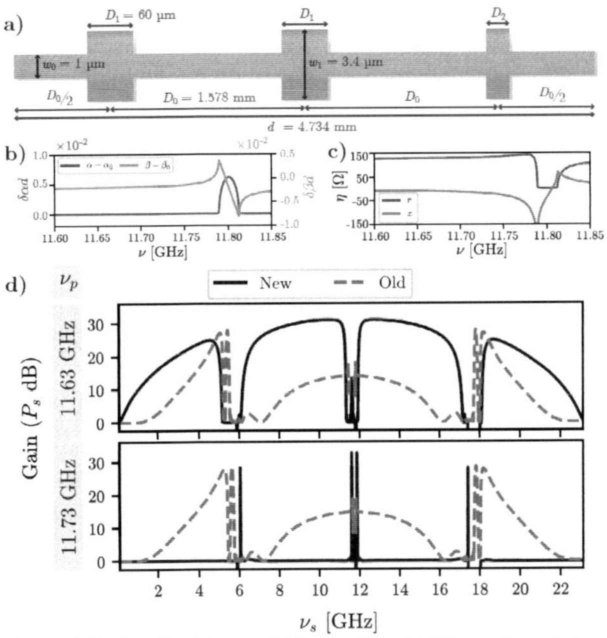

Fig. 1 (a) Unit cell of the used Floquet TL. (b) Dispersion relation near the 2nd stopband, where $\gamma_0 = \alpha_0 + j\beta_0$ is the propagation constant of the central line. (c) Complex impedance near the 2nd stopband. (d) Gain of the signal after traveling $z/d = 150$ unit cells for two values of pump frequency ν_p near the 2nd stopband, using an initial pump amplitude $A_p^0 = 0.2\, I_*$.

at first order, the terms in (1) must be balanced. The balancing depends on the magnitudes of C, R, G, L_0, the propagation constant γ, the frequency ν, and the initial magnitude of current, $|I(z=0)|$. However, the parameters of a Floquet TL depend strongly on ν, especially near the stopbands. This implies that the first order balancing of (1) gives different results at different frequencies. For achieving high amplification, the most relevant case occurs very close to the stopbands. Then, if the multiple-scales method is applied around this frequency and considering three propagating signals, *signal* (s), *idler* (i) and *pump* (p), we obtain

$$\frac{\partial A_{s(i)}}{\partial z} = jg_{s(i)}A_{s(i)} - 2\alpha_{s(i)}A_{s(i)} + \frac{jf_{s(i)}}{8I_*^2} \times \left[A_{s(i)}\left(\left|A_{s(i)}\right|^2 + 2\left|A_{i(s)}\right|^2 + 2\left|A_p\right|^2\right) + A_{i(s)}^* A_p^2 e^{j\Delta\beta z}\right], \quad (2.a)$$

$$\frac{\partial A_p}{\partial z} = jg_pA_p - 2\alpha_pA_p + \frac{jf_p}{8I_*^2}\left[A_p\left(2|A_s|^2 + 2|A_i|^2 + \left|A_p\right|^2\right) + 2A_p^* A_s A_i e^{-j\Delta\beta z}\right], \quad (2.b)$$

where $f_m = \frac{1}{2\beta_m}\left(\alpha_m^2 - \beta_m^2 - \frac{|\gamma_m|^2}{|\eta_m|^2}(r_m^2 - x_m^2)\right)$, $g_m = (\alpha_m^2 r_m^2 - \beta_m^2 x_m^2)/(\beta_m(r_m^2 + x_m^2))$, $m = s, i, p$, $\gamma = \alpha + j\beta$ is the propagation constant, $\eta = r + jx$ is the complex characteristic impedance, and $\Delta\beta = 2\beta_p - \beta_s - \beta_i$. Equations (2) differ from the model where α and x are neglected but, importantly, reduce to it.

III. Simulations and Discussion

We designed a CPW Floquet TL (Fig. 1.a) that has stopbands that allow suppressing undesirable harmonics and present a high non-linear dispersion relation $\delta\beta(\nu) \equiv \beta(\nu) - \beta_0(\nu)$ for operation near the stopbands. Fig. 1.b shows this effect around the 2nd stopband. Literature usually neglects x but Fig.1.c shows that it cannot be so since it significantly increases near the stopbands. Indeed, as shown in Fig. 1.d, very different gains of the target signal are obtained when the old (neglecting α and x) and the new model (given by equations 2) are compared. This figure shows two cases of different pump frequency. The case $\nu_p = 11.63$ GHz shows larger gain with the new model. This occurs because the total phase mismatch of the complex amplitudes stabilizes thanks to the jg_mA_m term in (2). The case $\nu_p = 11.73$ GHz, in contrast, shows almost no gain with the new model. The reason is the jg_mA_m term that dominates over the non-linear one, proportional to f_m.

These results show that amplification larger than predicted by the old model is possible, and that the pump frequency cannot be too close to the stopband in order to achieve amplification.

IV. Conclusions

We have presented a new set of amplitude equations for TKIPAs operating at frequencies near stopbands where, unlike other models, the effect of complex impedance and dispersion have been considered. To highlight the key differences between the models, simulations were performed for a CPW Floquet transmission line, showing that the new model can predict either larger or smaller gain than the traditional one, depending on how close to the stopband is the pump frequency. This happens because one of the new terms added to the amplitude equations is capable of stabilizing the phase mismatch, hence obtaining larger gain, but is also capable of dominate over the non-linear term responsible of amplification, since its magnitude depends on the frequency. Research to experimentally demonstrate these effects is underway.

References

[1] M. W. Pospieszalski, "On Extending the IF Bandwidth of ALMA Band # 6 SIS Mixers," Workshop 'The ALMA Vision: A Next Generation of Front End Receivers' (2021) [Online] Available: https://zenodo.org/record/5550951.

[2] B. H. Eom et al., "A wideband, low-noise superconducting amplifier with dynamic range," Nature 8, 623-627 (2012).

[3] S. Chaudhuri, J. Gao, and K. Irwin, IEEE Transactions on Applied Superconductivity **25**, 1 (2015).

[4] S. Zhao, S. Withington, D. J. Goldie, and C. N. Thomas, Journal of Physics D: Applied Physics **52**, 415301 (2019).

[5] A. H. Nayfeh and D. T. Mook, "Nonlinear Oscillations", John Wiley & Sons, Ltd (1995).

NOTES:

Gas Analysis with High-Resolution Fabry-Perot Spectrometry in the 550-650 GHz Range

Coralie Elmaleh*[1], Fabien Simon[1], Arnaud Cuisset[1], Anthony Roucou[1], Francis Hindle[1], and Gaël Mouret[1]

***Abstract*—The development of a high finesse THz cavity has enabled the construction of a Fabry-Perot spectrometer. This approach can be applied to the THz and submillimeter wavelength allowing the rotational transitions to be probed. The intra-cavity power levels have permitted the sub-Doppler Lamb-dip profiles of ethanol to be observed. The quantification of a trace of H2S is demonstrated with a detection limit around 60 ppb.**

Keywords—TeraHertz, Rotational Spectroscopy, CRDS, CEAS, Fabry-Perot, Lamb-dip.

I. Introduction

Cavity-Enhanced Absorption Spectroscopy (CEAS) and Cavity Ring-Down Spectroscopy (CRDS) are well established as very sensitive methods for the measurements in the infrared domain of gas phase compounds at trace level from their rovibrational signatures [1,2]. The recent successful development of a THz Fabry-Perot spectrometer shows that such techniques can be applied to the THz and submillimeter wavelength to measure rotational transitions. A high-finesse Fabry-Perot cavity working between 550 and 650 GHz has been built [3]. This 48 cm long terahertz resonator is composed of flat photonic mirrors with estimated reflectivity of 99,95% and a corrugated waveguide. Finesse values as high as 4000 have been obtained around 620 GHz. On the full spectral range, typical values are between 1600 and 4000 (Fig. 1).

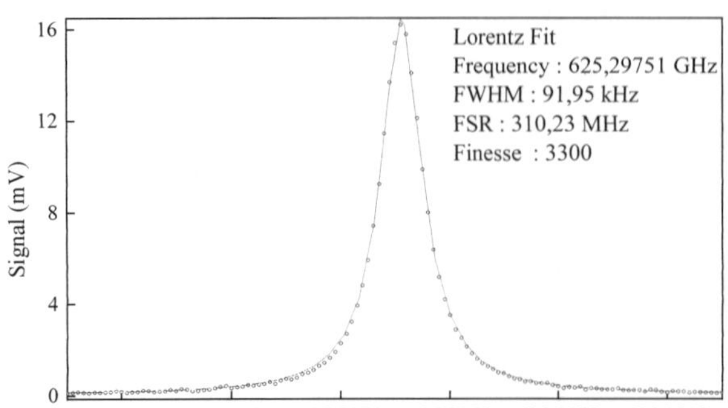

Fig. 1. Typical resonance mode of the THz cavity with a finesse of 3300 at 625.2975 GHz. (*Corresponding THz path* L_{eff} *~1.03 km*)

To obtain high-resolution spectra, the position of the cavity mode and emission frequency of the source must be simultaneously scanned. This is achieved by varying the cavity length by adjusting the position of photonics mirrors using piezo actuators. A PID control loop is established to lock the resonance of the cavity to the emission frequency. Then, two experimental protocols can be undertaken with this cavity assisted spectrometer, CEAS or CRDS mode as shown in the experimental setup (Fig. 2).

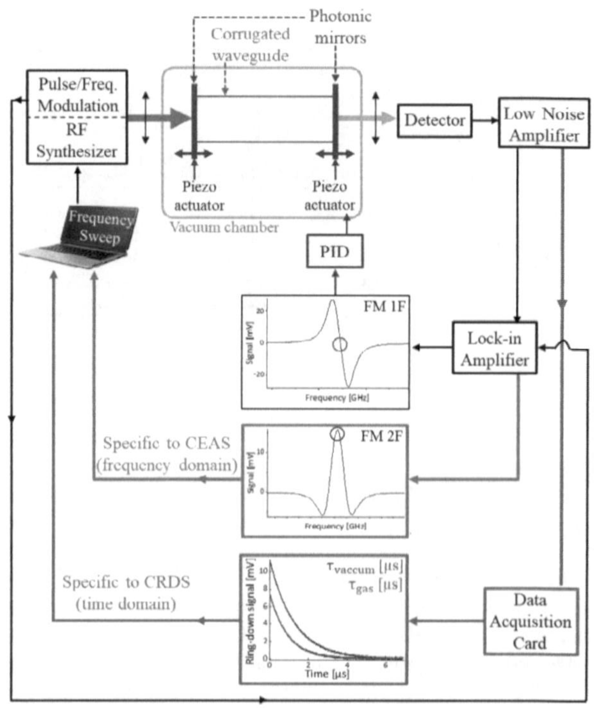

Fig. 2. Experimental setup (*in orange: setup specific to CEAS ; in blue: setup specific to CRDS*)

II. CEAS Results

With this setup, we scan the source frequency to measure the molecular absorption. The CEAS signal for a rotational transition of ethanol centered at 618.160 GHz is shown in Fig 3 (red line). This is confirmed by successive measurements of the cavity mode (black lines) that reduce in amplitude and broaden while passing the molecular absorption line. There is a strong correlation between the maximum of each resonance and the molecular signal obtained using a frequency modulation and detection at the second harmonic (FM-2F). We can also see that the finesse decreases once the absorption is observed. The finesse is representative of intracavity absorption.

The sub-Doppler Lamb-dip is visible close to the center of the frequency predicted at 618.160625 GHz by the Jet Propulsion Laboratory (JPL) for the $19_{06\text{-}14} \leftarrow 18_{05\text{-}13}$ rotational transition belonging to the transition state [4,5] (zoomed in Fig. 4). Instead of one individual rotational transition which is listed in the JPL database, a doublet of sub-Doppler lines is observed. The rotational-torsional

[1] *Université du Littoral Côte d'Opale, UR 4493, LPCA, Laboratoire de Physico-Chimie de l'Atmosphère, F-59140 Dunkerque, France.*
* *Corresponding author: coralie.elmaleh@univ-littoral.fr*

spectrum of ethanol is complicated with large amplitude motions involving -OH torsion [5] and $-CH_3$ internal rotation [6]. Such molecular motions produce tunneling splitting of rotational energy levels with different order of magnitude. Larger splittings are observed for -OH torsion with a barrier height of 400 cm^{-1}, leading to *gauche-*, *gauche+* and *trans* states. SubMHz line splittings are obtained for the torsion of $-CH_3$ with a barrier height of 1173.76 cm^{-1} for the *trans* state and 1331 cm^{-1} for the *gauche* states. As presented in Fig 3, subMHz line splittings associated to the methyl torsion were revealed by Lamb-dip measurements, completing the previous data obtained for $-CH_3$ internal rotation in *trans* state up to 600 GHz by J. C. Pearson *et al.*. These splittings have also been newly observed for the *gauche* states.

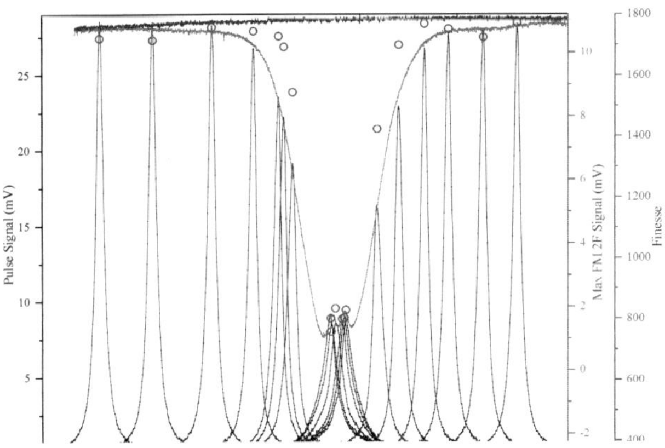

Fig. 3. CEAS of Ethanol (C_2H_5OH) P=3µbar, T=300K (red: molecular signal FM-2F; dark red: baseline; black: superposition of several resonances registered by moving the piezo actuators of the cavity; green: finesse of each resonance).

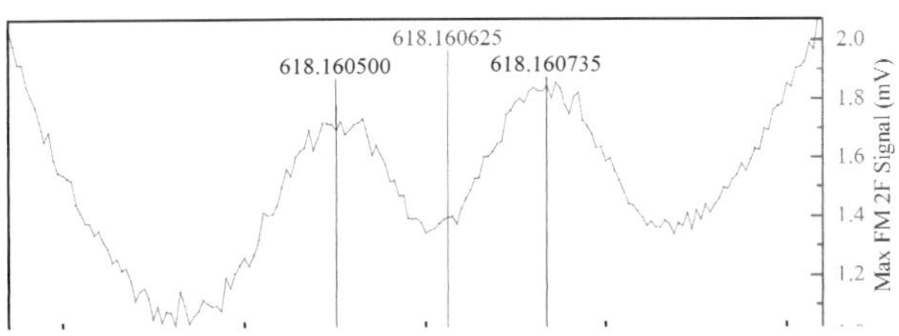

Fig. 4. Zoom on the absorption of the fig. 3. Predicted line frequency in the JPL spectroscopic database: f_{JPL}= 618 160 625 ± 5,3 kHz (transition 19_{06-14} <- 18_{05-13}, in red). F_{CEAS}: 618 160 500 kHz & 618 160 735 kHz (indicated in black)

III. CRDS Results

By recording the ring down time with and without gas sample, we can calculate the frequency-dependent absorption coefficient directly proportional to the absorbent concentrations. The Fig. 5 shows an example of an absorption profile for the $5_{05-00} \leftarrow 5_{04-01}$ rotational transition of trace of H_2S diluted in nitrogen. The H_2S concentration was determined to be 0.489 ± 0.059 ppm (2σ), and the gas input was 0.504 ± 0.010 ppm (2σ). It demonstrates that strongly polar compounds may be measured with a ppb level of concentration [7].

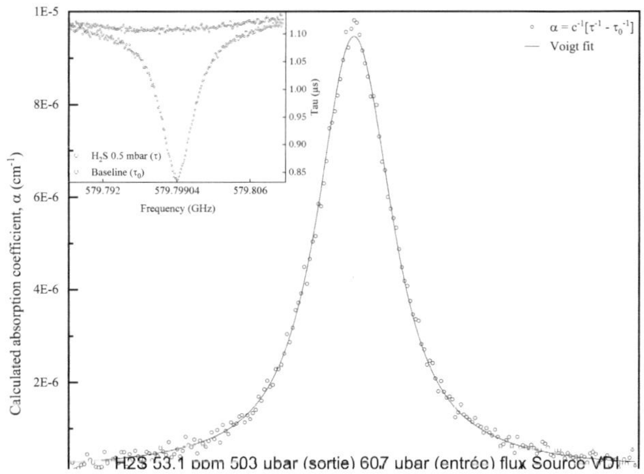

Fig. 5. Calculated absorption coefficient. Top-left: CRDS on the $5_{05-00} \leftarrow 5_{04-01}$ rotational transition of H_2S centered at 579,799042 GHz, measured in flux at a pressure of 504 microbars (and T=294K).

IV. Conclusion

We demonstrated the possibility to resolve hyperfine structures of rotational transitions using CEAS. High-resolution sub-Doppler saturated-absorption Lamb dips were measured at pressures of no more than 6µbar on rovibrational lines of C_2H_5OH from 550 to 620 GHz. With this technique, it is possible to completely resolve rotational lines with sub-MHz splitting. We also report the first THz-CRDS measurements above 600 GHz. A sub-ppm of calibrated H_2S mixture was easily detected.

With our setup we are able to quantify traces of gas at a level that can reach ppb, while being able to perform sub doppler measurements.

References

[1] E.R.T. Kerstel *et al.*, "A water isotope (^2H, ^{17}O, and ^{18}O) spectrometer based on optical feedback cavity-enhanced absorption for in situ airborne applications", *Appl. Phys. B* 85, 397–406, 2006.

[2] S. Kassi *et al.*, "Cavity ring down spectroscopy with 5 × 10^{-13} cm^{-1} sensitivity", *J. Chem. Phys.* 137, 234201, 2012.

[3] F. Hindle *et al.*, "Terahertz gas phase spectroscopy using a high-finesse Fabry–Pérot cavity", *Optica* 6, 1449-1454, 2019.

[4] H. M. Pickett, *et al.*, Submillimeter, millimeter, and microwave spectral line catalog. *J. Quant. Spectrosc. Radiat. Transfer*, 60: 883–890, 1998.

[5] J.C. Pearson, *et al.*, "The asymmetric top–asymmetric frame internal rotation spectrum of ethyl alcohol", *J. Mol. Spectrosc.* 251, 394-409, 2008

[6] J. C. Pearson, *et al.* "The Millimeter-and Submillimeter-Wave Spectrum of trans-Ethyl Alcohol." *J. Phys. Chem. Ref. Data,* 24(1), 1-32, 1995.

[7] C. Elmaleh *et al.*, "THz cavity ring-down quantitative gas phase spectroscopy," *Talanta*, 253, 2023.

A MKID-readout based on a heterogeneous, closely coupled architecture

Gerrit Grutzeck*[1], Ingo Krämer[1], Miroslaw Ciechanowicz[1], Nicolas Reyes[1], Carsten König[1], Andrey Baryshev[2], Stephen Yates[3] and Bernd Klein[1]

***Abstract*—Within this proceeding, we introduce the U-Board platform, a versatile platform for signal generation, acquisition and processing, based on a heterogenous processing architecture. Based on this platform we present a readout for Microwave Kinetic Inductance Detectors (MKIDs) for the A-MKID camera for APEX. In addition to the implementation of the readout on this heterogenous architecture, we also present a first comparison of the performance of the readout compared to the currently used readout of the A-MKID camera. Last but not least, we discuss how we plan to miniaturize the current prototype, which is based on commercial off the shelf components.**

Keywords—MKID, Backend, Readout, FPGA, GPGPU, Heterogeneous architecture

I. Introduction

Microwave Kinetic Inductance Detectors (MKIDs) have a great potential for large and sensitive detector arrays for use in sub-mm wavelength imaging and mid-resolution spectroscopy [1]. As these detectors are superconducting, notchfilters with a high-quality factor, they can be easily multiplexed in the frequency domain to allow for detector arrays with a high number of detectors [2]. Furthermore, the fabrication process is based on conventional optical or electron beam lithography, which leads to simple and reliable production. Also, the detectors can achieve photon noise limited sensitivity performance [3]. Due to the break-up of Cooper-Pairs by photons in the superconductor, the resonance frequency of MKIDs is a function of the photon-flux onto the detector.

To read out such detectors, the phase and amplitude changes of a readout tone placed close to the resonance frequency of the detector can be used. For this readout technique, the backend has to generate a tone for each detector, close to the detector's resonance frequency. The amplitudes and phases of these tones are then altered due to the interactions with the notch-filters of the detectors. At the end, each tone is analyzed again by the backend and the photon flux is determined based on the changes of signal, especially based on the phase change of the signal. Therefore, a backend for reading out MKIDs needs a phase coherent signal generation and acquisition combined with the processing power needed to generate and analyze the readout signal. The U-Board platform combines the strengths of General Purpose Graphical Processors (GPGPUs) and Field Programmable Gate Arrays (FPGAs), together with a phase stable signal generation and acquisition in the analog frontend. This platform is optimally suited to implementing a MKID readout.

The A-MKID instrument is a dual-color incoherent camera for the APEX telescope [4], which offers excellent observation conditions up to 300 µm based on MKIDs. The high frequency array is sensitive to 350 µm and the low frequency array is sensitive to 870 µm [5]. Both arrays are co-aligned and observe the same area of 15 arcmin x 15 arcmin of the sky. The nominal resonance frequencies of the detectors lie between 4 GHz and 8 GHz. In total the camera has 24 readout chains, where each chain hosts between 680 to 880 MKIDs.

The readout presented here is optimized for the A-MKID camera. Therefore, it aims for an analog bandwidth of 4 GHz for up to 1280 tones. The processing is based on a one million channel Fast Fourier Transformation (FFT), which is split between the FPGA and GPGPU. Furthermore, additional processing and reduction steps are migrated to the GPGPU to reduce the work-load on the controlling computer. It should replace the currently used readout of the A-MKID camera, which uses a heterodyne mixing scheme and lower resolution converters and therefore limits the performance of the camera.

II. The U-Board Plattform

In recent times, the performance of Digital-Analog-Converters (DACs) and Analog-Digital-Converters (ADCs) have not only improved strongly, but also the processing power of FPGAs and GPGPUs has increased a lot. This opens up new possibilities to build wideband readouts for MKID cameras, which can handle a thousand detectors per readout chain. Furthermore, the development of GPGPU modules for embedded systems, has removed the need for dedicated computers for hosting the GPGPU. This allows for compact heterogeneous architectures based on FPGAs and GPGPU modules. In the following, we will discuss the U-Board platform based on commercial off the shelf components as shown in Fig. 1, with an analog frontend optimized for the readout of the A-MKID camera.

A. Analog frontend and IF processing

To reduce the phase noise between a number of local oscillators (LOs), a homodyne mixing scheme is implemented. Therefore, the same LO is used for up and down conversion of the baseband signals. For this up and down converting IQ-mixers are used to reduce the required bandwidth of the DACs to 2.1 GHz, which allows the selection of 16 bit DACs. Furthermore, the IQ-mixers allow for a digital optimization of the sideband separation. The concept of the analog frontend and the IF processing is shown in the left part of Fig. 3.

For the signal generation, two DACs of the type AD9174 are used, each of which can generate a signal with a

[1]Max Planck Institute for Radio-Astronomy, Auf dem Hügel 69, 53121 Bonn, [2]Kapetyn Astronomical Institute, University of Groningen, The Netherlands, [3]SRON, Netherland Institute for Space Research, Groningen, The Nehterlands

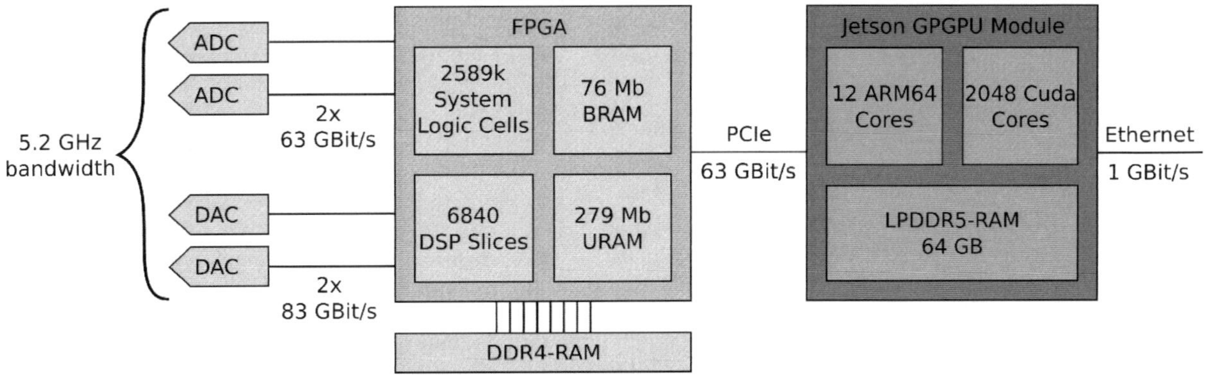

Fig. 1. The architecture of the U-Board prototype based on commercial off the shelf hardware. The DACs and ADCs are hosted on a custom PCB, which is connected to FPGA with high speed serial links (JESD204C). The FPGA (UltraScale+ Virtex 9) is from the high performance series of UltraScale+ FPGAs. Additional to the internal memory of the FPGA two 80-bit wide DDR4 memories are associated with the FPGA. The FPGA is finally connected to the Jetson GPGPU Module via a PCIe connection.

resolution of 16 bit with an instantaneous bandwidth of up to 3.08 GHz. The signal acquisition is done with a single ADC12DJ5200RF, which can sample an instantaneous bandwidth of up to 5.2 GHz with a resolution of 12 bit. The ADC is used in dual channel mode, where each channel has an instantaneous bandwidth of up to 2.6 GHz and typically an effective number of bits of 8.7.

B. Heterogeneous architecture

Both FPGAs and GPGPUs have their own strengths and weaknesses. FPGAs are per definition hard real-time capable and have flexible connection options. The combination of hard real-time capabilities and flexible interfacing options make possible to connect high-speed DACs and ADCs electrically and protocol wise. Furthermore, FPGAs have good integer processing power. But, their floating-point processing power and memory size are limited. Nether the less, the onboard memory of FPGAs has a bandwidth with very low latencies. GPGPUs on the other hand, do not have hard real-time capabilities, but need carefully designed buffers to ensure soft real-time capabilities. This behavior, is due to the fact, that GPGPUs can only be used in combination with an operating system. But, compared to FPGAs, GPGPUs deliver a higher floating-point processing power and have huge memory space. These memories also have a high bandwidth (not as high as the memory on FPGAs), but the latency of the memory is much higher. Last but not least, the GPGPU have a programming model, which is much closer to conventional software programming compared to the hardware description language used for FPGAs. This allows for faster implementations of new algorithms and ideas.

The idea of a heterogeneous architecture, based on FPGAs and GPGPUs, is to combine the strengths of both architectures. Two possible interconnections can be used to connect FPGAs and GPGPUs; high speed Ethernet (e.g. 100 Gbit/s) and PCIe. Ethernet can be set up for multicast applications, where one source sends data to multiple sinks, while PCIe is used for more point-to-point like connections. The disadvantage of high-speed Ethernet is, that the infrastructure costs (e.g. fiber links, switches, network cards) is significant. Therefore, high speed Ethernet has advantages only for applications, which rely on multicast data transfers (e.g. beam former). In heterogeneous architecture based on a PCIe interconnect, the FPGA and GPGPU are typically hosted in a powerful computer, as neither the FPGA nor the GPGPU are fully suited as PCIe host. This approach is again raising additional costs due to the requirements to host a FPGA and a GPGPU in a computer. Furthermore, a computer per FPGA is not compatible with the space requirements of typical backends, as the space close to the instrument is limited. An alternative to using a conventional computer to host both, the FPGA and the GPGPU, are the Jetson modules produced by Nvidia. These systems on a module include a multicore ARM CPU, a high performance GPGPU and a huge, high bandwidth memory region. These Jetson modules can be used as a PCIe host to which the FPGA is connected. This allows for a simple and fast point to point connection between the FPGA and the GPGPU.

To speed up the development time, we first developed a prototype mostly based on commercial off the shelf components, which is shown in Fig. 2. Only the analog frontend with the DACs and ADCs is hosted on a custom printed circuit board. This analog frontend is connected with a high-speed connector (FMC+) to the evaluation board VCU118 from Xilinx. From the hardware present on this evaluation board, we utilize FPGA, an UltraScale+ Virtex 9, as well as one of the DDR4-Memory banks for the MKID readout. The FPGA evaluation board is then connected to a

Fig. 2. The Picture of the prototype based on the commercial off the shelf hardware. In the foreground, the analog frontend, based on a custom PCB, is visible. This frontend is connected with a high-speed connector (FMC+) to the evaluation board VCU118 from Xilinx, which hosts the FPGA and the associated DDR4 memory. In the background, the GPGPU development Kit from Nvidia, with the Jetson GPGPU Module, is visible. This module is connected to the FPGA via PCIe.

development Kit from Nvidia with 8 lanes of PCIe 3. The development Kit from Nvidia is built around the Jetson Orin AGX module, which supplies the platform with 2048 CUDA cores (the GPGPU part of the module), 32GB of memory and 12 CPU cores, for housekeeping and controlling.

III. Readout

In the following we will discuss how the readout is implemented based on the platform described above, and then give a preliminary comparison to the currently used readout for the A-MKID camera.

A. Implementation of the readout

The readout is based on a one million point FFT, which is split into a first stage on the FPGA and a second stage on the GPGPU. Therefore, the possible frequencies of the readout tones are limited to a finite grid with an approximate spacing of 4 kHz. The concept of the readout and its division onto the FPGA and GPGPU is shown in the right part of Fig. 3.

The waveform for the signal generation via the DACs is streamed via the FPGA from the DDR4-memory to the DACs. While setting up the readout, this waveform is generated on the GPGPU based on the determined positions of the MKIDs. The generation of the waveform also takes into account a frequency dependent calibration of signal path from the DACs to the IQ up converter, to compensate for differences between the I and Q paths, as well as non-perfection in the IQ-mixer itself. Furthermore, blindtones are added to the waveform, which are used to correct for amplitude and phases changes due to changes of the system (e.g. drifting reference voltages, changes in propagation delay).

The real-time analysis of the digitized signal is split between the FPGA and the GPGPU. The first processing step is an 1M (2^{20}) point FFT. This FFT is split between the FPGA and the GPGPU, as described in the following. On the FPGA, 16 parallel FFTs of size of 64k are calculated in fixed point (see 16x 64k FFT in Fig. 3). Then the frequency bins of this FFTs, which contain the tones, are selected and transferred to the GPGPU via PCIe (see Channel Mask in Fig. 3). This reduces the data rate, which has to be transferred via PCIe, and further reduces the workload on the GPGPU. On the GPGPU the results from the 16 parallel FFTs are combined into the result of the 1M point FFT with a modified Discrete Fourier Transformation, which takes the calibration for the sideband separation of the down converter into account. This result corresponds to the complex S21 value of the system. The results of the consecutively calculated FFTs are interpreted as time-data with a sample rate of $\frac{4\text{ GHz}}{2^{20}} = 4.01\text{ kHz}$, which carry the changes of the photon flux on the detectors. In the next step, the data rate of this time-data is further reduced with a configurable filter (configurable length and coefficients) and a down sampling stage. The down sampling stage allows for an integer down sampling ratio; therefore, the possible sampling rates are $\frac{4.01\text{ kHz}}{n}$ with $n \geq 1$.

To reconstruct the phase change of the readout tone with respect to the MKID resonance, the FFT channels are first divided by the blindtones. This enhances the stability of the system as common, unwanted effects (e.g. drift of voltage references, changes of the propagation delay) are compensated for. In the next step, the calibration of the MKID is applied to map the complex S21 to the phase change with respect to the resonance. This phase change can be used to extract the frequency shift of the resonance, especially as the relationship is linear for small frequency shifts, which is the operation point of the A-MKID camera. All these reconstruction steps are implemented on the GPGPU. Finally, the data is streamed out of the Jetson Module to the controlling computer.

The readouts for one camera color are controlled by a central software instance on a computer (as shown in Fig. 5). All readout chains are connected via a high-speed Ethernet (e.g. 40 Gbit/s) to the controlling computer. Over this connection the chains are controlled, and the reduced data is sent from the chain to the computer, where all data streams are merged. The merged data stream is then sent out via another high-speed Ethernet connection to the telescope infrastructure.

B. Performance of the readout

In the following, we will briefly compare the performance of the readout based on the U-Board platform with the performance of the currently used readout for the A-MKID camera. The currently used readout uses a more complicated

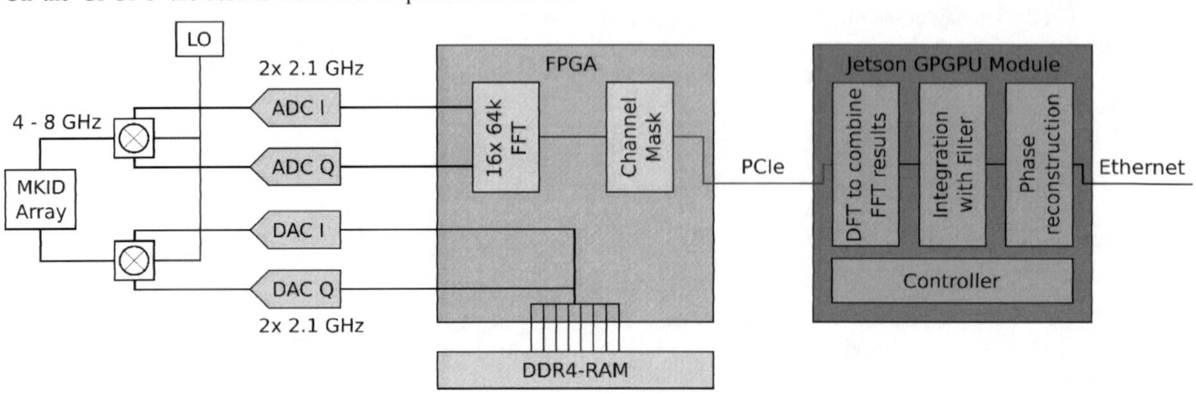

Fig. 3. The concept of the implemented MKID readout based on the U-Board for a single chain. The IF processing and analog frontend uses a homodyne mixing schema based on IQ-mixers to deliver 4GHz of usable, instantaneous bandwidth for reading out the MKIDs in the detector array. The signal with the readout tones is streamed from the DDR4-Memory associated with the FPGA. For the signal analysis, the first step is a one million point FFT split between the FPGA and GPGPU. Here only the spectral channels, which include a tone, are transferred and processed on the GPGPU. To reduce the data rate the complex results of the FFT are downsampled with a configurable antialiasing filter. Finally, the phase change due to the shifting of the MKIDs is extracted and send out via Ethernet.

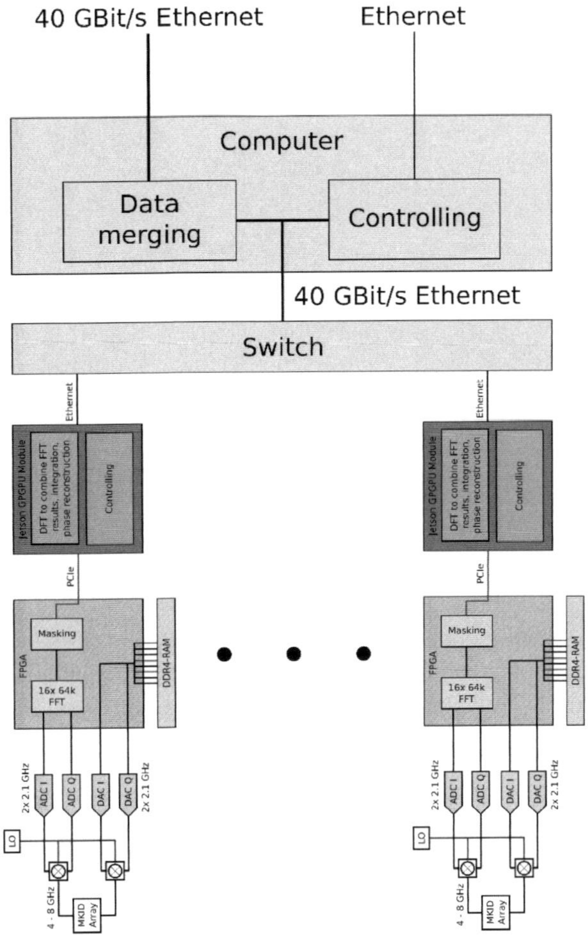

Fig. 5. The concept to combine a number of readout chains into a single instrument. For instruments with more than a single chain, like the A-MKID camera, the controlling of all chains and the data from all chains has to be centralized. As all the online reduction of the data is implemented on the heterogeneous U-Board, the data coming from the readout chains only has to be merged into a single data stream, which is send out to be saved. This reduced the requirements on the controlling computer enormously, as no demanding data reduction has to be done on it.

heterodyne mixing approach. Furthermore, the resolution of the DACs and ADCs is smaller. Particularly, the ADC only has 7.6 effective number of bits. Therefore, we expect a better performance of the presented readout compared to the readout currently under use.

In order to compare the performance of the two readout systems, measurements with the identical chain of detectors and the identical optical path are executed. The performance is measured with a noise scan, which is calibrated with a wirescan. For the noise scan, the camera is pointing at a bucket with liquid nitrogen, which is used to simulate the sky at APEX. The wirescanner moves two cables (one in x- and on in y-direction) in front of the liquid nitrogen through the field of view of the camera. These cables generate a signal of reproducible and known strength, which is used to calibrate the noise amplitude in the noise scan. As these measurements are done with a non-final optical filter stack and under laboratory conditions, the results are only preliminary.

In Fig. 4 the measured histograms, one for each readout,

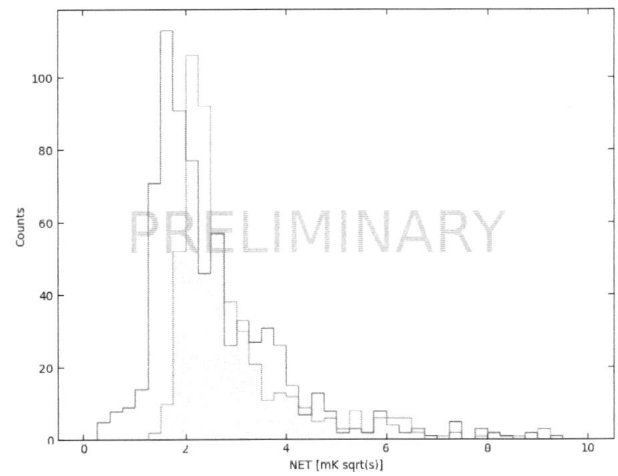

Fig. 4. Comparison of the sensitivities of the current readout with the sensitivities measured with the readout based on the U-Board platform. The green histogram represents the measured sensitivities with the readout based on the U-Board platform, and the blue one represents the sensitivities measured with the currently used readout.

of the sensitivities of one chain of the camera are shown. The green histogram represents the readout based on the U-Board platform, as presented here. The blue histogram shows the performance of the currently used readout. Two important aspects are visible in this figure. Firstly, the readout based on the U-Board platform gives a better sensitivity. This is expected, as the currently used readout is limiting the camera performance and the U-Board based readout uses ADCs with better performance. Secondly, the yield of MKIDs is higher. This is one of the benefits of using a side band separating, homodyne mixing scheme instead of a non-sideband separating, heterodyne mixing scheme. The separation of side bands reduces the number of MKIDs, which are close to each other in the baseband. These close MKIDs are not suitable for operation, as the crosstalk between them is significant.

IV. Outlook and Conclusion

We have presented a MKID readout, based on a 1M FFT, with a bandwidth between 4 GHz and 8 GHz, which is implemented on the U-Board platform, a heterogeneous architecture. The various processing architectures in the U-Board allow for a complete implementation of the online reduction pipeline on the backend. This reduces the requirements regarding the central, controlling computer. The choice of an FFT based approach, with 1M points, allows for a high maximal sample rate of 4 kHz and a fine grid of possible readout tone positions (also 4 kHz). Furthermore, the prototype based on the U-Board architecture for the A-MKID readout is performing better than the currently used readout, both in terms of sensitivities and in terms of yield. But the final performance of the readout compared to the current one has to be demonstrated on the sky.

As the U-Board platform is flexible, not only can a MKID readout be implemented on it, but also other types of backends, a high-resolution spectrometer for example.

The prototype based on commercial off the shelf components has no compact form, we are developing a more integrated solution. This new solution is based on a complete custom printed circuit board, which includes the analog frontend, an UltraScale+ Kintex 15 FPGA, the DDR4-memory associated with the FPGA and the connector for the Jetson module. As the analog frontend and the signal processing is the same on the new solution, as on the commercial off the shelf components prototype, we expect the same analog performance, with less space and power consumption.

REFERENCES

[1] P. Day, et al., "A broadband superconducting detector suitable for use in large arrays", *Nature*, vol 425, no. 6960, pp 817-821, Oct. 23, 2003.

[2] J. J. A. Baselmans, et al., "A kilo-pixel imaging system for future space based far-infrared observatories using microwave kinetic inductance detectors", *Astronomy & Astrophysics*, 29653, 2016.

[3] R. M. J. Janssen et al., "High optical efficiency and photon noise limited sensitivity of microwave kinetic inductance detectors using phase readout," *Appl. Phys. Lett.*, vol. 103, no. 20, 2013.

[4] R. Güsten, et al., "The Atacama pathfinder experiment (APEX), a new submillimeter facility for southern skies", *A&A*, 454(2): L13–L16, 2006.

[5] L. Esteras, "The optical system and the astronomical potential of A-MKID, a new camera using microwave kinetic inductance detector technology" Ph.D. dissertation. University of Bonn, 2014.

Development of an FPGA-Based, 1024-Channel Spectrometer with Individual Channel Lock-in Amplification

Jonathan R. Hoh[1], Mona Jarrahi[2], and Christopher Groppi[1]

Since the mid 1930's, the concept of a lock-in amplifier has been put into practice as an efficient and mathematically simple method for extracting weak signal amplitudes and phases from noisy environments. Using a periodic reference, lock-in amplifiers (LIAs from here on out) combine homodyne detection schemes with low-pass filtering to extract signals in a defined frequency band around the reference. In effect, this allows detection systems to reject all other frequency components that do not fall within the reference band, greatly reducing integrated noise power and boosting SNR.

Given the impressive ability to pick out faint signals from noisy measurements, it is no surprise that LIAs have found a great amount of use in the astronomy community, especially in higher frequency microwave systems where avoiding environmental noise is a constant battle. At their core, LIAs need only the ability to perform quadrature multiplication with their reference signals to downmix the tone of interest to baseband then apply an adjustable low-pass filter. The relatively simple engineering of LIAs has allowed them to maintain their place as a staple in mm-wave astronomy even as more complex methods of signal extraction such as the FFT and PFB have become the standard.

The primary downside of using LIAs as readout systems for astronomical observation is the fact that each LIA typically only extracts signal information from a single band located around a single reference tone. Analog LIAs, while effective and inexpensive, are not a space or power efficient means of examining a spectrum of information unlike the FFT which is able to output thousands of spectral amplitudes in a matter of milliseconds. At this point in time, LIAs and FFT-based spectrometers are typically seen as two different approaches to reaching the objective of extracting signal power from a noisy environment. Until now there has been little research done on combining the abilities of LIAs and FFT spectrometers into a single system.

Here we present the design, justification, and preliminary results for a 1024-channel spectrometer that employs lock-in methodology to further remove systematic noise from its output readings. By creating an FPGA-generated square tone which is converted to analog output and fed into a fast-switching RF diode, incoming noisy signals can be modulated by a precisely known chopping frequency. This up-converted signal then enters the FGPA through a digitizer and is passed through an FFT or PFB which separates spectral components into bins. The act of chopping the signal at a frequency slightly slower than the FFT output frequency (in this case, the ADC sampling speed divided by the FFT length) pushes all important information into a higher frequency band that is protected from systematic noise and 1/f pink noise.

Once the data leaves the FFT, it will be in one of two states depending on the phase of the chopping signal plus a small amount of digital delay. This results in each bin of the FFT to oscillate at the chop frequency which can easily be demodulated as the original chopping clock is created internal to the FPGA. Once demodulated, the resulting product is passed through an integrator to be low-pass filtered leaving only the DC portion of the remaining spectrum for each bin. Accumulated bins are stored in BRAM and fed to an external hard drive through ethernet where they can be used for spectral analysis of the signal of interest.

1. The Arizona State University, Tempe, AZ 85282 USA.
2. UCLA, Los Angeles, CA 90095 USA

NOTES:

Design and Characterisation of a 3.5-THz Fundamental Schottky Mixer

Divya Jayasankar[1], Nick Rothbart[2], Heiko Richter[2], Heinz-Wilhelm Hübers[2], Vladimir Drakinskiy[1], Peter Sobis[1,3], and Jan Stake[1]

Abstract—**Mixers operating at ambient temperatures are essential for realisation of THz heterodyne receivers for long lifetime space missions. In this paper, we report the design, circuit and block fabrication, assembly and characterisation of a 3.5-THz fundamental Schottky diode mixer. Design of planar, integrated mixer circuit realised using a suspended stripline technology on a 2-μm GaAs substrate and video signal detection results are presented.**

Index Terms—**Fundamental mixers, Heterodyne receivers, Schottky diodes, Terahertz electronics**

I. Introduction

High-resolution molecular spectroscopy at terahertz (THz) frequencies is a valuable tool for understanding the chemical composition of the Earth and other planetary atmospheres. In particular, the detection of gas species such as atomic oxygen (OI) at 4.7 THz [1] and hydroxyl radical at 3.5 THz in the least explored atmospheric regions can improve the climate and weather prediction models. With the recent unprecedented progress of quantum-cascade lasers (QCLs), the realisation of THz heterodyne spectrometers is now feasible. QCLs can provide a few mW of output power and operate in continuous-wave (CW) mode, thereby making it an ideal candidate as a local oscillator (LO) source for THz receivers [2]. Overall, there is a need for development of compact, reliable receivers up to 5 THz to enable high-resolution molecular spectroscopy.

To meet the scientific objectives and instrumentation requirements of future earth observation missions, frequency converters operating at ambient temperatures are preferable. Detectors based on Schottky diode technology are workhorses behind high-altitude balloons and space-borne missions such as ODIN, Microwave Instrument for Rosetta Orbiter (MIRO) and Jupiter icy moons explorer (JUICE) mission [3]. It offers a wide range of intermediate frequencies (IF), fast response time, and can work in ambient temperatures, thereby eliminating the need for cryocoolers [4].

Integrated and planar Schottky diode technology developed during the 90s opened up for more reliable and advanced terahertz mixers and multipliers compared to previous whisker-contacted devices [5]. In 1999, Siegel et al. demonstrated membrane-based technology that allows for low electrical parasitics and operation at terahertz frequencies [6]. Based on this monolithic membrane diode technology, the first planar 2.5 THz Schottky diode heterodyne receiver was developed for a NASA spaceflight mission using a CO_2-pumped methanol far-infrared laser. Based on this, we have developed a fabrication process for realised supra-THz integrated mixer circuits.

This paper reports the design, circuit and E-plane split block fabrication and assembly of the 3.5-THz planar, single-ended Schottky diode mixer in section II. Followed by the video detection experiment results in section III.

[1]Chalmers University of Technology, SE-412 96 Gothenburg, Sweden; [2]German Aerospace Center, Berlin 12489, Germany; [3]Low Noise Factory AB, SE-412 63 Gothenburg, Sweden.

II. Method

Fig. 1 shows the full 3D-EM model of the 3.5-THz fundamental mixer implemented in a finite element method (FEM) solver. The incoming radiation from the 3.5-THz QCL is coupled to the mixer by a diagonal horn integrated into a WM-64 rectangular waveguide (64 μm × 32 μm). The signal is coupled to the diode via an E-plane probe. To present the optimum RF/LO embedding impedance to the diode and to have maximum energy coupling, geometry optimisation and backshort tuning was carried out. To prevent RF signal leakage into the IF chain, a hammer-head filter was designed [7]. The rounded corners arising from the milling process is taken into consideration in the 3D-EM model design. To ensure only the fundamental quasi-TEM mode exists in the RF channel, the height and the width of both top and the bottom channels were optimised.

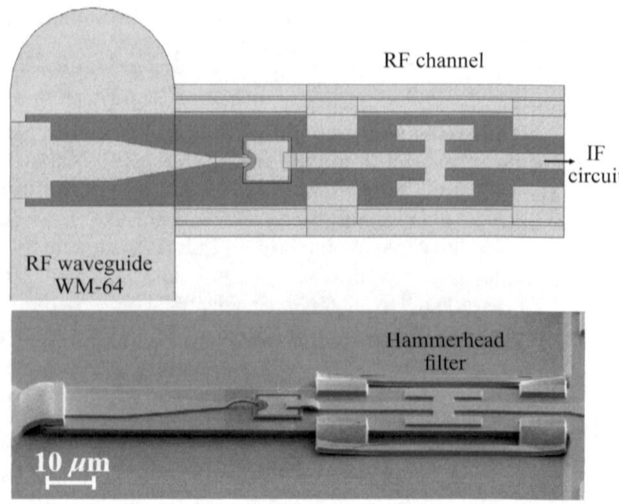

Fig. 1. 3.5-THz Schottky fundamental mixer. a) 3D-EM model of the fundamental mixer showing the rectangular waveguide WM-64 and the RF channel which houses the Schottky diode and the hammerhead filter. b) Scanned electron micrograph (SEM) of the integrated mixer circuit realised on a 2-μm GaAs substrate.

To realise THz components above 2 THz, minimising the circuit losses and parasitics is crucial. Hence, the 3.5-THz fundamental mixer circuit was realised using suspended stripline technology in a 2-μm thick GaAs membrane. An epi-layer doping concentration of about 6×10^{17} cm^{-3} was chosen to reduce parasitics and to operate below the cut-off frequency. It is followed by a highly doped buffer layer with a concentration of about 5×10^{18} cm^{-3}. E-beam lithography was used to define each pattern. The dc performance of the circuits were evaluated before the circuits were released and th ... Tl ... to ...

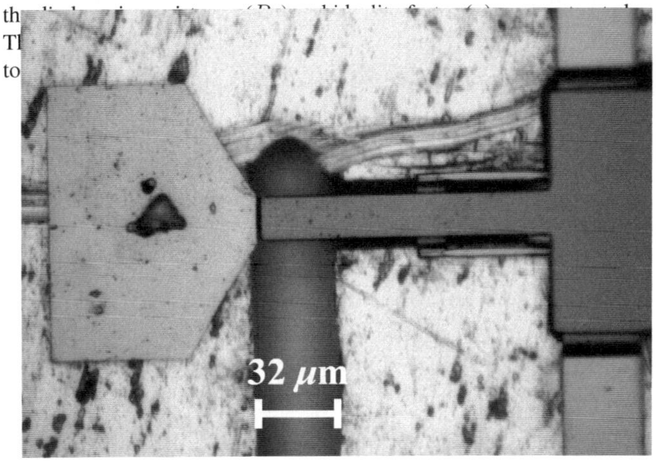

Fig. 2. Photograph of the integrated 3.5-THz Schottky diode mixer assembled in an E-plane Aluminum split block sputtered with 1-μm-thick gold. Note: The circuit is mounted upside down.

Fig. 2 shows the photograph of the fundamental mixer circuit assembled in an E-plane split-block machined using a high-precision Kern CNC machine at Chalmers University of Technology. The aluminum blocks were cleaned, etched and sputtered with 1-μm gold.

III. Results

Photograph of the experiment setup at the German Aerospace Ce...

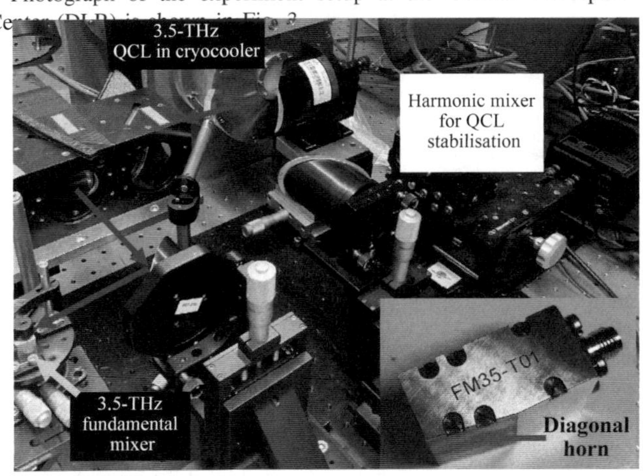

Fig. 3. Picture of measurement setup at the German Aerospace Center, Berlin. Inset: Fully assembled 3.5-THz fundamental mixer.

An off-axis parabolic mixer was used to focus the THz signal from the QCL. The signal was split using a beam splitter. Maximum power from the QCL was directed to the fundamental mixer via a Martin-Puplett diplexer which will be later utilised for noise temperature measurements. Whilst the other part of the signal will be sent to the Schottky harmonic mixer for QCL frequency stabilisation [8]. The total path length from the QCL to the fundamental mixer was about 70 cm and the QCL power at the mixer interface measured using a Thomas Keating power meter was about 5 mW.

For video detection, a mechanical chopper operating at 15 Hz was used for signal modulation and the video signal was observed using an oscilloscope and lock-in amplifier [9]. Fig. 4 shows the recorded video signal versus applied dc-voltage and the corresponding *I-V* response of the diode. The current safety limit was set to 1 mA. Noise temperature measurements and integration of harmonic mixer to this breadboard is currently under progress. We request for a full paper ...

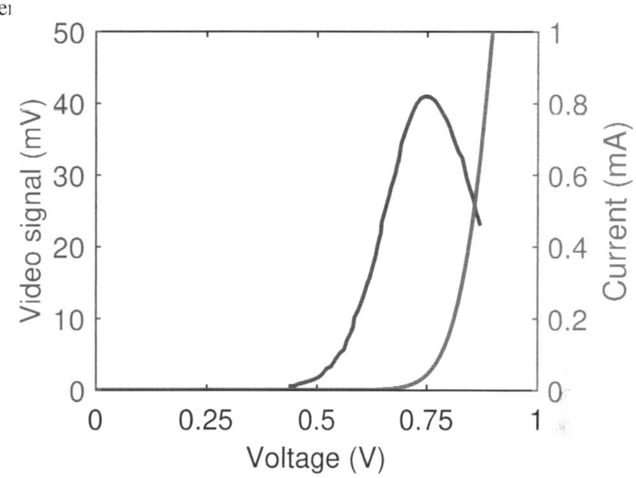

Fig. 4. Video detection. (Left axis) Video signal recorded using a lock-in amplifier and an oscilloscope. The peak signal was observed at 0.75 V. (Right axis) *I-V* response of the Schottky diode with 0.11 μm^2 contact area.

IV. Acknowledgement

The authors would like to acknowledge Dr. Holger T. Grahn, Dr. Lutz Schrottke, Dr. Xiang Lü, PDI for the QCL. Thanks to Dr. E. Saenz, ESTEC for fruitful discussions. Thanks to M. Myremark for precision machining of mixer blocks.

References

[1] H. Richter *et al.*, "Direct measurements of atomic oxygen in the mesosphere and lower thermosphere using terahertz heterodyne spectroscopy," *Commun. Earth Environment*, vol. 2, no. 1, Jan. 2021.

[2] L. Schrottke *et al.*, "High-performance GaAs/AlAs terahertz quantum-cascade lasers for spectroscopic applications," *IEEE Trans. THz Sci. Technol.*, vol. 10, no. 2, pp. 133–140, Mar. 2020.

[3] P. Sobis *et al.*, "SWI 1200/600 GHz highly integrated receiver front-ends," in *36th ESA Antenna Workshop on Antennas and RF Systems for Space Science, ESA/ESTEC ,Noordwijk, The Netherlands*, 2015.

[4] D. Cuadrado-Calle *et al.*, "Solid-state diode technology for millimeter and submillimeter-wave remote sensing applications: Current status and future trends," *IEEE Microwave Magazine*, vol. 23, no. 6, pp. 44–56, 2022.

[5] I. Mehdi *et al.*, "THz diode technology: Status, prospects, and applications," *Proc. IEEE*, vol. 105, no. 6, pp. 990–1007, Jun. 2017.

[6] P. H. Siegel *et al.*, "2.5-THz GaAs monolithic membrane-diode mixer," *IEEE Trans. Microw. Theory Techn.*, May 1999.

[7] D. Jayasankar *et al.*, "Design and development of 3.5 THz Schottky-based fundamental mixer," in *2020 50th European Microwave Conference (EuMC)*, 2021, pp. 595–598.

[8] D. Jayasankar *et al.*, "A 3.5-THz, ×6-Harmonic, Single-Ended Schottky Diode Mixer for Frequency Stabilization of Quantum-Cascade Lasers," *IEEE Trans. on Terahertz Science and Technology*, 2021.

[9] H.-W. Hübers *et al.*, "Video detection and mixing performance of GaAs Schottky-barrier diodes at 30 THz and comparison with metal-insulator-metal diodes," *Journal of Applied Physics*, vol. 75, no. 8, pp. 4243–4248, Apr. 1994.

Experimental characterisation of titanium nitride transmission lines for applications as kinetic inductance travelling wave parametric amplifiers

Joseph C. Longden*[1], Faouzi Boussaha[2], Christine Chaumont[2] and Boon-Kok Tan[1]

***Abstract*— Travelling wave parametric amplifiers (TWPAs) made from highly nonlinear reactive superconducting thin films have been demonstrated to be a potentially viable quantum-noise-limited amplifier technology for various fundamental physics platforms, including microwave/mm/sub-mm astronomy, dark matter search experiments, neutrino mass experiments and qubit readout. To date, only a limited number of successful kinetic inductance TWPA devices have been reported, with the majority fabricated from the same material, niobium titanium nitride (NbTiN), although in principle any highly nonlinear superconducting film can be used for kinetic inductance TWPA fabrication. Here, we present a detailed analysis of titanium nitride (TiN) transmission lines, to ascertain their suitability for use as kinetic inductance TWPAs. We will experimentally characterise our transmission line structures at cryogenic temperatures and compare the results with electromagnetic simulations. This characterisation and analysis would allow us to understand the advantages and limitations of TiN films, and whether they are suitable for applications as kinetic inductance TWPAs.**

Keywords— ***Low-noise, Amplifier, Superconducting, RF, Microwave, Travelling-wave***

I. INTRODUCTION

Kinetic inductance travelling wave parametric amplifiers (KITWPAs) [1, 2] are quantum devices, which can achieve high gain over broad bandwidth. They exhibit quantum-limited noise performance with negligible heat dissipation and their ease of fabrication makes them readily scalable to arrays for large pixel count applications, such as the readout of astronomical detector arrays.

The operation of a KITWPA, is reliant on the inherent non-linearity of the kinetic inductance of thin superconducting films. The majority of KITWPAs reported in the literature comprise a long superconducting transmission line patterned with a niobium titanium nitride (NbTiN) film [3], which is typically cooled to sub-Kelvin temperatures during operation. In principle, a KITWPA can be fabricated from any superconducting film that displays a high kinetic inductance, such as titanium nitride (TiN). TiN films have found applications in superconducting micro-resonator detectors, such as MKIDs, due to their high kinetic inductance, controllable critical temperature (T_c) and physical robustness [4]. TiN films have additionally been shown to display extremely low losses compared to other superconducting films [5]. Despite this, NbTiN films have proved a popular choice for KITWPA fabrication due to their high T_c making them better suited for applications at 4 K, such as heterodyne receivers. However, a plethora of applications at sub-Kelvin temperatures remain e.g., bolometric receivers and axion search experiments, and the negligible heat dissipation of KITWPAs allows for closer placement of the amplifier to the detector at the sub-Kelvin stage to further reduce signal loss. We, therefore, intend to explore the use of lower T_c TiN films for such applications and investigate additional possible advantages and/or limitations compared to the NbTiN film.

Before venturing into designing a KITWPA using the TiN film, it is important to fully understand the behaviour of the thin film. Therefore, in the first part of the paper, we present an experimental investigation of the various properties of our TiN film using a set of simple transmission line structures to probe their characteristics, which are related to their application as KITWPAs. We further investigate and report on how the transmission line geometries and topologies would affect the performance of the KITWPA. Finally, following comprehensive film and design geometry characterisation results, we conclude the paper by reporting on the preliminary measurement results of a TiN KITWPA operating from 4-12 GHz range.

II. RESULTS

In this paper, we focus our study on a 100 nm thick TiN film with T_c = 4.39 K and resistivity ρ_N = 140 $\mu\Omega$cm, deposited on a 500 μm thick sapphire substrate. Preliminary DC measurements of the thin film behaviour have already been performed to explore the IV characteristics of our TiN film. Preliminary results yield a maximum DC current allowed to transverse through the TiN strip before it quenches of about 1.8 mA at 400 mK, which is consistent with the expected critical current (I_c) value for our TiN superconducting film of ~1 mA. The maximum current as a function of bath temperature is shown in Fig. 1a. Fitting a linear trend to this data gives a max current of 2.2 mA at 0 K and zero current at 4.21 K, which are close to the expected I_c and T_c values, respectively. These preliminary DC measurement results suggest that our TiN films are behaving as we would expect,

[1]University of Oxford, Oxford, OX1 3RH, United Kingdom
[2]GEPI (CNRS UMR 8111), Observatoire de Paris, PSL Université, France
*joseph.longden@physics.ox.ac.uk

NOTES:

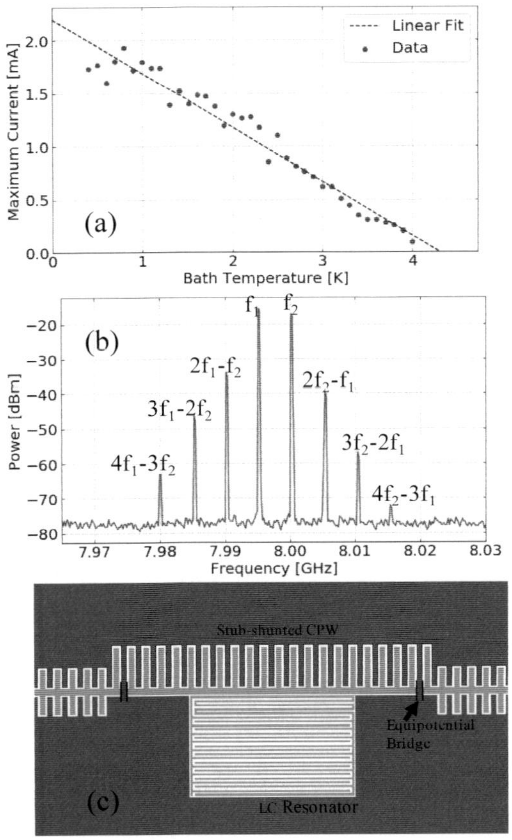

Fig. 1. (a) Maximum current through the superconducting strip as a function of the strip bath temperature. (b) IMD spectrum for TiN superconducting strip at 10 mK. (c) CAD image of an example resonator test structure.

allowing progression onto RF characterisation of our TiN transmission line structures.

In order to investigate the nonlinearity of the film, we have performed intermodular distortion (IMD) measurements, which involves sending two signals at different frequencies across a narrow band and measuring the non-linear cross terms. The signal levels are set below the I_c and T_c of the film. The results of this experiment are shown in Fig. 1b, which show the primary tones at frequencies f_1 and f_2 and the resultant third-order mixing terms. No gain was observed during this experiment and the peak heights of all tones were lower than expected. We can, therefore, confirm that there is nonlinearity present in our TiN films, but losses are higher than expected.

To further investigate the broadband behaviour and loss/nonlinearity performance of our TiN film, we have designed a series of simple test structures using a commercial 3D electromagnetic (EM) simulation software to obtain the required network parameters. An example test structure is a lumped-element LC resonator, shown in Fig. 1c, which is a particularly useful probe of the film properties, since the changes in the current dependence inductance would lead to a shift in the resonant frequency as the applied current is varied. The magnitude of the frequency shift can be used as a measure of the strength of nonlinearity, whilst the change in the depth of the resonance can be used to probe how lossy the film is. An ideal KITWPA film would exhibit a strong frequency shift with minimal change in the depth of the resonance.

The broadband performance of a KITWPA could also be affected by various design issues, such as the choice of transmission line and its corresponding dimensions. Most of the transmission line structures included in this study are based on co-planar waveguides (CPW) and include various combinations of the centre-strip and gap width ratio to explore the effect of geometry on KITWPA performance. To control the characteristic impedance of the transmission line, the CPW is shunted with additional capacitive stubs [6] (same width and gap dimensions as the main CPW), with their length optimised for a 50 Ω line. We additionally employ equipotential bridges for the suppression of parasitic transmission modes. A simple measurement of the S_{21} parameter will, therefore, confirm the feasibility of a particular dimension of test structure as a broadband transmission line and will also reveal the extent of losses in a long transmission line.

Our investigation also explores different transmission line topologies. Ideally, a KITWPA transmission line would be completely straight, however, the necessary long lengths make it impractical to achieve sufficiently high gains on a practical-sized chip. In practice, the transmission line must instead be wrapped into a compact arrangement, such as a spiral or hairpin-style pattern. The compactness of these arrangements may affect the performance of a KITWPA, due to cross talk between adjacent lines, so it is important that this effect is fully studied and understood.

Finally, based on the detailed analysis of our test structures, we conclude the paper with a KITWPA design, which includes its predicted gain-bandwidth product and preliminary experimental results including gain and noise temperature measurements.

REFERENCES

[1] B. Ho Eom, et al., "A wideband, low-noise superconducting amplifier with high dynamic range", *Nature Physics*, Jul. 2012.

[2] C. Bocksteigel, et al., "Development of a broadband NbTiN traveling wave parametric amplifier for MKID readout", *Journal of Low Temperature Physics*, Jan. 2014

[3] S. Shu, et al., "Nonlinearity and wide-band parametric amplification in a (Nb,Ti)N microstrip transmission line", *Phys. Rev. Research*, Jun. 2021

[4] H. G. Leduc, et al., "Titanium nitride films for ultrasensitive microresonator detectors", *Appl. Phys. Lett.*, Mar. 2010

[5] B. A. Mazin, "Superconducting materials for microwave kinetic inductance detectors", *arXiv*, Apr. 2020

[6] S. Chaudhuri, et al., "Broadband parametric amplifiers based on nonlinear kinetic inductance artificial transmission lines", *Appl. Phys. Lett.*, Apr. 2017

NOTES:

Design and Fabrication of All-metal Micromachined Finline Structures for Millimeter and Sub-millimeter Applications

Cristian López*[1], Vincent Desmaris[1], Denis Meledin[1], Alexey Pavolotsky[1], and Victor Belitsky[1]

***Abstract*— We present the design and fabrication of All-metal micromachined finline structures. These novel structures have all the advantages of conventional finlines, i.e., large fractional bandwidth and ease in the mounting process. However, the absence of a dielectric substrate prevents the excitation of substrate modes and the associated resonances, hence the chip area is not limited by substrate size. All-metal finlines could be integrated with existing finline-to-microstrip transitions. This opens up new possibilities for the development of planar circuitry in the millimeter and sub-millimeter range, e.g., integrating this structure with an SIS mixer design.**

***Keywords*— micromachining, finlines, waveguide-to-substrate, microfabrication.**

I. Introduction

Waveguides are widely adopted in modern THz systems for guiding electromagnetic waves and interconnecting different components and subsystems. However, certain key components are fabricated by employing thin-film technology, e.g., mixers and multipliers [1] [2]. Therefore, waveguide-to-substrate transitions play a crucial role in system performance. Among the variety of waveguide-to-substrate transitions reported in the literature, finlines stand out for their large fractional bandwidth and mounting tolerance when compared with other existing solutions [3], e.g., E-probes.

Typically, finline structures consist of a pair of conductive fins over a dielectric substrate. The substrate area and thickness are frequently minimized to reduce dielectric losses and avoid the excitation of substrate modes and resonances. Moreover, the substrate-waveguide interface introduces an impedance mismatch. In [3], the dielectric substrate between the fins was removed to reduce dielectric losses and improve matching over a larger bandwidth. Nonetheless, the dielectric underneath the fins was preserved to provide mechanical support to the structure. However, the substrate could be completely removed if the metallic fins are made sufficiently solid. All-metal finlines without any dielectric substrate constitute a promising solution where the chip area is not limited by the excitation of substrate modes and resonances. Moreover, the all-metal structure provides an excellent cooling capability, is not fragile, and tolerates stress during cooling better.

In this paper, we present the design and process development of all-metal micromachined finlines for millimeter and sub-millimeter applications. Such structures could be integrated with finline-to-microstrip transitions and employed as a new platform for prospective mixers SIS or Schottky mixers and multipliers.

*[1]E-mail: cristian.lopez@chalmers.se. The authors are with the Group for Advanced Receiver Development (GARD), Department of Space, Earth and Environment, Chalmers University of Technology, Gothenburg, Sweden.

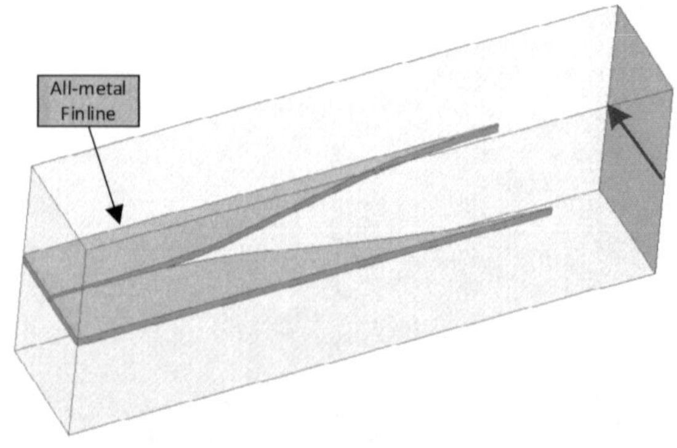

Fig. 1. HFSS 3D model of the proposed finline structure in a rectangular waveguide. The waveguide port is depicted in blue. The red arrow indicates the E-field excitation in the waveguide port. The design is intended for split-block technology.

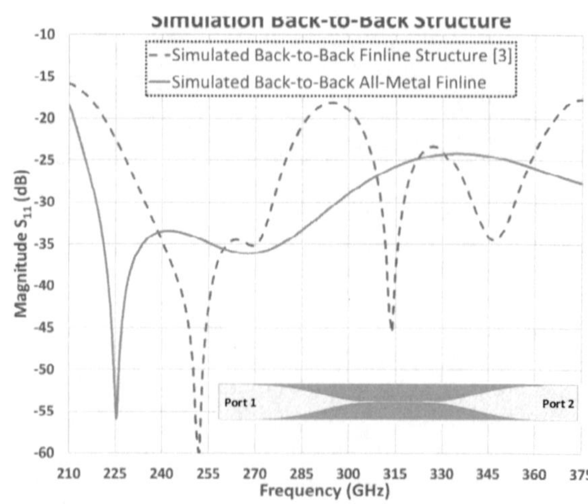

Fig. 2. Simulation of S11 Magnitude for 2 structures: finline structures with a thin dielectric substrate as presented in [3], and the proposed All-metal finlines in a Back-to-Back arrangement. It is clearly observed that the proposed All-Metal Finlines could achieve as broadband performance as their counterpart with thin dielectric substrate [3].

II. Design

The all-metal finline depicted in Fig. 1 was designed for the frequency range 210-375 GHz. In order to define the finline profile, the impedance vs width was analyzed following the procedure described in [3]. Afterward, the profile was optimized with the aid of Ansys HFSS. Fig. 2 depicts the performance comparison between the structure based on silicon substrate fins, presented in [3] and the proposed all-metal finlines. Both structures are arranged in a

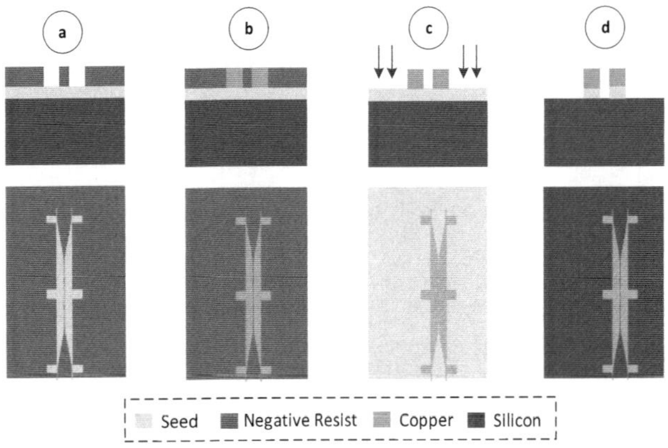

Fig. 3. Simplified fabrication process: (a) Sputtering of metallic seed and patterning of thick negative resist layer. (b) Copper electroplating. (c) Stripping of the resists and the remaining seed layer. (d) All-metal finlines over silicon. The silicon carrier is further etched away to release the structures.

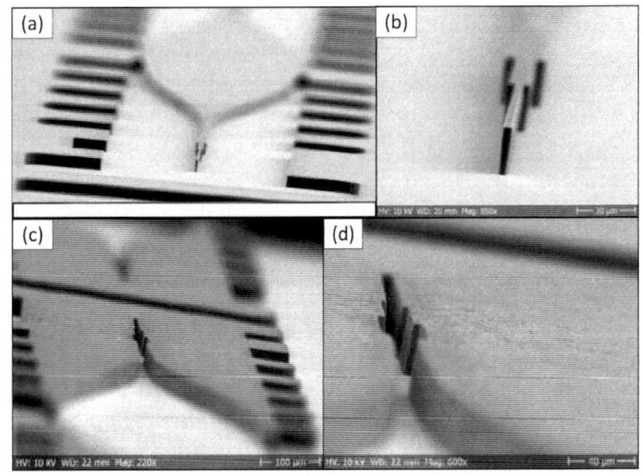

Fig. 4. SEM (Scanning Electron Microscope) pictures of the Fabricated samples (a) Patterned resist. (b) Detail of the central transformer in the patterned resist. (c) Fabricated structure fabricated on electroplated copper. The resist has been removed. (d) Detail of the central slotline transformer of the fabricated structure.

Back-to-Back configuration. It is seen that the proposed finlines could achieve broadband performance as their counterpart with a thin dielectric substrate.

A transition from finline to thin-film microstrip could be implemented with, e.g., a Marchand Balun which presents a large fractional bandwidth with minimum insertion loss [4].

III. Fabrication

The all-metal micromachining technique for waveguide components were demonstrated in [5] and [6]. A simplified process flow is depicted in Fig. 3. The process started with cleaning a 4" wafer followed by the deposition by magnetron sputtering of a metal seed bilayer of Ti/Au. Afterward, a negative thick photoresist was spin-coated, baked, and patterned by photolithography. This step was followed by the electroplating in copper plating solution. Structures with different copper thicknesses ranging from 30 µm to 50 µm were produced. Afterward, the resist layer was stripped by a wet remover. The remaining seed was dry etch using the copper structures as a hard mask. Finally, the silicon carrier was etched away to release the test structures. Fig. 4 depicts Scanning Electron Microscopes pictures of the resist layer and the fabricated test structures. Moreover, Fig. 5 shows the released structures where the smallest fingap is 4 µm.

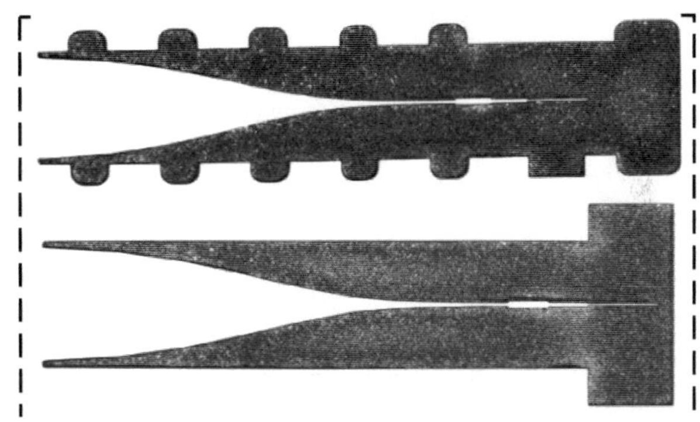

Fig. 5. Microscope picture of the fabricated test structures. The smallest fingap was 4 um.

IV. Conclusion

In this study, we have detailed the design, simulation, and fabrication of all-metal micromachined finlines, with the potential for integration with existing microstrip-to-slotline transitions. Moreover, all-metal finlines without a dielectric substrate are a promising solution to avoid limitations of,e.g., silicon substrate structures that usually present substrate modes, dielectric losses, and resonances. The high fractional bandwidth exhibited by the proposed structure enables its use in broadband SIS mixer design for future applications.

References

[1] V. Belitsky *et al.*, "ALMA Band 5 receiver cartridge - Design, performance, and commissioning, " *Astron. & Astrophys.*, vol. 611, pp.1-10, Mar. 2018, Art. no. A98, doi: 10.1051/0004-6361/201731883.

[2] H. Rashid, S. Krause, D. Meledin, V. Desmaris, A. Pavolotsky and V. Belitsky, "Frequency Multiplier Based on Distributed Superconducting Tunnel Junctions: Theory, Design, and Characterization," in *IEEE Transactions on Terahertz Science and Technology*, vol. 6, no. 5, pp. 724-736, Sept. 2016, doi: 10.1109/TTHZ.2016.2583201.

[3] C. López, V. Desmaris, D. Meledin, A. Pavolotsky, and V. Belitsky, "Waveguide-to-Substrate Transition Based on Unilateral Substrateless Finline Structure: Design, Fabrication, and Characterization," *IEEE Trans. THz Sci. Technol.*, vol. 10, no. 6, pp. 668-676, Nov. 2020, doi: 10.1109/TTHZ.2020.3020683.

[4] C. López, M. A. Mebarki, V. Desmaris, D. Meledin, A. Pavolotsky, and V. Belitsky, "Wideband Slotline-to-Microstrip Transition for 210-375 GHz based on Marchand Baluns," *IEEE Trans. THz Sci. Technol.*, 2022, doi: 10.1109/TTHZ.2022.3149413.

[5] V. Desmaris, D. Meledin, A. Pavolotsky, R. Monje and V. Belitsky, "All-metal micromachining for the fabrication of sub-millimetre and THz waveguide components and circuits," *Journal of Micromechanics and Microengineering*, 2008, doi: 10.1088/0960-1317/18/9/095004

[6] V. Desmaris et. al., "Terahertz components packaging using integrated waveguide technology," *2011 IEEE MTT-S International Microwave Workshop Series on Millimeter Wave Integration Technologies*, 2011, pp. 81-84, doi: 10.1109/IMWS3.2011.6061893.

Sub-mm Wave Schottky Mixer Pumped with 170µW Optically Generated Local Oscillator Power

Javier Martinez[1], Sumer Makhlouf[2], Marcel Grzeslo[2], Diego Moro[1], Oleg Cojocari[1], Andreas Stöhr[2].

***Abstract*—This paper presents a sub harmonic mixer at 270-320GHz, utilizing Low Barrier Schottky Diodes (LBSD). The novelty of this work is the use of a local oscillator (LO) power signal generated by an optical heterodyning system to pump the low barrier high-performance mixer. The Low Barrier Schottky Diodes (LBSD) used in this design can be pumped with a 170 µW LO signal over the band 135-160GHz.**

Keywords— Sub-Harmonic Mixer, Low Barrier Schottky Diodes, UTC-PD.

I. Introduction

The demand in commercial application such radar, communication, spectroscopy, and imaging of a high-power, compact and low power consumption THz devices is increasing day after day. The lack of available commercial power amplifiers in THz range makes necessary to implement a solution to lower down the high radio frequency (RF) signal to be easily amplified and handled. Mixers are a crucial component for such a purpose, being their main role to convert the high frequency received RF signal to an intermediate frequency (IF) signal. The amplitude of the IF signal is linearly related to the RF signal while featuring a much lower frequency. Due to the development of this technology, the frequency and bandwidth of many applications is increasing, leading to the research of a high-performance broadband mixers [1]. In this paper, the design, simulation, and characterization of a low barrier and high-performance Schottky Diode Mixer is presented, performing the measurements with an optical setup.

The state-of-the-art Sub Harmonic Mixers (SHMs) at 300GHz can achieve Conversion Loss between 6.5 and 10 dB using 2-4 mW of LO power featuring LO bandwidths between 20-60GHz [2-4].

The novelty of this research is the use of optical lasers and Uni-Traveling Carrie Photo Diode (UTC-PDs) optoelectronic device to create both the LO and RF signals used for the LBSD mixer. The aim of using optical pumping is the relative low phase noise they introduce and the simplification of the LO source. Usually, the electrically generated LO is up-converted, i.e., the phase noise is increased by 20*log(N) dBc/Hz, being N the multiplication factor [5]. The optically generated LO is down-converted, i.e., the phase noise does not increase, thus, the total noise of the system decreases considerably.

The results of joining the electrical mixer with an optical RF and LO signals will be explained and shown in the next sections.

II. 270-320 GHz Low Barrier Mixer

The low barrier high-performance mixer presented in this paper has been fully designed and fabricated in ACST GmbH. Figure 1 shows the mechanical block of the Mixer and the anti-parallel diodes used on it; the mixer architecture is like the one shown in [4].

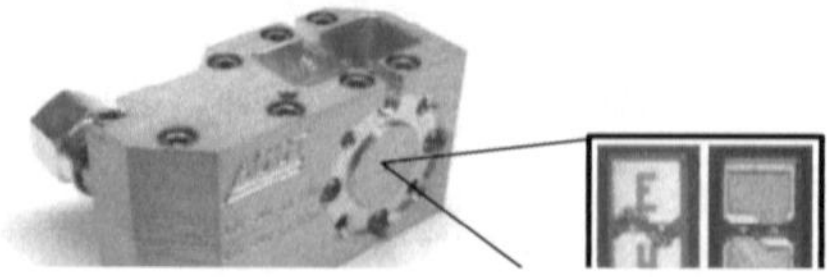

Fig. 1. 270-320 GHz Sub-Harmonic Low Barrier Mixer

The Mixer's features are described in Table 1:

TABLE I. Main Features of the 270-320 GHz SHM

	Min.	*Typ.*	*Max.*
LO Frequency (GHz)	135		160
LO Power (dBm)	-8	-7.5	-3
RF Frequency (GHz)	270		320
Input RF Power (dBm)			-17
IF Frequency (GHz)	0		18
DSB Conv. Loss (dB)	15	16	18

As Table 1 describes, the typical LO power requirement for the presented mixer is -7.5 dBm, which in comparison with the state-of-the-art achieved with GaAs SBDs it needs 10 dB less LO power. The typical DSB Conversion Loss in this LBSD is expected to be higher than in GaAs based Schottky mixers. The diode manufacture for the presented

[1]ACST GmbH, Hanau, 63457, Germany; [2]Universität Duisburg Essen (UDE), Duisburg, 47057, Germany

NOTES:

mixer wasn't as precise enough and the electrical properties were deviated from the expected values. The improvement in the diode properties fabrication can improve this mixer Conversion Loss (CL) in 4-5 dB. However, the performance of the presented mixer is still good enough for a wide range of applications. Although the Conversion Loss is higher, the optical pumping reduces the phase noise of the system, compensating those extra losses induced by the LBSD.

III. Measurement Setup

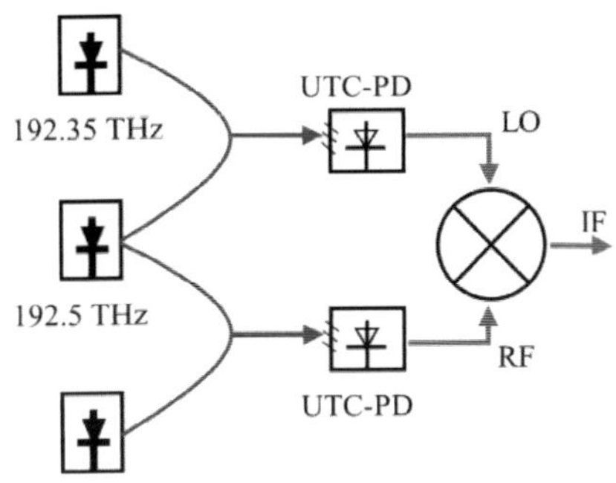

Fig. 2. Schematic of the measurement setup

The setup consists in three lasers, two are used as optical signal sources and the third one as a tuneable source. A commercial UTC-PD is used as RF source and a PD designed and fabricated in University Duisburg Essen (UDE) is used as LO source to pump the mixer. The optical signals generated by the lasers are coupled in pairs into 2 single optic fibers and amplified by 2 different Erbium-Doped Fiber Amplifiers (EDFAs).

Fig. 3. Mixer connected to the commercial WR3.4 UTC-PD. Also connected to the PD through a WR5 probe and WR5-to-WR6 transition.

After that, the RF and LO signals are coupled-out from the UTC-PD output and the PD to the WR3.4-mixer by means of WR3.4-flange and G-S-G WR5-probe and WR5-to-WR6 adapter, respectively. The IF output of the mixer is then connected to an Electrical Spectrum Analyzer to show the IF power.

IV. Results

Experimental results of the DSB Conversion Loss at room temperature are shown in Fig. 4. The measured values are from 18.2-25dB averaging 21dB.

Predicted CL were 16dB, these dB difference could mainly be due to the lack of power delivered by the UTC-PDs. The graph below doesn't show any ripples or drops, illustrating the flat behaviour of the mixer. No mismatches are induced by the PDs.

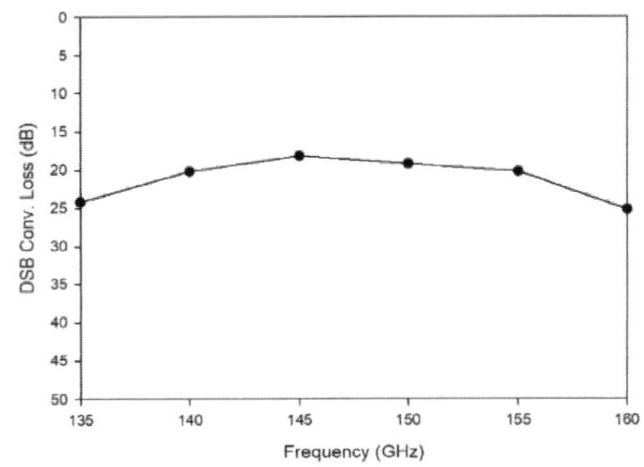

Fig. 4. DSB Conversion Loss using optical pumping

V. Conclusions and Future Work

A Low Barrier SHM has been demonstrated with only 170 μW LO power requirements and up to 18 dB DSB Conversion Loss using UTC-PD optical pumping. Improved manufacture LBSD are planned to improve the Mixer CL in 4-5 dB.

References

[1] Song Xiang., et. Al. "A 75-105GHz Broad-Band Balanced Mixer in a Novel Configuration," *International Conference on Microwave and Millimeter Wave Technology (ICMMT)*, May 2012.

[2] Cheng Guo., et. Al. "A 290–310 GHz Single Sideband Mixer With Integrated Waveguide Filters," *IEEE Transactions on Terahertz Science and Technology*, July 2018.

[3] Bertrand Thomas., et Al. "A Low-Noise Fixed-Tuned 300–360-GHz Sub-Harmonic Mixer Using Planar Schottky Diodes," *IEEE Microwave and Wireless Components Letters*, Dec. 2005.

[4] Zhe Chen., et Al. "Design of a Subharmonic Mixer with Plannar Schottky Diodes at 300-350 GHz," *International Workshop on Microwave and Millimeter Wave Circuits and System Technology*, April 2012.

[5] Xinlin Geng., et Al. "The Design of a 28GHz Mixer-Embedded Frequency Shifting PLL in 65nm CMOS with low In-band Phase Noise," *IEEE International Symposium on Circuits and Systems (ISCAS)*, May 2019.

NOTES:

CubeSounder: Development of Microwave Radiometer 3D Weather Imaging Sensors

K. D. Massingill[1], S. Bryan[1], C. Groppi[1], P. Mauskopf[1], R. Suresh[1], J. Greenfield[1], and J. Tinlin[1]

***Abstract*—Satellite based microwave radiometers are the most important driver of global weather forecasting (WMO 2012, NAS 2018). Current sensors rely on high-SWaP-C LO/mixer/RF components. We present ground testing and designs of a novel low-SWaP-C microwave sounding weather sensor, that will fly with a NASA technology demonstration flight on a stratospheric balloon. Our weather sensor utilizes parallel 60GHz/180GHz millimeter-wave filter bank spectrometers. The spectrometers work by amplifying a broadband signal with an LNA then channelizing with a millimeter-wave filter bank. Each channel of the filter bank is detected by a separate diode. We designed this sensor system for measuring 3D atmospheric water vapor and temperature by observing the pressure broadened 183 GHz water line and 60GHz oxygen line. The NASA tech flight will allow us to demonstrate our weather sensor technology in a relevant in environment, paving the way for either commercial adoption or space-based demonstrations. We will present progress on constructing and characterizing a flight ready prototype weather sensor system and demonstrate that the maturations and adoption of our novel sensor system would greatly improve weather forecasting.**

Keywords—Millimeter-wave, Remote Sensing, Weather, Spectrometer

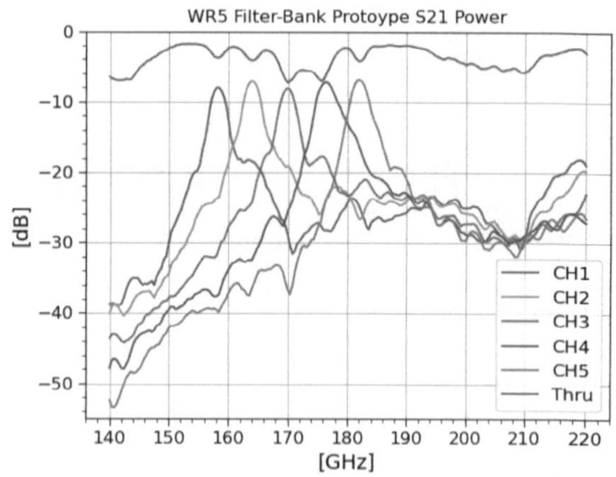

Fig. 1. The lab-measured scattering amplitude of a 180GHz WR5 filter bank prototype, showing the transmission of each channel.

References

[1] National Academies of Science, "Thriving on Our Changing Planet: A Decadal Strategy for Earth Observation from Space", *Decadal Survey for Earth Science and Applications from space*, 2018.

[2] World Meteorological Organization, "World Weather Watch: the fifth WMO Workshop on the impact of various observing systems on numerical weather prediction", *WMO Workshop Report*, May. 2012.

[1]Arizona State University, School of Earth and Space Exploration, Tempe, AZ 85287, USA

NOTES:

Proof-of-concept experiment on a novel microwave circulator based on frequency converters

Sho Masui[1, 2], Hideo Ogawa[1], Toshikazu Onishi[1], Kazumasa Makise, Takafumi Kojima[2], and Yoshinori Uzawa[2]

Development of a scalable integrated circuit toward a large format array is one of the major challenges in the superconducting electronics for applications such as radio astronomy and superconducting quantum computer. One of the critical components in the superconducting circuit that limit the realization of the large-scale system is a non-reciprocal cryogenic circulator based on a ferrite. In a receiver frontend of a radio telescope, it is used between a superconductor-insulator-superconductor (SIS) mixer and cryogenic low-noise amplifier to mitigate the multiple reflection and provides the ideal condition in order for the cryogenic active components to be independently operated in the system. Performance of the microwave ferrite-based circulators (isolators) reported in recent has greatly improved and shows acceptable low insertion loss over multi-octave bandwidths [1]. On the other hand, this conventional circulator is in principle physically large compared to scale of the superconducting millimeter/microwave integrated circuits. This makes it difficult to minimize physical size of the receiver frontend, and therefore, provides limitation of the number of pixels in multibeam heterodyne receivers.

To break through this limitation, we are developing a novel circulator that could be fully integrated on a superconducting planar circuit [2]. The circulator can be configured by using two identical 90-degree hybrid couplers and a gyrator inserted between them. A gyrator is a non-reciprocal device that has a 180-degree differential phase shift between waves propagating through it in opposite directions. It can be realized by employing two mixers as frequency up and down converters, and phase shifters (Fig. 1 (a)). In addition, if the SIS-based frequency converters can be used as the mixing element in the circuit, which is equivalent to a superconducting microwave amplifier being developed at the NAOJ [3, 4], the circulator with a possible gain or an unidirectional low-noise amplifier can be realized. Before designing and fabricating the on-chip circulator, we conducted a proof-of-principle experiment using commercial microwave components to clarify the design issues. As shown in Fig. 1 (b), the circulator in this experiment is consisted of two 90-degree hybrid couplers and two sets of frequency up and down converters: one set for the gyrator with the 90-degree phase shifter, and the other for compensation of the conversion loss in the gyrator. By combining them, an input signal from Port 1 can be propagated to Port 2, and an input signal from Port 2 can be propagated to Port 3. In this experiment, input/output frequencies are 200-500 MHz as the base band, and the local oscillator (LO) frequency is set to 2.0 GHz. This corresponds to the lower sideband: 1.5-1.8 GHz and upper sideband: 2.2-2.5 GHz in up-converted frequency bands. For the phase shifters used in the up-converted band, we adopted coaxial cables and adapters to achieve 90 degrees at around 2.0 GHz, since we could not obtain wideband characteristics with commercially available ones. The measurement results of the *S*-parameters for the gyrator composed of the above phase shifter and two commercial mixers are shown in Fig. 2. As a result, the phase difference between S_{21} and S_{12} is 180 degrees, indicating that it is operating as a gyrator. Regarding the characteristics as a circulator, although the phase shifters have frequency dependences, a typical isolation of 10 dB and

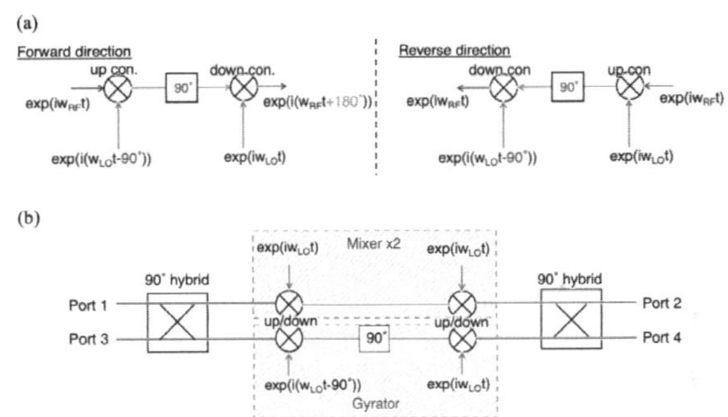

Fig. 1. (a) the schematic diagrams of a gyrator about forward direction and reverse direction. (b) the schematic diagram of circulator.

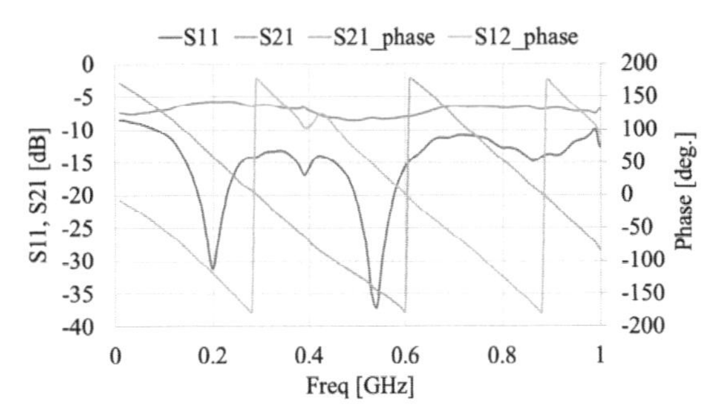

Fig. 2. The measurement results of *S*-parameter for the gyrator.

[1]Graduate School of Science, Osaka Prefecture University, Sakai, Osaka 599-8531 Japan; [2]Advanced Technology Center, National Astronomical Observatory of Japan (NAOJ), Mitaka, Tokyo 181-8588, Japan

NOTES:

the best isolation of 25 dB were successfully obtained in the base band.

In this symposium, we will report on the experimental results and the technical challenges in the circulator or isolator design. Besides, we plan to evaluate the gyrator using two SIS mixers as the frequency up and down converters, and we will present the progress of these measurements.

ACKNOWLEDGEMENTS

This work was supported in part by JST [Moonshot R&D] Grant Number [JPMJMS2067] and JSPS KAKENHI Grant Numbers JP18H03881 and JP19H02205.

REFERENCES

[1] L. Zeng, C. E. Tong, R. Blundell, P. K. Grimes and S. N. Paine, "A Low-Loss Edge-Mode Isolator with Improved Bandwidth for Cryogenic Operation," IEEE Transactions on Microwave Theory and Techniques, vol. 66, no. 5, pp. 2154-2160, May 2018.

[2] B. Abdo, M. Brink, and J. M. Chow, "Gyrator Operation Using Josephson Mixers", *Physical Review Applied*, Sep. 2017.

[3] T. Kojima, Y. Uzawa, and W. Shan, "Microwave amplification based on quasiparticle SIS up and down frequency converters", *AIP Advances*, Feb. 2018.

[4] Y. Uzawa, T. Kojima, Y. Kozuki, Y. Fujii, A. Miyachi, T. Tamura, S. Ezaki, and W. Shan, "An SIS-mixer-based amplifier for multi-pixel heterodyne receivers", *Proc. SPIE*, Dec. 2020.

NOTES:

A Cryogenic Scalable Small-Signal & Noise Model of GaN HEMTs

Mohamed Aniss Mebarki*[1], Ragnar Ferrand-Drake Del Castillo [2], Erik Sundin[1], Denis Meledin[1], Mattias Thorsell [2], Niklas Rorsman [2], Victor Belitskyi[1], Vincent Desmaris[1]

Abstract— **We present a detailed study of a scalable small signal and noise model at the cryogenic temperature of ~10 K of Gallium Nitride (GaN) - based High Electron Mobility Transistors (HEMTs). The results confirm a clear potential of the GaN technology for the cryogenic low noise applications as the model predicts a minimum noise temperature of ~ 4 K at the physical temperature of ~ 10 K. The improvement of the noise cryogenic performances is attributed to the decrease of the access resistances and the enhancement of the electron transport mechanisms. Moreover, the scalability of the model over the measured different gate peripheries is explored and provides new insights on the possibilities of further optimization of this technology for the cryogenic and low noise operation. Potentially, GaN HEMTs would provide enhanced dynamic range with the noise performance similar to InP devices.**

Keywords— ***AlGaN, cryogenic, GaN HEMT, LNA***

I. Introduction

Cryogenic low noise amplifiers (LNA) constitute a crucial part of the systems dealing with extremely weak signals, as it is particularly the case in the radio astronomy receivers. This however often takes place in the presence of disturbing signals from communication and other microwave applications, e.g. radars. Most recently, the GaN-based HEMTs were demonstrated as a potential novel candidate for the development of cryogenic LNAs in [1]. Indeed, even though the matching networks in the reported GaN-LNAs were exclusively optimized for room temperature (RT) conditions, they exhibited an impressive improvement of their noise performance by almost a factor of 10, achieving ~8 K noise temperature, when cooled down to ~ 10 K. These results opened up new perspectives for the development of the GaN-technology for cryogenic applications. Indeed, in addition to the potential fulfillment of the gain and noise requirements, the targeted systems could benefit from the superior linearity and robustness which is inherent to the GaN-material over the competing technologies. Furthermore, given the recently reported enhancement of the heterodyne receivers incorporating GaN and attributed to the latter material properties [2], the integration of GaN-LNAs could be of interest from the technological perspective.

Thus, in order to enable the full potential of the emerging cryogenic GaN-HEMTs for low noise applications, it is a prerequisite to determine small signal model (SSM) and noise models at the temperatures and frequencies of interest. Also, it is important to relate the physical dimensions of the devices to their noise characteristics. Indeed, the availability of such a relation allows the designer to choose a size of the transistor which is best adapted to the overall system requirements. Moreover, it can provide insights for the optimization of the technology itself. Therefore, in addition to allowing the assessment of the reliability, the scalability of the models helps to reduce the number of iterations and the costs required for the optimization of both the cryogenic GaN-HEMTs and the LNAs based on this technology. To the best of our knowledge, there is still a lack in the literature of a cryogenic model of GaN HEMTs dealing with the scalability of the data.

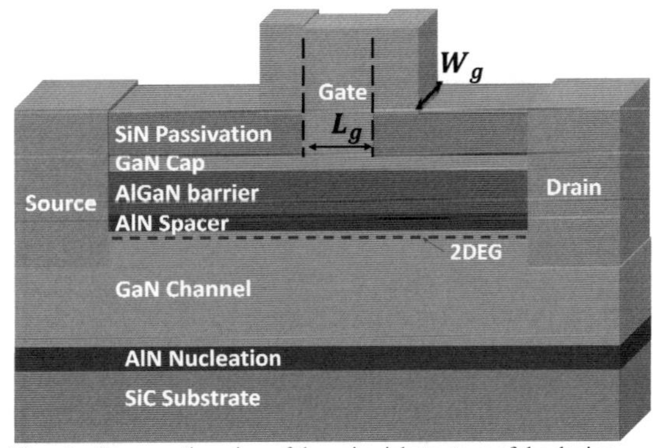

Fig. 1: Cross-section view of the epitaxial structure of the device.

This work proposes a scalable small signal and noise model of AlGaN/GaN HEMTs with a gate length of 0.2 μm at cryogenic temperatures (CT) of about 10 K. The validity of the model was investigated against the measurements for different total gate widths ranging from 50 μm to 150 μm. The HEMTs featured a 2 gate-fingers design. The typical cross-sectional view of the epitaxial structure of the devices is shown in Fig. 1. The static, microwave and noise characteristics of the same technology at cryogenic temperatures were recently studied in [3]. The details of the fabrication process are reported in [4].

II. II. Experimental Work

At both RT and CT, the devices were characterized on-wafer in a cryogenic probe station by using a parameter analyzer as well as a vector network analyzer. Fig 2 presents the typical measured I-V curves of the DUT at RT and CT, which feature a substantial decrease of the on-resistance of the HEMTs at CT.

For the high frequency noise tests, the devices were mounted in a fixture and measured in a cryogenic setup. The Y-factor technique and the cold-attenuator method [5], using a 20-dB attenuator, were implemented for the cryogenic noise characterization. We employed for this purpose a 50 Ω noise source and an Agilent MXA N9020A signal analyzer with its Noise Figure Measurement option. The noise measurements were performed within a frequency range of 4.5 to 6.5 GHz.

[1]E-mail: aniss@chalmers.se. The authors are with the Group for Advanced Receiver Development (GARD), Department of Space, Earth and Environment, Chalmers University of Technology, Gothenburg, Sweden

[2] Microwave Electronics Laboratory, Chalmers University of Technology, SE412 96, Gothenburg, Sweden.

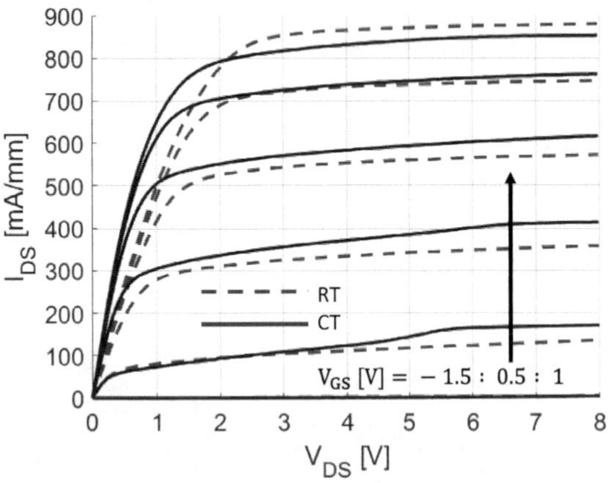

Fig. 1: Typical I_{DS}-V_{DS} curve of the DUT at RT and CT

The extraction of the noise parameters of the different HEMTs were performed by the de-embedding the contribution and noise temperature of the different elements in the cryogenic test system. Hence careful cryogenic calibration was performed prior to subtracting the noise contributions of the input and output chains of the system [6].

III. III. Results and Discussion

Relying on the measured S-parameters of the device, the extraction of the SSM followed the approach presented in [7]. Fig. 3 illustrates the equivalent small signal model, with the different noise sources associated with the lumped elements. The model revealed a consistent scalability of the SSM parameters with the transistor bias and its gate width, at both RT and CT, as can be seen in Fig. 4, 5 and 6 presenting respectively the variation of the access resistances, the intrinsic transconductance and gate to source capacitance. The figures reveal a significant increase of the intrinsic transconductance, g_{m-int}, together with the decrease of the access source and drain resistances, R_s and R_d respectively.

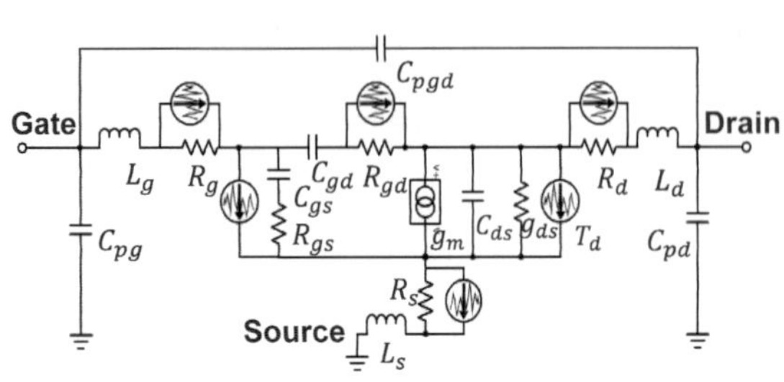

Fig. 2: Illustration of the equivalent small signal model, with the current noise sources associated with the different lumped elements

The inverse proportionality of R_s and R_d with the gate width reflects their dominating intrinsic nature. As recently reported in [8], using a similar technology, the gate resistance also reduces significantly at cryogenic temperatures. The gate resistance scales directly with the gate-width. The same applies to the gate-to-source capacitance, C_{gs}, impacting the input impedance of the device (Fig. 6).

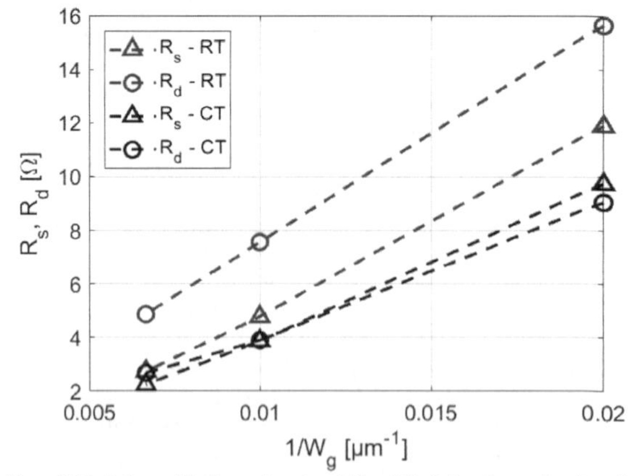

Fig. 3 Variation with the unit gate width of R_s (triangle markers) and (circle markers) R_d at RT (red) and CT (blue).

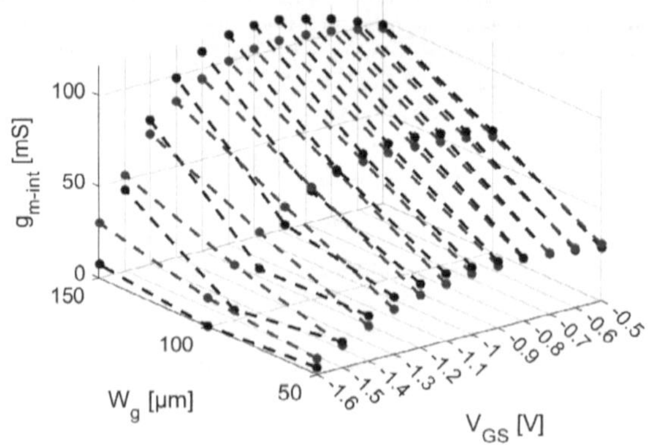

Fig. 4: Bias dependence of g_{m-int} and its variation with the gate width at RT (red) and CT (blue) at V_{DS} = 5 V..

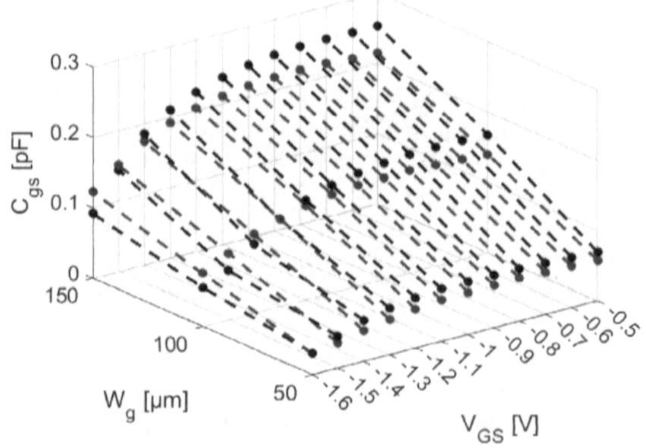

Fig. 5: Bias dependence of C_{gs} and its variation with the unit gate width at RT (red) and CT (blue) at V_{DS} = 5 V.

The combination of the decrease of the access resistances and the increase of the transconductance leads to a reduced impact of the thermal noise associated with the parasitic resistances of the device at cryogenic temperatures.

Once the cryogenic SSM and its scalability were determined, the Pospiezalski model [9] was implemented in order to extract the noise parameters of the devices. The model relies on the SSM parameters and the extraction from the noise measurements of the fitting parameter, T_d, accounting for the temperature related to the drain to source resistance. This latter is considered being different from that of the lattice, due to the confinement of the electrons in the 2-dimensional electrons gas (2DEG), and is proportionally dependent on the power dissipation of the device.

As presented in Fig. 7, the optimum noise temperature, $T_{min-opt}$ [K] at CT was found in the range of 4 – 5 K independently of the gate-widths. The observed weak variation of the minimum noise temperature with the gate-width is consistent with the conclusions from other works in the literature dealing with the scalability of the noise data of other FET and HEMT technologies [10]. The optimum bias point (V_{DS}, I_{DS}), at which $T_{min-opt}$ is obtained, was found to scale by approximately the same factor as the gate width (Fig. 7). As a consequence of the variation of of the small signal model with the gate-width, the other noise parameters scale with the gate width. Indeed, as it was demonstrated in [10], the radius of the constant noise figure circles varies with the gate width. Therefore, the radius of the constant noise circle can be optimized given an optimal gate width value. A broad noise circle facilitates the search of a good compromise on the gain and noise performances of the practical LNA. Fig. 8 shows the effects of this property in our model. The same plot shows the variation of the optimum reflection factor $\Gamma_{opt}^{W_g}$, with W_g.

The constant circles of the noise temperature ranging from 5 K to 10 K of the three studied transistor gate widths are plotted around the power dissipation level of 0.5 W/mm at the frequency f = 6 GHz. The transistor with the largest gate-width is found, in this case, to provide the broadest constant noise circles. In fact, this variation reflects also the decrease of the noise resistance R_n [Ω], with the gate-width.
The decrease of R_n leads to a reduced sensitivity of the minimum noise temperature towards the deviations from the optimum impedance. However, a transistor with a larger gate width will drive more current (Fig. 8). This will in turn affect the overall power consumption of the application system.
This might be especially critical when LNAs of multiple stages are implemented. The choice of the transistor size needs then to satisfy both the system requirements and the practicality of the design.

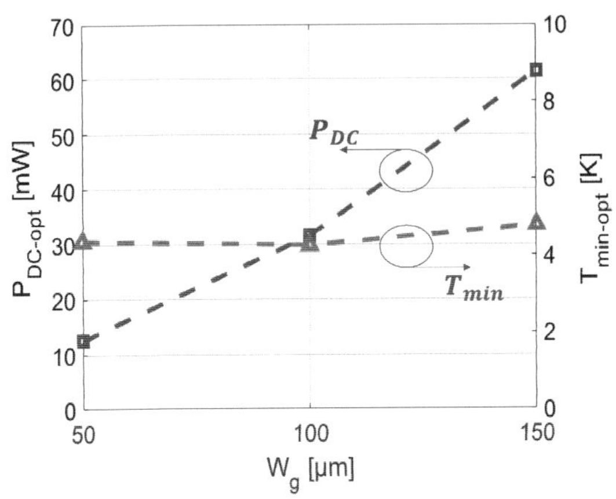

Fig. 7: Width-dependence of the predicted optimum noise bias (left vertical scale) and optimu noise temperature (right vertical scale) at CT.

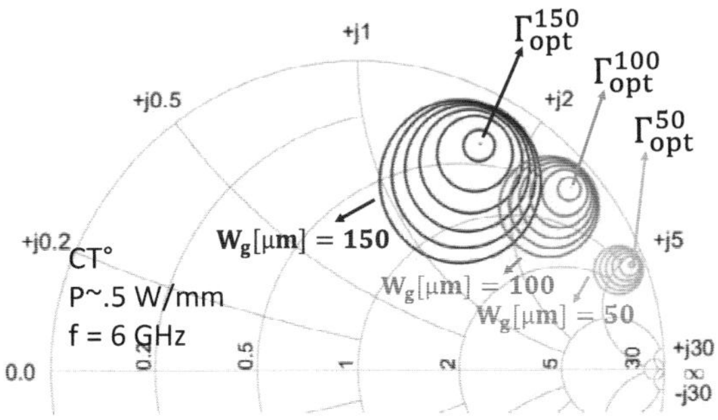

Fig. 8: Variation of the constant noise circles of T_e [K] = [5,6,7,8,9,10] for different gate widths: 2x25 um (orange), 2x50 um (red), 2x75 um (blue) at CT, at f = 6 GHz and power dissipation P ~ 0.5 W/mm.

IV. IV. Conclusion

This work experimentally verified for the first time the scalability with the gate periphery of a small signal and microwave noise model of GaN-HEMTs operating at the physical cryogenic temperature of ~10 K.

The results show a promising low-noise performance of GaN at cryogenic temperatures, as an optimum noise temperature 4.5 ± 0.5 K was obtained at CT from the different studied devices with the total gate-widths ranging from 50 up to 150 μm. This result is comparable to more established cryogenic low-noise technologies. Further optimization at the device level might lead to an enhanced competitivity of the GaN technology in this field. In addition, the extracted models provide the key elements for the future design of cryogenic GaN-based LNAs.

Acknowledgment

The research leading to these results has received funding from the Swedish Research Council (Vetenskapsrådet) under grant agreement 2018-065407.

References

[1] V. Desmaris, D. Meledin, E. Sundin, M. Thorsell, N. Rorsman and V. Belitsky, "Characterization of GaN-based Low Noise Amplifiers at Cryogenic Temperatures," in *ISSTT 2019 - 30th International Symposium on Space Terahertz Technology*, 2019.

[2] S. Krause, V. Desmaris, A. Pavolotsky, D. Meledin and V. Belitsky, "Suspended GaN beams and membranes on Si as a platform for waveguide-based THz applications," *Journal of Micromechanics and Microengineering,* vol. 28, p. 105007, 2018.

[3] M.A. Mebarki, et. al, "Noise Characterization and Modeling of GaN-HEMTs at Cryogenic Temperatures," *IEEE Transactions on Microwave Theory and Techniques,* 2022.

[4] D.-Y. Chen, A. Malmros, M. Thorsell, H. Hjelmgren, O. Kordina, J.-T. Chen and N. Rorsman, "Microwave performance of 'buffer-free'GaN-on-SiC high electron mobility transistors," IEEE Electron Device Letters, vol. 41, p. 828–831, 2020.

[5] J. D. Gallego and M. W. Pospieszalski, "Design and performance of cryogenically-coolable ultra low noise, L-band amplifier," in *1990 20th European Microwave Conference*, 1990.

[6] H. Rashid, D. Meledin, V. Desmaris, A. Pavolotsky and V. Belitsky, "Superconducting 4–8-GHz hybrid assembly for 2SB cryogenic THz receivers," IEEE Transactions on Terahertz Science and Technology, vol. 4, p. 193–200, 2014.

[7] N. Rorsman, M. Garcia, C. Karlsson and H. Zirath, "Accurate small-signal modeling of HFET's for millimeter-wave applications," IEEE Transactions on Microwave Theory and Techniques, vol. 44, p. 432–437, 1996.

[8] M. A. Mebarki, R. Ferrand-Drake Del Castillo, A. Pavolotsky, D. Meledin, E. Sundin, M. Thorsell, N. Rorsman, V. Belitsky and V. Desmaris, "GaN HEMT with superconducting Nb gates for low noise cryogenic applications," *physica status solidi (a),* 2200468, 2022.

[9] M. W. Pospieszalski, "Modeling of noise parameters of MESFETs and MODFETs and their frequency and temperature dependence," *IEEE transactions on microwave theory and techniques,* vol. 37, p. 1340–1350, 1989

[10] B. Hughes, "Designing FET's for broad noise circles," *IEEE transactions on microwave theory and techniques,* vol. 41, p. 190–198, 1993

Development of an instantaneous multiband digital 2SB receiver for the 67–116-GHz band

David Monasterio[1] and F. Patricio Mena[2]

Abstract— **In this work we present the initial development of an instantaneous multiband receiver with digital sideband separation for an extended W Band (67-116 GHz). In particular, we present the design and construction of a frequency diplexer that, thanks to the inclusion of a novel waveguide quadrature hybrid, shows an excellent behavior in the entire bandwidth. Moreover, we present a non-linear simulation of the entire receiver, including digital sideband separation. The results show complete RF coverage into four parallel IF outputs.**

Keywords— Heterodyne receivers, waveguide components, frequency conversion, frequency diplexers.

I. INTRODUCTION

The multiband heterodyne receiver proposed in [1] can be an alternative solution for the necessity of ultra-bandwidth systems in radio astronomy. It works with a frequency multiplexer that separates the RF bandwidth into a number of sub-bands that are down-converted independently. This architecture allows to obtain the relevant RF spectrum into several parallel IF outputs. Since the multiplexer is a key component in this type of receiver, the work presented in [1] also proposed a versatile design based in quadrature hybrids and filters. If several of them are concatenated, it is possible to achieve a multiplexer with any number of outputs.

Here we present a variation of this architecture that uses digital sideband (2SB) downconverters, and has been applied to an extended W band (67–116 GHz). The initial development includes the design and construction of a broadband frequency diplexer with relaxed mechanical constraints for easing the manufacture process. The diplexer uses a novel waveguide quadrature hybrid [3] with state-of-the-art performance.

II. PROPOSED MULTIBAND RECEIVER

Fig. 1 presents a general diagram of the multiband downconverter proposed in this work, which has the potential of covering the entire RF spectrum instantaneously. The complete RF coverage can only be implemented using a digital IF hybrid since, in contrast to its analog counterpart, it is not limited at lower frequencies. Then, by placing the LO frequency at the center each sub-band, total RF coverage can be achieved. Since this architecture divides the analog RF spectrum into two bands with half of the bandwidth, it offers two additional advantages besides the instantaneous access to the entire RF spectrum. (i) Reduction of the operational bandwidth needed for the electronics of each sub-band. (ii) As a consequence of this bandwidth reduction, further optimization can be archive in the performance of the receiver, reducing its overall noise temperature.

[1]Universidad de Chile, Santiago, 8370451, Chile; [2]National Radio Astronomy Observatory, Charlottesville, 22903, USA.

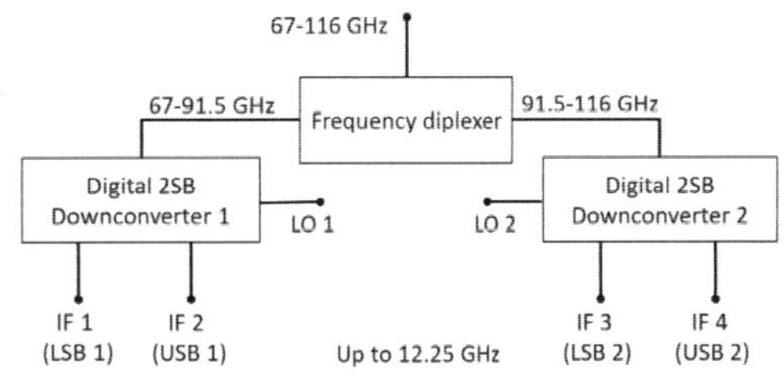

Fig. 1. Concept of a multiband heterodyne receiver with digital 2SB receivers for the extended W-band. The input RF signal is separated into two frequency bands with a diplexer, a device that is capable of selectively routing the signal depending on its frequency. The low (67-91.5 GHz) and high (91.5-116 GHz) RF ranges are transmitted to two independent digital 2SB downconverters, 1 and 2. Each downconverter, then, separates the signals even more to Upper sideband (USB) and Lower Sideband (LSB) with the use of a image-rejection configuration. In this way, the system allows instantaneous acces to the entire RF range.

Despite the possibility of solving various problems in radio astronomy, not much work has been done in developing this type of architecture. The main reason is the limitation of IF processing capabilities. However, in recent years, with the development of faster digital processing platforms, this limitation could be overcome in the near future.

III. WAVEGUIDE QUADRATURE HYBRID

In this section, we present a modified branch-line waveguide quadrature hybrid that besides achieving an excellent performance over a large fractional bandwidth, exceeding 50%, eases construction constraints. This breakthrough was accomplished by implementing the branch lines with a larger height than the main lines, which results in widening the former, and so, reducing the height-to-diameter ratio of the tool needed to mill them. We have applied this new concept to cover an extended W band because the frequency diplexer presented in this work needs a quadrature hybrid that covers its complete RF operational bandwidth at its input.

The modified waveguide branch-line coupler proposed in this work is presented in Fig 2. For this particular design, a tool with a ratio larger than 6.23 is needed, which is well below overmoded hybrids with similar bandwidths [3]. The proposed hybrid was simulated in ANSYS HFSS, showing a return loss over 20 dB, an amplitude imbalance below 0.5 dB, and a phase imbalance, $\Delta\varphi$, within $\pm 1°$. Measurements are in good agreement with simulations. They show a return

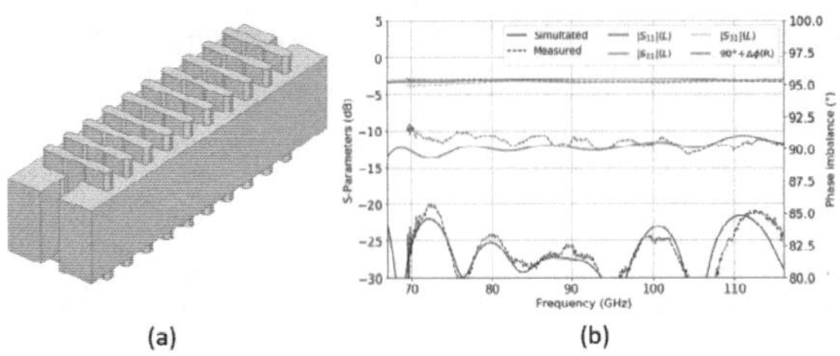

Fig. 2. Modified quadrature Hybrid. (a) CAD image. (b) Simulation and measurements results of the constrcuted prototype.

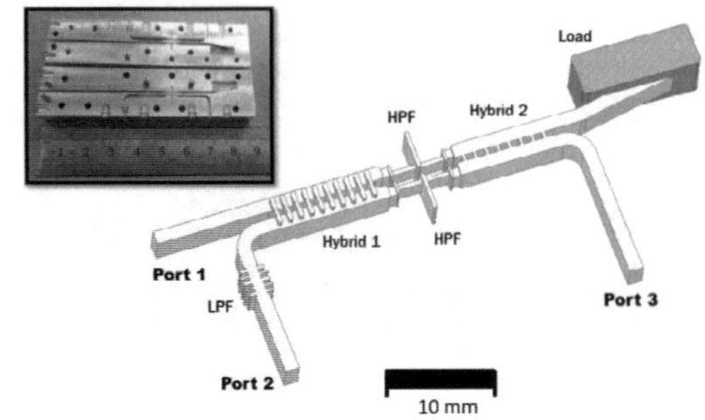

Fig. 3. CAD view of the complete diplexer, showing all of its components. The inset shows a photograph of its implementation as split-block.

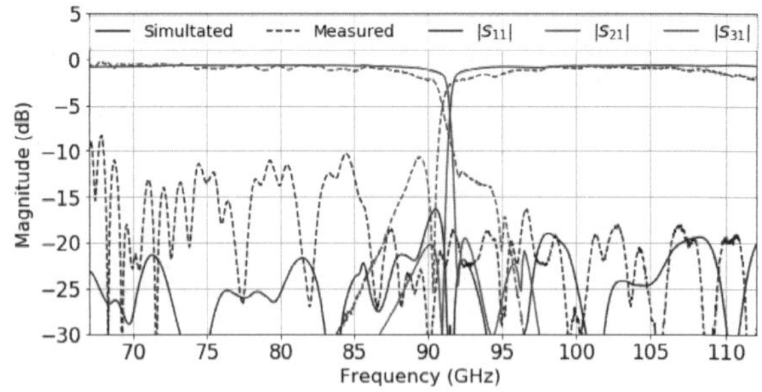

Fig. 4. Simulated and measured S parameters of the diplexer.

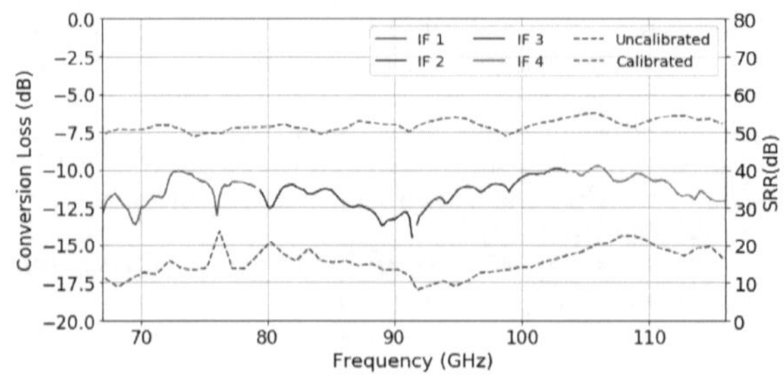

Fig. 5. Results of non-linear simulation of the convertion loss of the reciever and SRR with the digital sideband separation.

loss over 20 dB and amplitude and phase imbalances below 0.8 dB, and between -0.5° and $+2^{\circ}$, respectively.

IV. Frequency Diplexer

A frequency diplexer, based on [4], was designed and a prototype was manufactured. Fig. 3 shows the complete design. The components are an input quadrature hybrid that covers the entire input bandwidth (67–116 GHz), two high-pass filters with a frequency cutoff of 91.5 GHz, a second quadrature hybrid with an operational bandwidth of 91.5 to 116 GHz, and a low-pass filter to increase rejection between the two output bands.

The diplexer was simulated in ANSYS HFSS and the results are presented in Fig. 4. The simulations used a finite aluminum conductivity boundary condition and an ideal RF load. The $|S_{21}|$ and $|S_{31}|$ parameters show the transmissions of the lower band and the upper band respectively, with an interception point exactly at 91.5 GHz. The cutoff slope for each band matches the behavior of the bandpass filter. The $|S_{11}|$ parameter presents good behavior with a reflection below –20 dB in the majority of the band.

Fig. 4 also presents the scalar S-parameter measurements of the constructed prototype of the diplexer. In the $|S_{21}|$ and $|S_{31}|$ parameters, we can see two well-defined transmission bands for the lower and upper-frequency outputs. The interception point between the two frequency bands is at 90.98 GHz, approximately 500 MHz apart from its expected value. However, the shape of the cutoff slope of the $|S_{21}|$ and $|S_{31}|$ parameters differ from the simulation. We can explain these differences due to manufacturing errors in the high-pass filter sections of the constructed block. These errors could also partially explain the behavior of the $|S_{11}|$ parameter, as the lower-frequency band presents an overall higher value in comparison with the upper-frequency band. This difference is produced by asymmetries between the two band-pass filters, producing an ineffective cancellation of the reflected wave at the input port. Another possible cause for the increment in reflections is the poor performance of the low-pass filter placed at port 2. The diplexer losses possess an average value of 1.5 dB for the entire band and have good agreement with the simulations, only increasing at points of high reflection for the lower-frequency band and at the lower and upper edges in the upper-frequency band. The increase in the upper band can be also explained by asymmetries in construction, dissipating the remaining power at the load.

V. Receiver Simulations

In a three-step simulation, we have studied the response of a receiver using the previous diplexer in conjunction with realistic Schottky-based 2SB downconverters using a digital IF hybrid [2]. The first step uses ANSYS HFSS to simulate the majority of passive components such as the diplexer, waveguide-to-microstrip transitions, and microstrip IF paths. The second step is to use Cadence AWR in conjunction with the imported HFSS results to simulate the entire analog receiver using the harmonic balance nonlinear method, this includes MMICs I/Q mixers. The results (in Fig. 5) show that the receiver's conversion loss allows complete RF bandwidth coverage over the four IFs. In the final step, utilizing the information of harmonic content, the IF outputs were reconstructed as input data for a simulation utilizing the digital sideband separation method [5]. Results (also in Fig. 5) show an excellent performance after calibration, with a sideband rejection ratio (SRR) of 50 dB, allowing each IF channel to receive the signal for its specific sub-band with excellent isolation.

VI. Conclusions & Future Work

In this work, we have presented a multiband receiver architecture with digital 2SB that shows the feasibility of total broadband RF coverage through parallel IF outputs.

One of the key components, a fully functional waveguide diplexer, that utilizes a state-of-the-art waveguide hybrid, was also presented.

As a future work we plan to manufacture a new diplexer prototype without the low-pass filter to study its effect in the diplexer performance. Then, we propose to build the downconverter prototype and test it utilizing a second down-conversion stage with smaller IF bandwidth and a mobile LO, and implement digital sideband separation in a FPGA-based backend.

REFERENCES

[1] T. Kojima, et al., “Design and Development of a Hybrid-Coupled Waveguide Multiplexer for a Multiband Receiver”, *IEEE Transactions on Terahertz Science and Technology,* Jan. 2017.

[2] M. Morgan, et al., “Experiments with Digital Sideband-Separating Downconversion”, *Publications of the Astronomical Society of the Pacific*, March 2010.

[3] D. Monasterio, “Development of New Heterodyne-Downconverter Architectures for the Next Generation of Radio-Astronomy Instrumentation”, *Ph.D. thesis*, Universidad de Chile, Jan. 2023.

[4] A. Gonzalez, et al., “275–500 GHz Waveguide Diplexer to Combine Local Oscillators for Different Frequency Bands”, *IEEE Transactions on Terahertz Science and Technology,* Oct. 2017.

[5] R. Finger, et al., "A Calibrated Digital Sideband Separating Spectrometer for Radio Astronomy Applications", *Publications of the Astronomical Society of the Pacific*, Feb. 2013

Exploring the Limits of the Tunnel Junction Fabrication Technique for Josephson Junctions TWPA and the Preliminary Characterisation Results

Javier Navarro Montilla[1], Eduard F.C. Driessen[2], Arnaud Barbier[2], Faouzi Boussaha[3], Christine Chaumont[3] and Boon-Kok Tan[1]

***Abstract*— Travelling Wave Parametric Amplifiers (TWPAs) can potentially achieve quantum limited noise over a broad bandwidth in the microwave regime, with potential applications in the readout of millimetre (mm) and sub-millimetre (sub-mm) receivers to further improve the system sensitivity, among many other applications. TWPAs using embedded Josephson junctions (J-TWPA) have proven to exhibit noise performance approaching the quantum limit, however its compression point (P_{-1dB} ~ –100 dBm) is too low for reading out mm and sub-mm astronomical receivers. Therefore, we explored the design of higher dynamic range JTWPAs to match the power requirements, and to optimise the performances of the JTWPA for this specific application. Our aim is to adapt the well-established Nb-AlO$_x$-Nb tri-layer fabrication technique used routinely for developing high-quality Superconductor-Insulator-Superconductor (SIS) tunnel junctions to fabricate our JTWPA. Therefore, we present in this paper our investigation of the feasibility of such technique in fabricating large number of lower critical current density junctions embedded in a coplanar waveguide (CPW). The preliminary results on a 500-junctions device are in line with the expected behaviour, showing a measured gain consistent with theoretical calculations which demonstrates the potential use of the tri-layer tunnel junction technology for the fabrication of JTWPAs.**

Keywords—travelling waves, parametric amplifier, quantum limited, Josephson junctions, critical current density

I. INTRODUCTION

The travelling wave parametric amplifier (TWPA) is an emerging technology capable to achieve broadband high gain in the microwave regime with quantum limited noise performance. These appealing characteristics derive from the use of reactive properties of a superconducting transmission line formed using either high kinetic inductance superconducting film (KI-TWPAs) [1] or embedded Josephson junctions (J-TWPAs) [2], to transfer energy from a high amplitude pump tone to a weak signal. TWPAs may find applications in many fields ranging from quantum computing [3], dark matter axion search [4,5] to astrophysics, where it could replace the semiconductor-based readout amplifiers on millimetre (mm) and sub-millimetre (sub-mm) telescopes to decrease power dissipation and enhance the overall signal to noise ratio [6].

For applications requiring high power handling capability, including the readout of mm and sub-mm receivers' down-converted IF signal, KITWPAs are often preferred due to their higher compression point (–60 dBm < P_{-1dB} < –40 dBm). However, it requires long transmission lines and high pump powers, which may increase heating and induce Johnson noise, hence compromising the noise performance. On the contrary, JTWPAs have high inductance per unit length, therefore requiring shorter transmission lines and lower pump power to achieve similar high gain. Nevertheless, these advantages come at the cost of a lower compression point (P_{-1dB} ~ –100 dBm), restraining its application as readout for mm and sub-mm receivers beyond –100 dBm compression point.

The characteristic of a JTWPAs, including the dynamic range or power handling capability, is dictated by the various parameters of the "unit cell" (see Fig.1) comprising one or several Josephson junctions (JJs) embedded within a small section of the transmission line that is subsequently cascaded to form the amplifier. As the properties of the JJ can be easier to control compared to searching and fabricating exotic superconducting material in the case of KITWPAs, we opt to explore the possibility for engineering the behaviour of JTWPAs to realise a quantum amplifier that could match the power requirements of mm and sub-mm receivers (–90 dBm < P_{-1dB}< –80 dBm), while retaining the quantum-limit noise performance.

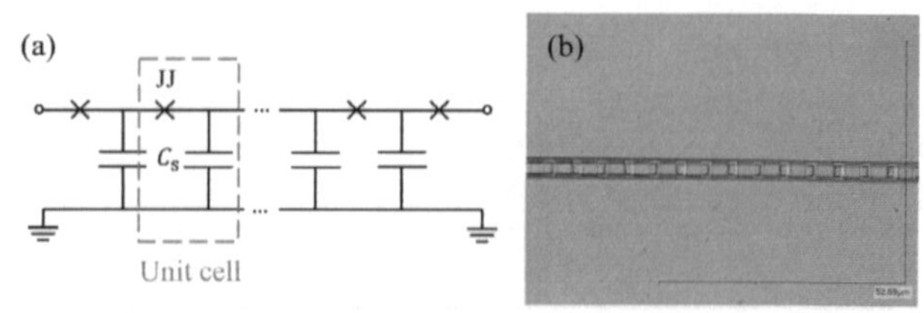

Figure 1. A unit cell cascaded repeatedly to form a full JTWPA. (a) Circuit diagram illustrating the unit cell where the cross represents the Josephson junction and C_s is the shunt capacitance to ground of the transmission line. (b) Optical microscope's image of a test device using false colours to identify the silicon substrate (grey), first niobium metallisation layer (red), SiO_2 dielectric layer (yellow), the Josephson junction (blue) and the second niobium metallisation layer (purple).

II. TEST DEVICES

A high gain JTWPA requires high inductance per unit length, which translates to the need for low critical current junctions around J_c < 0.3 kA/cm^2 with small area (< 1 um^2). This need for low critical current junction is in fact in contrast to the aim of Superconductor-Insulator-Superconductor (SIS) junctions' fabrication which often attempt to reach as high a J_c as possible. Given that the SIS fabrication procedure has been refined over many years of research and is more accessible for the astronomical community, we therefore intent to explore the know-how of such fabrication technique to engineer the junction properties that would suit to our objective. Furthermore, this opens up the opportunity to integrate the

[1]University of Oxford, Oxford, OX1 3RH, United Kingdom; [2]Institut de Radioastronomie Millimétrique (IRAM), Saint Martin d'Hères, 38406, France; [3]GEPI (CNRS UMR 8111), Observatoire de Paris, PSL Universite, Paris, 75014, France

JTWPA with the mm/sub-mm detectors in the near future e.g., fabricating the SIS mixer and the JTWPA on to a single chip using the same junction technology.

In order to understand how we can alter the well-established SIS junction fabrication recipe to realise the junction characteristic we need, i.e., lager number of junctions with low critical current density; we designed and fabricated test devices with 100, and 500 Nb-AlO_x-Nb junctions embedded in series within a niobium coplanar waveguide (Fig.1(b)). These devices are fabricated using four photolithography steps and an additional e-beam lithography step to define the junctions with different sizes (A_{JJ} = 0.25, 0.5 and 1 μm^2). We fabricate three sets of devices using 280-μm thick high-resistivity silicon wafers; and we altered the tri-layer deposition parameters for each set of devices to assess the junction's critical current density. The resulting measured J_c values (4, 3.5 and 2.5 kA/cm^2) as a function of the oxygen exposure $E = P_{O2} \cdot t$, where P_{O2} is oxygen partial pressure and t is oxidation time, are plotted in Fig.2; together with previous measurements at the same fabrication facilities (red dots) and data provided by our collaborators at GEPI, Observatoire de Paris (blue dots). The results clearly expose a $J_c(E)$ dependence proportional to $E^{-\beta}$, where the calculated $\beta = 0.4$ is consistent with results previously published in the literature [7].

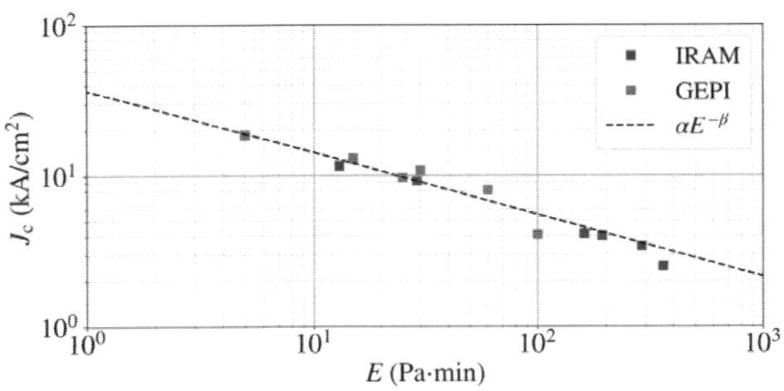

Figure 2. Measured critical current density J_c values as a function of the oxygen exposure E during the tri-layer fabrication process. The red dots with J_c = 4, 3.5 and 2.5 kA/cm^2 correspond to the set of devices fabricated at the Institut de Radioastronomie Millimetrique (IRAM) for this project; further data from previous fabrication rounds have been added to better estimate the trend. The blue dots correspond to data from our collaborators at GEPI, Observatoire de Paris. We measure a $J_c(E)$ dependence, where $\alpha = 36.8$ and $\beta = 0.4$.

III. Experimental Setup

We performed the DC and the RF characterisation of our devices at cryogenic temperatures (4K) to investigate the yield and accuracy of our refined fabrication recipe. First, we prepared the samples by mounting the chip on top of a cooper PCB board made using 2-mils thick Roger Duroid RO4350b material, patterned with a straight 50 Ω CPW line, except the area where the chip will be placed. The chip is bonded with aluminium bond-wires to make the electrical connections between the PCB and the chip, as shown in Fig.3(a). The PCB is then placed in a sample holder, as shown in Fig.3(b), and attached to a cooper bracket anchored to the cold plate of a 4K cryostat system.

The experimental setup inside the cryostat is displayed in Fig.4. We use a combination of a four-way splitter and a four-channels switch to measure up to 3 devices in a single cool-down. The DC characterisations were performed using a DC current source and a voltmeter, conducting a four-wire measurement of the chip using bias-tees on the RF lines. For the RF characterisation, we use a vector network analyser (VNA), and we calibrate the RF transmission measurement with a blank PCB sample inside the cryostat.

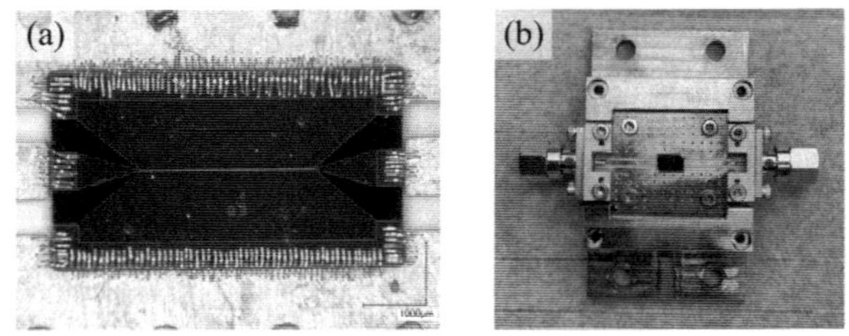

Figure 3. (a) Test device mounted and bonded to the PCB. (b) PCB mounted in the sample holder.

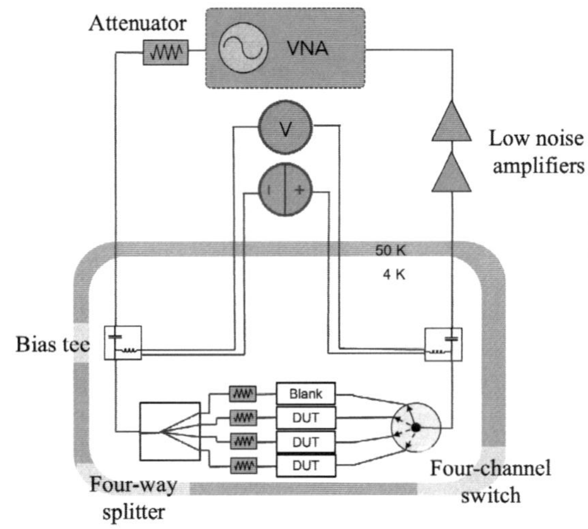

Figure 4. Experimental setup used to perform DC and RF characterisation of the test devices.

IV. Results and Discussion

Fig.5 shows the results obtained from a 100-junctions chip with a junction area A_{JJ} = 0.5 μm^2 and a critical current density J_c = 3.4 kA/cm^2. As shown in Fig.5(a), we managed to measure a clean RF transmission spectrum up to 10 GHz. We believe the deterioration of the transmission at higher frequencies is related with the way we mount our chip on the PCB, which is not optimal at the moment, as it requires long and bended bond-wires. Note that the observed feature around 7 GHz are artificial, originated from the calibration data using the blank sample. The unexpected dip at 3 GHz is under investigation at the moment, but should not affect our conclusion in this paper.

When we applied a DC current through the device, the RF spectrum barely changes until we reach I_{DC} = 140 μA, where the transmission starts to drop, reaching values lower than –15 dB for I_{DC} = 160 μA. This behaviour is consistent with the result obtained from the DC characterisation of the device, plotted in Fig.5(b) where the device exhibits no resistance until I_{DC} ~ 140-160 μA. Furthermore, when a single 3 GHz frequency tone (P_{in} > –25 dBm at the device's input port) is applied, we observe a 3rd harmonic generation without unwanted 2nd harmonic tone, suggesting that the transmission line exhibits the expected Kerr-3 (χ^3) non-linearity required for 4-wave mixing.

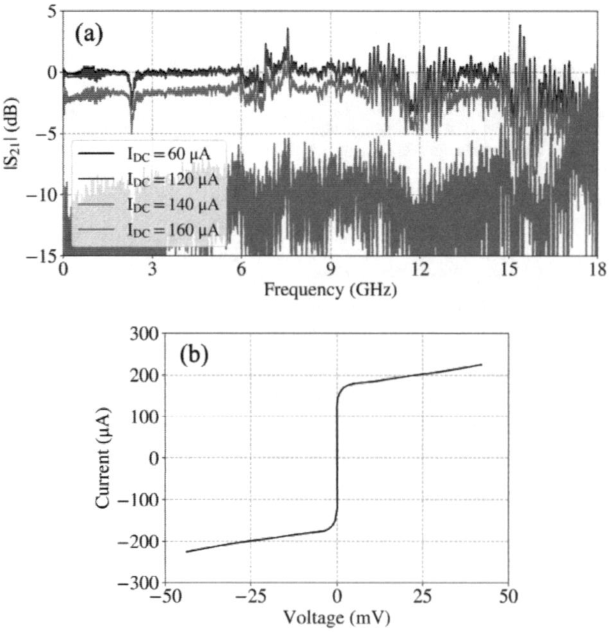

Figure 5. Characteristic for the 100-junctions device with A_{JJ} = 0.5 μm^2 and J_c = 3.4 kA/cm^2 measured at 4 K. We set the VNA signal power to P_{in} = −20 dBm with a 30 dB attenuator installed in the input line. (a) The S21 transmission profiles of the device under DC current biasing conditions. (b) Measured IV curve of the chip.

The DC and RF measured results for a 500-junctions sample with A_{JJ} = 0.5 μm^2 and J_c = 2.5 kA/cm^2 is plotted in Fig.6. The transmission measurements reveal a ripple effect as expected from the impedance mismatch generated by the CPW section loaded with junctions without additional features, which has a characteristic impedance higher than 50 Ω. The DC IV-curve characterisation plotted in Fig.6(b) shows a gap voltage V_g of about 500 times higher than the V_g of a single Nb-AlO$_x$-Nb junction, i.e., V_g = 2.8 mV; again, as expected. Fig.6(b) also shows an I_c value of 10 μA, very close to the expected value of I_c = 12 μA for the junctions with A_{JJ} = 0.5 μm^2 from this set of devices.

Given the encouraging results presented in Fig.6, we further tested this chip to search for possible gain, albeit with an expectation of a very small gain due to the short length of the device. Since the number of junctions is small and the critical current is higher than designated, we first estimate the expected gain for this device using the couple mode equations (CME) [8], assuming a shunt capacitance of C_s = 1 fF obtained from simulations of similar devices in Sonnet [9]. The results plotted in Fig.7(a) suggest that under this condition we should observe a gain up to 0.5 dB when pumping our device at 12 GHz, provided that the pump is very close to the I_c limit of the device. Similarly, a smaller gain below 0.1 dB should be observed when pumping the device at 6 GHz.

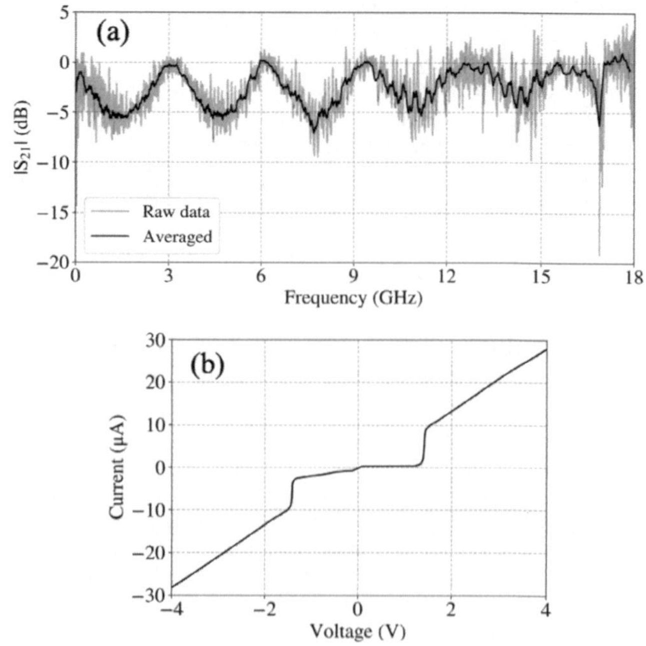

Figure 6. Measured results for a 500-junctions device with A_{JJ} = 0.5 μm^2 and J_c = 2.5 kA/cm^2 at 4 K. We set the VNA signal power to P_{in} = −20 dBm with a 30 dB attenuator in the input line. (a) RF transmission of the device. (b) Measured IV curve of the chip.

We performed the gain measurement of the device, where we input two signals using a power combiner at room temperature. One port of the power combiner is swept using a VNA with 50 dB attenuation, while the other port of the combiner is connected to a signal generator providing the pump tone. The output of the device under test is amplified using two low-noise amplifiers at room temperature and fed to the second port of the VNA for characterisation. Fig.7(b) shows the gain values measured when pumping the device at 6 and 12 GHz with P_{in} = −45 dBm and setting the power of the sweeping signal of the VNA to −20 dBm, i.e., −70 dBm at the power divider's port. Although the gain results are very noisy, given the extremely low power signal required to maximize the ratio between the pump amplitude and the signal for an optimal gain, we did indeed observe gain consistent with our theoretical expectation. Note that our theoretical calculations do not consider the impedance mismatch of the line, hence the lack of rippling effect shown in the simulated result. But the measured gain profile does indeed show a similar ripple periodicity as observed in the unpumped transmission profile of the chip, indicating that the gain is most probably real. More importantly, the peak gain is close to the value we expected from the simulation. The slightly lower gain values measured compared to our calculations is most probably caused by the pump depletion and additional losses of the line.

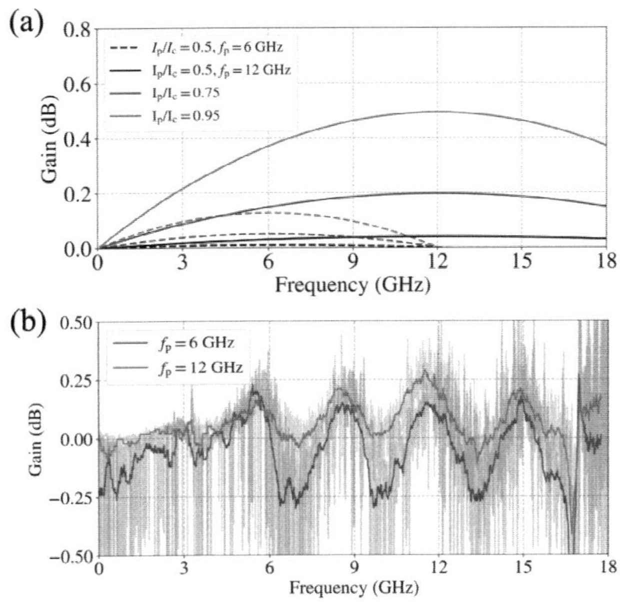

Figure 7. (a) Gain calculated using the analytical expression of the Couple Mode Equations (CME) for a 500-junctions device with $A_{JJ} = 0.5$ μm^2 and J_c = 2.5 kA/cm^2 when pumped with a pump at 6 GHz (dotted line) and 12 GHz (plain line). The different colours represent different amplitudes of the pump tone. (b) Measured gain for a 500-junctions device with $A_{JJ} = 0.5$ μm^2 and J_c = 2.5 kA/cm^2 when pumped with a strong tone at 6 GHz (red line) and 12 GHz (blue line). The raw data is plotted using light colours, while the darker lines correspond to a moving average of the data. We define the gain as the ratio of the transmission when the pump tone is on and off.

V. CONCLUSION

In this paper, we investigated the feasibility of utilising the tri-layer SIS junction fabrication technique for JTWPAs using a set of test devices. From the preliminary DC and RF characterisations, we observed a satisfactory transmission at frequencies below 10 GHz. We measured the DC and RF characteristics for a 500-junctions device, where the gap voltage shown to be 500 times the gap voltage of a single Nb-AlO_x-Nb junction as expected, and the critical current coincides with the value predicted. Finally, this device showed a measured gain consistent with theoretical calculations, demonstrating the potential use of the tri-layer tunnel junction technology for the fabrication of JTWPAs.

REFERENCES

[1] Byeong Ho Eom, et al., "A wideband, low-noise superconducting amplifier with high dynamic range", *Nature Physics,* Jul. 2012.

[2] C. Macklin et al., "A near-quantum-limited Josephson traveling-wave parametric amplifier", *Science,* October. 2015.

[3] L. Ranzani, et al., "Kinetic inductance traveling-wave amplifiers for multiplexed qubit readout", *Applied Physics Letters,* Dec. 2018.

[4] LC. Bartram, et al., "Dark Matter Axion Search Using a Josephson Traveling Wave Parametric Amplifier", *ArXiv:2110.10262,* Oct. 2021.

[5] C. Braggio, et al., "An Haloscope Amplification Chain based on a Travelling Wave Parametric Amplifier", *ArXiv:2205.02053,* May. 2022.

[6] M. Malnou, et al., "Performance of a Kinetic Inductance Traveling-Wave Parametric Amplifier at 4 Kelvin: Towards an Alternative to Semiconductor amplifiers", *Physical Review Applied,* Apr. 2022.

[7] S. K. Tolpygo, et al., "Properties of Unshunted and Resistively Shunted Nb/AlOx-Al/Nb Josephson Junctions With Critical Current Densities From 0.1 to 1 mA/μm2", *IEEE Transactions on Applied Superconductivity*, Jun. 2017.

[8] Kevin O'Brien, et al., "Resonant phase matching of josephson junction traveling wave parametric amplifiers", *Phys. Rev. Lett.*, Oct. 2014.

[9] Sonnet Software. https://www.sonnetsoftware.com/.

SIS photon detectors for THz observations beyond the gap energy

Ayako Niwa[1,2], Hajime Ezawa[2,3], Tomonori Tamura[2], Hiroshi Matsuo[2,3]

***Abstract*— Interferometry is advantageous for high-resolution observations, but the sensitivity of heterodyne receivers is limited due to quantum noises especially above 1-THz. Photon Counting Terahertz Interferometry (PCTI), we proposed, uses direct detectors to measure intensity fluctuations which are caused by photon bunching for aperture synthesis, and it keeps high sensitivity in terahertz band. Photon counting detectors for PCTI needs low-noise and high-time resolution and developments of them are undergoing with antenna-coupled SIS photon detectors. PCTI's first observation is planned in the later 2020s from Antarctica where water vapor is sufficiently small that enables ground-based terahertz observations. This paper shows a concept and a design of 1.5 THz photon counting detectors to observe 205 μm [NII] emission.**

Keywords—Intensity interferometry, Superconducting detector, Terahertz observation

I. Introduction

Terahertz observation provides information on thermal radiation and atomic emission lines of celestial bodies and holds the key to elucidating the state of the early universe and the process of galaxy and star formation. However, in heterodyne interferometers, the sensitivity is limited by the quantum noise, and high-angular resolution observations above 1 THz have not yet become a reality (see Fig. 1). Therefore, a new technology is required for high-resolution observation methods in the Terahertz band.

We propose Photon Counting Terahertz Interferometry (PCTI) for high-resolution observations from space in the Terahertz band. PCTI measures photon bunching by high-time resolution to determine delay time and performs aperture synthesis. Since PCTI's system needs direct detectors, PCTI performs below the quantum limit of the heterodyne receiver in the high-frequency region. We developed photon counting detectors for PCTI in the 500 GHz band with Nb/Al/AlOx/Al/Nb SIS junctions with about 1 pA leakage current at below 0.8 K[2]. As discussed below, the leakage current must be suppressed to measure photon bunching. Now we try to develop 1.5 THz photon counting detectors to demonstrate PCTI above 1 THz, which is our target.

PCTI's first test observations will take place in the late 2020s in Antarctica. Antarctic highlands have been found to have very good atmospheric transmission in the terahertz band compared to Chajnantor in Chile and Mauna Kea in Hawaii, due to the low water vapor content in the atmosphere[3]. PCTI will use two telescopes with an aperture of 30 cm (Antarctic 30-cm Submm Telescope). The first telescope was developed by University of Tsukuba, which are equipped with a 500 GHz heterodyne receiver. The telescope will be sent to Antarctica in late 2023 in cooperation with National Institute of Polar Research. We will assemble the second 30-cm telescope to follow the first and it will arrive in Dome Fuji as early as 2027. Then, both telescopes will be equipped with 1.5 THz photon detectors for test observations. Candidate observational objects are Carinae nebula and 30Dor which have strong 205 μm [NII] emission lines.

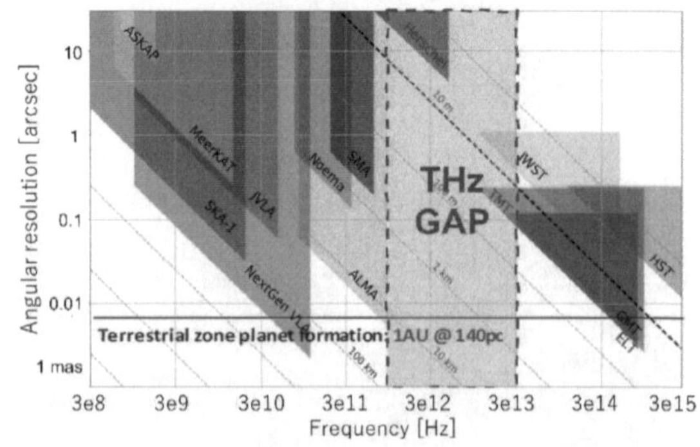

Fig.1 Observation target derived from Selina Robert et al., 2018[1].

II. Detector Concept

Our detectors must be low noise and fast response to measure photon bunching. Josephson junctions have fast response, but their leakage current will prevent identifying single photon. To suppress leakage current, our detectors use antenna coupled SIS junctions whose diameter is a few micrometers. We achieved a reduction in leakage currents of Nb/Al/AlOx/Al/Nb SIS junctions, and this study also uses the same structure.

A. SIS leakage current requirements

Thermal photons form photon bunching which is a signal for us by Bose-Einstein statistics which describe intensity fluctuation as eq. (1): P represents input power. If we observe from the Antarctica on summer, background intensity fluctuation could be estimated about $6 \times 10^{-16}\ \mathrm{W}/\sqrt{\mathrm{Hz}}$ for temperature of $T_B = 50$ K, bandwidth of $\Delta\nu/\nu = 10\%$. On the other hand, Noise Equivalent Power (NEP) of SIS junctions is described as eq. (2). So, background limit of leakage current is estimated that must be below 3 nA, assuming $\eta \approx 0.3$.

$$\mathrm{NEP} = \sqrt{2P(h\nu + kT_B)}\ [\mathrm{W}/\sqrt{\mathrm{Hz}}] \quad (1)$$

$$\mathrm{NEP} = \frac{h\nu}{\eta}\sqrt{\frac{2I_{leak}}{e}}\ [\mathrm{W}/\sqrt{\mathrm{Hz}}] \quad (2)$$

[1]University of Tsukuba, Tsukuba, Ibaraki 305-0006, Japan; [2]National Astronomical Observatory of Japan, Mitaka, Tokyo 181-8588, Japan; [3] The Graduate University for Advanced Studies, SOKENDAI, Mitaka, Tokyo 181-8588, Japan

B. Ground plane material

Our detectors consist of antenna coupled Nb/Al/AlOx/Al/Nb SIS junctions and detect photons by Cooper pair breaking. The double slot antenna is connected to the parallel connected twin junction (PCTJ) by the coplanar waveguide (CPW). PCTJ is a resonant circuit constructed by two SIS junctions. The CPW which is the same material used in the SIS junctions, causes transmission losses due to pair breaking. So, our detectors install an Al ground plane for the antenna and the CPW. Al is a superconductor at 0.8 K but behaves like a resistor for the 1.5 THz band, more than ten times higher frequency than superconducting gap energy.

C. Towards broadband observation

Previous work[4] achieved low leakage current with 0.3-kA/cm^2 critical current densities and 3 μm×3 μm square forms. Since these detectors are designed for the purpose of experiments in laboratory, so critical current density is relatively low and their bandwidth is limited. In this study, we try to expand the bandwidth keeping low leakage current by minimizing SIS junctions to 1 μm.

III. Detector Design

Fig.2 shows the design of 1.5 THz photon counting detectors. The double slot antenna and the PCTJ were tuned to 1.5 THz. These detectors use quartz wafers for lower stray capacitance. Impedance from antenna to PCTJ is matched to 100 Ω, where required critical current densities is about 1.3-kA/cm^2 for 1 μm circular junctions.

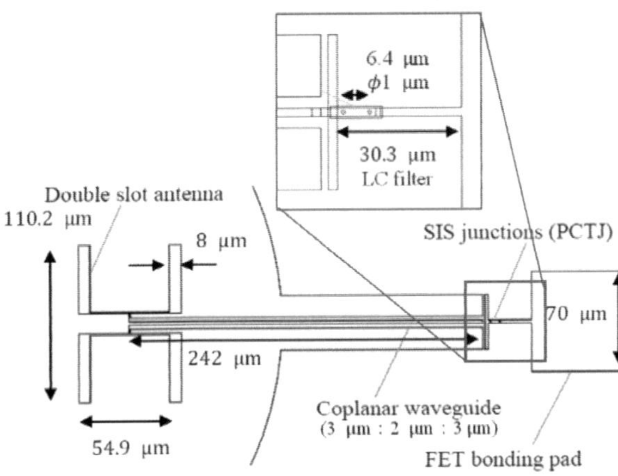

Fig.2 Design of 1.5 THz photon detector.

This study determined and fixed the ratio of antenna length to width referring to a previous paper[5] since this ratio affects directivity. Antenna parameters, length (and width) and slot line width were tuned by parametric analysis. Fig.3 is the calculation results of reflection coefficient of the antenna with final design parameters calculated by using FEKO, Altair. The reflection coefficient value is -38 dB at 1.5 THz. Fig. 4 shows the calculation results of 1.5 THz antenna beam pattern. This result is calculated by using HFSS, Ansys, by using the model whose antenna feed point is connected to wave port by CPW (see Fig.5). Calculation results of φ = 0 and 45 deg are slightly asymmetry due to CPW.

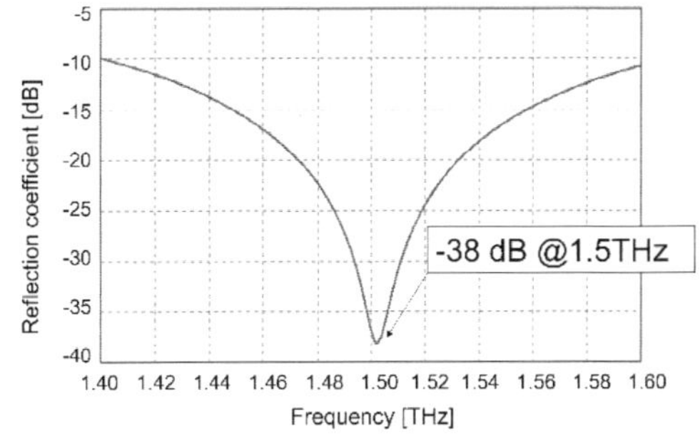

Fig.3 Calculation results of reflection coefficient of 1.5 THz double slot antenna.

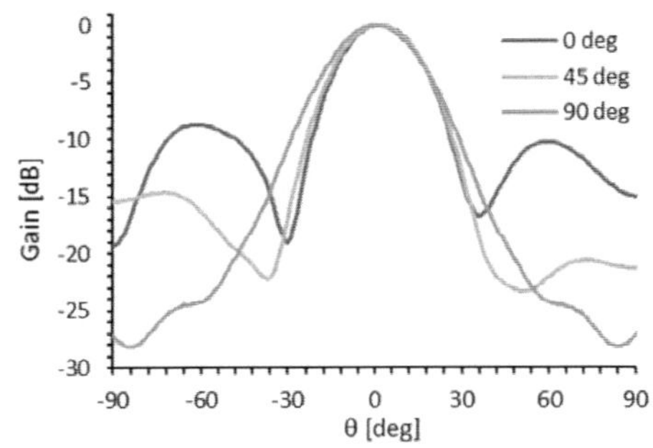

Fig.4 Calculation results of beam patterns for each azimuthal angle φ of 1.5 THz double slot antenna. The x-axis gives the polar angle θ.

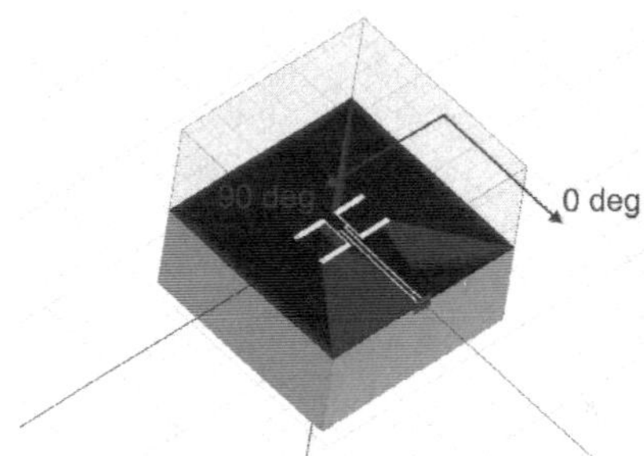

Fig.5 The double slot antenna model for calculating beam

Acknowledgments

Acknowledgments to JSPS KAKENHI Grant Numbers 20K20345 and 20K20346, and JAXA basic R&D for future space science missions. And this work is supported by Altair Engineering Inc. for the usage of FEKO for electromagnetic simulations.

Reference

[1] Selina Robert, Murphy Eric, McKinnon Mark, et al. 2018, Proc. SPIE 10700. https://doi.org/10.1117/12.2312089

[2] Hajime Ezawa, Hiroshi Matsuo, Masahiro Ukibe, Go Fujii, and Shigetomo Shiki. 2019, Journal of Low Temperature Physics, 194:426–432. https://doi.org/10.1007/s10909-019-02149-y

[3] P.Tremblin, N.Schneider, V.Minier, G.Al.Durand, and J.Urban. 2012, Astronomy & Astrophysics, 548, A65. https://doi.org/10.1051/0004-6361/201220420

[4] Hajime Ezawa, Hiroshi Matsuo, Masahiro Ukibe, Go Fujii, and Shigetomo Shiki. 2020, Journal of Low Temperature Physics, 200, 226-232. https://doi.org/10.1007/s10909-020-02513-3

[5] Daniel F. Filipovic, Steven S. Gearhart, Gabriel M. Rebeiz. 1993, IEEE Transactions on Microwave Theory and Techniques, Vol. 41, No. 10. https://doi.org/10.1109/22.247919

Design, Operation, and Characterization of a Laboratory Spatial-Spectral Fourier Transform Interferometer

Jeremy P. Scott*[1], Chris S. Benson[1], Locke D. Spencer[1]

***Abstract*—A spatial-spectral interferometer is an important milestone for far-infrared astronomy. Here we present the design, operation, and characterization of a laboratory-based spatial-spectral Fourier transform interferometer.**

Keywords*—*spectroscopy, far-infrared, interferometry, instrumentation

I. Introduction

The Far-Infrared (FIR; 0.3 – 30 THz) remains one of the least explored bands of the electromagnetic spectrum in astronomy despite how crucial FIR observations are for studying some of the most important topics in astronomy including the formation and evolution of planets, stars, and galaxies, and the conditions for life sustaining planets [1, 2]. Research in this area has been limited in-part by insufficient spatial resolution due to the diffraction limit of FIR radiation. Spatial resolution is limited by the diameter of the entrance aperture for a conventional telescope and by the separation of input apertures for a spatial interferometer. Thus, a spatial-spectral Fourier Transform Interferometer (2FTI) is a promising technology in future FIR astronomy missions, providing enhanced spatial resolution over a broad spectral range [2].

Though spectral and spatial interferometry individually are mature, their combination within a single instrument is novel. Our instrument (Fig. 1) consists of two input apertures which sample the electromagnetic field at two points in space. The field under observation is generated by a source incident on an off-axis parabolic mirror producing planar waves that simulate a distant source. One aperture is positioned on a linear translation spatial-stage which modulates the distance between the two apertures setting the baseline separation of the spatial interferometer. Light is then relayed through a series of flats to a double-rooftop configuration on the spectral-stage. The spectral-stage modulates the optical path difference between the two beams and provides the functionality of a Fourier Transform Spectrometer (FTS). The beams are then combined at a beamsplitter and focused onto the detector plane using an off-axis parabolic mirror. The resulting signal is an interferogram with intensity modulation dependent on the optical path difference between the two beams and the aperture baseline.

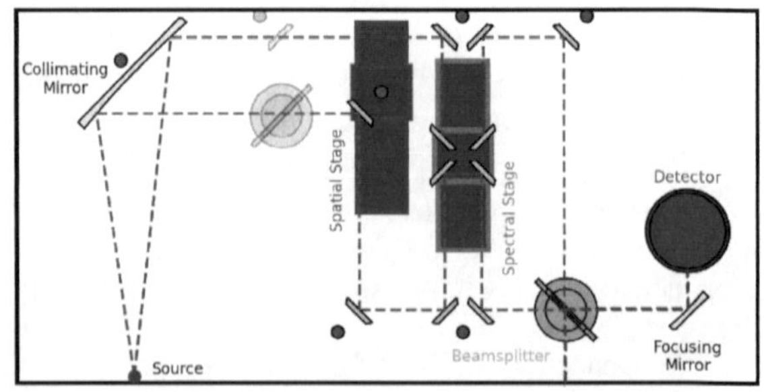

Fig 1. Diagram of 2FTI on laboratory optics table. FTS mode is obtained by adding the translucent components into the system.

II. Instrument Operation

An observation consists of setting a fixed baseline and slewing the spectral-stage in such a way that the resulting interferogram symmetrically samples both sides of zero-path-difference. Noise in the signals is averaged down either using a step and integrate method or by performing a series of rapid scans. The baseline is then changed and subsequent interferograms are recorded.

If the two input apertures overlapped, resulting in zero baseline, our instrument would be equivalent to an FTS. This configuration is achieved with our system by introducing the translucent components in Fig. 1. The resulting interferogram would then probe exclusively the temporal coherence of the source with its Fourier transform providing the spectral components (Fig. 2).

At non-zero baselines, a 2FTI probes the spatial coherence of the input radiation field in addition to temporal coherence. Spatial coherence manifests as the visibility of interference fringes and is determined by the intensity distribution of the source as related by the Van-Cittert Zernike theorem [3]. The Fourier transform of the resulting interferogram produces a spectrum with frequency dependent modulation determined by the spatial coherence of the radiation at the sampled baseline (Fig. 2). By measuring the modulation at a particular frequency at multiple baselines, it is possible to reconstruct a monochromatic intensity distribution of the source by use of the Van-Cittert Zernike theorem. Repeating this process for all frequencies produces a

[1]Institute for Space Imaging Science, University of Lethbridge, Lethbridge, Alberta, T1K 3M4, Canada

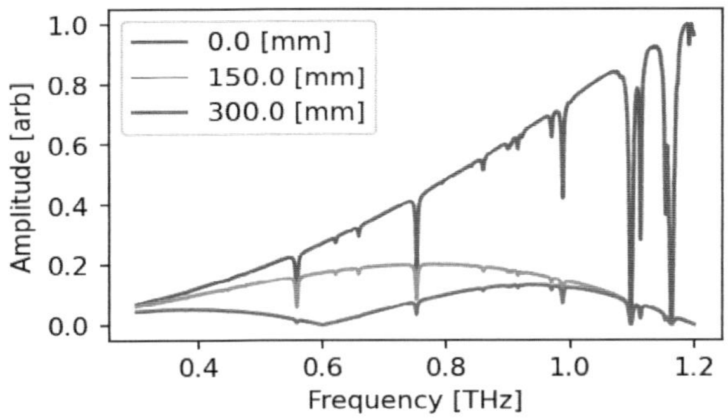

Fig 2. Model spectra of a 900 K blackbody with water absoption lines obtained for our 2FTI at various baselines viewing a 1 mm source. Modulation at non-zero baselines is due to the spatial coherence properties of the wavefront at the sampled baselines.

hyper-spectral image of the source with spatial resolution determined by the maximum baseline.

III. Demonstration of Visibility Measurement

We demonstrate visibility measurements using a tunable THz coherent source. Sinusoidal interferograms for a selection of frequencies are obtained at multiple baselines using the method outlined above. Amplitudes extracted from the resulting power spectra are used as unnormalized visibility measurements. The source emits though a 10 mm diameter circular lens, and we model the source as a circular intensity distribution. The expected visibility curve is then given by the absolute value of the besinc function. Using this model, we fit and normalize our visibility measurements as shown in Fig. 3. The fitted curves are good approximations of the data with deviations attributed to additional structure in the source shape.

IV. Instrument Characterization

The primary collimating mirror has an effective focal length of 1,200 mm and can provide a collimated beam 415 mm in extent. Circular 135 mm diameter flats are used as input apertures and optical relays. When placed at 45 degrees with respect to the collimated beam, the interferometer achieves a minimum and maximum baseline of 135mm and 310 mm, respectively, resulting in a maximum spatial resolution of 1 mm at 0.55 THz. The 5x5 transition-edge-sensor detector array is housed within a ~4 K cryostat. The array is a square with 18 mm linear dimension and coupled to the radiation field through a f/4 feedhorn array that is matched to the focusing mirror with an effective focal length of 240 mm. A magnification of -0.2 is obtained through the combination of the collimating and focusing mirror, and given the size of the detector array, a 100 mm field of view is obtained at the source plane. With a nominal optical path length of 3,400 mm and an aperture stop 60 mm in diameter at the focusing mirror, a 26 mm unvignetted field of view is obtained at the source plane. The maximum symmetric optical path difference for all possible baselines is 500 mm resulting in a spectral resolution of 0.3 GHz. The beamsplitter bandpass and detector spectral response limit the bandwidth to ~0.4 – 2.1 THz. Assuming a Nyquist frequency of 2.5 THz and using the measured detector cutoff frequency of 1.5 kHz, the spectral stage can be scanned at a maximum speed of 27 mm/s.

V. Conclusion

In this work we presented the design of a laboratory spatial-spectral Fourier transform interferometer and described its operation. The principles of visibility measurements were outlined, and a demonstration was provided for a circular laser source. Finally, we summarized the interferometer performance characteristics.

References

[1] Linz, Hendrik, et al., "Bringing high spatial resolution to the far-infrared." *Experimental astronomy*, Jun. 2021.
[2] Helmich, Frank P., et al., "FIRI—A far-infrared interferometer." *Experimental Astronomy*, Mar. 2009.
[3] Wolf, Emil. "A macroscopic theory of interference and diffraction of light from finite sources II. Fields with a spectral range of arbitrary width." *Proceedings of the Royal Society of London. Series A. Mathematical and Physical Sciences*, Jun. 1955.

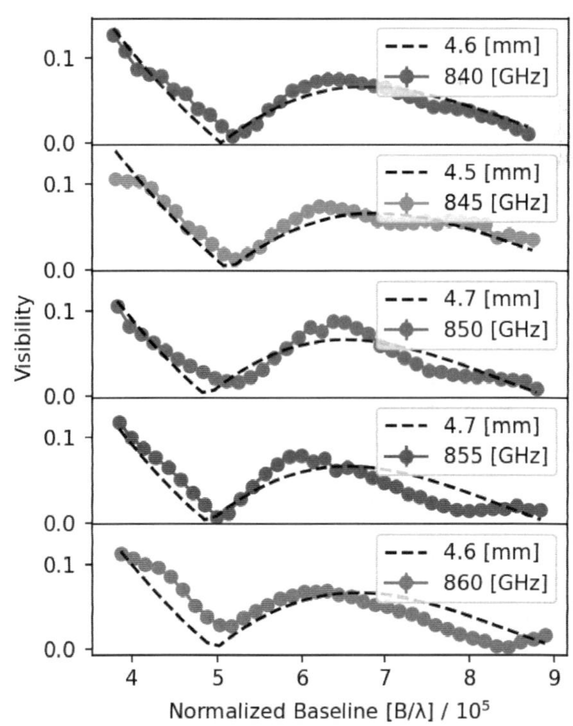

Fig 3. Measured (circles) and fit (dashed) visibility curves for a laser source at select frequencies. Values associated with fit indicate the radius of the circle that would produce the fit curve.

Development of the wide IF 230 GHz SIS mixer for KVN-Pyeongchang VLBI station

Naeun Shin [1,2], Jung-Won Lee [*,2], and Ming-Jye Wang [3]

***Abstract*—A 230 GHz SIS mixer is designed for the new Korean VLBI Network (KVN) telescope in Pyeongchang, South Korea. The design is focused on a good performance in the IF frequency range of 4-16 GHz over the RF band from 211 to 275 GHz. The designed mixer has series-connected junctions, and two tuning circuits are introduced. Type A uses transmission lines between the probe and the junction array to match the junction embedding impedance to the optimum source impedance. Type B uses junctions as a tuning element in transmission lines to replace the highly capacitive strip lines which increase capacitance from the tuning circuit. The capacitance of the whole circuits, required to be minimized for wider IF bandwidth, is 190 fF and 210 fF respectively. In the simulation, over the whole RF band, two designs show single sideband receiver noise temperatures less than two times the quantum limit, and the receiver temperatures vary less than 5 K within IF frequency from 4 to 16 GHz.**

***Index Terms*—receiver, SIS mixer, submillimeter**

I. Introduction

THE Korean VLBI Network (KVN) comprises three stations operating four receiver bands from 22 to 129 GHz. In addition to the current telescopes, the fourth telescope which is operable up to 230 GHz is being constructed in Pyeongchang, located in the eastern part of South Korea.

Through the recent outstanding works of the Event Horizon Telescope (EHT), the importance of the 230 GHz receiver is highlighted [1]. Optical depth at 230 GHz is about 0.3 (T_{sys} = around 200 K) in the cold winter season at Seoul. We expect similar atmospheric conditions at KVN-Pyeongchang station. The planned facility receivers of the new telescope are composed of a compact tri-band receiver (CTR) receiving 22/43/86 GHz [2] and 150/230 GHz receivers.

IF instantaneous bandwidth of KVN was just 2 GHz for all frontends in the beginning, constrained mainly by the processing bandwidth of the VLBI backends available around 2010. With the advent of a wide-band digitizer and data recorders working up to 32 Gbps, the mixers must be upgraded to provide a wider IF bandwidth of 16 GHz. At 230 GHz frequency, SIS mixers are still more sensitive than HEMT amplifiers and such requirement for wide IF bandwidth is the main motivation of this development.

In this paper, we present two wide IF bandwidth designs for KVN 230 GHz SIS mixers focusing on low noise performance at an IF frequency range of 4-16 GHz over the RF frequency from 211 to 275 GHz. A dominant factor that limits the IF bandwidth of an SIS mixer is often the RC time constant of the IF circuit in a mixer [4]. Although the lower normal resistance, R_N less than 10 ohms of a high Jc junction giving source impedance of several R_N at IF frequency helps for wide bandwidth by simpler matching to a following LNA, we adhere to normal Nb-Al/AlOx-Nb junctions for this development. It is because lowering capacitance at the IF frequency range using tri-layer-based junctions is another effective way to widen bandwidth by avoiding mixer gain roll-off.

An SIS mixer is generally composed of an RF tuning circuit to load optimum embedding impedance to a junction or junction array and an RF choke for transmission of IF signal. Therefore, limiting the total capacitance of such elements as an RF tuning network, RF choke, and junctions at the IF frequency range is required to achieve wider IF bandwidth. In our design, we utilized series-connected junctions for the lower total geometrical capacitance which still affects on a signal at the IF frequency range. We focused on lowering the capacitance of the RF tuning circuit and RF choke while increasing other mixer performances. In order to calculate the capacitance and the noise temperature, RF/IF circuit simulations were progressed using Supermix[5][6], a software package for the simulation of SIS mixers. The two types of circuits for the mixers are described in the following sections.

[1] Department of Physics and Astronomy, Seoul National University, Gwanak-ro 1, Gwanak-gu, Seoul 08826, Republic of Korea
[2] Korea Astronomy and Space science Institute, Daedeok-daero 776, Yuseong-gu, Daejeon 34055, Republic of Korea
[3] Academia Sinica Institute of Astronomy and Astrophysics, 11F of AS/NTU Astronomy-Mathematics Building, No.1, Sec. 4, Roosevelt Rd, Taipei 10617, Taiwan
Corresponding author: J.-W. Lee (email: jwl@kasi.re.kr).

TABLE I
Parameters used for the probe design

substrate material	quartz
waveguide size (μm)	940 x 470
probe radius (μm)	00
probe height (μm)	100
probe angle (degree)	45
substrate thickness (μm)	100
air gap height above substrate (μm)	50
air gap height below substrate (μm)	100
probe impedance (μm)	75 + 0.1 Ω

II. Mixer Designs

A. Probe

The size of the input waveguide is chosen as WR3.7 to cover the RF frequency range from 211 to 275 GHz. We adopted a single-sided probe [7] which has an RF choke at one side. The RF signal is transferred to the channel as shown in Figure 1. A channel where the substrate is located is perpendicular

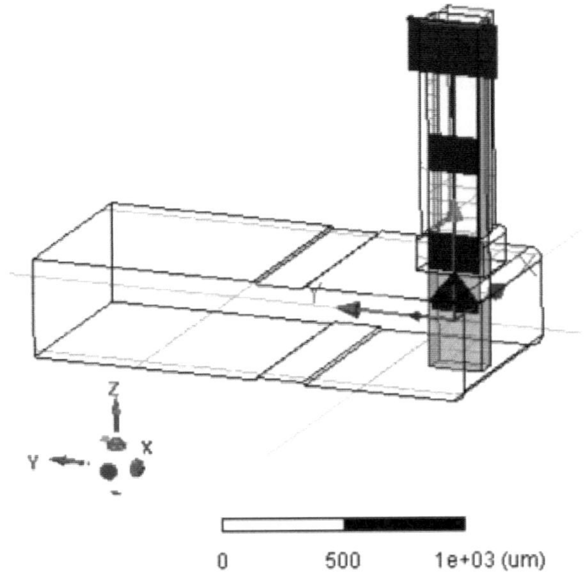

Fig. 1. A transition from a full height waveguide to probe simulated in HFSS. The substrate is extending out of the input waveguide and the thin-film microstrip circuit including junctions is patterned on the substrate. The probe has a home-plate shape; a combination of an isosceles triangle and a rectangle. In Table I, the length of two identical sides of the triangle is denoted as a probe radius, and the angle between them stands for a probe angle. A height of a rectangle is expressed as a probe height.

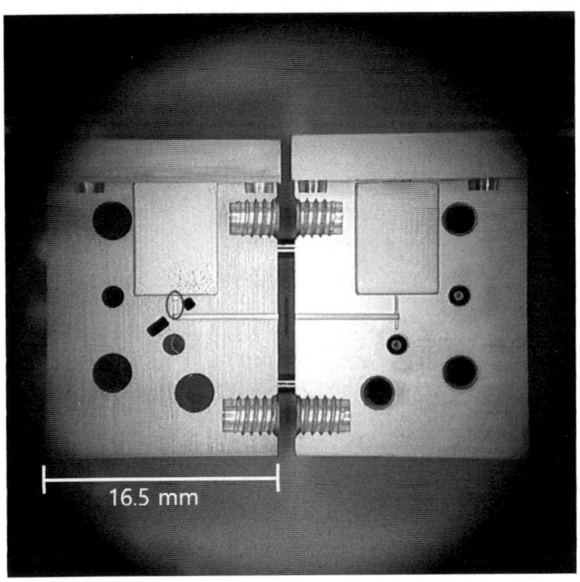

Fig. 2. Photo of the top and the bottom block of the mixer block (from left to right). The WR3.7 input waveguide is seen at the center. The mixer chip will be placed in the red-circled slot.

to the axis of the propagation in the waveguide. SIS junctions integrated with a superconducting tuning circuit and an RF choke are laid on a quartz substrate. We optimized a feed point impedance using Ansys HFSS, the 3D EM simulator. Several structures in the mixer block were used to optimize the feed point impedance having a compact and symmetric

TABLE II
JUNCTION PARAMETERS

area (μm^2)	1.1 × 1.1
number of junctions	3
normal resistance (Ω)	26
fractional bandwidth	0.26
specific capacitance, C_s (fF/μm^2)	80
critical current density, J_c (A/cm^2)	6700

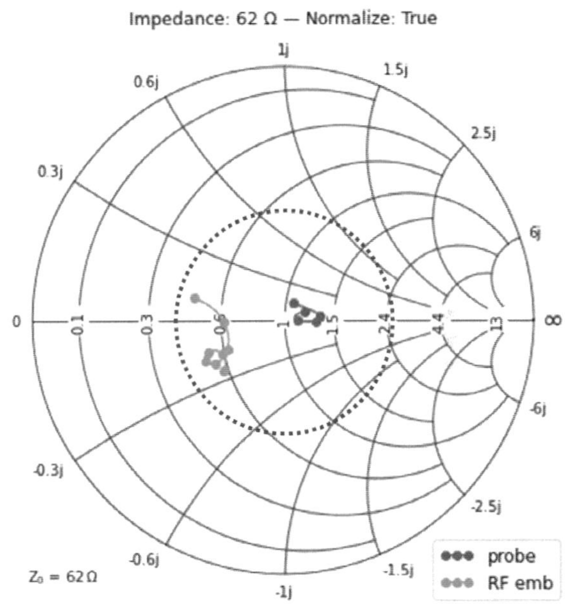

Fig. 3. The power from the feed point is transferred to the junction through tuning elements and it is finally seen to the junction as an RF embedding impedance. The RF embedding impedance of the Type 1 circuit from RF 211 to 275 GHz is located inside of the $|\rho| = 0.4$ circle indicated as a red dashed line. The Smith chart is normalized with an optimum source impedance.

tear-drop impedance trace on the Smith chart as shown in Figure 3. This method facilitates providing proper embedding impedance for junctions over the RF frequency range. First, the home-plate shape probe has more extra tuning parameters than the radial probe. The optimized parameters of the home plate probe are listed in Table I. In addition, widening the substrate channel under the first section of the RF choke decreases the imaginary component of the probe impedance. There are capacitive tuning steps on the long sides of the input waveguide near the probe reducing the height of the waveguide by 5%. Those structures decrease influences by non-propagating modes and return loss. At last, an air gap is added under the substrate to increase the cutoff frequency of the dielectric-loaded channel. As a result, the feed point impedance is 73+j0.1 Ω on average with a small variance. This probe impedance is shown in Figure 3. The mixer block containing the waveguide probe had been fabricated as seen in Figure 2. It will be combined with the mixer chip which consists of junctions, tuning elements, and an RF choke.

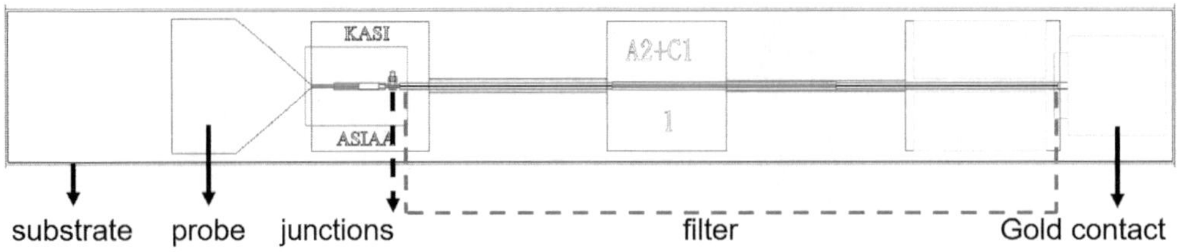

Fig. 4. The configuration of the mixer chip. The layers are denoted in different colors and the number in the center is information about the circuit and the number of columns on the wafer. The largest rectangle on the edge is a substrate. The probe at the left is a connection between the waveguide to the thin-film microstrip. The bonding pad is located at the last section of the RF choke followed by a gold contact. The gray dotted line is indicating a range of the RF filter composed of CPWs.

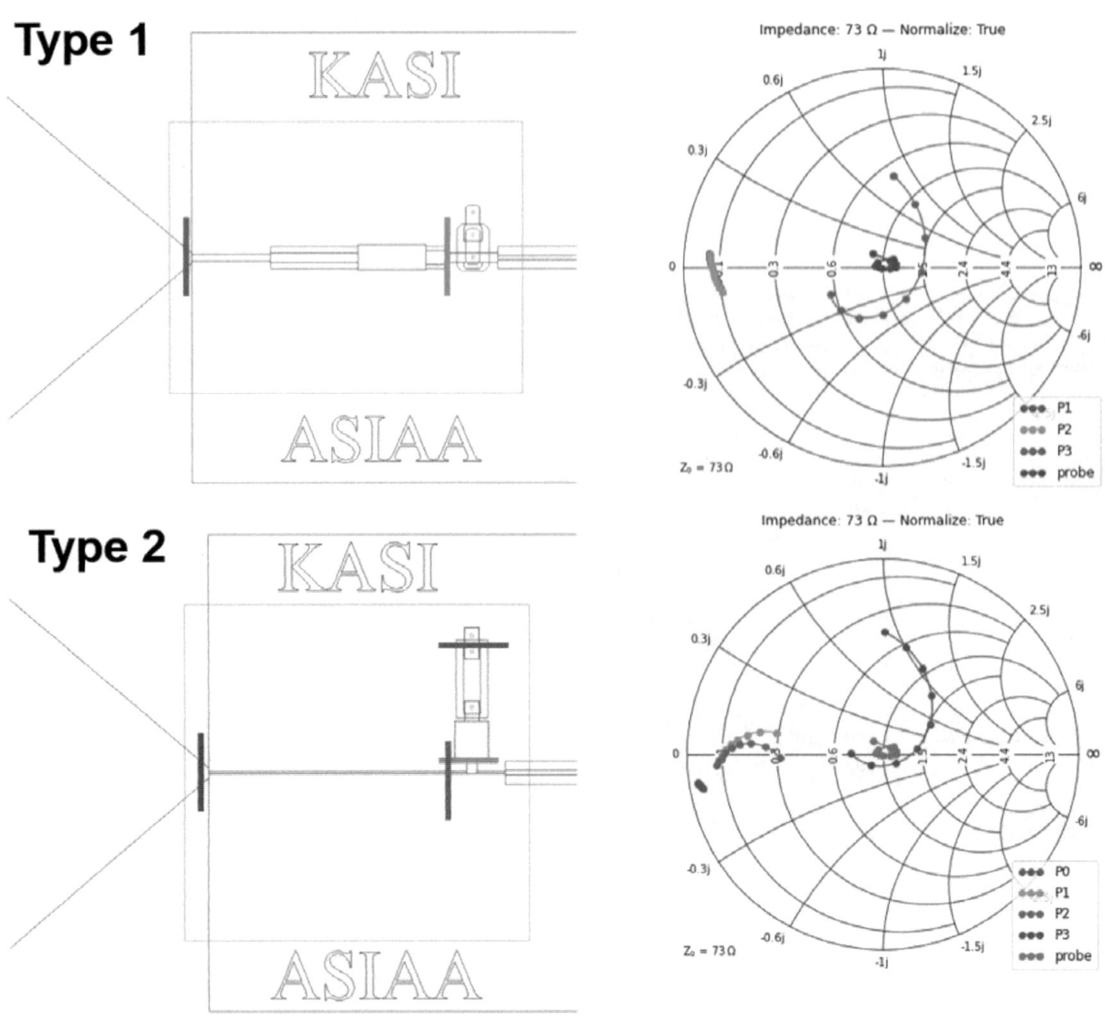

Fig. 5. (Top) An enlarged view of the Type 1 circuit between the probe and an island where the junctions are located. Thick lines are indicating the respective planes where the impedances are obtained. We can see the impedance is matched to the probe impedance from P1 to P3 through tuning elements. (Bottom) A maximized view of a Type 2 circuit showing an impedance matching from P0 to P3 through microstrips and CPWs.

B. Junctions and tuning circuits

The series-connected array of three Nb-Al/AlOx-Nb junctions is used in designs. The required critical current density is about 7000 $\mathrm{A/cm^2}$, estimated from the parameters in Table II. Two of the three junctions are placed on an island structure, as seen in Figure 4 and Figure 5, which is expected to have small series inductance to the capacitive junctions. To simulate the performance of the elements in the mixer chip, we used a program named Supermix, which performs harmonic balance for large signal analysis and small signal analysis based on Tucker's quantum mixer theory. The RF tuning elements composed of microstrips and coplanar waveguides (CPW) tune out the residual capacitance of the junction array.

There have been two approaches regarding junctions in a tuning circuit: that considers a junction as a lumped element as a whole [8] or an element used in distributed circuit[4]. Since the noise and conversion gain of the SIS mixer depends on the embedding impedance seen by the junction [8], the first approach focused on locating the embedding impedance nearby the optimum source impedance using only external transmission lines. The optimum source impedance is calculated by the Ke and Feldman formula based on the Tucker theory [3] and the boundary of the advantageous operating region is $|\rho| = 0.4$ where the $|\rho|$ is the reflection coefficient from the optimum source impedance. The RF source impedance combined with junction capacitance of Type 1 is shown in Figure 3, and it is well positioned near the optimum source impedance in RF bandwidth.

While the first one assumed the junction array essentially as a lumped junction, the second approach handled a junction as a series or shunt element in a transmission line. In this way, a low impedance tuning transmission line section that has a larger capacitance can be replaced with a junction partially tuned with a series inductive line and a less capacitive shunt line. In effect, the junction itself can be re-used as low impedance matching elements and the total capacitance of the junctions can be reduced in series connection [4].

After the tuning network and junctions, an RF choke composed of multiple CPWs follows, working as cascaded quarter-wave transformers. It is identical in two designs. We used scikit-rf, a Python package for linear RF circuit analysis, to calculate impedance transformation by the CPWs of the RF choke. The effective lengths of physical lines for the determined impedances are calculated using the package and implemented in the codes in Supermix. We confirmed most of the local oscillator (LO) power is transmitted to junctions and blocked by the RF choke from the simulation in Supermix. The additional surface impedance of a superconducting conductor is also taken into account.

As shown in Fig 5, both designs show coupled impedances with the feed point impedance in the RF bandwidth. Although the couplings are not fully optimized, it was the most enhanced result to match the impedance in the low and high frequencies at the same time with low capacitance tuning elements. The coupling can be advanced in the future with the improved design.

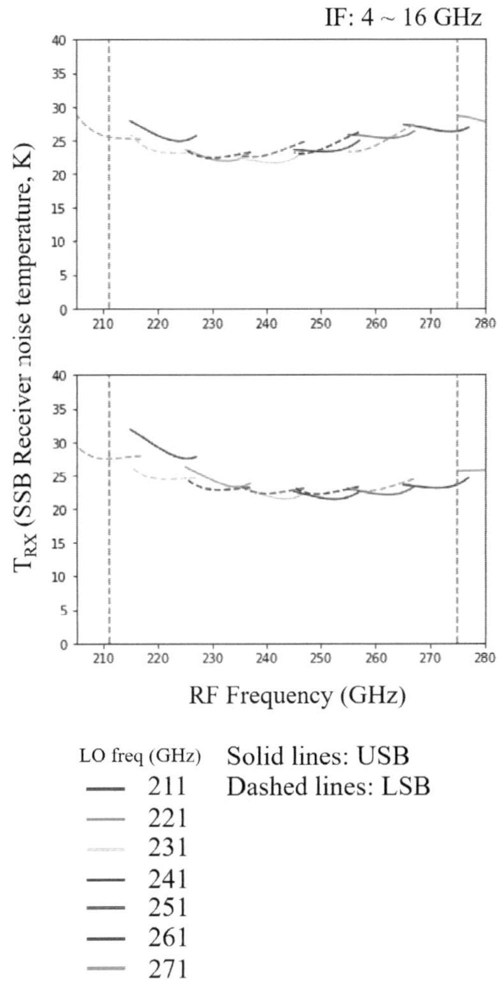

Fig. 6. The performance of designs simulated in Supermix. The top is for Type 1 and the bottom is about Type 2. T_{rx} is an SSB noise temperature and is calculated using a gain and a mixer noise temperature. Solid lines denoted upper sideband (USB) receiver noise temperatures with specific LO frequencies from 211 to 271 GHz and dashed lines are for lower sidebands (LSB).

III. Simulated Performance

A. Results from Supermix

The total capacitance of the circuit is critical to the IF bandwidth. We estimate the capacitance of the whole circuit by measuring the admittance at the IF output port using Supermix. Type A and B shows a total capacitance of 190 fF, and 210 fF, respectively, and these are compatible with previous 230 GHz mixers, which generally show capacitances value larger than 200 fF[9]. Two designs are expected to have an IF bandwidth of up to 16 GHz. For IF frequency range from 4 to 16 GHz, Type A and B show single sideband (SSB) mixer receiver noise temperatures less than two times the quantum limit, and T_{rx} varies less than 5 K throughout the RF frequency range from 211 to 275 GHz. Type A shows less variation in the receiver temperature than Type B in the whole RF band. Type B has flatter sensitivity in the IF band with a given LO frequency than type A except for LO at 211 GHz.

B. Analysis of the IF embedding impedance

The USB receiver noise temperature reaches the lowest point around IF frequency 14 GHz and increases a few K in other frequencies in case of LO 241 GHz as Figure 6 shows. The elevation of the noise temperature is originated from the decrease of the mixer gain at IF band edges. Since this trend appears in two types of RF tuning circuits at the same time, analysis of the IF circuit can explain the variance of the mixer gain. In general, optimizing the IF load admittance is required to obtain approximately constant mixer gain in wide IF bandwidth [10]. IF embedding impedance of 10 times R_N usually shows maximum gain, though the number can be different depending on LO and IF frequencies [11]. Roughly, the mixer conversion gain varies primarily with the real impedance and is affected less by changes in the imaginary impedance of the IF embedding circuit. As Figure 7 presents, the behavior of IF embedding impedance can be a reason for the gain variance. It will be confirmed after the lab test of the mixer chips which is under fabrication.

IV. Conclusion

We introduce designs of the wide IF bandwidth 230 GHz SIS mixer for the 4th KVN telescope under construction at Pyeongchang, South Korea. Covering the RF band of 211-275 GHz, the probe was optimized to get close to constant impedance. Two tuning circuits were designed to show similar noise temperature performance at the IF frequency range from 4 to 16 GHz. The mixer block has been made successfully, and we expect to have lab tests of the fabricated mixer chips in the near future.

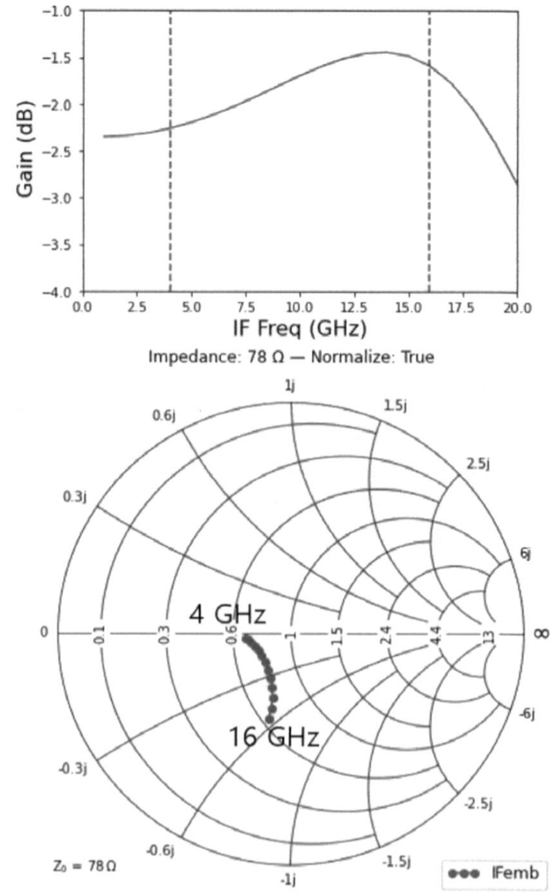

Fig. 7. The gain curve and IF embedding impedance of Type 1 from IF frequency 4 to 16 GHz with LO 243 GHz, which is normalized to 3 times of R_n. The gain curve has a peak at 14 GHz and the IF embedding impedance has the largest real part at the same point. It replies that the change of the gain is mainly contributed by the matching between the if impedance of the junction and the real part of the if embedding impedance.

References

[1] The Event Horizon Telescope collaboration et al., "First M87 Event Horizon Telescope Results. I. The Shadow of the Supermassive Black Hole", *ApJ*, 875, L1, 2019.

[2] Han, et al., A millimeter-wave quasi-optical circuit for compact triple-band receiving system. Journal of Infrared, Millimeter, and Terahertz Waves, September 2017.

[3] Ke and Feldman, "Optimum source conductance for high frequency superconducting quasiparticle receivers", *IEEE Trans. Microw. Theory Techn*, Apr. 1993.

[4] Tong, et al., "Wideband SIS receivers Using Series Distributed SIS Junction Array", *Proc. 29th IEEE International Symposium on Space THz Technology*, March 2018.

[5] Ward, et al., "Supermix: a Flexible Software Library for High-Frequency Circuit Simulation, including SIS Mixers and Superconducting Components", *Proc. 10th Int. Symposium on Space Terahertz Technology*, March 1999.

[6] Rice et al., "Fast Harmonic Balance of SIS Mixers with Multiple Junctions and Superconducting Circuits", *Proc. of 10th International Symposium on Space Terahertz Technology*, March 1999.

[7] Kooi et al., "A 275-425 GHz Tunerless Waveguide Receiver Based on AlN-barrier SIS Technology", *IEEE Transactions on microwave theory and techniques*, October 2017.

[8] Kerr, et al., "Development of the ALMA Band-3 and Band-6 Sideband-Separating SIS Mixers", *IEEE Transactions on Terahertz science and technology*, March 2014.

[9] Kerr et al., "MMA memo 205: A tunerless SIS mixer for 200-280 GHz with low output capacitance and inductance", *Proc. of 9th International Symposium on Space Terahertz Technology*, March 1998.

[10] Kooi et al., "Regarding the IF output conductance of SIS tunnel junctions and the integration with cryogenic InP MMIC amplifiers", *Proc. of 10th International Symposium on Space Terahertz Technology*, March 1999.

[11] Rice et al., "SIS Mixer Design for a Broadband Millimeter Spectrometer Suitable for Rapid Line Surveys and Redshift Determinations", *Proc of SPIE - The International Society for Optical Engineering*, November 2004.

4×2 Hot electron bolometer mixer arrays for detection at 1.46, 1.9 and 4.7 THz for a balloon borne terahertz observatory

José R. G. Silva[1,3*], Wouter M. Laauwen[1], Behnam Mirzaei[1,2], Nathan Vercruyssen[1,2], Matvey Finkel[1], Menno Westerveld[1], Nikhil More[1], Vitor Silva[1], Abram Young[4], Craig Kulesa[4], Christopher Walker[4], Floris van der Tak[1,3] and Jian Rong Gao[1,2*]

***Abstract*— We present a comprehensive characterization of three 4×2 HEB mixer arrays developed for the Galactic/Extra-Galactic ULDB Spectroscopic Terahertz Observatory (GUSTO), a NASA balloon borne terahertz observatory. These arrays were designed for operation frequencies of 1.46, 1.9 and 4.7 THz, respectively. The results regarding sensitivity, IF bandwidth and LO power requirements are presented, including their performance in the GUSTO instrument in the lab.**

Keywords*— *GUSTO, HEB, lens-antenna, mixer array

I. Introduction

GUSTO [1] is a NASA balloon borne THz observatory that is led by the University of Arizona and is scheduled to be launched from Antarctica in late 2023 for a flight duration of about 70 days. It aims at unveiling the physical structure and kinematics of star forming regions by mapping three THz fine structure lines: [NII] at 1.46 THz, [CII] at 1.9 THz and [OI] at 4.7 THz; along the galactic plane of Milky Way and a part of the Large Magellanic Cloud. With such detailed information one can unveil the dynamics and processes that dominate regions of star formation [2-3]. To achieve high spectral resolution, it will employ three 8 pixel arrays based on NbN hot electron bolometers (HEBs). Here we present comprehensive characterizations of the HEB mixer arrays developed for GUSTO.

II. HEB Mixers Arrays and Experimental setup

All HEB mixer arrays were designed to allow for eight pixels in a 4×2 configuration within a single metal block. All the pixels in an array share the same basic configuration that is shown in Fig. 1a. For each pixel THz radiation is collected on the surface of the elliptical Si lens. It is then focused, as it propagates through a lens and then an HEB chip substrate, to a spiral antenna, where the radiation is converted to an AC electrical current that is fed to the HEB. Through bonding wires, the HEB is connected to a co-planar waveguide (CPW) line that is used to both DC bias the device and collect the IF signal from the mixer. Each pixel is terminated with an IF connector that acts as the interface to a low noise amplifier (LNA). The lenses and the substrate of the HEB chips are made of pure, highly resistive Si (≥5 kΩ cm). Each HEB chip consists of a NbN bridge integrated with a planar logarithmic spiral antenna [4].

Two models of detector arrays were designed to accommodate the two types of lenses with different diameters. In Fig. 1b we show the completed B1 and B2 arrays, using elliptical Si lenses with 10 mm diameter and having a pitch size of 11 mm. The two arrays make use of the same model and were optimized for operation at 1.46 and 1.9 THz, respectively. Because these two frequencies are very close, it is difficult to separate them in the optical path of the instrument. Thus, B1 and B2 were designed to be placed side by side on the cold plate of the cryostat, mimicking a 4×4 array. The devices used have a NbN bridge of 2 μm in width, 0.15 μm in length, and 5 nm in thickness. The critical temperature of the NbN bridges is about 10 K. In Fig. 1c we show the completed B3 array that uses elliptical lenses of 5 mm diameter and has an 8 mm pitch size. This array is optimized for operation at 4.7 THz. In Fig. 1d we present a back side view of the B3 array, while partly assembled, where the eight detector chips, CPW lines and IF connectors are shown. The HEB devices used in this array are similar to the ones used in the other arrays (from the same wafer), however, the NbN bridge lengths are longer, being 0.2 μm instead.

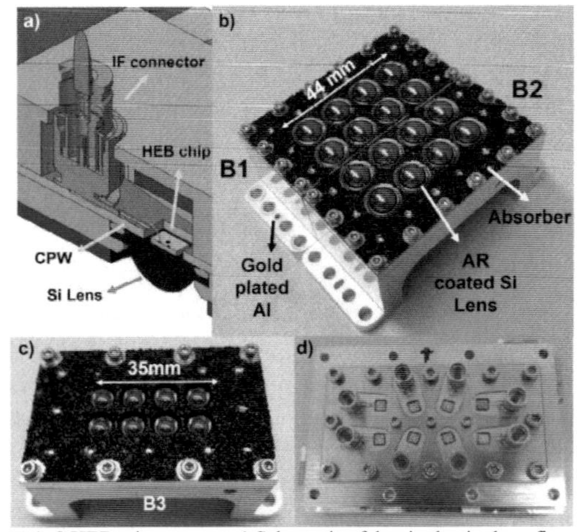

Fig. 1. 4×2 HEB mixer arrays. a) Schematic of the single pixel configuration used in all the arrays. b) Completed B1 and B2 arrays, for operation at 1.46 and 1.9 THz, respectively. c) Completed B3 array designed to operate at 4.7 THz. d) A back side view of the partly assembled B3 array, where the eight HEB chips, CPW lines and IF connectors are shown, from [5].

Each of the arrays has a different lens design, optimized to meet the GUSTO optical beam requirements, with details and verification reported elsewhere [4]. The methodology used to mount and align HEB antenna with the lens optical axis has also been described elsewhere [5]. With this methodology we were able to obtain beam pointing errors <0.1 deg relative to the array normal direction.

[1]SRON Netherlands Institute for Space Research, Landleven 12, 9747 AD Groningen and Niels Bohrweg 4, 2333 CA Leiden, the Netherlands. [2]Optics Research Group, Department of Imaging Physics, Delft University of Technology, Delft, 2628 CJ, the Netherlands. [3]Kapteyn Astronomical Institute, University of Groningen, Groningen, 9747 AD, the Netherlands. [4]Steward Observatory, University of Arizona, 933 N Cherry Ave, Tucson, AZ 85719, USA

We measure the DSB receiver noise temperature (T_{rec}^{DSB}), the receiver conversion loss (L_{rec}^{DSB}), and the required LO power (P_{LO}) for each pixel in the arrays. The IF noise bandwidth (NBW) was measured in the IF frequency range between 0.5 and 5 GHz for a few selected mixers.

T_{rec}^{DSB} of the different pixels in each array were measured at a mixer operating temperature of 4.4 K using a standard air setup and using 1.39, 1.63 or 5.25 THz FIR gas laser lines as LOs, respectively. These frequencies are slightly different from GUSTO's respective B1, B2, and B3 frequencies due to not having the same LOs as GUSTO available at SRON. The IF chain consists of a bias-T and a cryogenic SiGe low noise amplifier (LNA) at 4.2 K. The room temperature part of the IF chain includes two LNAs, a bandpass filter, and a microwave power meter. For T_{rec}^{DSB} measurements the IF was filtered by the bandpass filter centered at 2 GHz. For the noise bandwidth measurements, we replaced the components from the bandpass filter up to the power meter with a spectrum analyzer.

III. Results and Discussion

The T_{rec}^{DSB} is obtained using the Y-factor technique where the Callen-Welton blackbody temperatures are used to correct the physical temperature of both hot and cold load [6]. The L_{rec}^{DSB} is obtained using the U factor technique [7]. The P_{LO} is estimated using the isothermal technique [8]. For GUSTO we are interested in T_{mixer}^{DSB} and L_{mixer}^{DSB} which are defined as the noise temperature and conversion loss, respectively, after correcting for the effect of all optics in front of the Si lens. Additionally we also correct for the difference in frequency between LO used in the measurements and GUSTO target frequency.

To measure the NBW of an HEB mixer in our arrays we repeat the T_{rec}^{DSB} measurements over a wide IF range when it is biased at an optimal operating point. For both a B1 and a B3 pixels, operated at 1.39 and 5.25 THz, respectively, we obtained a NBW of 3.5 GHz.

We summarize the average and standard deviation value of T_{mixer}^{DSB}, L_{mixer}^{DSB} and P_{LO}, for each array in Table 1. Not only the arrays demonstrate a state of the art T_{mixer}^{DSB}, which is in line with the best reported so far in the literature [9-11], but also very good uniformity. For L_{mixer}^{DSB}, we find that it increases with the array operating frequency. Such an increase is confirmed even in the intrinsic L_{mixer}^{DSB} after removing the optical loss of the lens and coupling loss between antenna and HEB. This can be explained by existing artifacts that are present around the gold spiral antenna arms, which may introduce additional ohmic losses to the THz RF current. This effect would be stronger for a higher frequency. In terms of P_{LO} we also see a very good uniformity within a single array. Additionally, we note for the B3 array it is slightly higher than that for B1 and B2 due to the longer HEBs used in the B3 array.

TABLE 1

Performance summary of the three HEB mixer arrays averaged over the 8 pixels in an array. It includes the measured mixer noise temperature (T_{mixer}^{DSB}) and mixer conversion loss (L_{mixer}^{DSB}) at 2 GHz IF, the GUSTO receiver noise temperature ($T_{rec,GUSTO}^{DSB}$), measured at the instrument in the lab. and optimum LO power at HEB (P_{LO}). In parentheses are the standard deviations within the respective array.

Array	Operating Frequency	T_{mixer}^{DSB} (K)	L_{mixer}^{DSB} (dB)	$T_{rec,GUSTO}^{DSB}$ (K)	P_{LO} (nW)
B1	1.46 THz	330 (10)	5.7 (0.6)	≈870	210 (12)
B2	1.9 THz	420 (14)	6.9 (0.7)	≈1100	190 (10)
B3	4.7 THz	700 (26)	9.7 (0.7)	≈1920	240 (15)

The performance of the arrays in the GUSTO instrument is also shown in Table 1. These values are slightly above what we predicted, which is explained by extra coupling loss due to sidelobe spillover throughout the optics and especially beam vignetting in some optical elements for some of the mixers. The GUSTO instrument is currently undergoing final integration with the Gondola in preparation for the launch in December 2023. During the final commissioning, the ultimate performance can be determined and a more detailed paper on GUSTO's performance will be prepared.

IV. Conclusions

We have successfully demonstrated three 4×2 heterodyne HEB arrays for GUSTO, which will be operated at local oscillator frequencies of 1.46, 1.9 and 4.7 THz, respectively. These arrays represent, to date, the highest pixel count using the quasi-optical scheme at supra-THz frequencies. Our results demonstrate the heterodyne arrays with not only excellent sensitivity, but also good uniformity of the performance parameters.

Acknowledgment

We acknowledge the technical support from Jarno Panman, Rob van der Schuur, Erik van der Meer, Henk Ode, Duc Nguyen, Marcel Dijkstra. We also thank Yuner Gan, Axel Detrain, Geert Keizer, Gabby Aitink-Kroes, Brian Jackson and Willem Jellema for helpful discussions.

References

[1] C. Walker et al, Gal/Xgal U/LDB Spectroscopic/ Stratospheric THz Observatory: GUSTO, Proc. SPIE, vol. 12190, 2022, Art. no. 121900E.

[2] C. K. Walker, "THz coherent detection systems," in Terahertz Astronomy, 1st ed. New York, NY, USA: Taylor & Francis, pp. 159–227, 2016.

[3] C. Pabst, et al, "Disruption of the Orion molecular core 1 by wind from the massive star θ 1 Orionis C", Nature, vol. 565, no. 7741, pp. 618–621, 2019.

[4] J. R. G. Silva et al, " Beam waist properties of spiral antenna coupled HEB mixers at Supra-THz frequencies", IEEE Transactions on Terahertz Sci. and Tech., vol. 13, no. 2, pp. 167 – 177, 2023.

[5] J. R. G. Silva, et al, "High Accuracy Pointing for Quasi-optical THz Mixer Arrays", IEEE Trans. Terahertz Sci. Technol. vol. 12, no. 1, pp. 53-62, 2022.

[6] P. Khosropanah et al, "Low noise NbN hot electron bolometer mixer at 4.3THz", Appl. Phys. Lett. vol. 91, 221111, 2007.

[7] S. Cherednichenko et al, "1.6 THz heterodyne receiver for the far infrared space telescope", Phys. C: Supercond. and Its Applicat. Vol. 427, pp. 372, 2002.

[8] H. Ekström et al, "Conversion gain and noise of niobium superconducting hot-electron-mixers", IEEE Trans Microw. Theory Tech. vol. 43, no. 4, 938-947, 1995.

[9] W. Zhang, P. Khosropanah, J. R. Gao, E. L. Kollberg, K. S. Yngvesson, T. Bansal, R. Barends and T. M. Klapwijk, "Quantum noise in a terahertz hot electron bolometer mixer", Appl. Phys. Lett. vol. 96, pp.111113, 2010.

[10] K. M. Zhou et al, "1.4 THz Quasi-optical NbN Superconducting HEB Mixer Developed for the DATE5 Telescope", IEEE transactions on Applied superconductivity, vol. 25 no. 3, pp. 1-5, 2014.

[11] J. L. Kloosterman et al, "4-Pixel Heterodyne Receiver at 1.9 THz using a CMOS Spectrometer", Proc. of the 28th ISSTT, 74, 2017.

Orbiting Astronomical Satellite for Investigating Stellar Systems

Siddhartha Sirsi[1,2], Yuzuru Takashima[2], Art Palisoc[3], Henry Quach[2], Hyukmo Kang[2], Aman Chandra[1], Heejoo Choi[2,4], Daewook Kim[1,2,4], Jonathan Arenberg[5], Christopher Walker[1,2]

Orbiting Astronomical Satellite for Investigating Stellar Systems (OASIS) is a proposed space based observatory with a 14 m inflatable primary reflector that will perform high spectral resolution observations at terahertz frequencies[1]. An inflatable metallized polymer membrane with a large photon collecting area serves as the primary antenna (A1), followed by aberration correction mirror pair, and a FOV scanner that enables a large field of regards of 0.1 degrees while achieving diffraction limited performance over a wide terahertz wavelength ranging from 63 μm to 660 μm.

The shape of the primary antenna is a function of pressure, this offers a unique challenge with respect to optical design. The aberration correction mirror pair is designed to simultaneously tackle the deviation of the primary antenna from the desired ideal parabola and the off-axis aberrations like coma. A tip-tilt mirror system with a HRFZ Si field lens is designed to scan the FOV. To determine the best suitable design that meets all the science goals and system requirements, an analytical model is developed to perform a parametric design study and the results are represented as solution space contour plots.

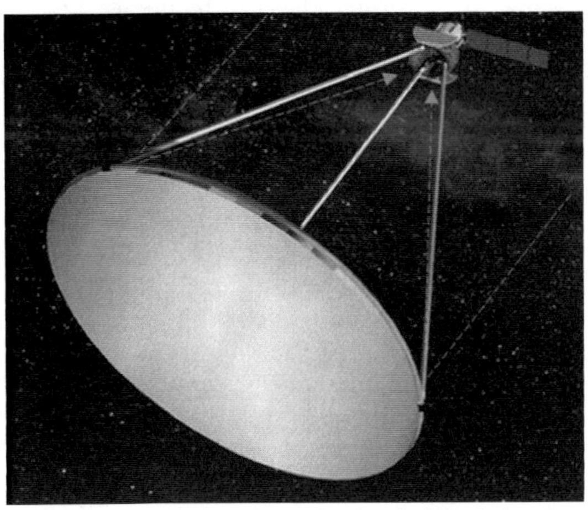

Fig.1 OASIS mission concept showing the fully deployed configuration.

The large diameter and irregular shape contribute to the challenges of accurately measuring the surface profile of the primary antenna. A 1m prototype of primary antenna (A1) is built to investigate whether the Nikon APDIS laser radar can be used to overcome these metrological challenges. The measured data is compared with Fichter solution of Hencky curve, and Finite Element Analyzer for Membranes (FAIM) developed by L'Garde Inc.

Fig.2 Nikon APDIS laser radar measuring the surface profile of 1m A1 prototype.

References

[1] Christopher Walker, et al., "Orbiting Astronomical Satellite for Investigating Stellar Systems (OASIS) Following the Water Trail from the Interstellar Medium to Oceans," SPIE Paper #11820-26, Proc SPIE, Optics+Photonics, San Diego, California, 1-5 August 2021.

[1] Department of Astronomy and Steward Observatory, University of Arizona, 933 N. Cherry Ave., Tucson, AZ 85721, USA
[2] Wyant College of Optical Sciences, University of Arizona, 1630 E. University Blvd., Tucson, AZ 85721, USA
[3] L'Garde, Inc., 15181 Woodlawn Avenue, Tustin, CA 92780, USA
[4] Large Binocular Telescope Observatory, University of Arizona 933 N Cherry Avenue, Tucson, AZ 85721, USA
[5] Northrop Grumman Aerospace Systems, 1 Space Park Blvd, Redondo Beach, CA 90278

NOTES:

Preliminary Characterisation of a Compact 240 GHz SIS Dual-Polarisation Receiver for Large Array Applications

Jakob Wenninger*‡, Faouzi Boussaha†, Christine Chaumont†, Boon-Kok Tan* and Ghassan Yassin*
*Department of Physics (Astrophysics), University of Oxford, Keble Road, Oxford, OX1 3RH, UK
†GEPI (CNRS UMR 8111), Observatoire de Paris, PSL Université, Paris, France.
‡Contact: jakob.wenninger@physics.ox.ac.uk

***Abstract*—We report the development of an on-chip, compact dual-polarisation Superconductor-Insulator-Superconductor (SIS) receiver covering the millimetre frequency band from 190 GHz to 290 GHz. All the required components for the receiver are integrated on-chip using planar circuit technology, except the feedhorn. We use a 4-probe orthomode transducer (OMT) to couple the radio frequency (RF) and local oscillator (LO) and to split the incoming signal into two RF polarisation states before they are routed to the twin junction $Nb/AlO_x/Nb$ mixers via a planar crossover, branch-line hybrids and bandpass filters. The entire receiver chip, only 4.0 mm by 4.1 mm in size, is mounted in a split block where the feedhorn and the magnetic biasing are located on the front side, and the intermediate frequency (IF) and DC biasing connectors are on the rear side, leaving the four adjacent sides of the block unobscured for a 2-dimensional extension into a multi-pixel array. In this paper, we present the receiver and mixer block design and preliminary experimental results obtained from testing the receiver.**

***Index Terms*—Heterodyne receiver, Superconductor-insulator-superconductor mixer, Dual-polarisation receiver, Focal Plane Array, Large format array;**

I. Introduction

Mapping large-scale structures in the Universe using heterodyne receivers, such as SIS mixers, requires long observation times due to the low pixel count in most instruments. Several multi-pixel focal-plane SIS arrays have already been developed, but most of these arrays only detect a single polarisation state of the incoming RF wave at a time [1]–[4].

The employment of single-polarisation receivers results not only in the loss of the polarimetry information but also in the loss of sensitivity since we are only detecting half of the incident signal power. Therefore, dual-polarisation receivers are favourable for their astronomical observation capabilities. However, most dual-polarisation receivers deployed to date are primarily a stack of two single-polarisation receivers with a waveguide OMT or a free-space optical element to separate the polarisation information [5]–[8]. The receiver setup, therefore, requires additional space, which does not provide a compact layout required for array extensions.

Here, we present a compact on-chip dual-polarisation receiver, which can be easily scaled to higher frequencies, as well as to form a large dual-polarisation heterodyne array. All the required RF components are fabricated on a single quartz chip using planar superconducting circuit technology.

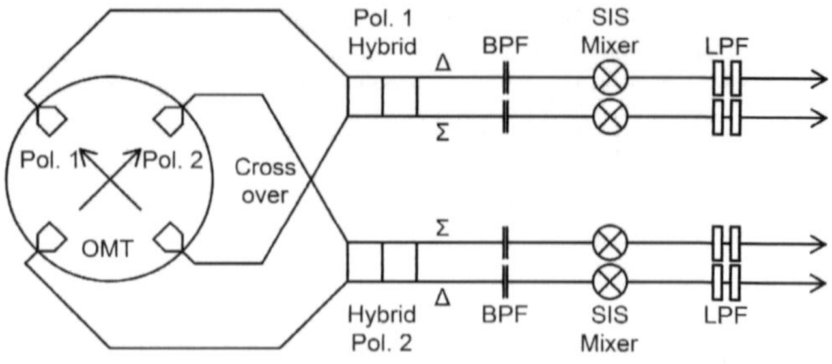

Fig. 1. A schematic of the receiver chip.

Consequently, the mixer chip can be housed in a split-block without any waveguide structure, apart from the feedhorn, which is drilled directly into the copper block, using the smooth-walled horn technology [9], [10].

II. The Dual-Polarisation Receiver Design

A detailed description of the receiver design has already been reported [11]. Hence, below, we will only give a brief account of the receiver design. Fig. 1 shows a schematic of the dual-polarisation receiver chip. A picture of the receiver chip is shown in Fig. 2. The receiver has all the front-end elements fabricated onto a single quartz chip and connected with microstrip transmission lines. A 4-probe OMT using two sets of opposing probes in a circular waveguide is used to split the LO and RF power incident at the feed horn into two orthogonal polarisations. The signals from each probe set, coupling one polarisation state, are then recombined at the quadrature hybrid Σ output port.

The microstrips connecting the OMT probes and the hybrids inevitably result in crossing transmission lines, which connect to OMT probes that couple different polarisations, as shown in Fig. 1. The crossover must be designed to minimise crosstalk between the two polarisations. Instead of interconnecting the microstrips in the wiring layer, a short coplanar waveguide (CPW) section is deposited in a gap in the ground layer where the Pol. 1 signal is routed underneath the Pol. 2 microstrip, as shown in Fig. 3. The CPW in the ground layer is capacitively connected to the feeding microstrip via broadside couplers, allowing for a closed dielectric layer throughout the chip. In summary, the crossover allows to connect the

Fig. 2. The thinned receiver chip mounted in the bottom block and bonded to the IF transformer. The chip consists of the OMT (1), the crossover (2) and two hybrids, where the hybrid for Pol. 1 is marked with (3). All hybrid outputs connect via a BPF (4) to a twin junction SIS mixer (5), which connects via the LPF and bond wires to the IF transformer (6). The ground is connected at (7).

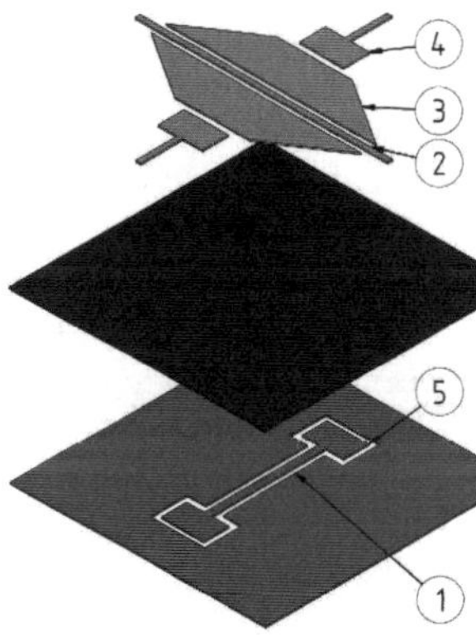

Fig. 3. An explosion view of the microstrip layers at the crossover. The ground plane is shown in red, the dielectric layer in blue and the wiring layer in green. Pol. 1 is routed in a CPW in the ground plane (1) underneath the Pol. 2 microstrip (2), which is bracketed to form a CPW in the wiring layer with two conductor patches (3). The Pol. 1 signal is connected via broadside couplers in the wiring layer (4) and the ground plane (5).

OMT probes to the hybrids for the respective polarisations to recombine the signals of each polarisation splitting between each probe set.

The hybrid has two outputs, the Σ output port containing the full strength of one of the polarised signals and the isolated Δ output. Both hybrid output ports connect to two identical mixer circuits for impedance symmetry reasons, although the Δ mixer circuit should couple only noise and can be terminated. The mixer circuits comprise a bandpass filter (BPF), a twin junction SIS mixer and a lowpass filter (LPF). The BPF is used before the SIS mixer to provide separate DC biasing and block the IF signal. The LPF after the mixer prevents RF signal leaking into the IF circuitry and ensures efficient RF coupling to the SIS junctions. The SIS mixer comprises two identical tunnel junctions in a conventional twin junction

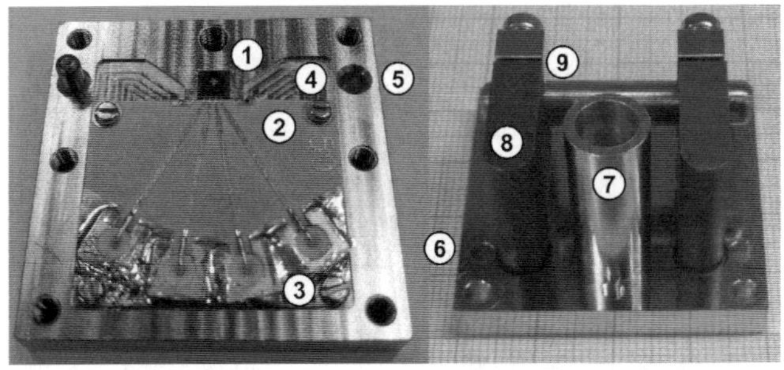

Fig. 4. The receiver is made from the bottom block on the left, the top block on the right, and the receiver chip on a pedestal in the bottom block (1). The receiver chip is wire bonded to the IF transformer (2), which ends in SMP connectors (3) to the back of the receiver. The bottom block has a recess for the IF transformer and coil shoes (4). An identically shaped but deeper counterpart is in the top block. Dowel holes in the bottom (5) and top (6) block align the feedhorn (7) with the receiver chip and the backshort section, which is milled into the bottom block. The magnetic assembly consists of a feed (8) connecting the shoes next to the receiver chip with the coil, which sits on a rod clamped on the feed (9).

arrangement to tune out the junction capacitance and maximise the coupling of power to the SIS mixer. As the mixer circuits are used to impedance match the Δ hybrid ports for both polarisations, each receiver chip contains eight SIS devices in four twin junction circuits.

Each on-chip circuit element is initially designed independently using the 3D electromagnetic software Ansys HFSS before interconnecting them with microstrip transmission lines to form the entire receiver chip. The full chip design is then carefully tweaked and optimised for bandwidth, identical power coupling between the two polarisation states, and minimised transmission losses. The detailed design description is published in [11].

The entire receiver is fabricated onto a 4.0 mm by 4.1 mm quartz substrate, hence the design of our split block is substantially simplified. As shown in Fig. 4, the mixer block is made up of two separate parts without the need for any complicated mechanical structure. The top block comprises a compact drilled horn and openings for the magnetic circuit assembly, and the bottom block has a large area milled out for hosting the mixer chip, magnetic shoes, and IF transformer with holes for the IF and DC connectors. The entire mixer block is only 40 mm by 40 mm, the size determined primarily by the size of the IF SMP connectors and the IF transformer. It is worthwhile noting that the block has plenty of space to accommodate different sizes of IF circuitries and magnetic bias options, so the block size can be further reduced if necessary.

The receiver is mounted in the cryostat and connected to the signal processing chain, as shown in Fig. 5. The four mixer circuits on the receiver chip connect via SMP connectors to the SMA connectors of the bias tees. SMP connectors are used due to the smaller footprint of the push-pull mechanism in comparison to the hex nut, which would need additional space for a spanner, keeping the receiver compact. The two bias tees connected to the Σ mixer circuits connect to circulators and low noise amplifiers to route the IF signal to room temperature, while the two bias tees connected to the Δ mixer circuits have terminated AC ports. The DC ports of the four bias

Fig. 5. The cryostat setup. The receiver (1) is connected via four SMP-to-SMA cables (2) to four bias tees (3). The outer two bias tees are terminated as they connect to Δ twin junctions, and the inner two bias tees connect via two isolators (4) and low noise amplifiers (5) to room temperature.

TABLE I
THE 5-STEP PLANAR-CIRCUIT THIN-FILM DEPOSITION PROCESS.

Step	Description	Material	Thickness
1	Deposition trilayer	$Nb/AlO_x/Nb$	400/1/200 nm
2	Etching junctions from trilayer	Nb Remains	400 nm
	Deposition of first dielectric	SiO	200 nm
3	Deposition of second dielectric	SiO	200 nm
4	Deposition of wiring layer	Nb	400 nm
5	Deposition of bonding pads	Ti/Au	10/150 nm

tees connect independently to room-temperature four-way bias modules and a data acquisition unit. Thus, the DC signal of all four mixers can be swept and read simultaneously. The linear arrangement of the signal processing chain behind and within the footprint of the receiver can be extended to an array in an ALMA-style cartridge.

III. Receiver Fabrication

The planar receiver chip is fabricated at the facilities of the Observatoire de Paris with a 5-step recipe, summarised in Table I. In the first step, the ground plane conductor is sputtered on the substrate in the form of a 14 kA/cm^2 $Nb/AlO_x/Nb$ trilayer, except for areas at the OMT, the IF bonding pads and CPW features of the crossover and BPF. The top layer of the trilayer is removed by reactive ion etching (RIE) in the second step and replaced with a first 200 nm SiO layer, except at the SIS junction locations. Thus, the area of the SIS junctions is defined, the top and bottom electrodes of the SIS junction are isolated by AlO_x and the remaining ground plane covered with SiO. A second 200 nm SiO layer is evaporated on the substrate in the third fabrication step to achieve the desired characteristic impedances and to minimise the likelihood of shorts between the ground conductor and the wiring layer. The second SiO layer leaves a 6 µm^2 window around each SIS junction with only 200 nm SiO to ensure contact between the trilayer top electrode and the wiring layer. The wiring layer is sputtered in the fourth fabrication step. The last deposition step is the evaporation of Ti and Au on areas used for wire bonding.

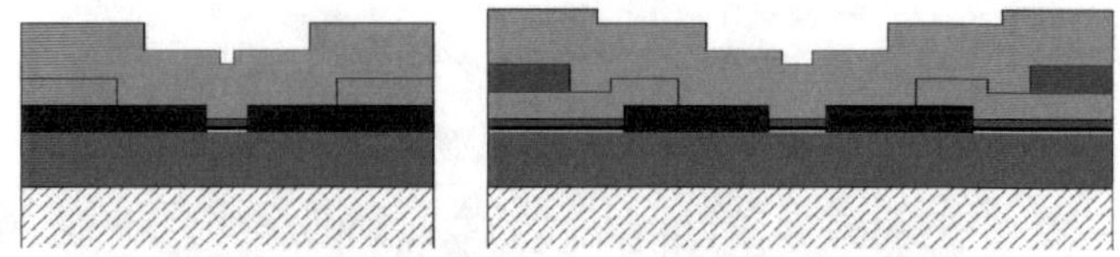

Fig. 6. Side views of the original mask set (left) and of the revised mask set (right) with larger SIS junctions and an additional SiO layer; The quartz substrate is shown hashed. The Nb of the trilayer is shown in red, which sandwiches the passivated Al in white and black. The first SiO layer is deposited where the trilayer has been removed, shown as dark blue. The second SiO layer is pale blue, and the third SiO layer is purple. The wiring layer Nb is shown in green.

A. DC Characterisation

After dicing the wafer and before thinning the receiver substrate to the thickness required for RF testing, we conducted DC screening tests. We have four device variations with junction areas of 1.0 µm^2 or 1.5 µm^2.

The DC screening results of the first fabricated batches showed a low fabrication yield, which is not unexpected, but too small for our devices requiring eight functioning SIS junctions. Consequently, we reviewed the recipe by also fabricating separate test device batches with 1.0 µm^2 to 10 µm^2 SIS junctions, and we found evidence that the large ground layer of our receiver causes issues during the RIE process. We suspect that the SF_6 etching of the trilayer might have polluted the photoresist defining the SIS junction area, preventing a clean lift-off. In the deposition of the wiring layer, these photoresist remnants were then sandwiched between the SIS top electrode and the wiring layer, causing a resistive response.

Accordingly, we modified the fabrication process by 1) increasing the SIS junction area and 2) reducing the area etched to define the SIS junctions. The mask of the second fabrication step now has a larger SIS junction area but only exposes the trilayer at a 10 µm^2 square around the SIS junctions to the RIE and subsequent SiO deposition, as shown in Fig. 6. Hence, the ground layer at most of the receiver chip consists of the trilayer, which is covered with 200 nm SiO and a 6 µm^2 square around the SIS junctions in the third fabrication step. An additional (3.5) fabrication step is inserted before the fourth step of Table I to deposit another 200 nm SiO with a 14 µm^2 square around the SIS junctions to achieve 400 nm SiO on most of the chip and, therefore, the desired characteristic impedances. With these modifications, the fabrication yield improved significantly.

The increased junction size inevitably affected the performance of the receiver. As shown in Fig. 7, the return loss of a signal incident at the circular waveguide is very high, but the ripples reach below −10 dB up to 240 GHz, offering us a chance to characterise the on-chip integrated circuit. In the initial design, most of the incident signal couples in the second SIS junction. Our simulation showed that the increased junction area causes the incident signal to couple mostly into the first SIS junction of the twin junction, leaving the second SIS junction with very little coupling. Despite this change, the total coupling in both SIS junctions is similar to that of the initial design at low RF frequencies.

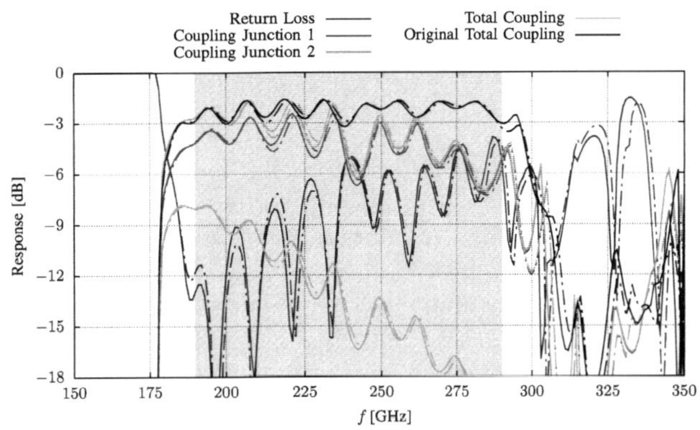

Fig. 7. The simulated overall performance of the receiver chip fabricated with the revised mask set and 3.0 µm² SIS junctions. The Pol. 1 responses incident at the circular waveguide of the model are shown as solid lines, and Pol. 2 responses as dash-dotted lines. Total coupling denotes the sum over the two SIS junctions of a twin junction circuit.

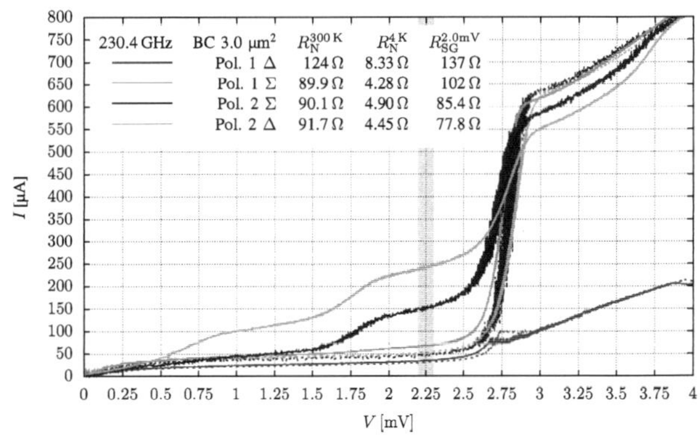

Fig. 8. The pumped IV curves and the unpumped IV curves (dotted) of the four twin junction circuits. The legend summarises the 300 K resistance, the normal resistance at 4 K and the subgap resistance measured at 2.0 mV. The grey shaded area is the (2.25 ± 0.05) mV interval used to gauge the direct detector response to the LO signal.

B. Receiver Preparation

We selected devices for thinning to the designed quartz thickness of 50 µm based on the fabrication yield measured by sampling devices across the wafer. The devices were thinned with a dicing saw at the Rutherford Appleton Laboratory. The thinned devices were mounted with a brush to their position in the bottom block and affixed with agar scientific ltd crystalbond 555, a low melting point (54°C) adhesive. Crystal bond is more versatile than the commonly used superglue since it can be applied without time restrictions and re-liquefied for adjustments.

A frequent source of difficulty was applying pressure on the receiver chip to remove excess adhesive between the chip and the bottom block. The excess adhesive would elevate the chip and consequently cause the chip to touch the top block, potentially destroying the receiver chip once assembling the receiver. In case the receiver chip protruded from the bottom block, we liquefied the crystal bond and adjusted the chip position. We followed the same procedure when we could identify an imperfect alignment of the backshort section underneath the transparent OMT. Once the receiver chip mounting in the bottom block was satisfactory, the receiver chip was wire-bonded to the IF transformer, and then the top block was mounted on the bottom block, forming the receiver. The receiver was mounted in the cryostat for RF characterisation.

IV. Optical Coupling Characterisation

The optical coupling of the receiver was tested using the twin junctions as direct detectors, with an LO source directly facing the receiver in the cryostat behind a Zotefoam vacuum window and a Zitex IR filter. The photon-step responses of the four twin junctions enabled us to infer the coupling efficiency of the circuit, particularly from the circular waveguide to the microstrips at the OMT and the power-combining capabilities of the hybrid.

The first thinned device we characterised has 3 µm² SIS junctions and has not been screened before thinning. The corresponding IV responses and parameters are shown in Fig. 8. The 300 K resistance test showed that the warm resistance of the Pol. 1 Δ twin junction has a resistance of 124 Ω, indicating that the twin junction device is faulty. The faulty response can be seen from the unpumped response with very low tunnelling current at the gap voltage, potentially caused by a defective SIS junction and a series resistance.

A. Power Measurement

The pumped IV responses in Fig. 8 show the highest photon step current we could achieve at 230.4 GHz, although the Gunn LO source was directly illuminating the horn, hence should have been strong enough to saturate the IV response. We conclude that the coupling of the SIS junctions to the LO source is poor, either due to faulty subcircuits or excessive losses. Furthermore, the LO couples to both twin junctions connected to the Pol. 2 hybrid. However, despite that the Δ twin junction was designed to be isolated, more LO signal couples to the Δ twin junction than the Σ twin junction at 230.4 GHz. In contrast, both twin junctions connected to the Pol. 1 hybrid are not at all responding to the LO. Following the design, a photon step should appear in the Pol. 1 Σ twin junction, considering that it has a good unpumped IV response. However, as the Δ twin junction is defective, we can not extract information to learn if any LO signal is incident at the Pol. 1 hybrid.

Fig. 9 shows the photon step current at a bias voltage centred at 2.25 mV as a function of LO frequencies. The data comprises two different cooldowns, a preliminary test and one cooldown designated for this frequency sweep, where most data points were recorded by sweeping the LO frequency. Photon steps appear in both twin junctions connected to the Pol. 2 hybrid at all frequencies. Since the power in the Δ and Σ ports are approximately equal, the observed behaviour resembles a single input signal into the hybrid, splitting with −3 dB and 90° phase difference between the outputs.

B. Polarisation Measurement

In order to confirm that the photon step responses in the Σ and Δ twin junction of the Pol. 2 hybrid originate from

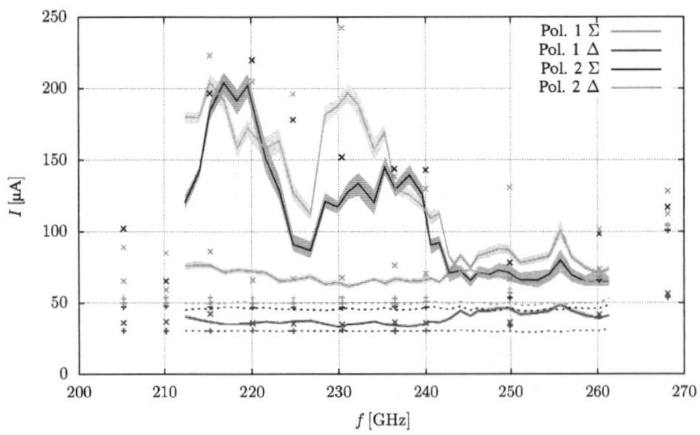

Fig. 9. The current of the pumped IV curve at a bias voltage of (2.25 ± 0.05) mV at various LO frequencies. The unpumped currents are shown as dotted lines. × and + show the pumped and unpumped data of the same experiment taken in a preliminary test.

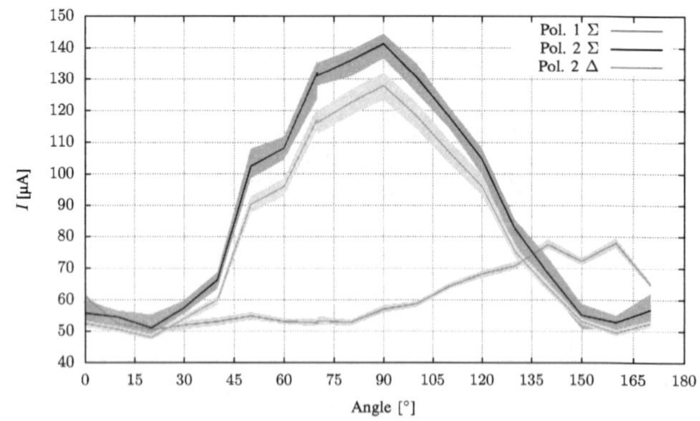

Fig. 10. The current of the pumped IV curve at a bias voltage of (2.25 ± 0.05) mV at different orientations of the wire grid at 216.8 GHz.

the same signal polarisation, we change the LO polarisation. Instead of rotating the LO source, we placed a wire grid between the LO source and the vacuum window. This reduces uncertainties and difficulties associated with the optical alignment as the LO source and wire grid are positioned only once. However, this method inevitably introduces losses due to the addition of the wire grid in the optical path.

The photon step currents centred at 2.25 mV are shown in Fig. 10 as a function of the wire grid orientation. We again recorded different LO powers for each data point and stepped the wire grid orientation by 20° to interleaf the data of a full rotation to the shown 10° resolution.

Both twin junctions attached to the Pol. 2 hybrid show photon-step currents varying sinusoidally with the wire grid angle and peaking together. This observation implies that the Σ and Δ twin junction responses originate from the same LO polarisation. Furthermore, since the power in the Δ and Σ ports is almost equal, it raises the possibility that the LO power is injected at only one hybrid input.

The current into the twin junction attached to the Pol. 1 hybrid Σ port is significantly less at the peak of the Pol. 2 photon step response compared to its maximum current. The peak of the Pol. 1 Σ twin junction occurs at a wire grid angle of 150°, providing evidence that the OMT, indeed, splits the two polarisations. However, the wire grid angles corresponding to the two polarisation peaks differ by only approximately 60°. This angle, however, was designed to be 90°, indicating some issues with the polarisation splitting capabilities of the OMT. Again, this observation is with the caveat of a defective Pol. 1 Δ twin junction.

We repeated this experiment for a set of LO frequencies shown in Fig. 11. Throughout, we observe that the photon steps of Pol. 1 and Pol. 2 peak at wire grid angle differences well below 90°. In spite of this discrepancy, the photon step currents of the two Pol. 2 twin junctions peaked at the same grid angle at all frequencies. An interesting observation is that the photon step currents peaked at different angles depending on the LO frequency. A reaffirming observation is that the photon step response varies sinusoidally with a period of 180°,

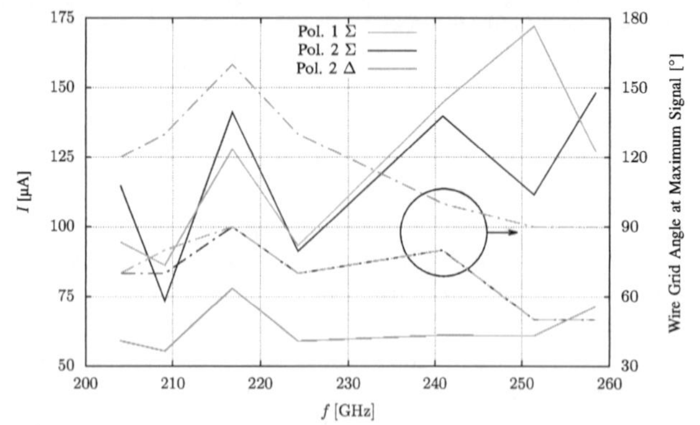

Fig. 11. The maximum current of the pumped IV curve at a bias voltage of (2.25 ± 0.05) mV in solid lines and the corresponding wire grid orientations in dash-dotted lines at various LO frequencies.

indicating that the experimental setup is fine.

C. Summary of RF Testing

We can summarise the RF testing in the previous section as follows: We have had difficulties pumping the twin junctions and were unable to saturate them at any frequency, although the Gunn LO has been facing the receiver directly. The characterisation of Pol. 1 has been compromised by the defective twin junction attached to the Pol. 1 Δ hybrid output. The Pol. 1 Δ twin junction response has been negligible throughout, and the Pol. 1 Σ twin junction response has shown rather small photon steps. In contrast, both twin junctions attached to the Pol. 2 hybrid showed clear photon steps. For Pol. 2, it depends on the LO frequency, which twin junction circuit, Σ or Δ, shows the larger photon step current. We confirmed that the responses of the twin junction circuits attached to the Pol. 2 hybrid originate from the same polarisation by rotating a wire grid inserted between the LO source and the cryostat vacuum window.

The optical coupling characterisation reveals two issues with the device:

1) The coupling of LO power from the Gunn LO source to the twin junctions is weak.

2) Both twin junctions attached to the Pol. 2 hybrid respond similarly to the LO with frequency dependence.

V. Investigation of the Causes for Weak Coupling

The weak coupling of LO power of the receiver could be caused by several factors, including:

1) Misalignment in assembling the top and bottom split blocks, where the feedhorn and OMT were aligned. This causes the waveguide containing the four OMT probes in the bottom block to be displaced with respect to the feeding circular waveguide in the top block. To investigate the influence of the displacement on the coupling, we simulated the behaviour with Ansys HFSS. The results are shown in Fig. 12, where we assumed that there is a misalignment of 10% of the waveguide diameter along the axis of a particular polarisation. The OMT probe shifting away from the circular waveguide couples significantly less than the opposing OMT probe, which shifts towards the centre of the circular waveguide. We also observe some cross-coupling into the OMT probes of the orthogonal polarisation. However, the coupling to this polarisation is affected very little by this misalignment.
   These simulations demonstrate that a displacement of the two circular waveguides could account for the poor coupling experienced in the experiment.
2) Fabrication defects of the receiver components on the chip. The on-chip receiver contains several components that require skilful fabrication for the receiver to operate as designed. We, therefore, inspected the receiver chip under the microscope and found the following:
   The crossover: The visual inspection showed that the transmission lines comprising the crossover are clean, and, therefore, assuming that no short to the ground plane exists, we don't expect the weak coupling to be caused by the crossover component.
   The hybrid: The transmission lines comprising the hybrids looked to be in good shape. Again, we expect the poor coupling to be caused by the hybrid fabrication. It is interesting, however, to emphasise that the Σ and Δ ports of Pol. 2 received the same amount of power as our test in the previous section demonstrated.
   The BPF: Inspection of the BPF under the microscope does show small defects in fabrication that could cause shorts to ground in both mixer circuits attached to the Pol. 2 hybrid, which could explain the weak coupling.

In order to find out the main reason for the poor coupling that we are experiencing, we are planning to perform the following experiments:

- Removing the wire grid and altering the polarisation by rotating the LO source. So far, we assumed that the LO power emitted from a rectangular waveguide and drilled feedhorn is polarised in an orientation that pumps each receiver polarisation with approximately 50% because our OMT probes are oriented by 45° with respect to the rectangular waveguide. Rotating the LO source, and therefore its rectangular waveguide, allows us to modify the coupling to the two sets of probes and will hopefully shed more light on the receiver behaviour.
- Removing the three horizontal connections of the hybrids, shown in Fig. 2. Each OMT probe will then connect to a twin junction circuit directly, allowing us to measure which OMT probe couples LO power. It will then also remove any potential problem in the hybrid design.
- Optical verification of the alignment of the top and bottom blocks, which could cause the displacement of the circular waveguide in the bottom block to the waveguide in the top block.

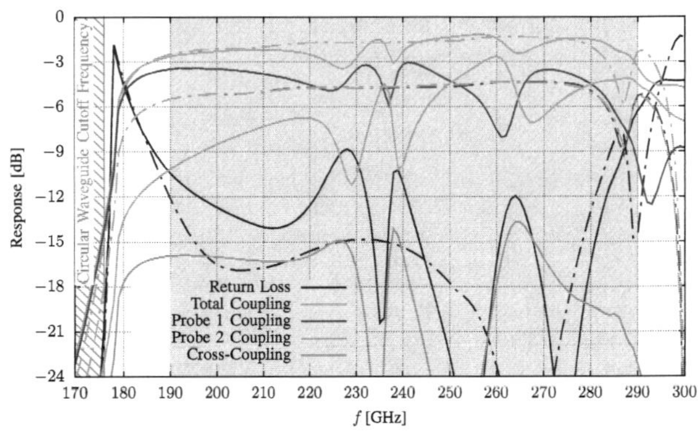

Fig. 12. The simulated OMT performance with the circular waveguide offset by 100 µm towards Probe 2 of the polarisation, which responses are shown as solid lines. The orthogonal polarisation, shown dash-dotted, shows largely similar responses to the designed performance, especially below 270 GHz, except for the excess cross-coupling in green.

VI. Conclusion

We have investigated the performance of an on-chip compact dual-polarisation receiver using SIS tunnel junctions in the frequency range of 190 GHz to 290 GHz. Our experimental test showed weak coupling to the SIS junctions, and we thoroughly experimentally investigated the polarisation behaviour of the receiver and concluded that the most likely reason for the poor coupling comes from a misalignment of the circular waveguide containing the four probes to the waveguide feeding the LO power. Further investigations are in progress to confirm the reason for poor coupling, including the employment of a different split block.

Acknowledgments

This research was funded by the UK-STFC Consortium Grant (ST/R000662/1). For the purpose of Open Access, the author has applied a CC BY public copyright licence to any Author Accepted Manuscript version arising from this submission. J. Wenninger's PhD studentship is supported by the UK Science and Technology Facilities Council and the Lincoln College's Kingsgate Scholarship. F. Boussaha and C. Chaumont acknowledge the supports from Paris Observatory and Paris Science and Lettre (PSL) University, of which they are the leader of the research and development group and clean room process engineer, respectively. The authors would like to thank Mark Merritt from the Rutherford Appleton Laboratory for thinning the devices and Florent Reix from the Paris Observatory for wafer dicing.

REFERENCES

[1] H. Smith, J. Buckle, R. Hills, G. Bell, J. Richer, E. Curtis, S. Withington, J. Leech, R. Williamson, W. Dent, P. Hastings, R. Redman, B. Wooff, K. Yeung, P. Friberg, C. Walther, R. Kackley, T. Jenness, R. Tilanus, J. Dempsey, M. Kroug, T. Zijlstra, and T. M. Klapwijk, "HARP: a submillimetre heterodyne array receiver operating on the James Clerk Maxwell Telescope," in *Millimeter and Submillimeter Detectors and Instrumentation for Astronomy IV*, 2008.

[2] C. Groppi, C. Walker, C. Kulesa, D. Golish, J. Kloosterman, S. Weinreb, G. Jones, J. Barden, H. Mani, and T. Kuiper, "SuperCam: A 64 pixel heterodyne array receiver for the 350 GHz atmospheric window," in *20th Int. Space Terahertz Technol. Symp*, 2009, pp. 90–96.

[3] C. E. Groppi and J. H. Kawamura, "Coherent detector arrays for terahertz astrophysics applications," *IEEE Transactions on Terahertz Science and Technology*, vol. 1, no. 1, pp. 85–96, 2011.

[4] J. Garrett, J. Leech, F. Boussaha, C. Chaumont, B. Ellison, and G. Yassin, "A 1x4 focal plane array using 230 GHz SIS mixers," in *2018 29th IEEE International Symposium on Space Terahertz Technology, ISSTT 2018*, 2018.

[5] R. Guesten, G. A. Ediss, F. Gueth, K. H. Gundlach, H. Hauschildt, C. Kasemann, T. Klein, J. W. Kooi, A. Korn, and I. Kramer, "CHAMP: The carbon heterodyne array of the MPIfR," in *Advanced Technology MMW, Radio, and Terahertz Telescopes*, vol. 3357. SPIE, 1998, pp. 167–177.

[6] K. F. Schuster *et al.*, "A 230 GHz heterodyne receiver array for the IRAM 30 m telescope," *Astronomy and Astrophysics*, 2004.

[7] V. Belitsky, I. Lapkin, B. Billade, E. Sundin, A. Pavolotsky, D. Meledin, V. Desmaris, M. Strandberg, R. Finger, and O. Nyström, "Prototype ALMA Band 5 Cartridge: Design and Performance," in *Proc. 20th Intl. Symp. Space Terahertz Tech*, 2009, pp. 2–5.

[8] J.-Y. Chenu, A. Navarrini, Y. Bortolotti, G. Butin, A. L. Fontana, S. Mahieu, D. Maier, F. Mattiocco, P. Serres, and M. Berton, "The front-end of the NOEMA interferometer," *IEEE Transactions on Terahertz Science and Technology*, vol. 6, no. 2, pp. 223–237, 2016.

[9] P. Kittara, A. Jiralucksanawong, G. Yassin, S. Wangsuya, and J. Leech, "The design of potter horns for THz applications using a genetic algorithm," *International Journal of Infrared and Millimeter Waves*, vol. 28, no. 12, pp. 1103–1114, 2007.

[10] J. Leech, B. K. Tan, G. Yassin, P. Kittara, and S. Wangsuya, "Experimental investigation of a low-cost, high performance focal-plane horn array," *IEEE Transactions on Terahertz Science and Technology*, vol. 2, no. 1, pp. 61–70, 2011.

[11] J. Wenninger, F. Boussaha, C. Chaumont, B. K. Tan, and G. Yassin, "Design of a 240 GHz On-Chip Dual-Polarization Receiver for SIS Mixer Arrays," *Superconductor Science and Technology*, 2023.

Study of anti-reflection layer on dielectric lens for the new 72–116 GHz 7-beam receiver of the Nobeyama 45-m telescope

Yasumasa Yamasaki[1], Yutaka Hasegawa[1], Toshikazu Onishi[1], Hideo Ogawa[1], Atsushi Nishimura[2], Ken'ichi Tatematsu[2], Alvaro Gonzalez[3], Hiroaki Imada[3], Keiko Kaneko[3], Ryo Sakai[3], and Takafumi Kojima[3]

We have developed a new 7-beam heterodyne receiver "7BEE" (7BEamEquipment) for the 72–116 GHz radio frequency (RF) band and installed in the Nobeyama 45-m telescope. This receiver could become an upgrade of current receivers, such as "T70" covering the 74–89 GHz band and "FOREST" (FOur beam REceiver System on the 45-m Telescope) covering the 80–116 GHz band [1]. The main scientific purpose of this development is to understand the early stage of star formation by deriving the deuterium fraction of dense cores.

The optics for this receiver employs wideband corrugated feed horns covering the 67–116 GHz band [2] and dielectric lenses to couple the beams from the feed to those from the 45-m antenna. One of the important aspects in the lens design is the anti-reflection (AR) layers used to mitigate reflections at the lens surfaces. Several types of AR layers using different geometrical shapes have been studied in [3]. In the comparison between the concentric grooves and the straight grooves, it was confirmed that the concentric grooves caused distortion of the wavefront of the incident wave and degraded the beam symmetry, cross-polarization characteristics and aperture efficiency. In addition, when the spacing between straight grooves was compared between 1.2 mm and 1.7 mm, almost the same results were obtained for the beam pattern symmetry and noise contribution. On the other hand, the aperture efficiency was several percent higher in the case of 1.7 mm. Therefore, straight grooves with a spacing of 1.7 mm were determined as the AR structure of the lens.

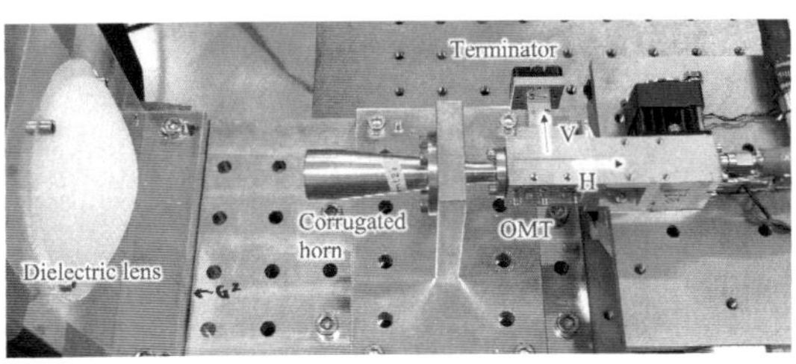

Fig. 1. Photograph of the beam measurement system.

Next, we have investigated the polarization characteristics of the AR structure of the lens. We have constructed the optical system consisting of the lens, the corrugated horn, the double-ridged waveguide type Orthomode Transducer (OMT), and the waveguide transition from circular to square waveguide with a width of 0.73 mm for the evaluation (Fig. 1). The far-field beam patterns were measured by using the beam measurement system [4]. The OMT has high performances with the insertion loss of –0.2 dB or better, the return loss of 22 dB or better, and the cross-polarization separation of –40 dB or better. We have also measured the beam patterns of the optical system in which the waveguide transition from circular to rectangular waveguide was used instead of the OMT to investigate the effect of the OMT on the beam pattern.

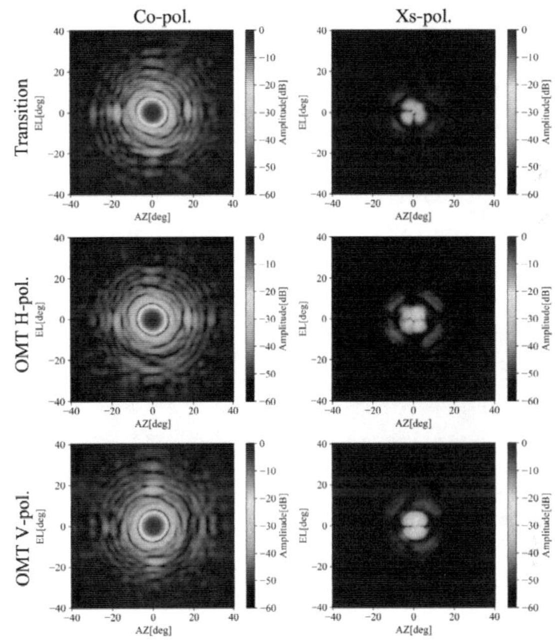

Fig. 2. Results of the far-filed beam patterns in the three cases as follows: (1) lens+horn+waveguide transition (2) lens+horn+OMT H-pol. (3) lens+horn+OMT V-pol.

The results of the far-filed beam pattern at 95 GHz were shown in Fig. 2. The aperture efficiency was calculated with the results. In the same way as the method calculated in [3], the optical systems except the lens and horn were modeled in the physical optics software GRASP (TICRA), the beam radiated from the antenna was calculated using the measurement results as the source, and the aperture efficiency was obtained from the peak gain. The results of aperture efficiency were 0.760 (transition), 0.763 (OMT, H-pol.), and 0.754 (OMT, V-pol.). The differences were within 1 %.

As for the beam pattern of cross-polarization, we analyzed it in detail because we found differences in intensity and

[1]Osaka Metropolitan University, Sakai, Osaka, Japan; [2] Nobeyama Radio Observatory, National Astronomical Observatory of Japan, Minamisaku, Nagano, Japan; [3] National Astronomical Observatory of Japan, Mitaka, Tokyo, Japan

distribution. The maximum cross-polarization intensities relative to the co-polarization were summarized in Table 1. In the system without the lens, an increase of about 3 dB in the cross-polarization level was observed at 115 GHz for V-pol. due to the addition of the OMT. As for the distribution of cross-polarization, the similar results were obtained for H-pol.. On the other hand, the higher-order mode of the circular waveguide may dominate the distribution on the high-frequency side of the V-pol. When the lens was installed, there was an increase in cross-polarization levels. In addition, differences were found between V-H in OMT. In particular, for the pattern of V-pol, we obtained results with two peaks rather than four. It was found that the beam characteristics after passing through the lens were changed by the nature of the electric field incident on the lens, especially in cross-polarization. However, it has little effect on aperture efficiency, which is fine observationally.

Table 1. Maximum cross-polarization intensity of the far-field beam pattern in dB.

Frequency [GHz]	75	95	115
horn+transition	-28.0	-26.3	-28.5
horn+OMT-H	-28.3	-26.9	-27.3
horn+OMT-V	-26.2	-26.9	-24.9
lens+horn+transition	-29.6	-32.6	-21.7
lens+horn+OMT-H	-27.9	-24.5	-18.8
lens+horn+OMT-V	-19.3	-22.7	-19.4

Acknowledgement

This work is supported by JSPS KAKENHI Grant Numbers JP20H05645. This work is also supported by Japan Science and Technology Agency, the establishment of university fellowships towards the creation of science technology innovation, Grant Number JPMJFS 2138.

References

[1] Tetsuhiro Minamidani, et al., "Development of the New Multi-Beam 100 GHz Band SIS Receiver FOREST for the Nobeyama 45-m Telescope", *SPIE*, 2016.

[2] Alvaro Gonzalez, et al., "Development of receiver optics for ALMA bands 1 and 2, and possible synergies with ngVLA", *SPIE*, Dec. 2020.

[3] Yasumasa Yamasaki, et al., "Development of seven-beam optics using dielectric lenses for a new 72–116 GHz receiver in the Nobeyama 45 m telescope", *Publications of the Astronomical Society of Japan*, Mar. 2023.

[4] Alvaro Gonzalez, et al. "Reconfigurable Near-Field Beam Pattern Measurement System From 0.03 to 1.6 THz", *IEEE Trans. Terahertz Sci. Technol.*, 2016.

Tunable Antenna-Coupled Intersubband Terahertz (TACIT) Mixer Integrated with Self-complementary Antenna

Changyun Yoo[1], Jonathan H. Kawamura[2], Ken W. West[3], Loren N. Pfeiffer[3], Boris S. Karasik[2], and Mark S. Sherwin[1]

Abstract—**We are developing a novel THz hot-electron bolometer (HEB) mixer based on a high-mobility 2-dimensional electron gas (2DEG) confined in a single GaAs/AlGaAs quantum well. Named as Tunable Antenna-Coupled Intersubband Terahertz (TACIT) mixer, the device takes advantages of intersubband transitions of the confined 2DEG for efficient, tunable absorption of THz radiation. Here we demonstrate a TACIT mixer integrated with a self-complementary antenna suitable for the broadband operation of the device. We study the tunability in both direct detection and heterodyne detection responses and show that the observed tunability behavior is consistent with our device model based on intersubband transitions. In addition, we study the DC-bias dependence of the noise at 1.4 GHz and show that the noise behavior follows that of a HEB device.**

Keywords— THz heterodyne detector; high-mobility 2-dimensional electron gas; intersubband transitions

I. Introduction

Heterodyne detection is an important technique for high-resolution spectroscopy at THz frequencies for astrophysics and planetary applications. Above 1 THz, superconducting hot-electron bolometers (HEBs) are the heterodyne detectors of choice for most astrophysics applications because of their excellent sensitivity. For applications that cannot afford active cryogenic cooling (i.e., deep-space planetary applications), Schottky-diode mixers are the only option so far because of their higher operating temperature. However, compared with superconducting HEBs, Schottky-diode mixers suffer from much less sensitivity and much higher required local-oscillator (LO) power (~1 mW), the latter of which especially limits the use of the technology for multi-pixel heterodyne array applications.

We are developing a novel THz heterodyne detector based on intersubband transitions of a high-mobility 2-dimensional electron gas (2DEG) that may offer the best of both worlds [1]. Named as Tunable Antenna-Coupled Intersubband Terahertz (TACIT) mixers, our devices can be as sensitive as superconducting HEBs (with predicted single-sideband noise temperature $T_{SSB} \sim 1{,}000$ K) at higher operating temperature (40-60 K) with moderate LO power requirement (~1 μW) and with a useful intermediate frequency (IF) bandwidth (~10 GHz). In addition, TACIT mixers can offer wide in-situ tunability in the detection frequency (2-5 THz) with small (<1 V) DC voltage biases. These impressive mixer characteristics along with the tunability makes TACIT mixers an attractive THz mixer technology for low-noise, multi-pixel THz heterodyne spectrometers, especially for applications in which either active cryogenic cooling or available THz LO power is limited.

[1] Physics Dept. & Inst. for Terahertz Science and Technology, University of California at Santa Barbara, Santa Barbara, CA 93106, USA; [2] Jet Propulsion Laboratory, California Institute of Technology, Pasadena, CA 91109, USA; [3] Dept. of Elec. Eng., Princeton University, Princeton, NJ 08544, USA.

II. Results

In our previous work, we demonstrated a proof-of-concept TACIT device integrated with a single-slot antenna and confirmed the tunability in the detection frequency (2.5-3.1 THz) as well as THz mixing capability at 60 K with an IF bandwidth exceeding 6 GHz [2]. Here we report on another version of a TACIT mixer integrated with a self-complementary Maltese-cross antenna (Fig.1. (a)). In this version of the TACIT mixer, the source and the drain of the device contain additional metal contacts that capacitively couple out the IF response from the 2DEG (Fig.1. (a) inset and Fig.1. (b)). With the top and the bottom gate metals, these capacitive contacts also form the self-complementary Maltese-cross antenna structure suitable for the broad tunability of the device (Fig.1. (a) inset).

Direct detection measurements in response to a monochromatic radiation at 2.52 THz and 3.11 THz at 10 K show that the tunability behavior of the device is consistent with the previous results and in good agreement with the device model presented in [1] (Fig.1. (a)). Heterodyne measurements in response to two monochromatic sources near 2.52 THz at 10 K also resulted in the tunability behavior consistent with the direct detection result at 2.52 THz. In addition, noise power measurements at 1.4 GHz show that the noise power peaks at the source-drain DC biases near the knee of the current-voltage (IV) curve in which the contribution of the thermal fluctuation noise is expected to be maximum. By measuring the noise power at zero source-drain DC bias with and without 2.52 THz, we observed the change in the noise power that corresponds to ~ 20 K of heating of the 2DEG with THz.

Heterodyne measurements at higher temperatures (40-60 K) and noise temperature characterizations are currently underway.

NOTES:

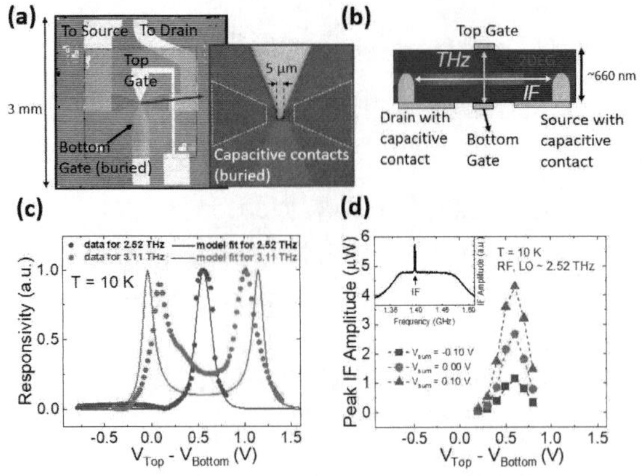

Fig. 1. (a) TACIT mixer integrated with a Maltese-cross antenna on a high-resistivity Si substrate. The inset shows the four-terminals of the device near the active region (5 μm × 5 μm patch) that form the self-complementary antenna for broadband THz absorption. (b) schematic showing the vertical profile and the four-terminals of the device. The two Schottky contacts (top and bottom gates) sandwiching the active region orient the THz electric fields perpendicular to the 2DEG plane (required to satisfy the selection rule for the intersubband transition) for efficient THz absorption. The two Ohmic contacts with capacitive contacts (source and drain) couple out the IF response in the in-plane resistance of the 2DEG. (c) direct detection response at 10 K with theoretical model curves (d) gate-voltage dependence of the IF peak amplitude for the heterodyne detection of ~ 2.52 THz at 10 K. The inset shows the IF signal near 1.4 GHz for the heterodyne detection.

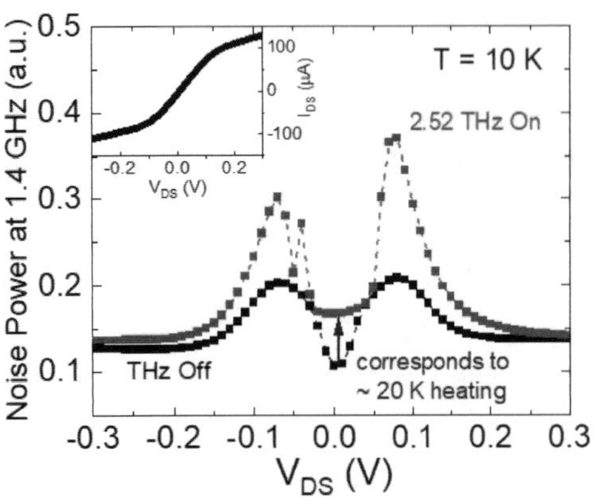

Fig. 2. DC-bias dependence of the noise power at 1.4 GHz with and without 2.52 THz at 10 K. The inset shows the device current-voltage (IV) curve without THz illumination at 10 K.

REFERENCES

[1] M. S. Sherwin et al., "Tunable Antenna Coupled Intersubband Terahertz (TACIT) Mixers: The Quantum Limit Without the Quantum Liquid," *Proc. Far-IR, Sub-mm and mm Detector Technology Workshop*, 2002; https://arxiv.org/abs/1909.10664

[2] C. Yoo et al., "Demonstration of a Tunable Antenna-Coupled Intersubband Terahertz (TACIT) Mixer.", *Appl. Phys. Lett*, Jan. 2020.

NOTES:

325 GHz and 650 GHz Dual-polarisation receivers Concept

O. Auriacombe*[1], P. Sobis[2,3], V. Drakinskiy[2], K. Kempe[1], J. Embretsén[1], A. Emrich[1], D. Cuadrado-Calle[4], M. van der Vorst[4], J. Stake[2]

***Abstract*— The integrated dual-polarisation receivers utilize a dual probe concept, efficiently integrating the antenna and MMIC package environment which allows for polarisation discrimination without the use of bulky and lossy external orthomode transducers. This concept increases the sensitivity of the instrument and reduces its size, enabling the development of future earth observation arrays. Omnisys Instruments AB (Sweden) and Chalmers University of Technology (Sweden) are working to demonstrate state-of-the-art dual polarisation capability with two integrated receiver modules at 325 GHz and 650 GHz.**

Keywords*— *Radiometer, receivers, Polarisations, remote sensing, Schottky diodes, Low noise amplifiers, Microwave,

I. Introduction

Integrated dual polarisation receiver demonstrators at 325 GHz and 650 GHz are an advantage for weather and climate applications such as, constellation of small weather satellites, cost efficient atmospheric science missions and future cubesat instruments, due to the strict volume accommodation and performance requirements. This article describes the architecture of two integrated Schottky – MMIC front-end receivers operating at 325 GHz and 650 GHz, respectively.

The schematic of the spaceflight intended dual-polarisation receivers is shown in Fig. 1. A synthetiser device with integrated voltage control oscillators (VCO) can deliver a tone from 40 MHz to 15 GHz. The selected VCO frequency is 10.156 GHz. A first stage of LO driver with active multiplier and power amplifier provides around 180 mW at a frequency of 81.25 GHz. A Schottky multiplier (x2) converts the 81.25 GHz frequency to 162.5 GHz with input and output powers of 125 mW and 50 mW, respectively. This multiplier provides the LO signal to the 325 GHz front-end. For the 650 GHz front-end, another Schottky multiplier (x2) is added to the LO multiplier stage to reach an output frequency of 325 GHz with 10 mW output power. Each front-end is integrated in one single block with similar architecture. One front-end is composed of a spline horn antenna, an integrated orthomode transducer (OMT), two sub-harmonic mixers (SHM), two IF low noise amplifier MMICs, and a LO splitter. The dual probe concept consists of the integration of an orthogonal probe layout with the two mixers on a single substrate. The design is a 3µm thick suspended GaAs membrane with 1µm thick gold strip. It is suspended in between the split blocks interconnected with on circuit integrated waveguide probe transitions and freestanding gold beam leads (used for mechanical support), RF/DC ground connections, and IF circuit interconnects. This configuration has previously been demonstrated at 424 GHz [1]. Anti-parallel Schottky diodes are provided by Chalmers University of Technology. They have been used as mixer diodes with state-of-the-art performances in several instruments in the past [2], and at frequencies up to 3.5 THz [3]. Low noise MMIC amplifier chips from Low Noise Factory AB (Sweden) are integrated with their matching networks inside the front-end block. A single chip has an RF bandwidth of 1-15 GHz and gain of 37 dB. Finally, spline horn has been designed for high cross-polarisation discrimination.

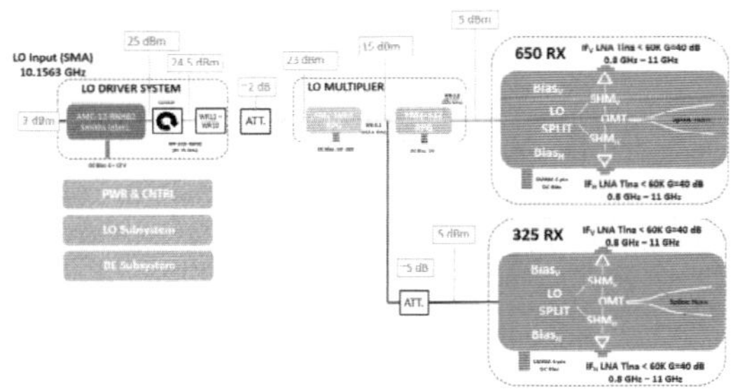

Fig. 1. Schematic of the dual-polarisation receivers at 325 GHz and 650 GHz

II. Results

Two dual-polarised receivers operating at 325 GHz and 650 GHz were modelled using Ansys HFSS and Advanced design System (ADS) and simulated with a non-linear analysis. The preliminary simulated performance of a single polarised sub-harmonic DSB mixer with an integrated OMT RF probe structure is shown in Table I. The OMTs for the 325 GHz and 650 GHz receivers have a simulated cross-polarisation discrimination better than -25 dB across the band and low transmission losses (around 0.2 dB). The mixer simulated conversion loss was simulated at around 9 dB with a 2 mW LO power at 325 GHz and 10 dB with 2.5 mW LO power at 650 GHz. Assuming a 180 K diode output noise, the receiver noise temperature with the integrated MMIC would range between 737 – 1024 K for the 325 GHz receiver and 982 – 1424 K, for 650 GHz receiver.

TABLE I. Simulation Results

Frequency (GHz)	*Mixer Conv. Loss (dB)*	*Trec.(K)*	*OMT X-pol. Discrimination (dB)*
314 – 336	8.67 – 9.14	737 - 1024	-26.3 ± 1.0
639 – 661	9.92 – 10.57	982 - 1424	-25.3 ± 0.2

References

[1] P. Sobis, et al., "Membrane Integrated Asymmetric Dual E-plane Probe Ortho Mode Transducer at 424 GHz", *29th IEEE International Symposium on Space Terahertz Technology*, 2018.

[2] P. Sobis, et al., "Discrete GaAs Schottky beamlead mixer diodes for space-borne receiver applications", *29th IEEE International Symposium on Space Terahertz Technology*, 2018.

[3] D. Jayasankar et al., "A 3.5-THz, ×6-Harmonic, Single-Ended Schottky Diode Mixer for Frequency Stabilization of Quantum-Cascade Lasers," in *IEEE Transactions on Terahertz Science and Technology*, vol. 11, no. 6, pp. 684-694, Nov. 2021.

[1]Omnisys Instruments AB (AAC Clyde, Ltd.), SE-42132, Västra Frölunda, Sweden; [2]Terahertz and Millimetre Wave Laboratory, Chalmers University of Technology, SE-41296, Göteborg, Sweden; [3]Low Noise Factory AB, SE-41263, Göteborg, Sweden; [4]European Space Research and Technology Centre (ESA ESTEC), 2201, Noordwijk Netherlands.

SIS technology development to serve Next Generation receivers for ALMA

Alexey Pavolotsky*[1], Takafumi Kojima*[2], Sho Masui[2], and Victor Belitsky[1]

***Abstract*— Modern radio astronomy demands for broadband receiver systems. For SIS mixers, this translates into objective to employ superconducting tunnel junctions with a very low R_nA and low specific capacitance. The traditionally used $Nb/AlO_x/Nb$ junctions have largely approached their physical limit of minimizing those parameters. It is commonly recognized that it is AlN-barrier junctions, which are needed for further progressing of the broadband SIS mixer instrumentation for radio astronomy. In this work, we present the progress in development of the process for high quality Nb/Al-AlN/Nb superconducting tunnel (SIS) junctions' fabrication and their characterization in terms of their specific capacitance.**

Keywords—SIS junctions, AlN tunnel barrier, specific capacitance

I. Introduction

We are developing SIS process technology capable of fabricating mixer chips for Next Generation ALMA receivers. The requirements for such SIS process technology stems from the ALMA 2030 Development Roadmap document [1] calling for twice or triple enhancement of the RF and IF bandwidths of the receivers and consequently the SIS mixers. This in turn translates into the requirement for SIS junctions to have a smaller specific capacitance, C_s, (i.e. junctions with AlN tunnel barrier) and having smaller size.

We have earlier reported on the process development for high-quality Nb/Al-AlN/Nb junction fabrication based on microwave plasma nitridation [2]. We showed that the Nb/Al-AlN/Nb junctions with R_nA product down to ~5 Ohm·μm^2 demonstrate excellent quality. Also, we showed that the produced junctions were quite stable against the thermal annealing, at least up to 200°C, thus allowing for thermal impact during almost any possible fabricating or packaging technology processes.

In this manuscript, we present results of the Nb/Al-AlN/Nb junction process development aiming for fabricating of smaller area junctions, as well as the specific capacitance measurements following the approach similar to that reported in the papers [3] at GARD and in [4] in NAOJ, by means of on-wafer capacitance measurements with 4K probe station.

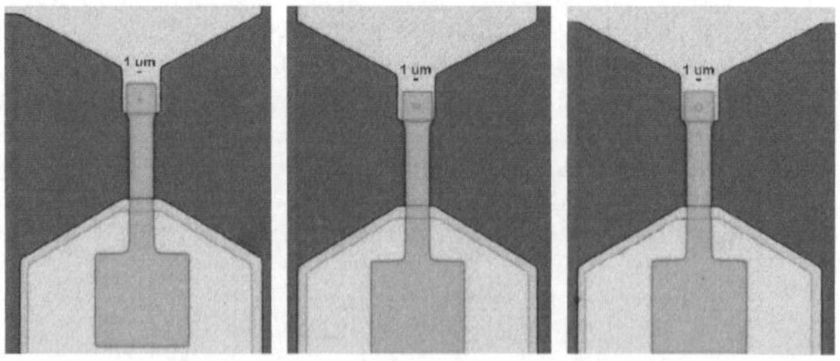

Fig. 1. Micrographs of the Nb/Al-AlN/Nb SIS junctions of different sizes and shapes.

II. Results

We have developed the process capable of fabricating the Nb-based SIS junctions with the junction size achieving junction dimensions $\leq 1\mu m^2$. We have employed direct laser writing for definition of the junction pattern. After setting up the direct laser exposure process with the AR-N 4340 negative tone resist, the rest of the small SIS junction process has been integrated and the test junctions have been fabricated. The test wafer included the junctions of different shapes and sizes (Fig. 1), as well as the bigger sized round junctions for extracting of the R_nA product along with the junction size offset due to the lithography and etching processes (Fig. 2). The presented on the Fig. 3, the current voltage characteristics of the fabricated submicron Nb/Al-AlN/Nb junctions demonstrate their excellent quality.

Earlier, we reported on the Nb/Al-AlN/Nb junctions specific capacitance measurements [2]. The latter was performed in the test cryostat with long and inevitably lossy stainless-steel coaxial cables connecting device under test at 4K physical temperature with VNA. Use of the cables is largely affects the accuracy of the measurements because the losses and electrical length of the cables are changing while cooling and make the VNA calibration compromised.

Direct measurement of the junction capacitance in cryogenic probe station at NAOJ [4], [6] is seen as a very attractive alternative option both from the point of view of a quicker throughput and - importantly - higher accuracy.

[1]Group for Advanced Receiver Development, Department of Space, Earth and Environment, Chalmers University of Technology, Gothenburg 412-96, Sweden, [2]Advanced Technology Center, National Astronomical Observatory of Japan, Mitaka, Tokyo 181-8588, Japan

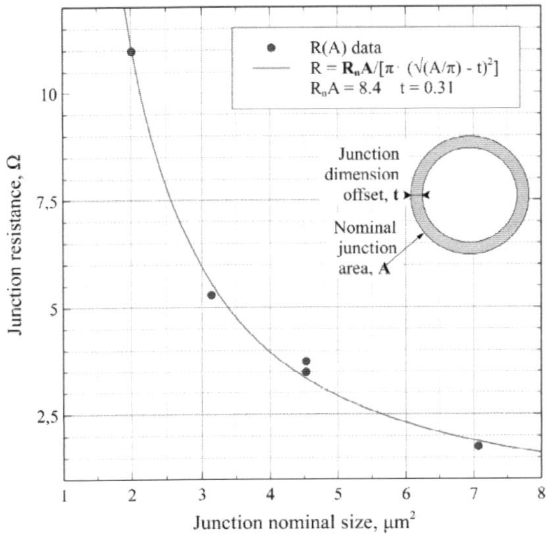

Fig. 2. True R_nA and dimension offset t is extracted from from the scaling of junctions' Rn vs nominal area of the test round junctions.

Shown on the Fig. 4, a cryogenic probe station at NAOJ allows to calibrate out all embedding circuitry down to the junction itself. Moreover, through careful design of the test structures on-wafer, we have all three standards needed for 1-port Short-Open-Load, SOL calibration technique integrated right on the chip to be applied. This way, every measurement inherently includes fresh calibration in the same cooling cycle.

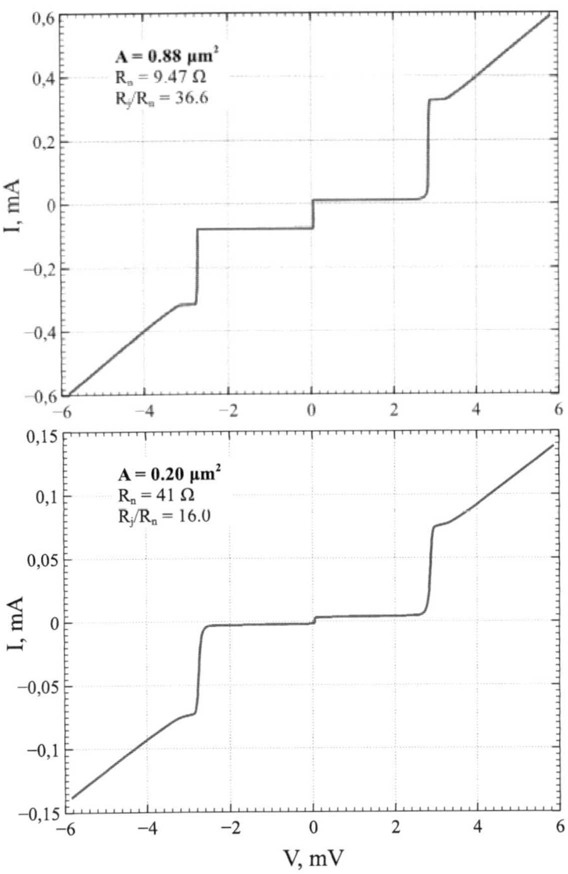

Fig. 3. Typical current-voltage characteristics of the Nb/Al-AlN/Nb junctions.

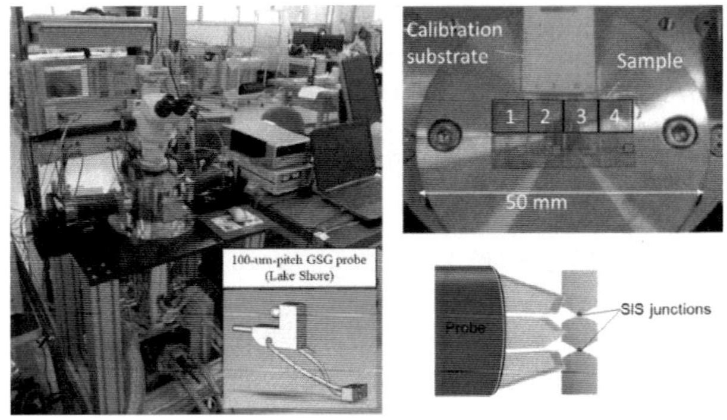

Fig. 4. Photographs of 4K probe station at NAOJ and device sample stage. The probe station allows to perform on-wafer characterization of SIS junction lots of various sizes at frequencies ranging from dc to microwave.

The dedicated test SIS wafer was designed and fabricated (Fig. 5). The test SIS wafer included single SIS junctions of different sizes for dc-IVC measurements (placed closely adjacent to the big external contact pads) and 112 test structures dedicated for capacitance characterization.

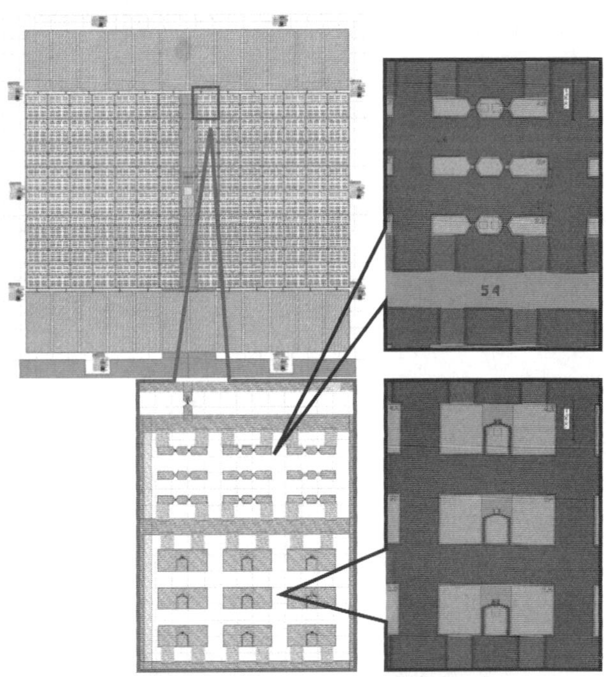

Fig. 5. The wafer (*upper left*) contains 112 test structures for capacitance characterization as well as single SIS junctions of different sizes for dc-IVC measurements (placed closely adjacent to the big external contact pads). Magnified view of the test structures (*lower left*) include structures with SIS junctions, as well as calibration patterns (open, short and load) – all for single and twin junctions' arrangements. Optical micrographs of fabricated structures are shown to the right

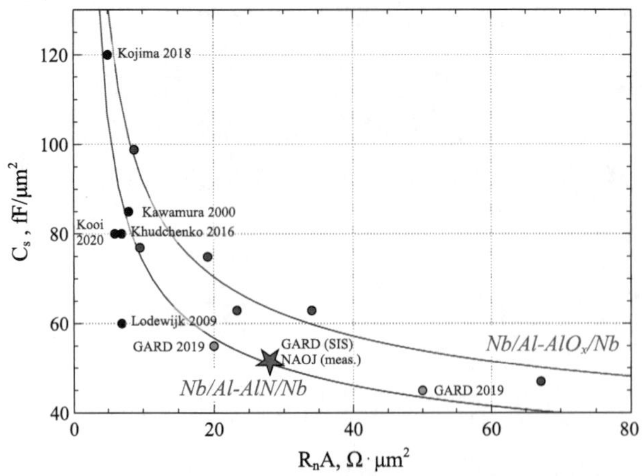

Fig. 6. Specific capacitance of Nb/Al-AlN/Nb junctions (*blue*) as compared with that of Nb/Al-AlO$_x$/Nb junctions [5] (*red*). The capacitance data for the junctions are approximated with semi-empirical relation $Cs = a / \ln(R_n A)$ [7], where a is equal to 211 [5] for the Nb/Al-AlO$_x$/Nb junctions and to 170 for the measured Nb/Al-AlN/Nb junctions. The *dark-blue dots* are the data communicated by the other groups.

Each test chip contained structures with SIS junctions, single and twin junctions' arrangements, as well as standard patterns (open, short and load).

The measurement result (Fig. 6) has confirmed that specific capacitance of the Nb/Al-AlN/Nb junctions is noticeably lower than that reported for the Nb/AlO$_x$/Nb junctions, e.g. in [5], as well as the agreement between the measurements performed independently at GARD and at NAOJ is very good.

III. Conclusion

We have presented the status of SIS process developing capable of fabricating mixer chips for Next Generation ALMA receivers. The fabricated Nb/Al-AlN/Nb SIS junctions demonstrate excellent junction quality for the R_nA down to ~5 Ohm·μm^2 and down to submicron junctions' area defined by means of direct laser writing. Direct measurements of the Nb/Al-AlN/Nb SIS junctions' specific capacitance at GARD and NAOJ give the consistent results and confirm that the specific junction capacitance is ca. 20% lower than that for the Nb/Al-AlO$_x$/Nb junctions.

IV. Acknowledgement

This work was partially supported in the framework of the European Organisation for Astronomical Research in the Southern Hemisphere (ESO) "Advanced Study for Upgrades of the Atacama Large Millimeter / submillimeter Array (ALMA)".

References

[1] J. Carpenter, et al., "The ALMA Development Program: Roadmap to 2030", *arXiv preprint arXiv:2001.11076. 2020,* Jan 29.

[2] A. Pavolotsky, et al., "Specific capacitance of Nb/Al-AlN/Nb superconducting tunnel junctions", *Proceedings of ISSTT 2019*, April 2019.

[3] P. Y. Aghdam, H. Rashid, A. Pavolotsky, V. Desmaris, D. Meledin, and V. Belitsky, "Direct Measurement of Superconducting Tunnel Junction Capacitance," *IEEE Trans. Terahertz Sci. Technol.*, 2015.

[4] T. Kojima, et al., "On-wafer capacitance Measurement of Nb-Based SIS Junctions with a 4-K Probe Station", *IEEE Transactions on Applied Superconductivity*, June 2017

[5] P.Y. Aghdam, et al., "Specific Capacitance Dependence on the Specific Resistance in Nb/Al–AlO$_x$/Nb Tunnel Junctions", *IEEE Transactions on Terahertz Science and Technology*, September 2017.

[6] T. Kojima, et al., "Contribution of Quantum Susceptance in SIS Junction Capacitance Measurement," *IEEE Trans. Appl. Supercond.*, 2019.

[7] V. Belitsky, et al., "100 GHz mixer with vertically integrated (stacked) SIS junction array," *Int. J. Infrared Millim. Waves*, 1993.

Integrated Schottky Technology for Supra-THz Applications

Vladimir Drakinskiy[1], Divya Jayasankar[1], Peter Sobis[1,2], and Jan Stake[1]

Abstract—**This paper presents the technological development of e-beam lithography based monolithically integrated supra-THz Schottky diode process at the Chalmers University of Technology. Schottky diodes with sub-micron contact area integrated with mixer circuits were realised on an ultra-thin 2-μm GaAs substrate. The fabrication process, challenges and the yield of planar integrated mixer circuits are presented.**

Index Terms—**Schottky diodes, GaAs, Mixers, Terahertz electronics**

I. Introduction

Schottky diodes are employed as primary detecting element as well as in the local oscillator chain for frequency stabilisation of quantum-cascade lasers (QCLs) [1]. They operate in wide intermediate frequency and ambient temperature, which is crucial for long lifetime space/air-borne missions to study Earth and other planetary atmospheres. Detection of molecular lines such as OI and OH at terahertz (THz) frequencies is essential. Our group has demonstrated successful frequency stabilisation of quantum-cascade lasers (QCLs) at 3.5 THz, and 4.7 THz using a Schottky-based harmonic mixer [2]. This paper reports the development of integrated sub-micron Schottky diode circuits on ultra-thin GaAs substrate.

II. Fabrication Process

The realisation of terahertz integrated circuits demands the alignment of patterns with high accuracy and precision in the order of sub-microns. Hence, we have developed a fabrication process entirely based on electron-beam lithography. The GaAs wafer consists of a 50-nm-thick n-doped epi-layer with a doping concentration of $6 \times 10^{17}\,\mathrm{cm}^{-3}$. Followed by a 500-nm-thick, heavily doped n^{++} buffer layer with a doping concentration of $5\times10^{18}\,\mathrm{cm}^{-3}$. The wafer consists of two (Al,Ga)As etch stop layers as given in [3].

The detailed fabrication process is as follows, to protect the active layer, SiO_2 was deposited using PECVD. Later, the ohmic (Pd/Ge/Au/Pd/Au) and anode (Ti/Pt/Au) contacts were formed and followed by the formation of the mesa by etching both epi- and buffer layer. High anisotropy of the dry etching process compared to wet etching allows vertical sidewalls and minimises the risk of over-etching under the resist mask. In the next step, air bridge contacts and mixer circuit elements such as hammer-head filter were formed [4]. Finally, to provide mechanical support and heat transfer, 1-μm gold beamleads are formed. Fig. 1a shows the scanned electron micrograph (SEM) of the fabricated single-ended planar Schottky diode fundamental mixer circuit with sub-μm anode contact area on a 2-μm GaAs substrate.

III. Yield Analysis

Before releasing the circuits, on-wafer dc tests were carried out using Kelvin probes in dark conditions. Fig. 1b shows the scatter plot of extracted diode parameters: series resistance (R_s) and ideality factor (η) consisting of about 30 integrated mixer circuits with three different contact areas. The mean and standard deviation of R_s and η are summarised below in Table. I. We believe the variation in diode parameters is caused due to a small deviation in Schottky contact areas.

[1]Chalmers University of Technology, SE-412 96 Gothenburg, Sweden.
[2]Low Noise Factory AB, SE-412 63 Gothenburg, Sweden.

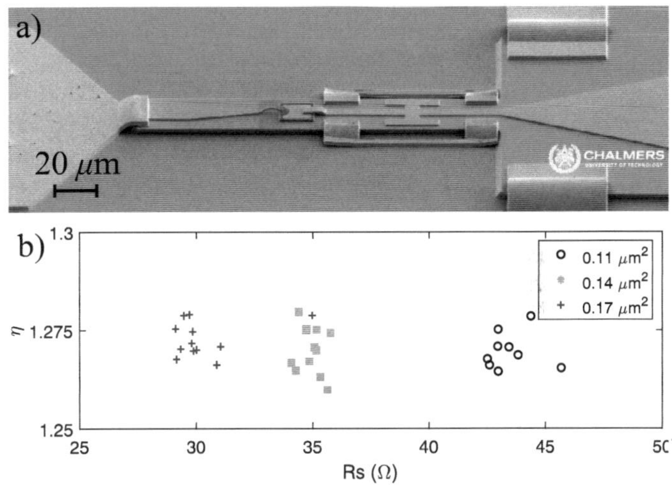

Fig. 1. Supra-THz Schottky mixers. a) SEM of the integrated 3.5-THz fundamental mixer circuit realised on a 2-μm GaAs substrate. b) Extracted R_s and η for three Schottky diode contacts (0.11, 0.14, and 0.17 μm^2).

TABLE I
Extracted DC-Parameters of Integrated 3.5-THz Fundamental Mixer Circuits

Area (μm^2)	Series resistance (Ω)		Ideality factor (η)	
	Mean	Std.Dev	Mean	Std.Dev
0.11	44	2.7	1.27	6.0×10^{-3}
0.14	35	0.5	1.27	5.9×10^{-3}
0.17	30	1.6	1.27	4.4×10^{-3}

References

[1] D. Jayasankar *et al.*, "A 3.5-THz, x6-Harmonic, Single-Ended Schottky Diode Mixer for Frequency Stabilization of Quantum-Cascade Lasers," *IEEE Trans. on Terahertz Science and Technology*, vol. 11, no. 6, pp. 684–694, 2021.

[2] H. Richter *et al.*, "Phase-locking of 3.5-THz and 4.7-THz Quantum-Cascade Lasers using a Schottky diode harmonic mixer," in *46th IRMMW-THz*, Sep. 2021.

[3] P. H. Siegel *et al.*, "2.5-THz GaAs monolithic membrane-diode mixer," *IEEE Trans. Microw. Theory Techn.*, vol. 47, no. 5, pp. 596–604, May 1999.

[4] D. Jayasankar, V. Drakinskiy, M. Myremark, P. Sobis, and J. Stake, "Design and development of 3.5 THz Schottky-based fundamental mixer," in *2020 50th European Microwave Conference (EuMC)*, 2021.

Estimation of Input Power Handling Capability of Next-Generation GaN Schottky Diodes for Millimeter Wave Frequency Multipliers

Priyanka Mondal*[1], Giuseppe Di Gioia[2], Thibaut Vacelet[1], Alexandre Feret[1], Yvon Cordier[3], Guillaume Ducournau[2], Yannick Roelens[2], Mohammed Zaknoune[2] and Jeanne Treuttel[1]

***Abstract*—This abstract is on the estimation of input power handling capability of GaN Schottky diodes developed till date. The destructive reverse breakdown voltage is characterized as -69.5 V for the diodes developed with the epitaxy having active epilayer thickness 590 nm and doped at 6.6×10^16^ cm^-3^ on 6H-SiC substrate. Simulations based on a 2-anode varactor doubler with optimized efficiency show that an input power as high as 500 mW can be applied to a single GaN Schottky diode until the diode voltage swing reaches the limit of reverse destruction.**

Keywords*—doubler, GaN Schottky, local oscillator, millimeter wave, power handling, reverse breakdown voltage.

I. Introduction

Schottky diodes play a key role in local oscillator (LO) and detector systems in the instruments for planetary and Earth sciences [1]. GaAs technology is the solution so far for the millimeter wave and sub-millimeter wave bands. But, exploration of GaN Schottky technology is a promising one for the LO power generation at millimeter wave band in view of excellent power handling capability [1], [2].

Recently, GaN Schottky diodes with small anode sizes are developed and characterized on 6H-SiC substrate after rigorous and successful development on Sapphire substrate through e-beam lithography [3]. Destructive reverse breakdown voltage is characterized to be -69.5 V for an epitaxy of active epilayer thickness 590 nm and doping density of the n^- layer as 6.6×10^{16} cm^{-3}. Mobility of the epitaxy is determined as ~ 400 $cm^2/V.s.$ which is substantially low. This low mobility is attributed to the high defect density of the crystal inherent to the large mismatch between GaN and the substrate [4]. In homoepitaxy context, the GaN electron mobility can reach 1000-1200 $cm^2/V.s.$ and better results can be expected on GaN substrate particularly for the breakdown voltage.

Parameters of the diode after DC measurements are listed in Table 1. Aim of the present abstract is the estimation of input power handling capability of the developed GaN Schottky diode. The frequency of operation is chosen as 105-120 GHz to fit the atmospheric absorption bands in addition to the potential application for the low frequency LO stages, for an example 1.06 THz LO source [5]. A 2-anode varactor doubler configuration is considered for the simulations due to limited commercial availability of driver source having power level 1 W or more.

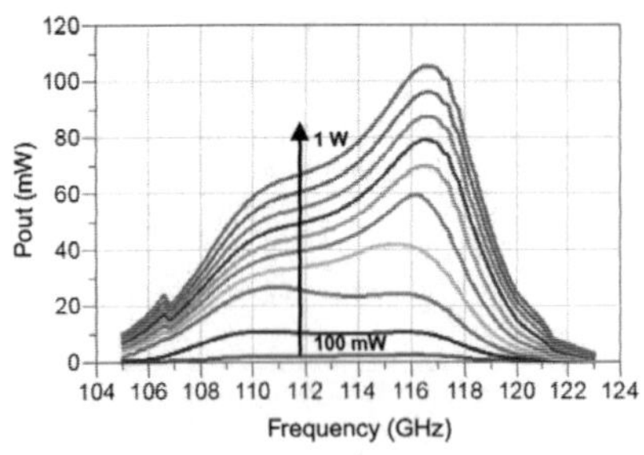

Fig. 1. Simulated performance of a 2-anode GaN Schottky varactor.

TABLE I. Parameters of GaN Schottky Diode

Diameter (μm)	*Destructive Reverse Breakdown Voltage (V)*	C_{j0} *(fF)*	R_S *(Ω)*
6	-69.5	27.3	34.7

II. Doubler Simulation

Simulated performance of the 2-anode GaN Schottky doubler in electromagnetic environment for different input power level is shown in Fig. 1. Reverse bias of the varactor doubler is optimized to achieve the best efficiency with the experimentally obtained diode parameters. Input power is increased but to keep the voltage swing across each diode well below the destructive reverse breakdown voltage. It is seen that an input power as high as 500 mW/anode can be applied. However, the best efficiency of the doubler is achieved over a certain input power level. Further GaN Schottky diodes for multiplier application with the objectives of input power handling capability and efficiency while operating at multiple frequency bands are under fabrication for experimental verifications.

Acknowledgements

This work was supported by the French Agence Nationale de la Recherche (ANR) and the French network RENATECH, under the 'SchoGaN' project. The authors are thankful to Lytid team for their participation in thoughtful discussions.

References

[1] I. Mehdi, et al, "THz Diode Technology: Status, Prospects, and Applications," Proceedings of IEEE, Jun. 2017.

[2] J. V. Siles, et al, "Capabilities of GaN Schottky Multipliers for LO Power Generation at Millimeter-Wave Bands," 19th International Symposium on Space Terahertz Technology, Apr. 2008.

[3] G. D. Gioia, et al, "GaN Schottky Diode on Sapphire Substrate for THz Frequency Multiplier Applications," Superlattices and Microstructures, in press.

[4] Nils G. Weimann et al, Scattering of electrons at threading dislocations in GaN, Journal of Applied Physics, Apr. 1998

[5] P. Mondal, et al, "Development Status of Millimeter Wave GaN Schottky Doublers above W-band for the Implementation of European Terahertz Sources for Astronomy and Astrophysics" Nano, Meso, Micro: Science and Innovation for Radio and Photonic, URSI JS22, Mar. 2022.

[1]LERMA, Observatory of Paris, Paris, 75014, France; [2]IEMN, University of Lille, Villeneuve d'Ascq Cedex, 59652, France; [3]Université Côte d'Azur, CNRS, CRHEA, Valbonne, 06560, France. *pmondal@ieee.org

Characterization of a Wide-Band Microstrip Parametric Amplifier Utilizing 4-Wave-Mixing Techniques

E. Linden,[1] N. Klimovich,[2] S. Shu, B. Eom,[3] C. Kim,[3] R. DeLuc,[3] and P. Day[3]
[1]Department of Physics, The College of Liberal Arts and Sciences, Arizona State University, Tempe, AZ, 85281
[2]The Division of Physics, Mathematics and Astronomy, California Institute of Technology, Pasadena, California 91125, USA
[3]Jet Propulsion Laboratory, California Institute of Technology, Pasadena, California 91109, USA

Kinetic Inductance Traveling Wave Parametric Amplifiers (KI-TWPA) are being developed that can achieve quantum-limited sensitivity over wide fractional bandwidths, while being able to amplify signals of the level needed for sensitive astronomical measurements. We present initial measurements on the KI-TWPA through 4 Wave Mixing (4WM) of a superconducting NbTiN microstrip transmission line using the nonlinear kinetic inductance inherent in the NbTiN material, measured at 1K. The target of our design is to produce gain in a signal bandwidth of 4-12GHz for use as a detector readout amplifier, or as an IF amplifier in a heterodyne receiver system. Because both the pump tone and the generated idler tones lie at higher frequency than the amplified region, the entire 4-12 GHz range will be useful for signal amplification without introduction of image tones.

Analysis of the low temperature behavior of GaN-on-SiC Schottky barrier diodes

B. Orfao*[1], G. Di Gioia[2], B. G. Vasallo[1], S. Pérez[1], M. Zaknoune[2], Y. Cordier[3], Y. Roelens[2], J. Mateos[1] and T. González[1]

***Abstract*—GaN-on-SiC Schottky Barrier Diodes have been characterized from 33 to 475 K. A theoretical model accounting for image force lowering is used to extract the parameters of the diodes and predict the reverse-bias current. Deviations from the ideal behavior are observed for low temperatures both in forward and reverse bias.**

***Keywords*—GaN Schottky barrier diodes, low temperature characterization.**

I. Introduction

In the last years, GaN has attracted great interest for the fabrication of Schottky Barrier Diodes (SBDs) for high-frequency high-power applications [1], such as frequency multipliers in the sub-THz range overcoming the power output performance of the standard GaAs technology [2]. The potential use of GaN SBDs in systems for space applications motivates the study of the diodes at cryogenic temperatures reported in this work.

II. Results

C-V and *I-V* measurements were carried out from 33 to 475 K on a (221 μm diameter) GaN-on-SiC SBD fabricated at IEMN. *C-V* measurements provide an epilayer doping level of 1.05×10^{17} cm^{-3}, used for the fitting of the *I-V-T* curves shown in Fig. 1(a), from which the barrier height and the ideality factor were extracted (inset). A thermionic-emission (TE) model including image force lowering and series resistance (around 10 Ω for all *T*) was used. In Fig. 1(a), the expected decrease of the TE current when lowering *T* is observed, jointly with a non-ideal current mechanism preventing a good fitting of the *I-V* curves at the lowest temperatures, confirmed by the decrease of the barrier height and increase of the ideality factor (inset). Indeed, a kink in the *I-V* curves appears at low *T*, which could be attributed to the presence of defects or inhomogeneities in the barrier.

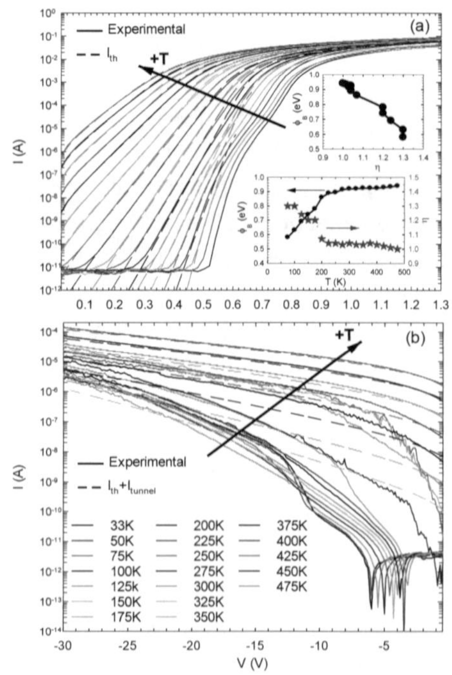

Fig. 1. Comparison of the (a) forward- and (b) reverse-bias *I-V-T* characteristics obtained experimentally (solid lines) and with the model (dashed lines). Insets: barrier height and idelity factor vs. *T*.

For reverse bias, Fig. 1(b), our model, which includes also the tunneling current (using the same barrier height as in forward-bias), correctly fits the measurements for the highest temperatures. However, below 400 K other tunneling current mechanisms become significant.

References

[1] F. Roccaforte, et al., “Nitride Semiconductor Technology: Power Electronics and Optoelectronic Devices”, Wiley, August. 2020.

[2] S. Liang, et al., “A 177-183 GHz High-Power GaN-Based Frequency Doubler With Over 200mW Output Power”, *IEEE Electron Device Letters*, 2020.

[1]University of Salamanca, Salamanca, 37008, Spain; [2]IEMN, Villeneuve d’Ascq, 59650, France; [3]CRHEA, Valbonne, 06560, France

NOTES:

LO power division circuits for the CCAT-prime Heterodyne Array Instrument (CHAI)

I. Barrueto[1], K. Jacobs[1], S.Wulff[1], M.Schultz[1], M. Justen[1], C.E. Honingh[1] and J. Stutzki[1]

***Abstract*—We present power dividers to be used in the local oscillator distribution network in CHAI. The power divider is based on superconducting 90° CPW hybrid with waveguide antennas on a thin Silicon substrate. We present preliminary experimental results and comparison with simulations to gain insight about the feasibility of the selected power dividing scheme.**

***Keywords*—Heterodyne mixer, superconductive circuits, 90° hybrids, power division.**

I. Introduction

The scientific goal of CHAI is to map the narrow spectral lines of [CI], 3P1-3P0 and 3P2-3P1 as well as the medium/high excitations of CO.
In order to accomplish this task, CHAI is conceived as a dual colour heterodyne receiver intended for simultaneous observations of the Lower Frequency Array 460-500 GHz (LFA) and the Higher Frequency Array 780-820 GHz (HFA). Each band consists of a square array of 8x8 pixels in a superconductor –insulator-superconductor (SIS) balanced mixer configuration. For each band the cryogenic (4K) 64-pixel array consist of 2x8 vertically stacked 1x4 pixel waveguide blocks. These blocks are fabricated in CuTe split block technology with 460 μm x 230 μm (480 GHz) and 240 μm x 140 μm (800 GHz) rectangular waveguides connecting the 3 LO power dividers and 4 mixers in one block.

II. Low frequency array and block

The LFA is shown in Fig.1 with the three main components depicted plus the bottom half of a 1x4 pixel block with the RF and IF components therein. This block receives the local oscillator (LO) power at his back and then distributes it to the 4 mixers by a cascade of 3 dB power dividers (white circles). Mixers and couplers are made in the same superconducting technology, consisting of Niobium coplanar waveguide transmission lines on 9 μm thick Silicon substrates, contacted by Gold beamleads. The balanced mixer design is based on the design by Westig et al [2] (**Fig. 4 top**). The LO power divider is a 3 dB 90° CPW branch line coupler with the isolated port terminated with a thin film Titanium Nitride load of 13x20 μm^2, with a DC resistance of 43 Ohm, equal to the CPW line impedance (**Fig. 2.b**). For test purposes, 2 other terminations, a resonant termination and a short circuit have been fabricated as well (**Fig. 2.c-d**). To test the LO power divider we use a simpler 2-pixel block which represents the half of the 4-pixel block with one power divider and 2 mixers (**Fig. 2.a**). Two 2-pixel blocks were fabricated in out in-house workshop, the difference between them being that one of the blocks has 5 μm deep pockets in the top half that fit just over the beam lead ends (**Fig. 3**).

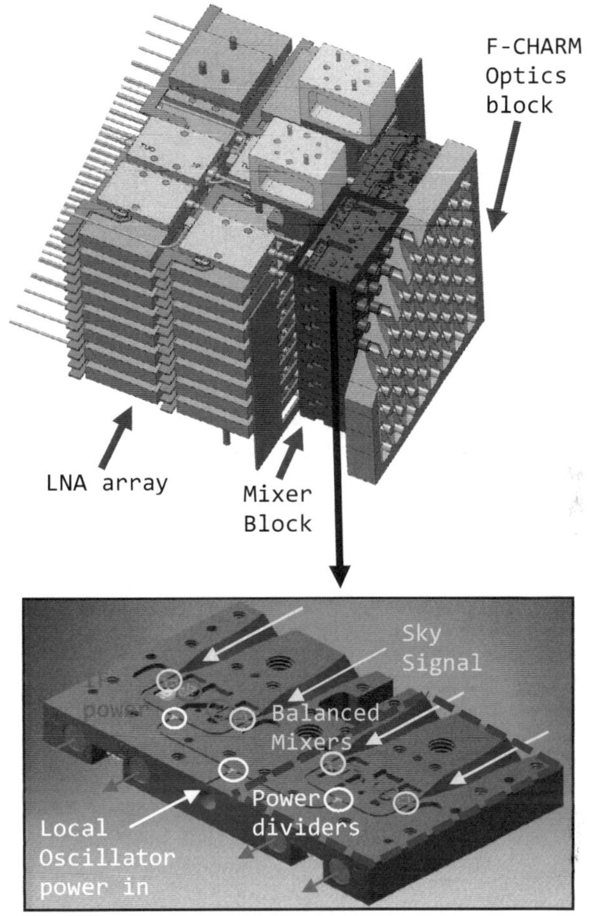

Figure 1: 3D CAD of the Focal plane unit array (Courtesy of U.U. Graf) with its main components labelled (**Top**) and a detailed image of a half 4-pixel block with its core components (**Bottom**). The dotted red line marks the which part of the 4-pixel block which is being tested with the debugging 2-pixel block

III. Methods & results

To measure the LO coupling to the mixer, we use the SIS junctions as a power detector. We record the current on the first photon step below the gap voltage induced by the

[1]I. Physikalisches Institut, Universität zu Köln, Köln, 50937, Germany.

NOTES:

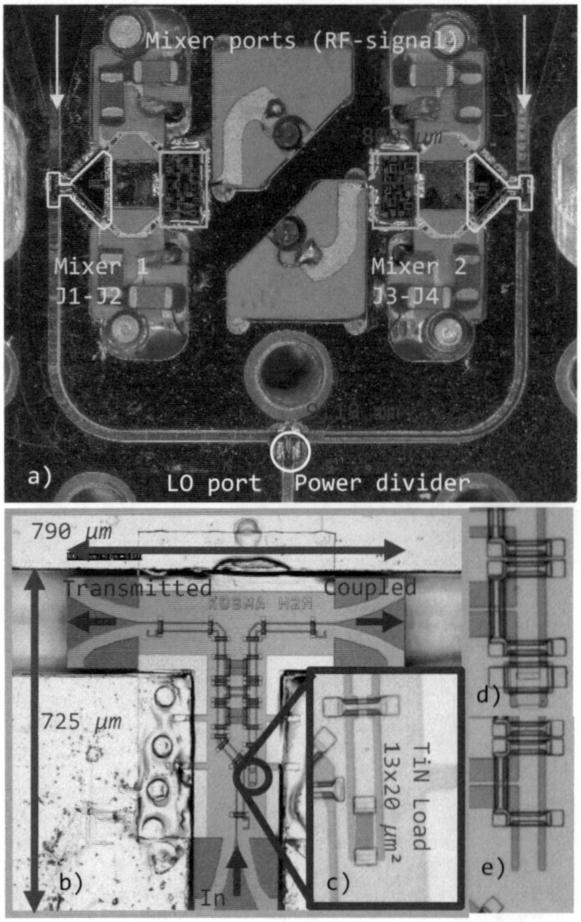

Figure 2: **a)** Half block of the 2-pixel block with the position of the main components therein. **b)** Image of the fabricated power divider with the lumped element TiN Load marked in violet. **c)** Lumped element load **d)** Lossy resonator load. e) Short circuit termination.

incident LO power at a fixed bias voltage of about 2 mV. We use a VDI signal generator extender module SGX 494 WR 6.5, with a range from 110 to 170 GHz plus a frequency Tripler that allows it to reach the 330-510 GHz band. In order to characterize the mixers in the present setup we inject the CW signal through the Mixer ports (**Fig. 2.a** yellow arrow) and record the induced current in the junction. The experimental results and simulations are presented as relative response, meaning one of the junctions in the mixer (**Fig. 4 Top**) is selected as reference and it is used to normalize the response of the second one. This format allows for a reduction of external influences, such as variations in the available signal power, making the evaluation of the performance of the mixer easier. Three facts jump out immediately. First, is that the measured mixer seems to be shifted down in frequency by about 20 GHz compared to simulation results. Second, the expected standing wave behaviour is similar to the measured one, However, this similarity ends about 470 GHz. Third, there is a

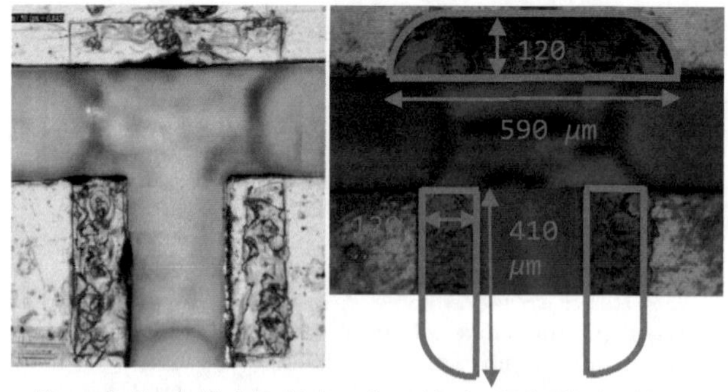

Figure 3: Top half of the fabricated test blocks. **Left:** Block with no beamlead pockets milled. The effects of pressing the beamleads is visible after the removal of the device. **Right:** This block has 5 μm deep pockets in the top half that fit over the beamleads of the device. The purpose of these pockets is to decrease mechanical strain on the beamlead ends when the block is closed, facilitating the mounting and demounting of devices in the block, preventing damage in the remaining devices.

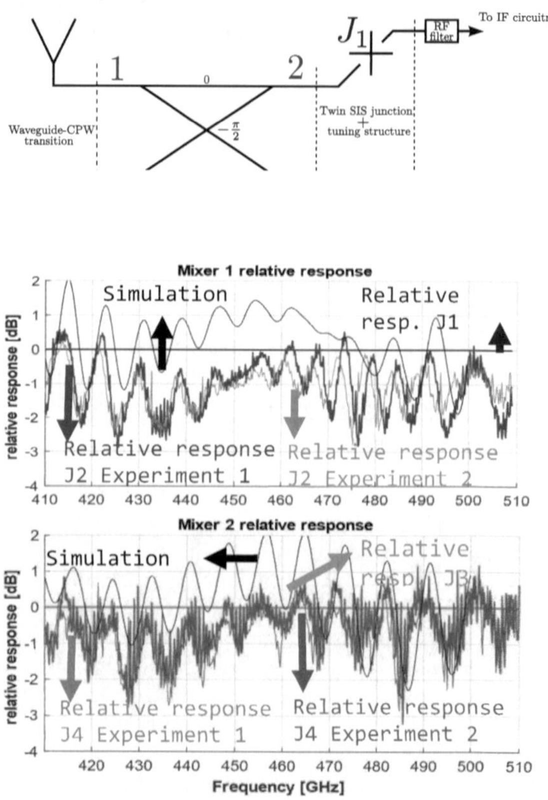

Figure 4: Schematics and mixer response for both mixers. Given the mirrored placement of mixers M1 and M2 in the 2-pixel block M1 receives the CW signal from the port 1, meanwhile, M2 receives it from port 4. For M2 the port junction correspondence is J3- Port 3, J4-Port 2. The experiment is repeated with different LO power dividers, a lumped element load and a resonant respectively, without significant differences. A 20 GHz downshift is present in both mixers, when comparing to simulations **Top:** Simulated and measured relative response of mixer 1. **Bottom:** Simulated and measured relative response of the mixer 2.

NOTES:

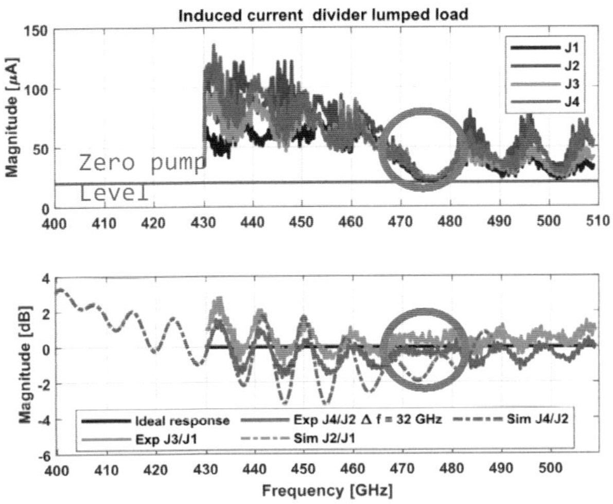

Figure 5: Results for a power divider with lumped element load. **Top:** Measured induced current in all junctions across frequency. Between 470 and 480 GHz there is little to no LO coupled the junctions. **Bottom:** Normalized response plus CST simulation (30 GHz downshifted). The induced current of J3 and J4 is normalized by the induced current of the corresponding mirrored junction in mixer one, that being J1 and J2 respectively.

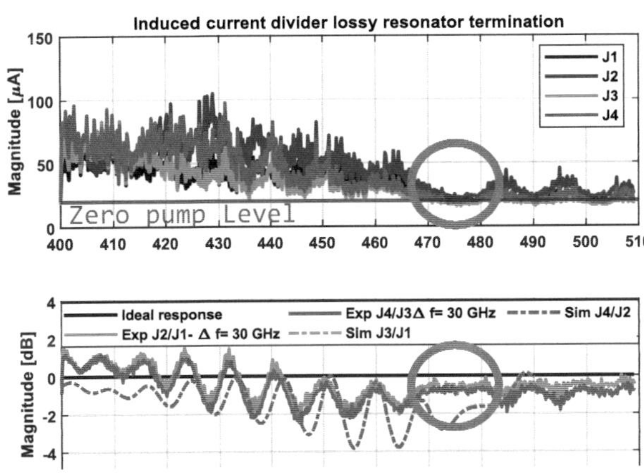

Figure 6: Results for a Hybrid with a resonant circuit load. **Top:** Measured induced current in all junctions across frequency. The frequency range where no power is coupled to the junctions (Blue circle) is also observed in the test of this power divider. **Bottom:** Normalized response plus CST simulation (30 GHz downshifted). The induced current of J3 and J4 is normalized by the induced current of the corresponding mirrored junction in mixer one, that being J1 and J2 respectively.

disagreement between simulation and experimental measurements with regards to the power balance between the two mixers, around 1 and 0.7 dB for mixers M1 and M2 respectively.

After the mixers have been characterized, power divider devices with the three different loads are tested by injecting the CW signal through the LO port, reaching mixers M1 and

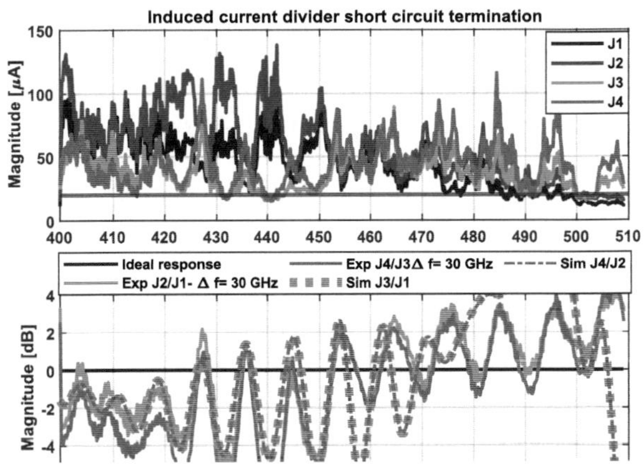

Figure 7: Results for a Hybrid with a short circuit load. **Top:** Measured induced current in all junctions across frequency. There is no frequency range where power is not coupled to the junctions **Bottom:** Normalized response plus CST simulation (30 GHz downshift). The induced current of J3 and J4 is normalized by the induced current of the corresponding mirrored junction in mixer one, that being J1 and J2 respectively.

M2 (**Fig.2**). Once again, junction J1 from mixer M1 is chosen as the reference junction for normalization. In addition to the normalized results, the direct measured power division results for the lumped element load, the lossy resonant load and the short circuit are displayed in **Fig.5**, **Fig.6** and **Fig. 7** respectively. A 30 GHz downshift in frequency of the measured results with respect to the simulations is observed for the three power dividers tested. This 30 GHz shift is believed to be caused by characteristics of the power divider, as it is seen in the comparison between junctions from different mixers, where the properties of the power divider come into play. In **Fig.5**, **Fig.6** and **Fig. 7** comparison between J1/J2 or J3/J4 only grants information about the mixer characteristics instead of the power divider. For the power dividers with the lumped element and resonant load, the graphs of the induced current show a frequency range between 470 and 480 GHz where there is hardly any power coupled to any junction followed by, at higher frequencies, a standing wave-like behaviour dissimilar from what is expected from simulations. Aside from this phenomenon, it can be seen that the measurements of the power divider with the lumped element load and the resonant load have similar features. This is not observed for the power divider with the short circuit load. Repeated experiments with different power dividers in the two available testing blocks (**Fig.3**) yields the insight that the phenomenon seen in **Fig.6** and **Fig.7** was only seen in power dividers tested in the block which had beamlead pockets for the power dividers (**Fig.3**). Detailed simulations of individual components did not show any indication of a behaviour similar to the observed one, suggesting that a complete 3D integrated system simulation is necessary for an accurate prediction of

NOTES:

the performance. For the power divider with a short circuit there is a good agreement between simulation and measurement up to 460 GHz, save for the 30 GHz downshift. Above 460 GHz there is a change in the period of the standing waves and the agreement is lost.

IV. Summary & future work

In this work we have presented the status of the LO power divider devices for the Low Frequency Array of CHAI. Three power divider models were tested in two different housing blocks. Each power divider had a different load termination at the isolated port, with the termination being a lumped element load, a lossy resonator and a short circuit (For testing purposes). Experimental results of power dividers tested on a block with beam lead pockets suffered from a frequency range (470-480 GHz) where no power was coupled to the junctions, the exact cause is currently under examination.

There is a frequency downshift of around 30 GHz observed in all three power dividers results in comparison with simulation. After correcting for the frequency shift, there is a reasonable agreement between simulations and experimental results up to about 460 GHz. The power divider with a lumped element load and a resonant load have similar power division characteristics. Unfortunately, from the experimental data it is concluded that neither of the two power divider models is suitable for large array operations in the present configuration. The investigation of the present devices is currently ongoing, as well as the development of new ones.

References

[1] http://www.ccatobservatory.org

[2] M. P. Westig, K. Jacobs, J. Stutzki, M. Justen, and C. E. Honingh, Supercond. Sci. Technol. 24, 085012 (2011).

NOTES:

Characterization of monolithically integrated lithium niobate ring resonator for a high sensitivity room temperature radiometer

Jessica César Cuello*[1], Alberto Zarzuelo[1], Robinson C. Guzmán[1], Luis Gonzalez-Guerrero[1], Gabriel Santamaria[2], Enderson Falcón-Gómez[1], Luis E. García[1], and Guillermo Carpintero[1]

***Abstract*—Radiometry in the sub-millimeter and THz region is required for applications as spectroscopy, earth observation, space missions and radio astronomy. A high sensitivity room temperature radiometer, based on the upconversion of the RF signal to the optical domain, has been already proposed. High-Q resonators, made of nonlinear crystals, are require for realizing such upconversion process. Since small physical footprint and low power consumption is necessary for space applications, the miniaturization of the components becomes relevant. In this work we present the study of an integrated ring resonator (RR), the first step of the integration, in a Photonic Integrated Circuit (PIC), of the upconversion radiometer. The device was fabricated in a thin-film lithium niobate PIC.**

Keywords—photonic integrated circuit, lithium niobate ring resonators, radiometer

I. INTRODUCTION

Upconversion of millimeter-wave or THz signals to the optical domain, where room temperature photonic detectors are highly sensitive, is a novel detection approach proposed in [1]. In this approach, a laser pump is modulated by millimeter-wave or THz signal in a high-Q whispering-gallery mode resonator (WGMR), which is used as an efficient electro-optic modulator (EOM), upconverting the signal detected by the antenna to optical sidebands. Detection can be then performed with conventional photodetectors in a direct or coherent detection scheme. For this approach, using a high-Q resonator as the EOM is necessary to critically couple the pump mode, enhancing intracavity power and thus photon conversion efficiency, leading to a radiometer sensitivity comparable to conventional cooled receivers while working at room temperature as described in [2].

Our goal is to develop such architecture [1] into a photonic integrated circuit (PIC) to achieved low size, weight and power consumption (SWaP), three critical parameters for a satellite payload application. In this work, we present the characterization of a key component for this application, an integrated Lithium Niobate (LiNbO3) resonator. The transmission spectrum of the resonator, and several figures of merits, are obtained. Due to the need of nonlinear characteristics of the chip, the device was fabricated on a thin-film LiNbO3 substrate. The conclusions obtained from this work have been used as input for the development of a second PIC, where the complete photonic-based radiometer architecture is being fabricated.

[1]University Carlos III of Madrid, Leganes, 28911, Spain; [2]University of Colorado at Boulder, Boulder, CO 80309, USA
*jecesarc@ing.uc3m.es

II. DEVICE DESCRIPTION

Thin-film Lithium Niobate-on-insulator (LNOI) has emerged as a promising material platform for different applications due to its high refractive index, large piezoelectric response and large Pockels electro-optic efficiency [3]. Furthermore, two of the advantages that makes LNOI suitable for the develop of our photonic integrated circuit is its low propagation loss and high-Q on-chip ring resonators [4].

The integrated system is shown in Fig. 1. The upconverter is based on a WGMR for optical upconversion and an asymmetric Mach-Zehnder interferometer (AMZI) to increase the upconversion bandwidth [5]. By phase matching the optical carrier and the signal detected by the antenna inside the resonator, the upconversion of the RF signal to the optical domain is achieved, generating two optical sidebands around the laser pump v_p by sum-frequency v_s and difference-frequency v_d generation. These optical sidebands represent the signal that we want to detect. The detection can be performed with conventional photodetectors in a direct or coherent detection scheme.

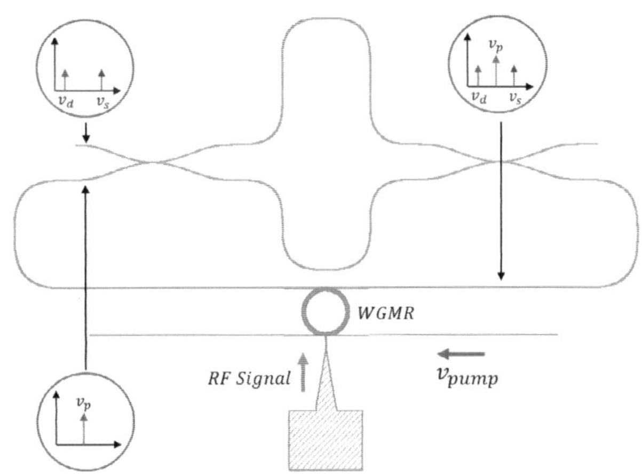

Fig. 1. Integrated photonic-based radiometer architecture.

For testing purposes, a first set of devices was fabricated to individually characterize the WGMR and AMZI. The physical footprint of the PIC is 10 mm x 4 mm. A microscope image of the structures under test is presented in Fig. 2.

To study the influence of the separation between the bus waveguide and the resonator on the transmission spectrum of the resonator, eight structures with different separation (gap) were fabricated. The gap values are: 1.4 μm (RR1), 1.26 μm (RR2), 1.11 μm (RR3), 0.97 μm (RR4), 0.829 μm (RR5), 0.686 μm (RR6), 0.543 μm (RR7), and 0.4 μm (RR8). A resonator radius of 50 μm was used for all eight structures.

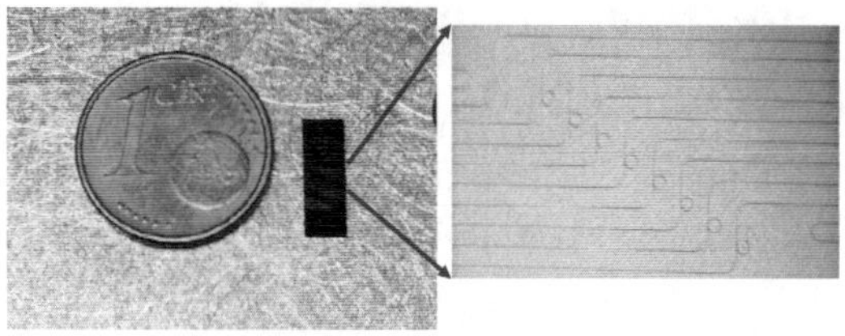

Fig. 2. Image of the physical footprint of the photonic integrated circuit (PIC) and a microscope image of the LiNbO3 resonators.

III. Experimental Results

The characterization of the structures was performed using the Component Analyzer feature of the High-Resolution Optical Spectrum Analyzer (BOSA) from Aragon Photonics. The light is injected and collected through edge coupled optical waveguides, using lensed fibers. The insertion loss measured at each facet, working with a lensed fiber of 2 μm spot diameter and 1.8 μm wide waveguides, was found to be around 5.5 dB.

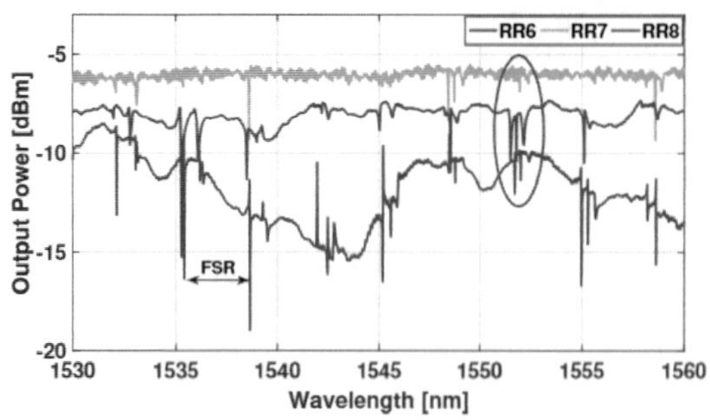

Fig. 3. Transmision spectra of RR6, RR7 and RR8.

After testing all the structures, we observed that resonances begin to appear on RR6, which suggests that for waveguide gaps larger than 0.7 μm there is no appreciable coupling between the access waveguide and the resonator.

The transmission spectra of RR6, RR7 and RR8 are presented in Fig. 3. In the transmission spectrum, three different sets of resonant modes are identified, each with similar Free Spectral Range (FSR) around 3.2 ± 0.2 nm (400 ± 25 GHz).

A modulation of the optical spectrum was also observed in the transmission response of all the structures resulting from the Fabry-Perot cavity created between the facets of the PIC, which can be reduced by using anti-reflection coating at both edges of the chip.

From the transmission spectra we can define that the optimum waveguide-ring resonator gap for the final radiometer design is around 0.543 μm (RR7), due to the flat response of its spectrum. By fitting the experimental results of the under-coupled RR7 (Fig. 4), we estimate a loaded quality factor of $0.8x10^5$. The intrinsic Q-factor (Q_i) of the resonator is related to the measured loaded Q-factor (Q_L) by:

$$Q_i = 2Q_L/\left(1 \pm \sqrt{T_0}\right) \tag{1}$$

where T_0 is the normalized transmission value and + and - correspond to under- and over-coupled regimes, respectively. At critical coupling regime, $T_0 = 0$ corresponding then to an intrinsic Q-factor $Q_i = 2Q_L$.

The measured intrinsic Q-factor of the monolithically integrated LiNbO3 ring resonator is not as good as the state-of-the-art demonstrations on the same platform [4] due to the additional bending loss added by working with a smaller resonator radius.

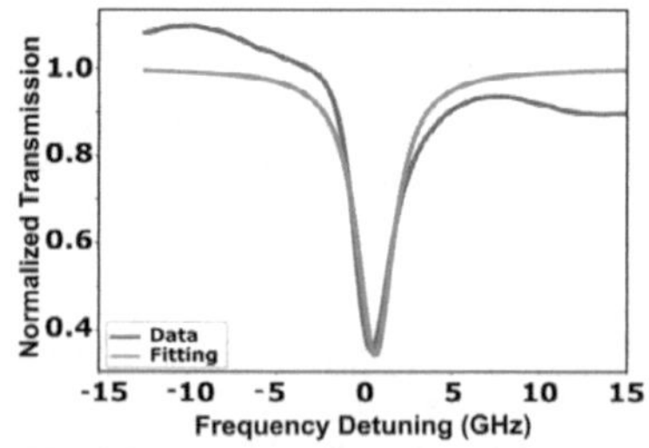

Fig. 4. Fit of the resonance dip to Lorentzian function at wavelength of 1551.5 nm

IV. Conclusions

In this work we have presented the characterization results of a Lithium Niobate ring resonator to be part of a Cubesat payload radiometer. For waveguide/resonator gaps larger than 0.7 μm, almost no light is coupled to the resonator. High-Q LiNbO3 resonators have been demonstrated, with an intrinsic Q-factor of around $1x10^5$ in the best structure (RR7). It is expected that the quality factor of the resonators will increase in future fabrication runs by increasing the radius of curvature of the rings to reduce losses.

Furthermore, photonic integration has allowed the reduction in terms of size, weight and power consumption of the photonic-based radiometer for satellite payload, and the LNOI platform has proven to be the right choice for the development of our high-Q WGMR. The power consumption of the integrated radiometer, associated to the electronic part of the final system, is expected to be around 6W.

Funding. This work has been partly supported by the Comunidad de Madrid through grant S2018/NMT-4333 MARTINLARA-CM Project, by the Spanish Ministerio de Ciencia e Innovación through Programa Estatal de Generación de Conocimiento y Fortalecimiento Científico y Tecnológico del sistema de I+D+i orientada a los Retos de la Sociedad convocatoria 2019 (grant PID2019-109984RB-C42, Circuitos Integrados Fotónicos para radiómetros avanzados en rango milimétrico para instrumentos de nueva generación en observación terrestre) and by MINISTERIO DE ASUNTOS ECONÓMICOS Y TRANSFORMACIÓN DIGITAL by the project "6G-Xtreme" within the Plan de Recuperación, Transformación y Resiliencia. Mecanismo de Recuperación y Resiliencia Perspectiva de género en los materiales de difusión (Unión Europea-NextGenerationEU).

Acknowledgments. The authors thank M. Lonĉar, J. Holzgrafe and H. Warner for valuable discussions and PIC fabrication, which was performed at the Center for Nanoscale Systems (CNS) at Harvard University.

References

[1] L. E. G. Muñoz, et al., "Room temperature radiometer based on an upconversion process for CubeSats applications", invited paper *Terahertz Photonics*, 2020.

[2] G. Santamaría-Botello, Z. Popovic, K. A. Abdalmalak, D. Segovia-Vargas, E. R. Brown, and L. E. García Muñoz, "Sensitivity and noise in thz electro-optic upconversion radiometers," Sci. Reports 10, 1–13, 2020.

[3] C. Wang, et al., "Nanophotonic lithium niobate electro-optic modulators," Opt. Express 26, 1547–1555 (2018).

[4] M. Zhang, et al., "Monolithic ultra-high-Q lithium niobate microring resonator," Optica 4, 1536–1537 (2017).
[5] G. Santamaría-Botello, et al., "Broadband millimitre-wave to optical up-conversion for room-temperature high sensitivity radiometers", *39th ESA Antenna Workshop*, 2018.

Characterization of Microwave Properties of Superconducting NbTiN Films using TDS

Fedor V. Khan[*, 1, 2], Elena S. Zhukova [2], Boris P. Gorshunov [2], Lenar. S. Kadyrov [2], Artem M. Chekushkin [1], Andrey V. Khudchenko [3], and Valery P. Koshelets [1]

***Abstract*—In this work we perform comprehensive study of electrodynamic properties of superconducting films of NbTiN at frequencies up to 2.5 THz and in temperature range from 4 up to 15 K using commercial time-domain spectrometer TeraView TPS Spectra 3000. A set of NbTiN films with different content controlled by the pressure of nitrogen in magnetron chamber was fabricated. We found out that there is a trade-off between low normal-state resistivity, small London penetration depth and high critical temperature. By characterizing the films, the optimal manufacturing conditions were determined. Two models, with and without taking intragap states into account were used to describe experimental data and both show quantitative correspondence with the experiment.**

***Keywords*—thin films, superconducting materials, terahertz measurements**

I. Introduction

Devices of superconducting electronics have been extensively used in fundamental research and number of practical applications. The devices based on superconductor-insulator-superconductor (SIS) tunnel junctions are the main element of receiving systems in terahertz (THz) range having the lowest noise temperature only few times higher than quantum limit. The operating frequency of the superconducting SIS-receivers is limited by the gap, being 750 GHz for Nb which is used in fabrication of the majority of superconducting devices nowadays. In order to expand the operating range to frequencies higher than 1 THz, other materials, e.g. Nb compounds, like NbN and NbTiN, with higher gap frequency should be used.

This work is dedicated to characterization of the parameters (normal state resistivity ρ_0, critical temperature T_c, London penetration depth λ_L) of superconducting NbTiN films fabricated at different technological conditions at frequencies close to 1 THz. The thicknesses of the films are around 330 nm, exceeding London penetration depth, which turned out to be more than 280 nm for all the samples.

II. Fabrication Processes

NbTiN films were sputtered on 535-µm-thick high-resistivity silicon substrates at room temperature using cluster magnetron system Kurt J. Lesker. The sputtering was performed from 3" NbTi (78 % of Nb and 22 % of Ti) target in mixture of nitrogen and argon. Power of magnetron was around 500 W. The pressure of nitrogen in magnetron chamber was varied from $0.35 \cdot 10^{-3}$ to $0.65 \cdot 10^{-3}$ mbar in order to investigate its impact on critical temperature of the film T_c and ρ_0 which is DC conductivity at normal state near T_c, and also determine the conditions at which the highest T_c and lowest ρ_0 are obtained. The DC-parameters measured by four-probe technique of all the samples are listed in Table I.

III. TDS-Measurements

The study of the samples at THz frequencies was performed using TDS-spectrometer TeraView TPS Spectra 3000. Transmission spectra of the superconducting films on substrates were measured for all the films in temperature range from 4 to 15 K. Conducting medium can be expressed in terms of complex permittivity $\varepsilon^* = \varepsilon' - 4\pi i \sigma_1/\omega$, where ε' is the real part of permittivity and σ_1 is the real part of conductivity. Measured spectra together with the fitting curves are shown in Fig. 1.

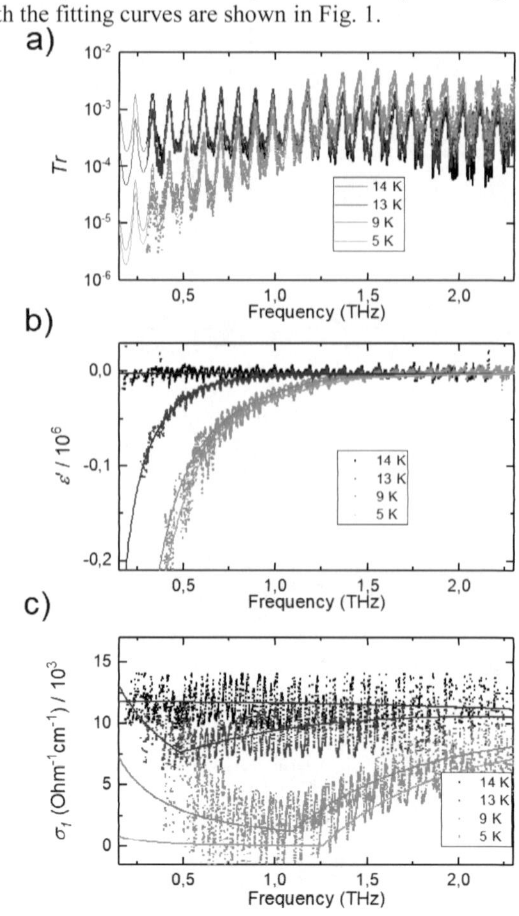

Fig. 1. Spectra of transmission coefficient (a), real part of permittivity (b) and conductivity (c) of superconducting NbTiN film (sample #5).

[*] *Corresponding author; e-mail: khanfv@hitech.cplire.ru*

[1]Kotelnikov Institute of Radio Engineering and Electronics of RAS, Moscow, 125009, Russian Federation; [2]Moscow Institute of Physics and Technology (National Research University), Dolgoprudny, Moscow Region, 141701, Russian Federation; [3]Astro Space Center of Lebedev Physical Institute of Russian Academy of Sciences, Moscow, 119991, Russian Federation

IV. Theoretical Models

We used the expressions from the paper by Zimmermann *et al.* [1] in order to fit our data. This model not only allows to calculate the complex conductivity of the superconductor, but also takes into account finite quasiparticle scattering time. The additional parameter, quasiparticle scattering rate γ, is reciprocal to scattering time τ: $\gamma = \hbar/\tau$. From fitting the experimental data, we obtain ρ_0, Δ and γ. Since the permittivity in superconducting state is many times higher than in normal state due to Cooper pairs, T_c is determined directly from Fig. 1 b as the lowest temperature where ε' takes finite value at zero frequency.

As is known, Nb compounds including NbTiN are superconductors with strong coupling. Related effects lead to that the ratio between the critical temperature and the gap at $T = 0$ $\Delta_0 = 1.76k_BT_c$ predicted by BCS-theory is no longer valid. The coefficient at k_BT_c determined from our measurements turned out to take values from 2 up to 2.2 for all the samples.

In recent papers on studying the properties of NbTiN films it was claimed that the density of states is also different from that predicted by BCS for whatever reasons [2]. Singularities in the density of states at the energies near the gap vanish and the gap frequency itself is slightly reduced. These effects can be treated by more complicated models, e.g. described in [3].

As can be seen from Fig. 1, theoretical curves calculated using the expressions from [1], where the BCS-like density of states is considered, are in good correspondence with the results of experiment in our work. Therefore, in order to understand the difference between the models we used both of them to process the data from [2], where the spectra of relative change in reflection coefficient upon transition from superconducting to normal state were measured. Multiple reflections were suppressed by adding additional silicon plates to substrate. Both models allow to obtain reasonable agreement with the experimental data (see Fig. 2). However, a small discrepancy between the curve corresponding to model [1] appears at frequencies near the gap which is around 1.1 THz. This discrepancy can be eliminated using model [3] by adjusting parameter Γ_s that corresponds to scattering rate on magnetic impurities, though the accuracy of the fit at frequencies higher than 1.3 THz becomes worse.

It still remains unclear if the discrepancy near the gap frequency is caused by intrinsic or extrinsic effects (such as surface roughness or contamination).

V. Conclusions

The resultant parameters of all the films are listed in Table I. It should be noted, that there is a trade-off between the lowest ρ_0, smallest London penetration depth at zero temperature λ_0 and highest gap and T_c. It was found out that ρ_0 monotonously grows with increasing the pressure of nitrogen in chamber; λ_0 has minimum near 4.5 mbar and T_c reaches its maximum at 5.5 mbar. Thus for particular purposes one should use the corresponding conditions.

Moreover, the values of the parameters, obtained from DC-measurements and using TDS differ by approximately 10 %, what is likely to be caused by the grain structure of the films. At temperatures close to T_c, there are areas in film both in superconducting and normal state. Direct current flows through superconducting pattern, which is not the case for

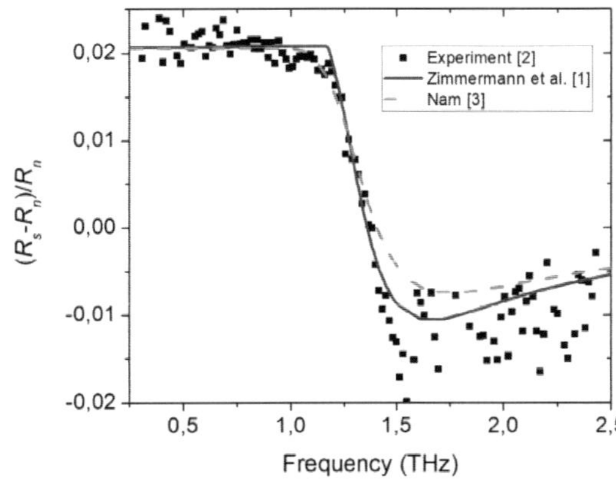

Fig. 2. Relative difference of reflection coefficients of NbTiN film in superconducting and normal states. Experimantal points were taken from [2]; solid and dashed curves represent the fits using expressions considering BCS-like and modified densities of states, respectively.

the current induced by the AC-radiation that flows through both the grains and the boundaries. Therefore, T_c of the films at high frequencies is determined by the transition temperature of the boundaries, which is lower. Therefore, the parameters obtained by TDS are those that should be used in design and modeling of superconducting devices.

TABLE I. Parameters of the Films

Sample no.	#1	#2	#3	#4	#5	#6
N_2 pressure, 10^{-2} mbar	6.5	5.9	5.3	4.7	4.1	3.5
Thickness, nm	333	332	328	339	338	325
$\rho_{0,DC}$, μΩ cm	127	114	104	98	93	92
$T_{c,DC}$, K	14.9	15.2	15.2	15.3	15.2	14.8
$\rho_{0,TDS}$, mΩ cm	110	105	95	91	89	85
$T_{c,TDS}$, K	14.1	14.4	14.6	14.4	13.9	13.8
$2\Delta_0$, meV	4.9	5.1	5.1	5.1	5.2	4.7
λ_0, nm	330	300	290	280	280	290

References

[1] W. Zimmermann, et al., "Optical conductivity of BCS superconductors with arbitrary purity", *Physica C: Superconductivity*, Nov. 1991

[2] B. N. R. Lap, et al. "Characterization of superconducting NbTiN films using a dispersive Fourier transform spectrometer." *Applied Physics Letters*, Sep. 2021

[3] S. B. Nam, "Theory of electromagnetic properties of superconducting and normal systems I", *Physical Review*, Apr. 1967

A Low Loss Diplexer for Submillimeter-wave Sideband Separating Receivers

Subash Khanal[1], Alain Maestrini[1], Robert Lin[1], and Goutam Chattopadhyay[1]

***Abstract*—This work presents a design of a low loss waveguide-based diplexer which enables sharing of a common antenna between two distinct frequency bands. The diplexer works as a passive high/low pass filter for the incoming RF signal. The filtered RF bands are mixed with the LO signal on two sub-harmonically pumped mixers realizing the sideband separating receiver architecture at submillimeter-wave frequencies. Usually, sideband separation is implemented mixing the incoming radio frequency (RF) signal with a quadrature feed local oscillator (LO). However, any changes in gain or phase of either the I or Q path will critically affect the sideband balance and will degrade the receiver performance significantly. In contrast, the diplexer designed in the work is a fully passive (waveguide) structure and is insensitive to such variations leading to an improved overall rejection and receiver stability.**

***Keywords*—Diplexer, sideband separation, sub-harmonic mixes, submillimeter wave.**

I. Introduction

At submillimeter-wave, radiometry/spectrometry are generally the main drivers for the development of highly sensitive detectors and receivers. This band constitutes a wealth of absorption and emission lines [1] for several gaseous molecules such as oxygen, ozone, water vapor and many more whose physical parameters, such as temperature and pressure, can be extracted by observing the shape of the spectral lines. The vertical distribution of key trace gases on earth's atmospheric layers is a critical observation to help our understanding on the processes that control the overall climate. This can be achieved by limb sounding technique and is commonly implemented with space born instruments. Room temperature GaAs Schottky diode technology-based receivers have been an attractive choice at submillimeter-wave frequencies for limb sounding due to their high reliability, in addition to lower mass, cost, and complexity compared to the cryogenic receivers. Sideband separation technique is known to provide better sensitivity and calibrated measured data compared to double side band counterpart when it comes to the observation of spectral signature of the molecules [3]. This is mainly due to the reduced noise on the single side band system obtained by suppressing/rejecting the image band. Usually, sideband separation is implemented with quasi-optical diplexers and image rejection filters [4]. However, these are lossy, bulky and expensive. A waveguide based low loss, compact and easy to integrate sideband filtering technique is presented in this work. The designed diplexer consists of a waveguide structure comprising of a T-junction, and a series of waveguide steps and stubs which are optimized such that only certain frequencies are allowed in each branch rejecting all other unwanted frequencies. The merit of this design is that the waveguide structures are optimized considering the intrinsic impedances of the mixers such that all the unwanted high frequency signals on lower side band (LSB) branch are reflected back to the upper side band (USB) branch and vice versa. This means, we do not pay the 3 dB penalty from the T-junction split and therefore a low loss RF path can be realized.

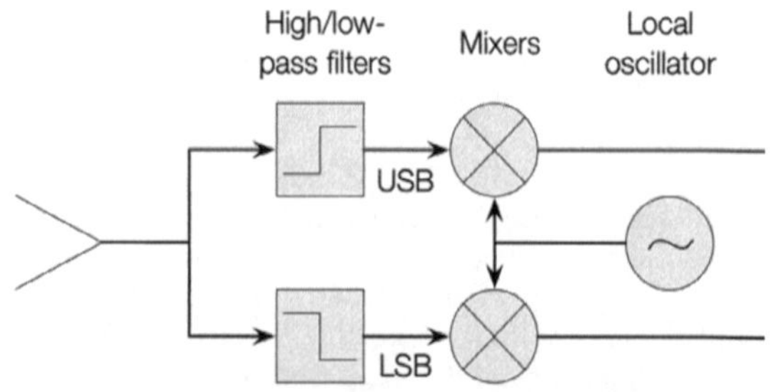

Fig. 1. Block diagram of the 340 GHz sideband separation receiver architecture using RF filtering approach.

References

[1] Jet Propulsion Laboratory, California Institute of Technology. http://spec.jpl.nasa.gov/

[2] P. H. Siegel, "THz instruments for space," *IEEE Transactions on Microwave Theory and Techniques*, vol. 55, no. 11, pp. 2957–2965, November 2007.

[3] P. R. Jewell and J. G. Mangum, "System temperatures single versus double sideband operation and optimum receiver performance", International Journal of Infrared and Millimeter Waves, Vol. 20, No 2, pp. 171-191, May 1999.

[4] N. Erickson, "A Very Low-Noise Single-Sideband Receiver for 200-260 GHz," IEEE Transactions on Microwave Theory and Techniques, vol. 33, no. 11, pp. 1179–1188, Nov. 1995.

[1]Jet Propulsion Laboratory, Pasadena, CA 91109, USA

NOTES:

Design of RF waveguide structure for 2SB SIS receiver for Millimetron 211-275 GHz VLBI channel

Andrey Khudchenko*[1], Ivan Tretyakov[1], Valery P. Koshelets[1,2], Andrey M. Baryshev[3], Roman A. Cherniy[1]

***Abstract*—The waveguide structure of a Sideband-separating (2SB) receiver based on superconductor-insulator-superconductor (SIS) mixers strongly influences the Sideband Rejection Ratio (SRR), one of the key parameters of the receiver. During development of a RF waveguide structure for a 211-275 GHz 2SB SIS receiver it was found that SRR degrades from 26 dB to about 18-20 dB due to reflections from SIS mixers and from RF loads. Additionally, the imbalance of an LO signal is found and described. This imbalance can cause up to 0.5 dB difference in pumping level of SIS junctions and additional phase shift up to 5 degrees. This effect strongly depends on LO frequency. As result, the SRR can degrade further down by 1..2 dB.**

Keywords*—*SIS receivers, sideband separation, sideband rejection ratio

I. Introduction

The 1.3mm band corresponding to ALMA Band 6 is the prime observational range for the current and for the future Event Horizon Telescope activities. The VLBI instrument for Millimetron space project includes a 211-275 GHz receiver, which is planned to be based on superconductor-insulator-superconductor (SIS) mixers. For the best possible sensitivity of the Space-Earth interferometer the onboard receiver should have the same sideband configuration as the receivers installed in the ground telescopes, i.e. it should be sideband separating (2SB) one. Using a DSB mixer on board of Millimetron would cause degradation of the interferometer sensitivity by a factor of square root of 2.

II. Design

In our development of 2SB SIS receiver for the frequencies 211-275 GHz we have designed a waveguide block comprising of RF hybrid, 3dB LO splitter and LO couplers, see Fig.1. The design has modular approach same as in [1] and the elements structures are similar to [2]. The RF hybrid was designed with a gain balance within 1 dB and with a phase imbalance not exceeding 0.3 degrees (see fig.2). The corresponding Sideband Rejection Ratio (SRR) is better than -26dB in the entire frequency band. The prime focus in the design of the waveguide structures was dedicated to the reduction of reflections, which are strongly distorting the sideband rejection ration (SRR) [3].

Using 3D-simulator and we have analyzed the entire waveguide structure of the RF block and estimated the SRR efficiency for different levels of reflection form RF loads and from the SIS mixers.

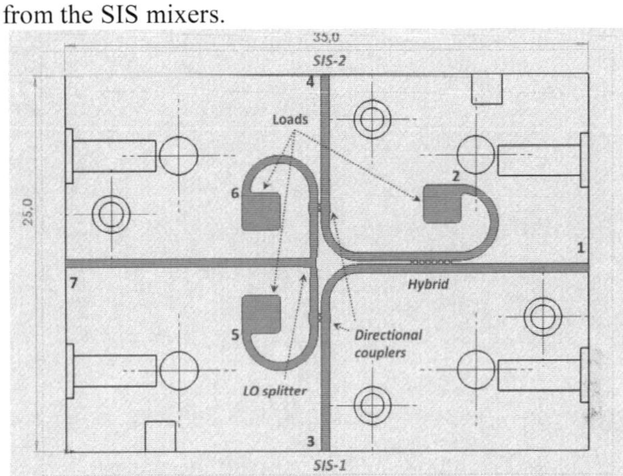

Fig.1 Drawing of the hybrid block showing the RF waveguide structure. Ports 1 and 7 show the inputs for the LO and RF signals correspondingly. Ports 3 and 4 are associated with signs ``SIS-1" and ``SIS-2", which indicate the location of the SIS junctions after the backpieces are installed. Ports 2, 6, 5 indicates the loads: 2 - RF load to suppress parasitic reflections, 6 and 5 - LO loads dump the unused LO power. The waveguide block size is 25\,mm x 35\,mm.

III. Results

The influence of the reflections form SIS mixers and from the RF load is demonstrated in Fig.2 and Fig.3. The reflection level for SIS mixers is varying here from -3.4dB to -9.2dB, at the same time reflection from RF load is fixed at -20dB. A strong degradation of amplitude and phase balance leads to deterioration of the SRR level, as example for SIS reflection of about -7dB it reaches 19 dB in the worst point, while for the pure RF hybrid it was not higher than -26 dB.

Additionally, the imbalance of an LO signal is found (see Fig. 3). It takes place also due to reflections from the SIS mixers and the RF load. This imbalance is periodic and the period is determined by the distance between the mixers and the RF load. For the given above levels of reflection, the LO amplitude unbalance can be up to 0.5 dB, causing different

[1]Astro Space Center of P.N. Lebedev Physical Institute RAS, Moscow, 119991, Russia; [2]Kotel'nikov Institute of Radio Engineering and Electronics RAS, Moscow, 125009, Russia; [3]Kapteyn Astronomical Institute RUG, Groningen, 9747AD, Netherlands

NOTES:

pumping level and, as result, different gain of the mixers. Meanwhile, more critical is the phase shift, which can reach the level of 5 degrees. This effect strongly depends on LO frequency.

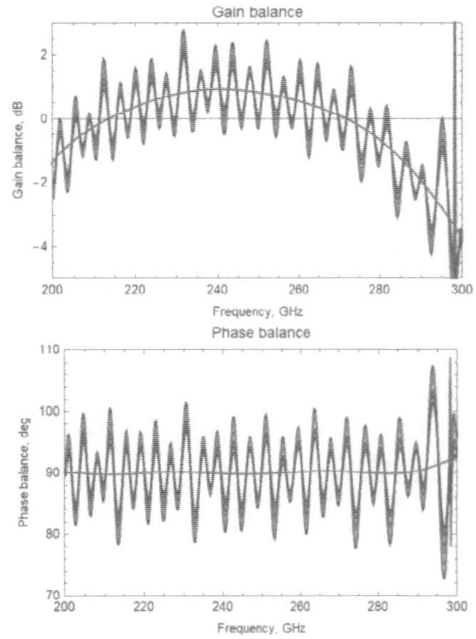

Fig.2 The top plot shows the simulated amplitude balance between two SIS mixers then reflections are added – oscillating curves; the smooth blue line represents gain balance of the single hybrid. The bottom plot similarly shows the simulated phase balance with reflections are added – oscillating curves; the smooth blue line is a phase balance for the single hybrid.

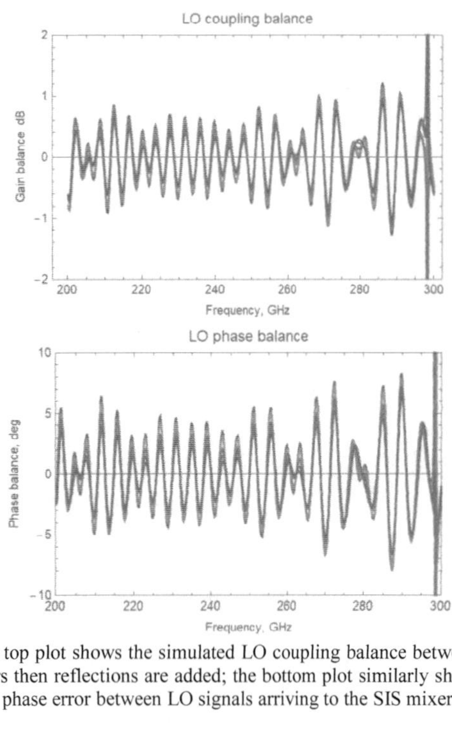

Fig.3 The top plot shows the simulated LO coupling balance between two SIS mixers then reflections are added; the bottom plot similarly shows the simulated phase error between LO signals arriving to the SIS mixers.

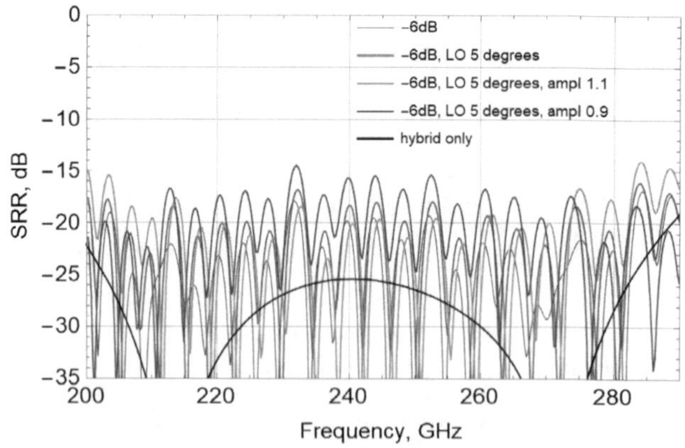

Fig.4 SRR vs frequency.

As seen from Fig.4 SRR degrades from 26 dB (black curve) to about 18..20 dB (blue curve) due to reflections from SIS mixers and from RF loads. Here SIS reflection is -6.7 dB, RF Load reflection -20 dB. For this design the hybrid isolation is 26…30 dB.

Taking into account the LO phase imbalance reduces the SRR level further by about 2 dB in the worst points (green curve vs blue one). The mixer gain imbalance caused by difference of mixers and by various LO pumping level could be estimated as 10% in amplitude or 0.8dB in power. This imbalance can be in-phase or contra-phase to amplitude imbalance causing degradation or improving of SRR, as shown by red and by orange curves in Fig.4. This give motivation to try swapping SIS mixer in 2SB block for better performance.

As conclusion, the influence of reflections on SRR level shows necessity to improve RF load, hybrid isolation and to pay attention to SIS mixer reflections for better performance of the entire 2SB receiver.

This work was supported by a grant from the Russian Science Foundation (project No. 23-79-00061, https://rscf.ru/project/23-79-00061/). The equipment of USU #352529"Cryointegral" was used to perform the simulations.

References

[1] A. Khudchenko et. al, "Modular 2SB SIS Receiver for 600-720 GHz: Performance and Characterization Methods", *IEEE Tr. on TST*, vol. 7, No. 1, pp. 2-9, 2017.

[2] R. Hesper et. al, "A new high-performance sideband-separating mixer for 650 GHz," Proc. SPIE, vol. 9914, 2016, Art. no. 99140G.

[3] R. Hesper, et al., "A High-Performance 650-GHz Sideband-Separating Mixer. Design and Results", *IEEE Tr. on TST*, vol. 7, No. 6, pp. 686-693, 2017

NOTES:

A 67-116 GHz MMIC-based dual-polarized 2SB down-converter for the ALMA Band 2 Warm Cartridge Assembly prototype development

B. Thomas*[1], R. Hesper[2], S. Mohr[1], J. Barkhof[2], A. Busch[1], and J. Adema[2]

***Abstract*—We present here the development of a dual-polarized 2SB down-converter covering the extended frequency range 67-116 GHz for the ALMA Band 2 Warm Cartridge Assembly prototype. The 2SB down-converter operates at room temperature and relies on MMIC technology for the ultra-wideband mixers and LNAs as well as powerful PAs. The gain measurements show a flat response with <5dB p-p average variation over 2-20 GHz IF ranges. The sideband ratio ranges between 17 and 30 dB in average in the nominal configuration.**

Keywords—MMIC, sideband separating, down-converter, dual-polarized, ultra-broadband

I. Introduction

ALMA Band 2 receiver is one of the key milestones in the ALMA 2030 development roadmap [1]. Prototypes for the Cold Cartridge Assembly (CCA) and Warm Cartridge Assembly (WCA) are currently under development in Europe. One of the key features of this band is the use of semi-conductor MMICs for amplification at cryogenic temperatures in the CCA, and for down-conversion at room temperature in the WCA. A pre-prototype of the ALMA Band 2 Receiver has demonstrated very broad instantaneous RF & IF bandwidths [2]. The 2SB down-converter prototype developed here is an evolution of the original pre-prototype presented at the ISSTT 2018 and described in [3].

II. Development and Preliminary Test Results

The 2SB down-converter prototype relies on a full-MMIC approach covering the extended 67-116 GHz range. A careful optimization of the RF, LO and IF components has been performed in order to maximize the side-band ratio, and improve the gain flatness response over a 2-20 GHz IF range. The gain and sideband characterization for each channel and polarization has been done with a Vector Network Analyzer and associated extender heads, using a novel UOSM calibration method (Unknown-Open-Short-Match) which allows a full S-parameters measurement to be performed on the 2SB down-converter. This results in very accurate amplitude and phase gain and sideband measurements from 60 GHz to 120 GHz, and can be scaled up to 1100 GHz with the corresponding VNA extenders. An example of the measured gain response and sideband ratio is shown in Fig.1.

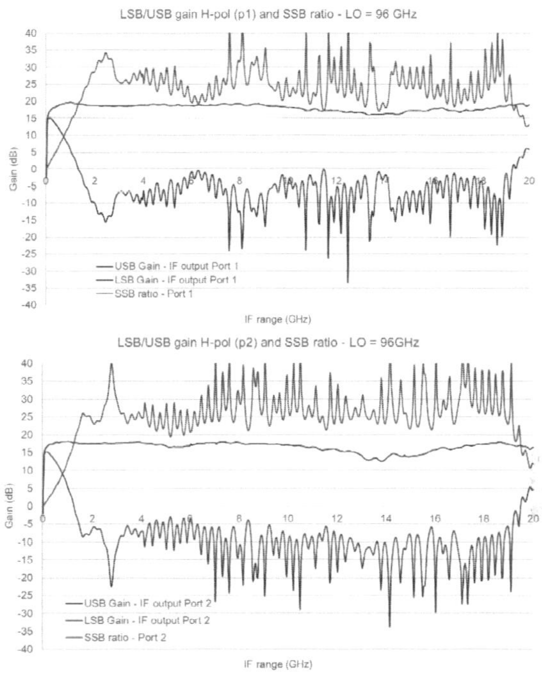

Figure 1. Measured gain response and sideband ratio for the Port 1 (top graph) and Port 2 (bottom graph) vs IF output range. Configuration: H-Polarization 1 - LO= 96 GHz - RF = 76-116 GHz – IF = 10 MHz- 20 GHz.

A summary of the 2SB down-converter test results will be presented, along with the novel UOSM measurement method used. Suggestions for further improvements such as highly integrated IQ-mixer modules under development and LO power control capabilities will be addressed.

References

[1] "The ALMA Development Program: Roadmap to 2030", J. Carpenter et al.. Website (stand: 15.03.2022): https://science.nrao.edu/facilities/alma/science_sustainability/URSI.pdf

[2] "Wideband 67−116 GHz receiver development for ALMA Band 2", P. Yagoubov et al., Astronomy & Astrophysics, Volume 634, A46, pp-22, February 2020.

[3] "A dual-polarization sideband separating Schottky based receiver for ALMA Band 2+3 Warm Cartridge Assembly", B. Thomas et al., proceedings of the 29th IEEE Int. Symp. on Space THz Tech., March 26-28th 2018, Pasadena, CA, USA.

[1] Radiometer Physics GmbH, Werner-von-Siemens-Strasse 4, 53340 Meckenheim, Germany; [2] NOVA, Kapteyn Instituut RUG, Landleven 12, 9747 AD Groningen.

NOTES:

SEPIA345: a dual polarization 2SB cartridge receiver for APEX telescope: Design and Performance

D. Meledin*[1], I. Lapkin[1], S-E. Ferm[1], V. Desmaris[1], E. Sundin[1], A. Pavolotsky[1], M. Fredrixon[1], M. Strandberg[1], P. Bergman[2], K. Torstensson[3], C. Durán[4], F. M. Montenegro-Montes[5,6], and V. Belitsky[1]

***Abstract*—A new receiver channel covering the 271-377 GHz frequency band has been installed into the SEPIA receiver at the APEX telescope. The receiver channel was designed and built in an ALMA-compatible cartridge layout. The receiver has a dual polarization layout with OMT and employs 2SB SIS mixers featuring an extended 4-12 GHz IF band, providing 32 GHz instantaneous IF bandwidth for two polarizations and two sidebands.**

Keywords—sub-mm receiver, SIS technology, dual-polarization

I. Introduction

The frequency band between 200-425 GHz has been marked as the highest priority for the new generation of ALMA receivers in the "ALMA 2030" roadmap document because of its most direct impact on the new ALMA science goals [1]. The upper part of this band, corresponding to the ALMA Band 7, has quite many astrophysical important molecular transitions. The new generation of state-of-the-art receivers has to provide a wide IF band, which opens the possibility of performing science observations more effectively.

The Atacama Pathfinder Experiment (APEX) telescope placed next to ALMA at the Chajnantor plateau in northern Chile has offered outstanding millimeter- and submillimeter-wavelength observing possibilities [2]. A major component of the current set of APEX facility receivers is provided by the Swedish ESO PI Instrument for the **A**PEX telescope (SEPIA), which is a multiband heterodyne instrument developed, designed, and built by the Group for Advanced Receiver Development (GARD), Onsala Space Observatory, Chalmers University of Technology, in collaboration with ESO [3]. This instrument features a cartridge layout that is fully compatible with ALMA technologies. The SEPIA instrument, therefore, provides a flexible platform for observations but also for testing different receivers.

One of three SEPIA receiver channels, the SEPIA345 receiver, operates over the frequency range of 271–377 GHz and was installed in 2020. The primary use of the receiver is to observe important molecular transitions (CO, CH_3OH, and N_2H^+) as part of the single-dish spectroscopic observing program of APEX. Moreover, the SEPIA345 receiver also expands APEX participation in Very Large Baseline Interferometry (VLBI) observations for the Event Horizon Telescope (EHT) from 230 GHz to 345 GHz. In April 2022 the SEPIA345 receiver channel was successfully used in EHT test observations at 345 GHz. The scientific importance of EHT telescopes such as APEX moving to a higher observing frequency is that this gives longer VLBI baselines in terms of wavelengths, allowing higher-resolution images to be made of Super Massive Black Holes in the center of galaxies [4].

In this paper, we present the SEPIA345 receiver channel that covers 271-377 GHz and provides the IF bandwidth of 4-12 GHz in 2SB and dual polarization configuration.

II. Receiver Design

The SEPIA instrument occupies one of the facility positions inside the Nasmyth cabin A at the APEX telescope. The optics design implements a frequency-independent illumination of the secondary for all SEPIA receiver channels (159–722 GHz) with an edge taper of about −12 dB [5]. To be able to provide beam coupling in the limited cabin space, the SEPIA optics has a total of ten mirrors, including five warm mirrors. As part of the receiver verification, measurements of the beam at several signal frequencies were performed in the lab as near-field scans over the cryostat windows. The calculated Gaussivity was found to be about 96%.

The SEPIA345 receiver is built for dual polarization operation and utilizes the 2SB receiver architecture, as

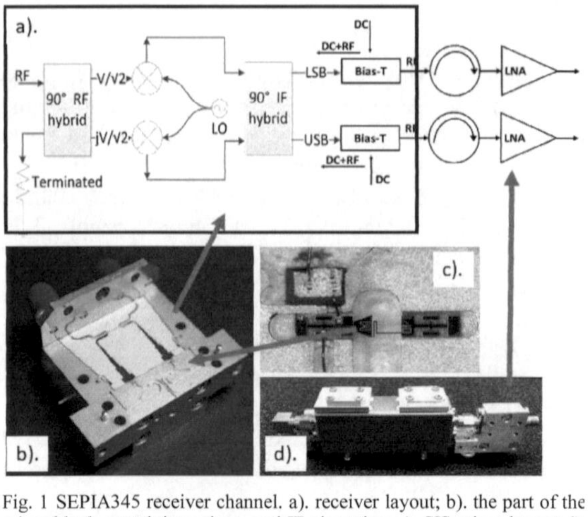

Fig. 1 SEPIA345 receiver channel. a). receiver layout; b). the part of the mixer block containing mixer- and IF circutries; c). SIS mixer layout d). IF assembly of one of the sidebands consisting of a cryogenic isolator and a low noise amplifier (LNA).

[1]Group for Advanced Receiver Development, Chalmers University of Technology, SE412 96, Gothenburg, Sweden.

[2] Department of Space, Earth and Environment, Chalmers University of Technology, Onsala Space Observatory (OSO), SE439 9243992 Onsala, Sweden.

[3] Joint ALMA Observatory, Alonso de Córdova 3107, Vitacura 763-0355, Santiago de Chile, Chile.

[4] European Southern Observatory, Alonso de Córdova 3107, Vitacura, Casilla 19001, Santiago de Chile, Chile.

[5] Departamento de Física de la Tierra y Astrofísica, Universidad Complutense de Madrid, 28040 Madrid, Spain

[6] IPARCOS-UCM, Instituto de Física de Partículas y del Cosmos, E-28040, Madrid, Spain.

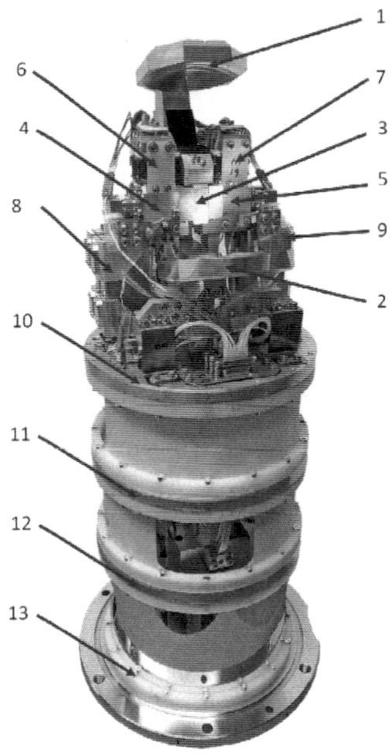

Fig.2 SEPIA345 Cold Cartridge Assembly (CCA) layout [6]. The follow-ing components are marked: 1- first cold mirror; 2 - second cold mir-ror; 3 - orthomode transducer (OMT); 4 - 2SB mixer block at Pol0; 5-2SB mixer block at Pol1; 6 - magnetic coil assembly at Pol0; 7 - magnetic coil assembly at Pol1; 8 - LSB and USB IF assemblies at Pol0; 9 - LSB and USB IF assemblies at Pol1; 10 - 4K plate; 11 - 15K plate; 12 - 110K plate; 13 - room temperature plate.

shown in Fig.1a. An orthomode transducer (OMT) is employed for polarization split. Since both polarizations share the same feed horn, they are co-aligned on the sky. Lab measurement results confirmed a cross-polarization lower than −25 dB over the entire operational band.

As any ALMA receiver channel, the SEPIA345 consists of a Cold Cartridge Assembly (CCA) and a Warm Cartridge Assembly (WCA). The WCA is the standard ALMA Band 7 WCA produced by NRAO. The SEPIA345 CCA was designed from scratch and consists of the cold optics with two elliptical mirrors, corrugated feed horn, OMT, 2SB mixer blocks, IF chains, and DC bias circuitry with the LO multipliers installed at 110 K of the CCA. Details of the CCA are presented in Fig.2 [6].

In the 2SB mixer block, we integrated two Superconductor-Isolator-Superconductor (SIS) mixer chips, along with the RF and IF circuitries. The bottom part of the 2SB mixer block is shown in Fig. 1b. As in the commonly used 2SB mixer configuration, the RF circuitry comprises a signal waveguide quadrature 3 dB hybrid and the LO injection directional couplers (with a coupling factor of -18dB). Mixer- and IF circuitries are also located inside each of the two mixer blocks. Waveguides' dimensions of 380x760μm demand the strict fabrication accuracy of the receiver components.

The mixer chips (shown in Fig. 1c) employ Nb/Al-AlO_x/Nb SIS in the twin-junction configuration [7] and have an area of 2.3 μm^2; the SIS junctions have been fabricated in-house and the mixer chip employs 65μm thick quartz substrates [8]. The twin-junction tuning circuitry with a single-step transformer compensates for the reactance of the junctions in the RF signal band of 270-375 GHz. The transformer is formed on 360 nm of SiO_2 sputtered on top of the RF choke, serving as a ground layer for the tuning circuitry. The extraction of the IF signal occurs at an additional port situated close to the SIS mixers and via a landing capacitor (Fig. 1c) and enhances the flatness of the IF response across the 4–12 GHz band.

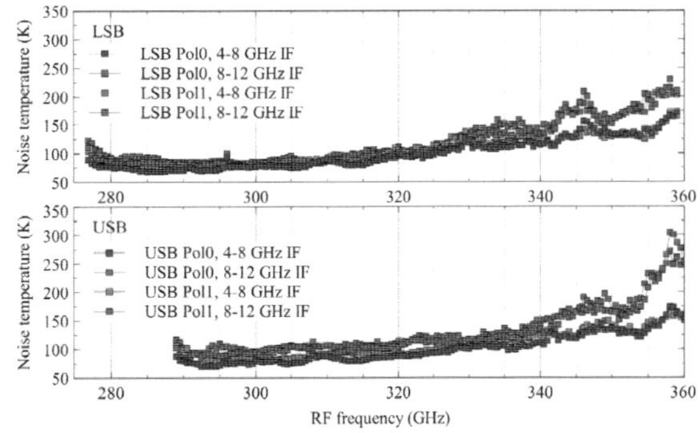

Fig.3 SSB receiver noise temperature measurements performed for each polarization and sideband averaged over each 4.0-GHz-wide FFTS IF sub-band. The temperature in the higher IF section (8–12 GHz) is illustrated with blue (Pol0) and violet squares (Pol1). The results in the inner segment (4–8 GHz) are shown with red (Pol0) and green squares (Pol1).

The IF circuitry comprises two IF matching transformers, two bias-Ts, and an IF 90-degree hybrid integrated on the same 20 mils thick alumina substrate as depicted in Fig. 1b, enabling a broader IF bandwidth of 4-12 GHz for each sideband. Further amplification at IF is performed by low-noise cryogenic amplifiers (LNAs) preceded by 4–12 GHz isolators, as shown in Fig.1d. As a result, 16 GHz (2 × 8 GHz) of instantaneous IF bandwidth is reached per polarization.

The LO (283–365 GHz) is an integral part of the WCA produced by the National Radio Astronomy Observatory (NRAO) and uses a direct multiplication LO chain (x18), following a Yttrium Iron Garnet (YIG) oscillator. The last frequency multipliers (×3) for each polarization are mounted inside the CCA at the 110K stage. The YIG oscillator, LO chain components, as well as warm IF amplifiers are located in the Warm Cartridge Assembly (WCA), which is attached to the CCA. Amplified IF signal is processed further with the help of the APEX IF processor and the (Fast Fourier Transformer Spectrometer) FFTS. As a result, 4–12 GHz IF bandwidth for each side-band and polarization is divided into two 4-GHz-wide bands. The FFTS provides spectral resolution of up to 64000 channels per 4 GHz of input bandwidth.

III. Characterization of the Receiver at the Telescope

The technical verification of SEPIA345 at the APEX telescope took place in 2020. The science commissioning of the receiver band was then continuously performed until the middle of 2021.

The receiver noise temperatures were measured over the full RF band by tuning the LO in steps of 0.5 GHz with help of the facility calibration unit (FCU) following a standard COLD-HOT-SKY calibration procedure per sideband and polarization. Results of such measurements averaged over each 4-GHz-wide FFTS section are presented in Fig.3 [6].

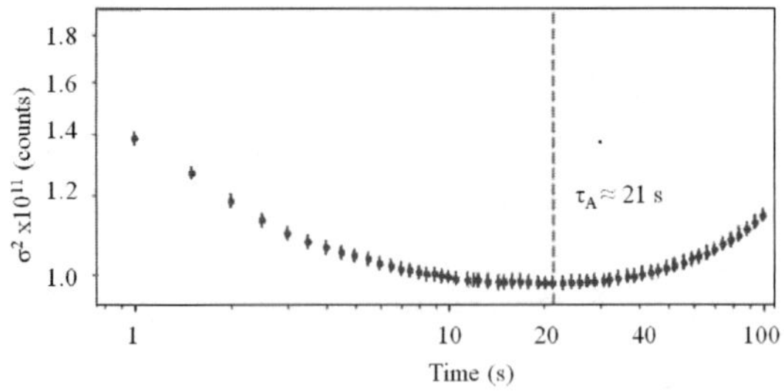

Fig.4 Example of total power Allan variance measurement performed for Pol0 LSB within a 125-MHz-wide sub-band in the 4–8 GHz FFTS segment at the LO frequency of 297 GHz.

The sideband rejection ratio (SBRR) plays an important role in the ability to perform high-quality science observations since these require as little contamination as possible in the signal band from lines in the image band. This is especially true for the single-dish observations. In the case of astronomical sources with high spectral line density, a poor SBRR may, for instance, hamper line identification. From our measurements, the average SBRR is as low as –18 dB. The Pol0 mixer demonstrates SBRR average values around –20 dB, while Pol1 channel average values lie around –15 dB [6]. These results are in line with measurements performed on-sky by observing the WB 947 source with strong and narrow CO J = 3 → 2 line [6]. In this experiment, the CO line was placed at different parts of the IF band, by retuning the LO. For each LO setting, the SBRR was evaluated by measuring the intensity ratio of the weakly appearing image line to the signal line.

Stability is one of the most important receiver parameters. A standard method to characterize receiver stability is performing measurements of the Allan variance [9]. These measurements took place while the telescope was standing still (stowed). The cold load of the FCU was providing more stable conditions than the cabin ambient temperature while data was being collected over several hours. The total power Allan time was measured in small sub-bands combining consecutive FFTS channels, spanning a total bandwidth of 125 MHz per sub-band. In Fig. 4, we show the total power Allan variance that was obtained using the Pol0 USB within a 125-MHz-wide sub-band within the 4–8 GHz FFTS segment at the LO frequency of 297 GHz. This frequency is recommended for continuum observations as it is relatively free of atmospheric features. The spectroscopic Allan times are calculated by creating smaller sub-bands combining a few consecutive FFTS channels, spanning 1.0 MHz in total. Then, the signals of two of these 1.0 MHz sub-bands are subtracted, and their noise characteristics are inspected to have a more complete overview.

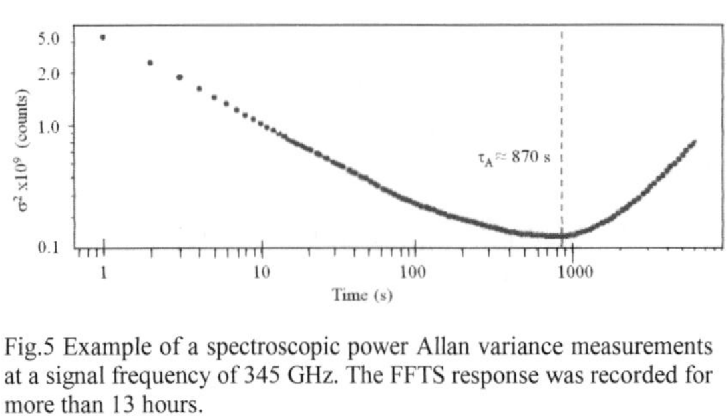

Fig.5 Example of a spectroscopic power Allan variance measurements at a signal frequency of 345 GHz. The FFTS response was recorded for more than 13 hours.

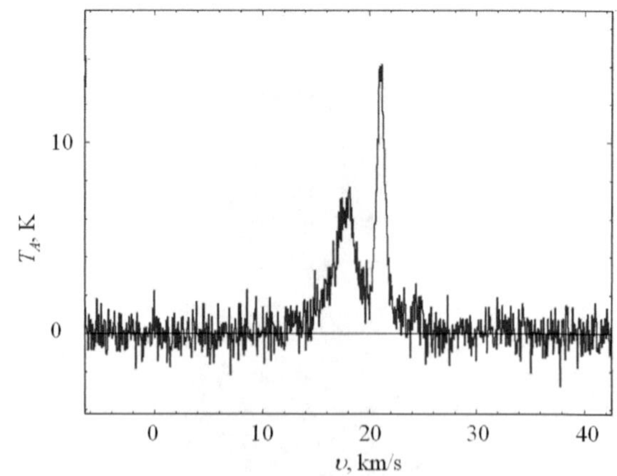

Fig.6 The observed H_2O 321 GHz maser spectrum towards the red supegiant VY CMa

An example of the spectroscopic Allan variance measurements is displayed in Fig. 5. These measurements were performed with Pol1 mixers at the 345 GHz USB in the 4–8 GHz FFTS IF section using the difference between two different 1.0-MHz-wide sub-bands, some 1.8 GHz apart. The Allan time is estimated to be larger than 850 s.

In a frame of science verification, among many other tests, observations of the water maser at 321 GHz toward VY CMa have been performed. The water maser emission originates from energy levels near 2000 K in excitation energy. It can be considered as a point source concerning the APEX beam size at 321 GHz. Fig.6 shows this water maser line.

IV. Conclusion

The SEPIA345 receiver channel, installed at the APEX telescope in February 2020, covers a signal band of 271–377 GHz. The dual polarization receiver employs 2SB SIS mixers and offers an instantaneous IF bandwidth of 4×8 GHz. The receiver has an average SBRR of –20 dB and –15 dB measured at the telescope for the Pol0 and Pol1 mixer channels, respectively. The receiver stability measurements demonstrate typical total power Allan times on average larger than 10 s in the 125 MHz effective noise bandwidth and better than 60 s spectroscopic Allan times in the 1 MHz effective noise bandwidth. The SEPIA345 receiver become one of the APEX facility instruments and is now available to all APEX observers.

V. References

[1] J. Carpenter, D. Iono, L. Testi, et al., ALMA Memo 612, 2019, https://arxiv.org/abs/1902.02856.
[2] Güsten, R., Nyman, L.-A., Menten, K., Cesarsky, C., Booth, R. S., and Schilke, P. F., "The Atacama Pathfinder EXperiment", *The Messenger*, vol. 124, p. 12, 2006.
[3] V. Belitsky et al., "SEPIA - a new single pixel receiver at the APEX Telescope", *Astron. & Astroph.*, vol. 612, A23, 2018.
[4] Event Horizon Telescope Collaboration, "First Sagittarius A* Event Horizon Telescope Results. I. The Shadow of the Supermassive Black Hole in the Center of the Milky Way", The *Astrophysical Journal*, vol. 930, no. 2, 2022. doi:10.3847/2041-8213/ac6674.
[5] I. Lapkin et al.," New Optics for SEPIA- Heterodyne Facility Instrument for the APEX Telescope," *30th ISSTT*, Gothenburg, 2019, pp.150-154.

[6] D.Meledin et al., "SEPIA345: A 345 GHz dual polarization heterodyne receiver channel for SEPIA at the APEX telescope", *Astronomy and Astrophysics*, vol. 668, 2022. doi:10.1051/0004-6361/202244211.

[7] V. Belitsky, M. Tarasov, "SIS junction reactance complete compensation," *IEEE Transactions on Magnetics*, v.27(2), pp. 2638-2641, 1991.

[8] A. Pavolotsky, D. Dochev, and V. Belitsky, "Aging and annealing-induced variations in Nb/Al-AlOx/Nb tunnel junction properties," *J. Appl. Phys.*, vol. 109, no. 2, p. 024502, 2011.

[9] Schieder, R. and Kramer, C., "Optimization of heterodyne observations using Allan variance measurements", *Astronomy and Astrophysics*, vol. 373, pp. 746–756, 2001. doi:10.1051/0004-6361:20010611.

ALMA Band 2 Receiver Automated Test System

J.Barkhof*[1], R.Hesper[1], S.Realini[1], M.Bekema[1], A. Koops[1], R.de Haan[1], J.Adema[1], A.Baryshev[1], M.Rodenhuis[1], V. Belitsky[2], M. Fredrixon[2], I. Lapkin[2], A. Pavolotsky[2], M. Strandberg[2], S. Ferm[2], C. Lopes[2], F.Cuttaia[3], S.Ricciardi[3], L.Terenzi[3], F.Villa[3], R.Nesti[4], N. Phillips[5], P. Yagoubov[5]

***Abstract*—As part of the ALMA Band 2 project, an automated test system was developed to fully qualify state-of-the-art Band 2 receivers, based on heritage of the Band 9 and Band 5 productions. The RF range of this receiver is 67 - 116 GHz, with a goal IF band of 4-18 GHz. Each receiver will undergo a thorough acceptance testing to verify its operation and performance prior to delivery. The core of the test system are a single-cartridge test cryostat, a dual-channel intermediate frequency signal processor, a vector near-field test system, and script-based measurement and control software that enables automated testing. We present details of the test system and measurement results of the first Band 2 receivers.**

***Keywords*— automatic test equipment, Sideband rejection ratio, Noise temperature, Sub-mm receivers**

I. Introduction

In contrast to the higher ALMA bands (3-10), which use cryogenic mixers for heterodyne downconversion, the Band 2 receiver employs cryogenic waveguide amplifiers in the Cold Cartridge Assembly (CCA) [1], while the sideband-separating downconversion happens in the air-side Warm Cartridge Assembly (WCA). The CCA is constructed by a consortium of NOVA (Netherlands), GARD (Sweden) and INAF (Italy) [2], while the WCA is built at NOVA using a wide-band downconverter developed by RPG [3].

The key receiver specifications to be qualified before delivery are: heterodyne sensitivity and cartridge output power as a function of local oscillator (LO) frequency and intermediate frequency (IF), sideband rejection ratio, aperture and polarization efficiency, output power and signal path phase stability, and gain compression. The two most important ones are shown in the tables below.

Noise Specifications	
Range (GHz)	*Tnoise (K)*
67-90	≤ 47
67-90	80% ≤ 30
90-116	≤ 43

Sideband Rejection Ratio	
Range (GHz)	*SRR (dB)*
67-116	> 7 dB
67-116	90% of the IF > 10

An example of the measured noise temperature is shown in Fig. 1. The noise temperature is measured over 4-18GHz IF at 5 GHz LO spacings and, corrected for residual image response and averaged over bins of 4GHz. Fig. 2 shows the sideband rejection ratio measured over 4-18GHz at the LO frequencies 79, 84, 89,… 104 GHz.

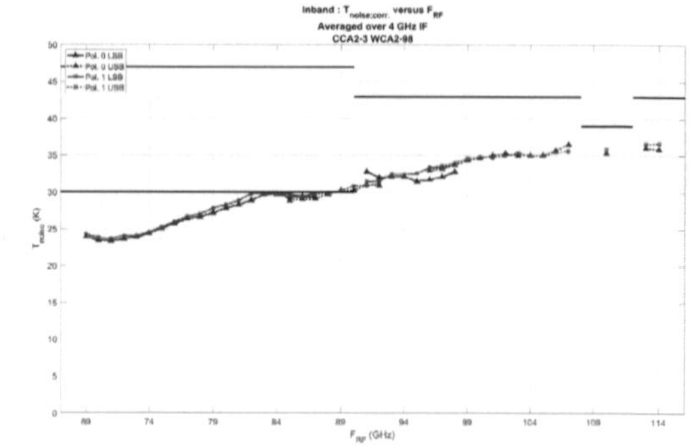

Fig. 1. Corrected Receiver noise temperature vs. RF frequency averaged over 4GHz bins, including specification markers.

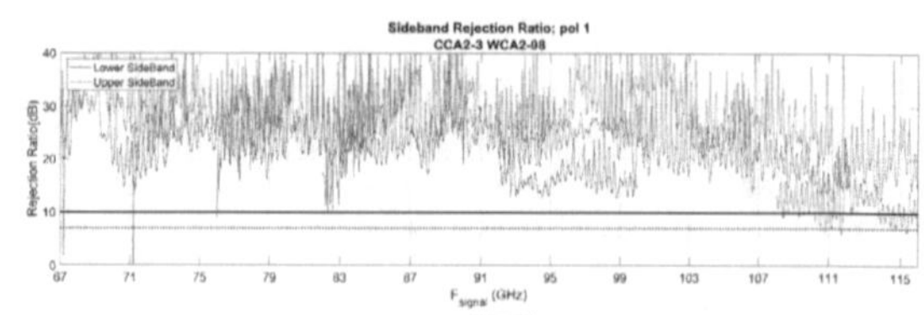

Fig. 2. Cartridge sideband rejection ratio vs. RF frequency. The upper graph shows polarization 0 and the lower graph polarization 1. The black line indicates a rejection ratio of 10 dB and the dotted black line 7 dB.

References

[1] P. Yagoubov, et al., "Wideband 67−116 GHz receiver development for ALMA Band 2", *Astronomy & Astrophysics*, February 2020.

[2] V. Belitsky et.al, "ALMA Band 2 Cold Cartridge Assembly Design", ISSTT2022

[3] B. Thomas, et al., "A dual-polarization sideband separating Schottky based receiver for ALMA Band 2+3 Warm Cartridge Assembly", *proceedings of the 29th IEEE Int. Symp. on Space THz Tech.*, March. 2018.

[1]Kapteyn Astronomical Institute, University of Groningen, 9747 AV Groningen, The Netherlands; [2]GARD, Chalmers University, Gothenburg, 41296, Sweden; [3]INAF-OAS, Via P. Gobetti 101, 40129 Bologna; [4]INAF-OAA, Largo Enrico Fermi 5, I - 50125 Firenze; [5]ESO Headquarters, 85748 Garching bei München, Germany

NOTES:

Development of ALMA Band 1 Corrugated Horns based on Metal 3D Printer

K. Kaneko*[1], R. Sakai[1], T. Kanzawa[1], K. Mitsui[1], H. Kang[2], T. Kojima[1], Y. Uzawa[1], A. Gonzalez[1], C.-D. Huang[3], and Y.-D. Huang[4]

Additive Manufacturing (AM) technology has grown dramatically and is now used in a wide variety of industrial applications. In the aerospace industry, it has been applied to fabrication of antennas and waveguide components and is now being put to practical use, as is the case with the remarkable example of the development of the satellite antenna by Boeing [1]. The application of the AM technology to radio astronomical instrument is also under consideration. This technology benefits shorter development periods, cost reductions, realization of complex shapes that are difficult to be fabricated with conventional technology, and reduction in the number of parts by integrating multiple parts. On the other hand, since a component is fabricated by melting and solidifying materials, verification of material properties is an important item.

In order to verify feasibility of this AM technology, we have tried to manufacture corrugated horns used in a frequency range of 35 to 50 GHz in the period of 2016-2018, and successfully obtained promising performance for the radio astronomical application [2]. Based on the result, we have developed the corrugated horn to be implemented into the Band 1 receiver (35-50 GHz, up to 52 GHz as best effort) for the Atacama Large Millimeter/submillimeter Array (ALMA) [3]. One of the major conventional manufacturing methods of corrugated horns is milling or electro-forming that are mature technologies in recent, while they provide long lead time, high costs and design limitation. Through this development, we have tried to fabricate the 3D-printed corrugated horns that met ALMA technical specifications, but also verified whether if the AM technology could mitigate the risks of existing technology. In addition, for the practical use, it is important not only for the horn performance but also the material properties. Since the component is made by repeatedly melting and solidifying material powder by laser irradiation, the physical properties may have changed from the original ones. Besides, the corrugated horn is located and operated in the cryogenic stage at around 15 K, and thermal cycling between room temperature and operation temperature is assumed for the receiver maintenance. Thus, we have investigated material properties at both warm and cryogenic temperatures, and obtained the beneficial information for the component design and practical use.

In this symposium, the development process, benefits and risk of 3D-printed components, measured material properties and the beam performance of the horn individual and the receiver implemented 3D printed horn will be presented.

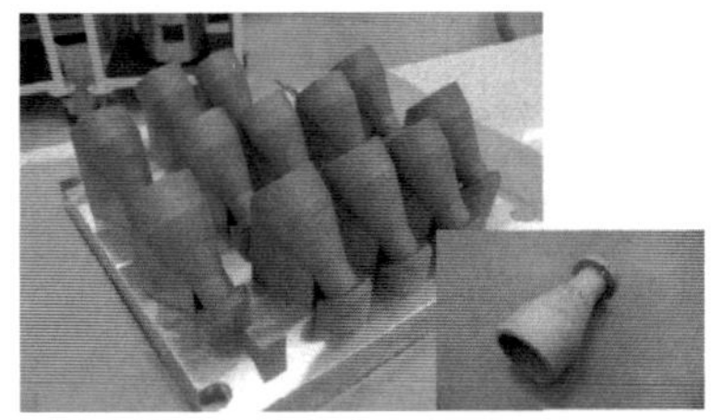

Fig. 1. ALMA Band 1 corrugated horn fabricated by 3D printer

Fig. 2. ALMA Band 1 receiver with 3D printed horn

REFERENCES

[1] https://3dprint.com/239651/boeing-creates-the-first-3d-printed-metal-satellite-antenna-saves-on-mass-time-and-costs/
[2] A. Gonzalez, et al., "Metal 3D-Printed 35–50-GHz Corrugated Horn for Cryogenic Operation", JIMTW, 42:920-973, 2021.
[3] Y.-J. Hwang, et al., "Band 1 receiver front-end cartridges for Atacama Large Millimeter/submillimeter Array (ALMA): design and development toward production", SPIE conference on Astronomical Telescopes + Instrumentation 2016, 2016

[1]National Astronomical Observatory of Japan, Mitaka, 181-8588, Japan; [2] The University of Tokyo, Bunkyo, 113-8654, Japan; [3] National Chung-Shang Institute of Science and Technology, Taichung, 40741, Taiwan; [4] Academia Sinica Institute of Astronomy and Astrophysics, Taipei, 10617, Taiwan

NOTES:

Demonstrator of Cryogenic Multibeam Receiver with MMIC LNAs for 75-116 GHz

P. Serres*[1], A. Navarrini[2], O. Garnier[1], F. Schaefer[3], S. Türk[3], F. Thome[4], P. Ortu[2], G. Valente[2], A. Saba[2], AL. Fontana[1], D. Maier[1], A. Barbier[1], D. Billon[1], E. Driessen[1]

Abstract— **We describe the design of a nine-element multibeam receiver demonstrator for the 75-116 GHz (W-band). The design consists of a cryogenic 3×3 dual-polarization focal plane array (FPA) based on individual reimaging optics for the corrugated horns, and employs an innovative "active OMT" with two cascaded mHEMT MMIC LNAs packaged in each of the polarization channels of the OMT waveguide structure. The RF signals from the cryogenic array active OMT outputs are linked, through WR10 waveguides, to three-pixel sideband separating mixer (2SB) modules located at room temperature, attached to the baseplate of the cryostat. One of such modules converts six RF signals from one line of the array (1×3×2 pols) down to the 4-12 GHz IF band, thus providing a total of 12 IF outputs (USB and LSB). The optics of the FPA demonstrator was designed and optimized for high-efficiency broadband operation on the IRAM 30-m telescope at Pico Veleta, Spain.**

Keywords—Receiver, optics, OMT, MMIC, LNA, mixer, sideband separation, cryogenics.

I. Introduction

Under the auspices of the AETHRA (Advanced European Technologies for Heterodyne Receiver for Astronomy) WP5.1 project, funded by the European Union's Horizon 2020 research and innovation programme, we developed a multipixel W-band FPA demonstrator [1] composed of cryogenic modules and room temperature down-conversion modules based on metamorphic HEMT semiconductor MMIC LNAs [2] and MMIC mixers [3] for the 75-116 GHz band. Here, we present the design of the receiver.

II. Receiver Design

Fig. 1 shows a 3D model of the FPA receiver cryostat with room-temperature reimaging optics designed for the IRAM 30-m radio telescope. The details of the inner part of the receiver cryostat, including the full 3×3 cryogenic array of pixels, are shown in Fig. 2. The optical design is based on nine corrugated feed-horns and individual fully reflective frequency-independent optics (Fig. 3) that couple the signals from the telescope sub-reflector. The HDPE vacuum window consists of 3×3 antireflection-coated sub-windows. A PTFE Infrared Filter at ~70 K and a Goretex filter at ~15 K reduce the thermal loading on the coldest stage, at ~15 K. The cryostat internal details are shown in Fig. 4.

Two of the key modules of the receivers are the cryogenically cooled active OMT (described in [4], see Fig. 2), based on four W-band MMIC LNAs (two MMICs per polarization channel), and the room temperature three-pixel downconverter module, shown in Fig. 5.

Each of the nine active OMT modules has a small footprint, with a size compatible with the 44 mm spacing between the pixels. An active OMT module is based on a square waveguide input and on two standard WR10 waveguide outputs, in line with the input, each delivering a total gain of ≈45 dB for the two independent linear polarization channels.

The three three-pixel downconverter modules adopt a dual-sideband separating mixers (2SB) downconversion scheme, where each 2SB subassembly delivers two 4-12 GHz IF output signals, the USB (Upper Sideband) and the LSB (Lower Side Band). A tuneable 14.50-17.33 GHz baseband LO is injected into each of the three-pixel downconverter modules. The downconverter is based on fully-integrated InGaAs MMIC sub-harmonic mixers that include LO-multiplier and IF-LNA.

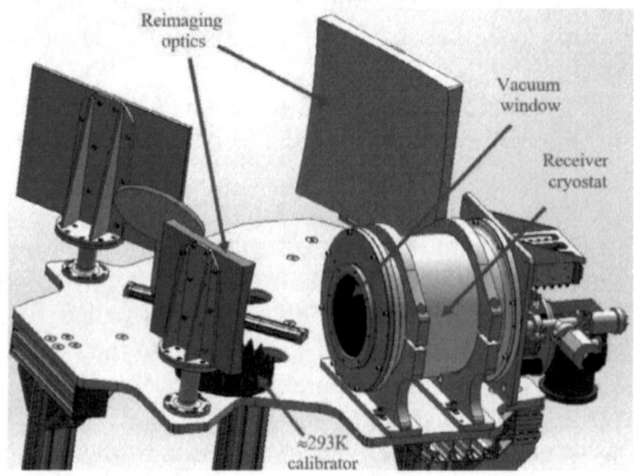

Fig. 1. Design of the receiver array demonstrator on the mechanical frame for the IRAM 30-m telescope.

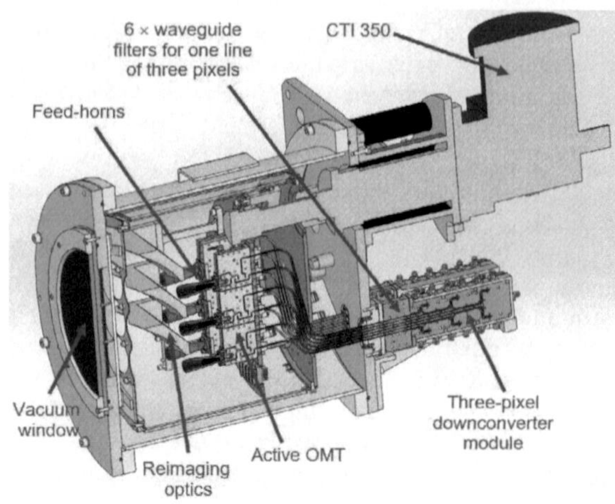

Fig. 2. Cross-cut view of the dual-polarization nine-pixel cryogenic MMIC receiver array with room-temperature sideband separating three-pixel downconverter modules. The reimaging optics and active OMT modules are cryogenically cooled at ~15 K inside the cryostat by a commercial cryocooler (Cryodyne CTI 350).

[1]IRAM, Saint Martin d'Hères, 38406, France; [2]INAF-OA Cagliari, Selargius, 09045, Italy, [3]MPIfR, Bonn, 53121, Germany; [4]Fraunhofer IAF, Freiburg, 79018, Germany.

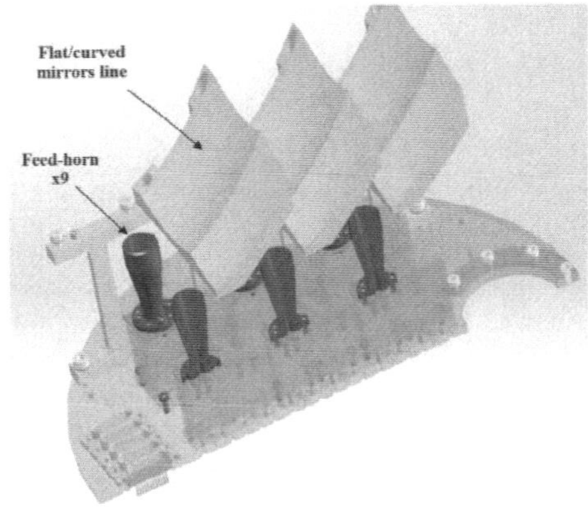

Fig. 3. Cross-cut view of the cryostat 15 K stage with fully reflective cryogenic optics for the nine individual pixels.

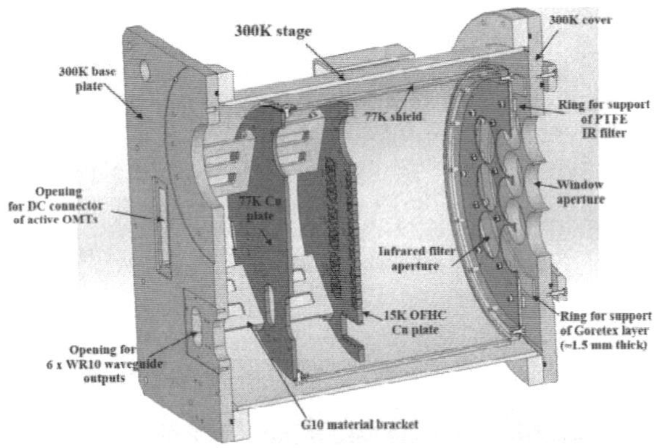

Fig. 4. Cross-cut view of the cryostat showing the 15 K and 77 K cryogenic stages, the location of the IR filters, and the G10 brackets for thermal isolation of the stages.

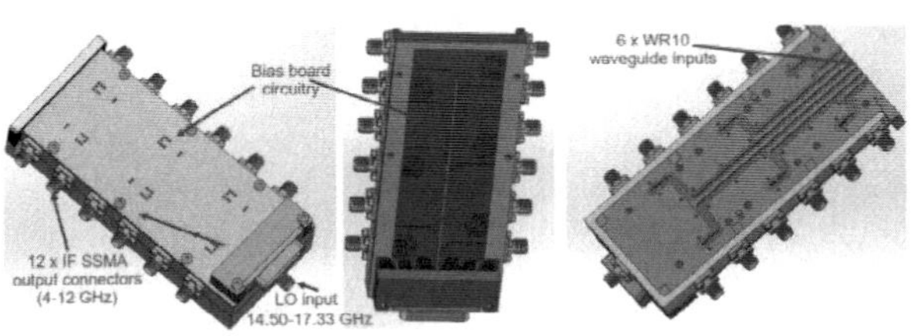

Fig. 5. Three views of the three-pixel dual-polarization dual-sideband separating mixers downconverter module.

The receiver is biased by an M&C (Monitor and Control) unit specifically developed for this project. The M&C unit adopts independent boards for biasing the cryogenic LNAs and the downconverter modules.

The receiver subassemblies were constructed and tested. Then, the array demonstrator was assembled using only one line of three active OMT modules (the one containing the central pixel) and one of the three single-pixel modules. Laboratory tests of the system were successfully carried out at the room and cryogenic temperatures, resulting in low-noise broadband performance close to predicted values across the full 75-116 GHz band. Then, the receiver was installed on the IRAM 30-m telescope, Pico Veleta, Spain, and successfully used to conduct radio astronomy observation tests.

III. Conclusion

We presented an innovative design of a cryogenic focal plane array receiver suitable for radio astronomy observation in W-band. The array utilizes active OMT incorporating low-noise MMIC LNAs, three-pixel dual-polarization sideband separating modules, and a dedicated M&C and bias module for the LNAs and the downconverters. The receiver was assembled and successfully tested on the IRAM 30-m telescope.

References

[1] O. Garnier, A. Navarrini, F. Schaefer, P. Serres, F. Thome, S. Türk, "Multipixel W-band FPA demonstrator composed of cryogenic module and down conversion module," AETHRA – WP5.1 Semiconductor LNAs and MMIC Receivers, Final report H2020-INFRAIA-2016-2017/H2020-INFRAIA-2016-1, Jan 29, 2021. Available: https://radiowiki.mpifr-bonn.mpg.de/lib/exe/fetch.php?media=jra:730562_radionet-d5.3_210129.pdf.

[2] F. Thome, F. Schaefer, S. Türk, P. Yagoubov, and A. Leuther, "A 67–116-GHz Cryogenic Low-Noise Amplifier in a 50-nm InGaAs Metamorphic HEMT Technology," IEEE Microw. Wirel. Compon. Lett., Vol. 32, No. 5, May 2022.

[3] F. Thome, E. Ture, R. Iannucci, A. Leuther, F. Schaefer, A. Navarrini, P. Serres, "Frequency Multiplier and Mixer MMICs Based on a Metamorphic HEMT Technology Including Schottky Diodes," IEEE Access, vol. 8, no 1, pp 12697-12712, 2020, ISSN: 2169-3536, DOI: 10.1109/ACCESS.2020.2965823.

[4] A. Navarrini, G. Valente, P. Serres, F. Schaefer, F. Thome, O. Garnier, "Compact Dual-Polarization Cryogenic Receiver Module for the 75-116 GHz Band," 2018 IEEE International Conference on Electromagnetics in Advanced Applications (ICEAA), Cartagena de Indias, Colombia, pp 479-482, Sep. 10-14, 2018.

Study of ALMA Band 2 receiver optical design of ACA 7-m antenna

Hiroaki Imada *1, Keiko Kaneko, Ryo Sakai, Takafumi Kojima, and Alvaro Gonzalez

***Abstract*—ALMA Band 2 receiver optical designs for ACA 7-m antenna are studied with ray-tracing software prior to heavy electromagnetic full-wave simulation. We have obtained good candidates for our application.**

***Keywords*—optical design, ray tracing, EM simulation**

I. Introduction

The ALMA Band 2 receiver is being developed [1]. ESO leads the development project with NAOJ, whose activity covers the receiver optical designs for the 12-m antennas and the 7-m antennas. We have made the component designs for the 12-m antennas. The poster shows the receiver optical design study to the corrugated horn.

The requirements are as follows:

- Aperture efficiency at sub-reflector: > 0.8,
- Polarization efficiency at sub-reflector: > 0.995,
- Forward efficiency: > 0.95,
- Focus efficiency: > 0.98.

The specific constraints for the Band 2 receiver optical system in the 7-m antenna are as follows:

- Same horn position as in 12-m antenna,
- Same lens (shape and material),
- Same IR filter (shape, material, and position).

II. Methods

We employed ray tracing to obtain designs. The goals of ray tracing were 1) to image the sub-reflector on the horn aperture and 2) to make rays converge on the phase center of a horn beam. Point 1) ensures that the electric field on the sub-reflector is copied on the horn aperture geometrically [2]. Point 2) achieves the phase matching between the beams from the sky and the receiver. The phase center of the horn beam seems located around 100 mm from the horn aperture (Fig. 1). Once we obtained a design candidate that satisfied those points, PO+MOM simulation was carried out to see its performance. The efficiencies were calculated with the following equations:

$$\eta_{\mathrm{ap}} = \frac{G}{D_{\mathrm{std}}}, \qquad \text{ap. eff. at sky (PO+MoM)}$$

$$\eta_{\mathrm{ap,sub}} = \frac{\eta_{\mathrm{ap}}\eta_{\mathrm{spill,rec,sub}}}{\eta_{\mathrm{spill,ent}}\eta_{\mathrm{spill,rec,pri}}}, \qquad \text{ap. eff. at subref. (PO+MoM)}$$

$$\eta_{\mathrm{spill,ent}} = \frac{\pi\left(R_{\mathrm{ent}}^2 - R_{\mathrm{cas}}^2\right)\cos\theta_{\mathrm{inc}}}{\pi R_{\mathrm{pri}}^2}, \qquad \text{ent. spill. eff. (geometrically)}$$

$$\eta_{\mathrm{pol}} = \frac{\iint_{\Omega_{\mathrm{sub}}} |E_{\mathrm{co}}(\theta,\phi)|^2 \,\mathrm{d}\Omega}{\iint_{\Omega_{\mathrm{sub}}} \left(|E_{\mathrm{co}}(\theta,\phi)|^2 + |E_{\mathrm{cx}}(\theta,\phi)|^2\right) \mathrm{d}\Omega}, \qquad \text{pol. eff. (PO+MoM)}$$

$$\eta_{\mathrm{fwd}} = \frac{\eta_{\mathrm{spill,rec,pri}}}{\eta_{\mathrm{spill,rec,sub}}}, \qquad \text{fwd eff. (PO+MoM)}$$

$$\eta_{\mathrm{foc}} = \frac{\left|\iint_S E_{\mathrm{horn}} E_{\mathrm{inc}}^* \mathrm{d}S\right|^2}{\left(\iint_S |E_{\mathrm{horn}}|\,|E_{\mathrm{inc}}|\,\mathrm{d}S\right)^2}. \qquad \text{focus eff. (MoM+ray tracing)}$$

The right column describes each efficiency with calculation methods.

III. Results

We made two designs: only the lens was adjusted (Fig. 2), and both the lens and sub-reflector were adjusted (Fig. 3). The coma aberration dominates the lens-only case performance. On the other hand, the spherical aberration by the sub-reflector dominates the lens + sub-reflector case performance. Both cases show the entrance pupil spillover efficiency [3] of 0.89 and 0.98, respectively. Interestingly, the aperture efficiency at the sub-reflector does not necessarily ensure maximum antenna gain, or aperture efficiency at the sky. Fig. 4 shows the lens-only case satisfies the requirements on the aperture, polarization, and forward efficiencies. Fig. 5 shows the focus efficiency for the lens-only case, which also satisfies the requirement. Fig. 6 demonstrates the beam patterns on the sky just for reference.

IV. Conclusion

We made receiver optical designs for the ALMA Band 2 with ray tracing simulation. Electromagnetic simulations confirmed that adjusting the sub-reflector position achieves the requirements. We also found that aperture efficiency at the sub-reflector does not necessarily reflect the antenna gain on the sky, or aperture efficiency on the sky.

References

[1] P. Yagoubov, et al., "Wideband 67—116 GHz receiver development for ALMA Band 2", *A&A*, 2020.

[2] H. Imada, et al., "Condition of optical systems independent of frequency for wide field-of-view radio telescopes", IEEE TST, 2015.

[3] M. Nagai, et al., "Factorization of Antenna Efficiency of Aperture-type Antenna: Beam coupling and Two Spillovers", IEEE TAP, 2021.

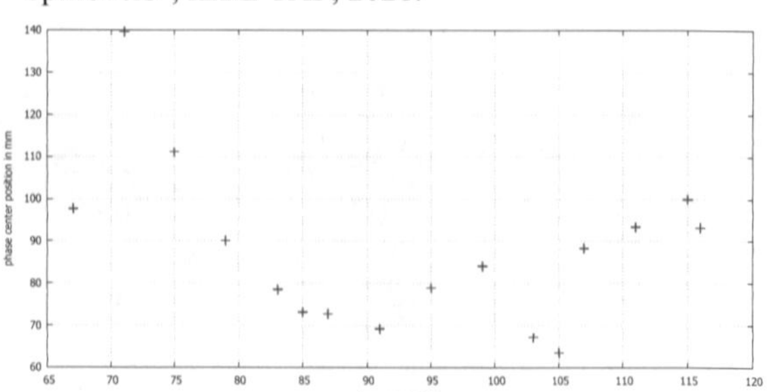

Fig. 1. Phase center position from the horn aperture.

[1]National Astronomical Observatory of Japan, Mitaka, 181-8588, Japan.

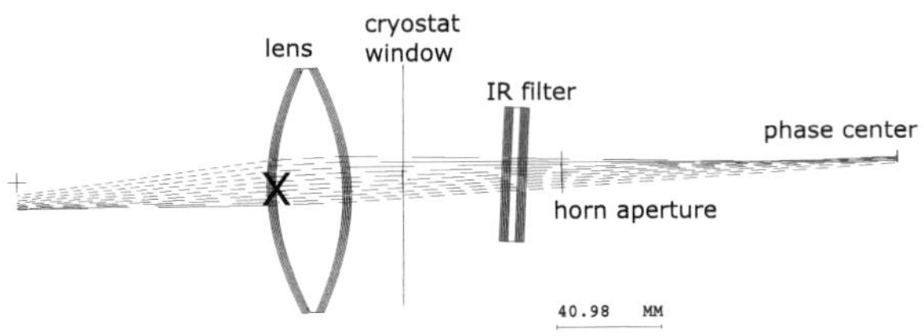

Fig. 2. Adjusting lens case.

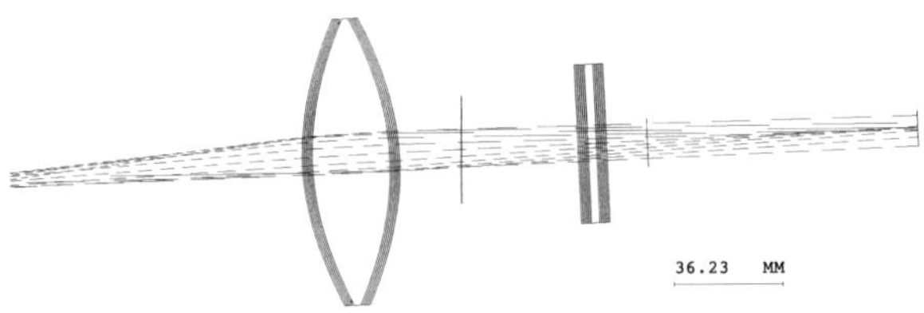

Fig. 3. Adjusting lens and sub-reflector case.

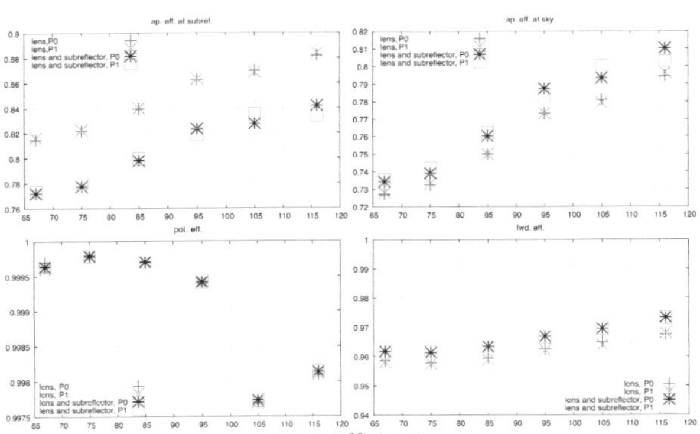

Fig. 4. Efficiencies.

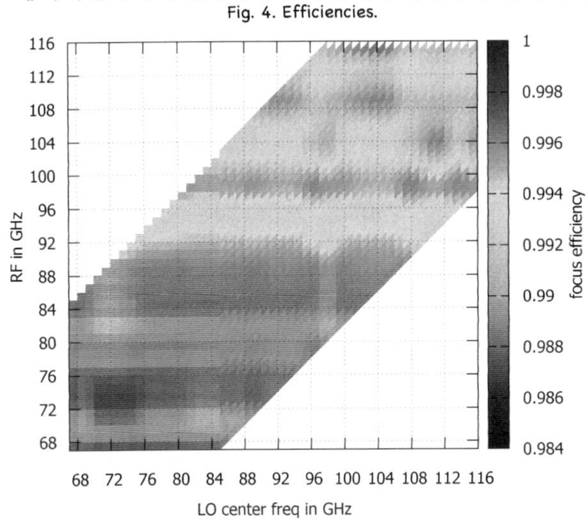

Fig. 5 Focus efficiency for the lens-only case.

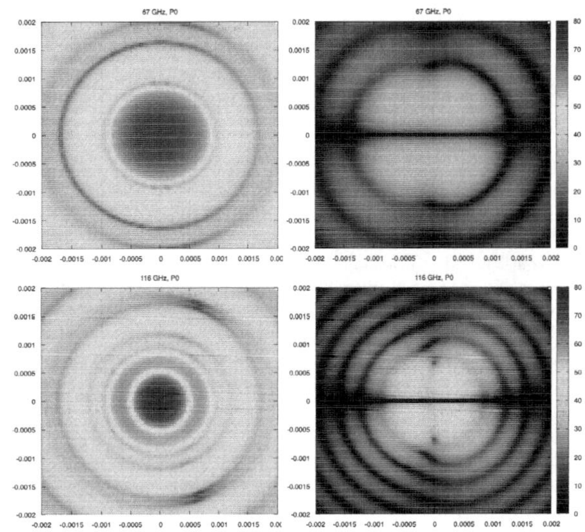

Fig. 6 Beam patterns on the sky for the lens-only case. Upper panels show the 67-GHz results and the lower the 116-GHz ones.

Design of a 480GHz Metamaterial Flat Lens

Cassandra N. Whitton[1], Jose V. Siles[2], Christopher E. Groppi[1], Philip D. Mauskopf[1], Adrian J. Tang[2]

The measuring of water vapor column densities and temperatures is relevant to many fields of study, including astrophysics, planetary science, and Earth science. For astrophysical and planetary science purposes in particular, such measurements are virtually impossible to perform without a space-based observatory due to the high attenuation introduced by water in the Earth's atsmophere. Observations of the 557 GHz line of ortho-water have been performed previously with large space-based instruments, but the ability to perform such observations with a cubesat would significantly increase the cost-effectiveness of such missions and thereby opens up previously unavailable science applications which would be financially impractical to perform with a larger platform.

For these reasons, we are designing a cubesat capable of spectroscopically resolving the 557 GHz and 1113 GHz lines of water vapor. Measurements performed with such a cubesat would allow us to gain information about the temperature, structure, and density of water vapor. Such a cubesat requires the miniaturization of several crucial technologies due to the small form factor and low weight requirements inherent in the cubesat standards. One such structure is the primary observing aperture. A metamaterial lens acting as the primary aperture is ideal for this purpose. Our metamaterial lens is a lens in which the focusing of light is achieved by using patterned elements of conductive material embedded in the lens structure as opposed to a traditional curved lens or Fresnel geometry. Such a lens can be made lightweight, flat, and can be designed to have near-zero reflections to reduce observing time.

As an intermediate step, we here present a design for a lightweight and flat metamaterial lens which operates at 480 GHz. The design has been simulated to measure its focusing ability as well as its reflectivity and attenuation at the necessary frequencies. Overall, we find very low reflective and optical losses at the design frequency (less than 5% each) and less than 20% conductive/dielectric loss.

***Keywords—terahertz, cubesat, metamaterial, lens, optics** (key words)*

I. Introduction

Flat metamaterial optics may offer substantial improvements in the size, weight, and power (SWaP) of optics in space-based applications. This is of particular interest for Smallsat applications where SWaP restrictions are of paramount importance. The main characteristic of the proposed optics that results in dramatic improvements in SWaP resource requirements is the thin geometry of the metamaterial structure. In a traditional bulk lens, focusing is achieved through the transformation of a planar wavefront to a spherical wavefront with a dielectric of varying thickness. The required shape of the surface of the lens is governed by Snell's law to provide a focus at a given location. This shape results in a minimum required thickness of bulk dielectric. In the sub-mm and THz, this thickness is at a minimum several mm and increases when a lens with a small f/D ratio is required. Zone plate lenses (e.g. a Fresnel lens) can help to address this problem, with a bandwidth penalty caused by the wavelength dependent design of the lens [3]. A gradient index lens would instead achieve this phase shift by altering the index of refraction of the lens over its surface, again creating a varying path-length over the lens. A mirror provides a similar transformation but through a reflective surface governed by the law of specular reflection. Again, this results in a material thickness of several mm at a minimum to create and support the reflective mirror geometry.

By contrast, a metamaterial flat lens can be less than a fraction of a mm thick, with a corresponding decrease in weight.

In addition, multi-layered metamaterial structures can effectively have anti-reflection coating built into them from their initial design. This can result in low reflective loss and thus significantly greater efficiency than a traditional curved lens. The lower thickness of such a flat lens can also result in significantly lower dielectric losses than an equivalent curved lens.

II. Methods

Our design process is primarily based on the work of Pisano (Ref. [1]), with some key differences. Our overall design process is as follows:

1) Choose the primary optical characteristics of the lens: diameter, focal length, operating frequency.
2) Choose materials and secondary characteristics: dielectric substrate material, conductive material, thickness of dielectric and metal layers, and number of layers.
3) Subdivide the lens surface into pixels based on the operating frequency.
4) From the desired optical characteristics, determine the desired phase transformation that each pixel will apply as a function of frequency.
5) In HFSS (or equivalent EM simulation software), simulate varying sizes of individual conductive elements embedded in dielectric.
6) Using the results from HFSS, create a model which simulates the S-parameters of a single pixel, using transmission line theory.
7) In a scripting language, run an optimization script which tweaks the sizes of the conductive elements in order to find the best element sizes for each pixel in the lens.

For our design, we chose a diameter of 124mm, due to working with a manufacturing process which uses 150mm wafers with an exclusion zone around the edge. We chose a focal length of 150mm since we are interested in using these

[1] School of Earth and Space Exploration, Arizona State University, Tempe, AZ 85281 USA (email: cassandra.whitton@gmail.com) [2] Jet Propulsion Laboratory, Pasadena, CA 91109 USA

lenses for cubesat applications where short focal lengths could be advantageous to save space. We chose an operating frequency of 480GHz in order to be able to test the lens with 500GHz VNA extenders that we have access to. In principle, our design process can be used to optimize across a bandwidth rather than a single frequency. We chose a single frequency for simplicity.

In step 2, we chose to use polyimide as our dielectric and aluminum as our conductive material, as well as a metal thickness of 0.5um. Aluminum was chosen over copper because it can be dry-etched, which makes it easier to manufacture. We found the simulated difference in loss between copper and aluminum was less than 1%. The metal thickness was chosen to make the metal multiple skin-depths thick at our chosen frequency. We chose to use 10 metal layers (in between 11 polyimide layers). We chose a dielectric thickness of 10um per polyimide layer. This ensures that the total lens thickness is roughly 110um, which is approaching the maximum thickness that the manufacturing process can accept.

In step 3, we chose to subdivide our lens into square pixels with side dimension 120um, or one fifth of a free-space 500GHz wavelength. This follows Pisano's recommendation.

In step 4, we differ from Pisano's process. While Pisano uses the Gaussian beam approximation to generate the phase pattern on the lens, we found that this approximation failed to accurately predict our lens focal length when we tested a 20GHz lens prototype [6]. We found the actual focal length was 10% longer than expected. Instead, we use the Gaussian beam approximation [3] to estimate the beam waist of a gaussian beam coupled to the lens from its focal point, based solely on the lens diameter and focal length. Then we use Fourier optics (as described in Ref. [2]) code written in matlab to simulate propagating the beam from the focal point to the lens surface. Then we use the simulated beam phase to give us the desired phase pattern for the lens.

In step 5, we simulated varying sizes of aluminum squares embedded within polyimide. Floquet ports were used to simulate a plane wave entering and then exiting the aluminum square, with periodic boundary conditions. Additional polyimide material (roughly a half wavelength) had to be added on both sides of the element and then de-embedded to ensure accurate results. Full S-parameters from 450-500GHz were simulated for roughly 100 different sizes of squares, ranging from 1% to 99% of a pixel length dimension.

In step 6, we model the S21 of each pixel. Note that each pixel is an alternating stack of 11 dielectric layers and 10 conductive elements. We combine the results simulated in step 5, and a model for the polyimide dielectric ($\varepsilon_r = 3.5$ and $\tan\delta = 0.015$). This polyimide model is taken from THz measurements in Ref. [4] and verified by our own independent measurements of the material. To model a pixel, we convert the S-parameters of the thicknesses of polyimide layers and given sizes of conductive element to ABCD parameters, and then multiply them together, alternating polyimide-conductor-polyimide-conductor, etc. We take the characteristic impedance of the S-parameters to be the impedance of free space, since the lens inputs and outputs will be free space. A full explanation of this technique may be found in section 4.4 of Ref. [5].

Fig. 1. A zoomed-in image of a single layer of the lens manufacturing drawing. Each pink square will be a conductive element embedded in polyimide.

In step 7, we use the pixel model from step 6 in combination with the desired phase pattern found in step 4. The optimizer is given the freedom to alter the size parameter of each conductive element in the pixel. The goal function is created from the S21 of the pixel, as the product of the magnitude of S21 and the cosine of the phase difference between the desired and achieved S21. This ensures that both the transmittance and phase of the pixel are optimized for. Note that the optimization may be subject to additional constraints. In our case, the manufacturer's minimum feature size of 5um limited the largest and smallest possible conductive element sizes that we could use.

Once each pixel in the lens has been optimized, we can convert each layer of the lens to a manufacturing file. One layer of the lens manufacturing drawing is depicted in Fig. 1.

We can also simulate the performance of the lens. In our case, we first use the S21 of the lens to see what a plane wave input to the lens would be transformed to in its output plane. We can then use the Fourier optics code described earlier to propagate that output wave to the lens's focal plane, then use this focal plane information to analyze the lens performance.

III. Simulation Results

The results of simulating a plane wave propagated through the lens can be seen in Fig. 2. As a figure of merit, we use the Strehl ratio of the lens, which is defined as the peak intensity of the actual lens at its focal point divided by the peak intensity of an ideal lens of the same diameter at its focal point [7]. This helps to define the performance of the lens without requiring a specific feed source to couple to. A Strehl ratio of 1 is ideal.

We simulated the Strehl ratio of the lens at 480GHz to be 0.78. The Strehl ratio can be broken down into three inefficiencies: reflective loss, conductive/dielectric loss, and optical loss. We found the reflective efficiency to be 96.5%: that is, 3.5% of power incident on the lens is reflected. We found the dielectric/conductive efficiency to be 82%: that is, 18% of power that is not reflected from the lens is absorbed by the lens as heat. Finally, the optical efficiency of the lens is 98.1%: 1.9% of the power that is output from the lens is scattered due to phase inaccuracy.

Performing these simulations over a wider frequency range, we find that the 3dB bandwidth (i.e. the bandwidth in which the Strehl ratio is at least half its maximum) is roughly 5%. The reflective and dielectric loss stay relatively constant while the optical loss drops quickly away from the center frequency.

IV. Conclusion

We have designed a flat lens that performs well at 480GHz. From our simulations, we can see that our lens has solid performance with a Strehl ratio of 0.78. The breakdown of inefficiencies demonstrate that, at least theoretically, our metamaterial's built-in anti-reflection coating is very effective, resulting in only 3.5% reflective loss. At the design frequency, optical efficiency is also very strong, leaving dielectric and conductive efficiency as the primary sources of loss.

It is possible, if additional performance is necessary, that the lens can be made thinner, or out of different dielectric material, to reduce the dielectric loss. Polyimide has a relatively high loss tangent of 0.015 at our design frequency, while some plastics such as high-density polyethylene or polypropylene have a loss tangent nearly an order of magnitude lower [3]. However, the plastics we can use are limited by the capabilities of our manufacturing partners.

The low frequency bandwidth of the lens is a downside, but not surprising given that the lens was optimized for a single frequency. Additional study will be done to see if we can increase the bandwidth by using wider optimizations without significantly decreasing the center-frequency performance. We are hopeful given successful experiments with this in, e.g., [1], where a roughly 30% fractional bandwidth is obtained.

We have recently received the first batch of manufactured lenses from our manufacturing partners and are currently working to characterize its performance with real measurements.

References

[1] Pisano, Giampaolo, et al., "Dielectrically Embedded Flat Mesh Lens for Millimeter Waves Applications", *Applied Optics*, Apr. 2013.

[2] Davis, Kristina, et al., "Analysis Techniques for Complex Field Radiation Pattern Measurements", *Proceedings of SPIE,* 2018.

[3] Goldsmith, Paul F., "Quasioptical Systems", 1998.

[4] Sahin, Seckin, et al., "Dielectric Properties of Low-Loss Polymers for mmW and THz Applications", *International Journal of Infrared and Millimeter Waves*, Mar. 2019.

[5] Pozar, David M., "Microwave Engineering: Fourth Edition", 2012.

[6] Whitton, Cassandra N., et al. "Measurements and Simulations of a 20 GHz Metamaterial Lens", *31st IEEE International Symposium on Space Terahertz Technology,* Mar. 2020.

[7] Strehl, Karl W. A. "Theory of the telescope due to the diffraction of light", 1894.

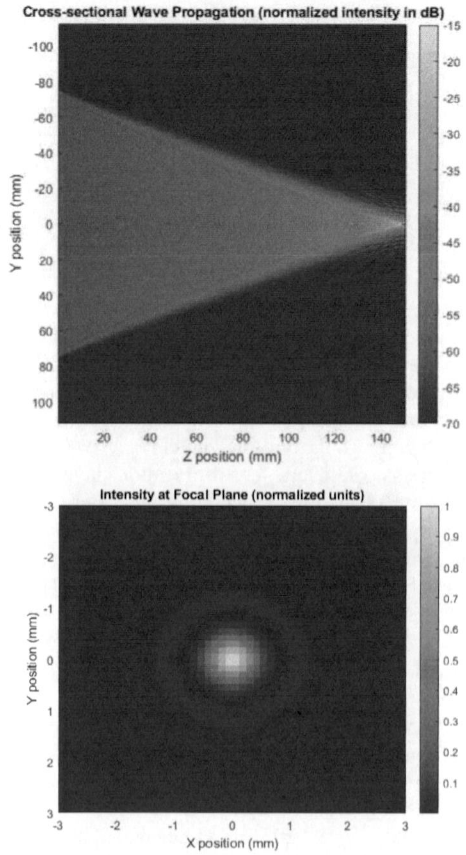

Fig. 2. Top: A cross-section of a Fourier-optics-simulated wave being propagated from the lens output (left) to the focal plane (right). Bottom: The intensity plotted at the lens focal plane, demonstrating an Airy pattern.

A Tunable Linear to Right-Handed Circularly Polarized THz Antenna Based on Graphene Switch

Mohammad M. Fakharian[1], Mohammad Alibakhshikenari[2], and Mariana Dalarsson[3]

[1]Faculty of Engineering, University of Garmsar, Garmsar, Iran

[2]Department of Signal Theory and Communications, Universidad Carlos III de Madrid, 28911 Leganés, Madrid, Spain

[3]School of Electrical Engineering and Computer Science, KTH Royal Institute of Technology, SE 100-44 Stockholm, Sweden

fakharian@fmgarmsar.ac.ir, mohammad.alibakhshikenari@uc3m.es, mardal@kth.se

***Abstract*— We propose a linear to right-handed circularly polarized tunable terahertz antenna. Polarization-switchable graphene is used to enable tunable polarization conversion and circular polarization. Polyimide is used as a substrate material for designing THz components due to its low absorption. By adjusting the graphene chemical potential to be between 0 eV to 0.8 eV, the polarization state in the 0.55 THz band can be obtained without changing the physical geometry. The proposed antenna has a significant potential for use in tunable terahertz devices and related applications.**

Keywords—Terahertz, antenna, tunable, polarization converter, graphene.

I. Introduction

The electromagnetic radiation in frequency range 0.1–10 terahertz (THz) is known as the THz band of the electromagnetic spectrum [1]. Wave manipulation in the THz band is a technology of great significance for research and applications of wireless communication, space exploration and other related fields [2–3]. In recent years, manipulating the polarization state of terahertz waves has also become an important research area in high resolution imaging, terahertz communication, remote sensing and other mid-infrared applications [4–5]. On the other hand, as one of the key THz devices, a polarized tunable antenna can effectively reduce signal loss caused by polarization mismatch and multipath effects, which has attracted increasing attention in recent years. Traditional antennas' polarization transformation methods include birefringent materials, crystals, optical gratings [6], controlling the on-off of a PIN diode [7], and rotating converters to achieve different polarization states [8], which have limitations such as high loss, large volume, and requirements to change the physical structure. Recently, polarization conversion using graphene switches in terahertz band has become a new research hotspot [9].

Graphene is a two-dimensional material built up by carbon atoms in a honeycomb lattice [10]. The Fermi energy levels of graphene can be altered by using chemical doping or electrostatic gating [11]. Such procedures change the electrical conductivity of graphene, which makes graphene a promising candidate for the design of tunable devices. Therefore, graphene has been widely used in micro-nano devices including sensors, absorbers, and antennas [9]. One idea is to combine the antenna with graphene switches to expand the polarization conversion functionality, in order to enable tunable linear polarization and circular polarization characteristics at the same working frequency band.

In this work, a tunable linear to right-handed circularly polarized THz antenna is designed and studied. Graphene is employed as a switch to achieve a reconstruction of the polarization state. Polyimide is selected as a suitable dielectric substrate for broadband THz components due to its low absorption. By adjusting the chemical potential of graphene between 0 eV and 0.8 eV, the working states of the circularly polarized antenna and linearly polarized antenna can be switched for operating frequencies around 1 THz (0.7–0.75 THz) without changing the physical geometry of the antenna. This combination can easily create a multistate THz antenna that can be used for the design of circularly polarized antennas.

II. Conductivity Model Of Graphene

The conductivity of graphene is provided by Kubo's equation [12], which is determined by both intra-band and inter-band transitions.

$$\sigma_S = \sigma_{\text{intra}} + \sigma_{\text{inter}} \tag{1}$$

$$\sigma_{\text{intra}} = -j\frac{e^2 K_B T}{\pi\hbar^2(\omega - j2\Gamma)}\left(\frac{E_F}{k_B T} + 2\ln\left(e^{-E_F/k_B T} + 1\right)\right) \tag{2}$$

$$\sigma_{\text{inter}} = \frac{-je^2}{4\pi\hbar}\ln\left(\frac{2|E_F| - (\omega - j2\Gamma)\hbar}{2|E_F| + (\omega - j2\Gamma)\hbar}\right) \tag{3}$$

In these equations, k_B is Boltzmann's constant, e is the electron charge, E_F is the Fermi energy, $\omega = 2\pi f$ is the angular frequency, and $\hbar = h/(2\pi)$ is the reduced Planck's constant. In the simulations, T is the environmental temperature, which is fixed at 300 K, $G = 1/2\tau$ is the phenomenological scattering rate, where $\tau = 1$ ps is the electron-phonon relaxation time. The main advantage of graphene is that its surface conductivity can be tuned by changing the Fermi energy. By applying a transverse electric field through a bias gated structure, E_F can be adjusted over a wide range of energies (between ±1.0 eV), so the conductivity of graphene can be controlled by a DC bias voltage. An approximate closed-form expression relating the Fermi energy E_F to the bias voltage V_g is given by Ref [13].

$$E_F \approx \hbar v_f \sqrt{\frac{\pi \varepsilon_r \varepsilon_0 V_g}{e t_S}} \tag{4}$$

In Equation (4), v_f is the Fermi velocity, which is fixed at 1.1×10^6 m/s, V_g is the bias voltage, which can be artificially controlled, ε_0 and ε_r are the permittivity of the vacuum and dielectric, respectively, and t_s is the thickness of the insulating spacer. In summary, Equations (1)–(4) provide an effective solution of the equations that dynamically control the polarization state of the terahertz wave by the bias voltage. Actually, by loading the graphene units as a switch on the antenna configuration, the working modes of the designed graphene-based antenna can be tuned by the adjustable surface conductivity of the graphene.

III. StructureOf The Proposed THz Antenna

The configuration of the designed THz antenna is shown in Figure 1, which is printed on a polyimide substrate with relative permittivity 3.5, thickness 6.25μm, tangential loss factor 0.008. It consists of a circular radiating patch as a driven element in the center of the substrate and two parasitic elements around it. In order to increase the capacitive load, the distance between the parasitic elements and the driven element is reduced, and thus the overall radius of the antenna is reduced. The parasitic elements are connected to the driven element through graphene switches. Metal layers in the two sides of graphene can be used to provide a tunable graphene complex conductivity through the electrostatic field effect and produce varying chemical potential on the graphene stubs by applying a gate bias voltage. The modified geometrical parameters of the antenna are shown in Table I. The spectra and the results in the present paper are obtained by using the commercial 3D software CST Microwave Studio, based on the Finite Integration Technique.

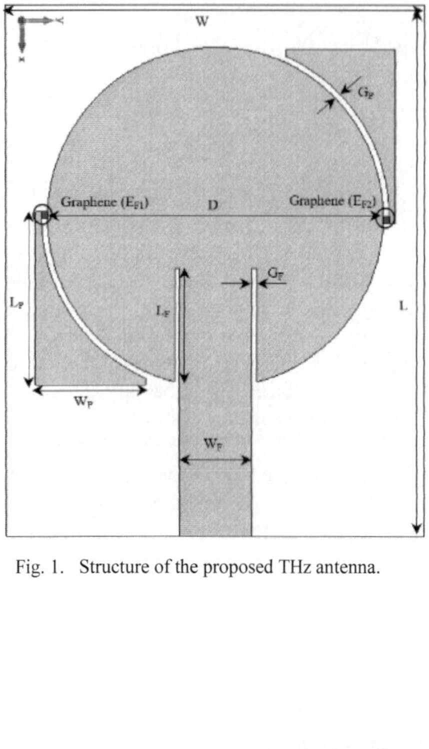

Fig. 1. Structure of the proposed THz antenna.

TABLE I. The Detail Parameters Of The Proposed Antenna.

Parameters	Dimension (μm)	Parameters	Dimension (μm)
W	220	W_F	37
L	273	L_F	65
D	175	G_F	3.2
W_P	57.3	G_P	3.5
L_P	89.5	-	-

IV. Numerical Results and Discussions

The purpose of the proposed design is to control the polarization of a linearly polarized incident wave. The designed structure can convert the polarization from linear to right-hand circular polarization. This feature can be shown by examining the axial ratio (AR) of the wave passing through the structure. AR is the ratio of the cross-polarized wave to the co-polarized wave, so for a perfect circular polarization conversion, the AR should be equal to one. A 3-dB point can be considered as the reference point for polarization conversion. The performance of the proposed antenna has been studied in the range of 0.1– 1.1 THz with the reflection coefficient ($|S_{11}|$), AR, and radiation pattern responses, as shown in Fig. 2. The chemical potential of graphene switches varies from 0 to 0.8 eV and the corresponding characterizes for the linear and polarization agilities are illustrated in Fig. 2(b). For RHCP, impedance bandwidth fully covers the corresponding 3-dB AR bandwidth at the center frequency of 0.55 THz. The minor polarization Z <0 is the LHCP and the major is the RHCP for Z >0 for the proposed antenna. Moreover, the antenna can be operated as a multi-characterized antenna with different radiation patterns at the same operating frequency, as verified in [7]. It is however not studied in this work. The gain attained by the proposed antenna is more than 0 dB with an efficiency of 40% at different modes.

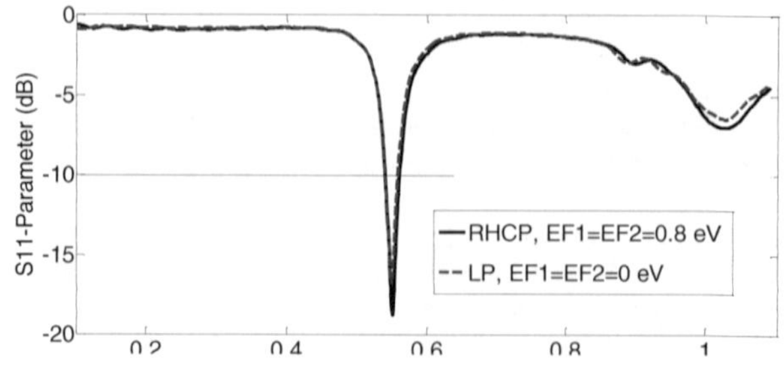

(a)

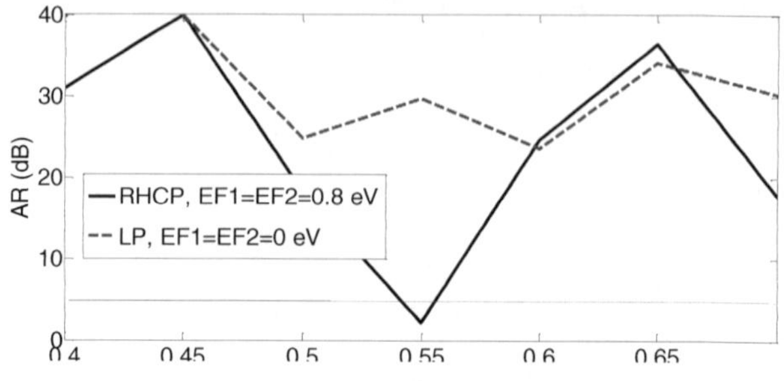

(b)

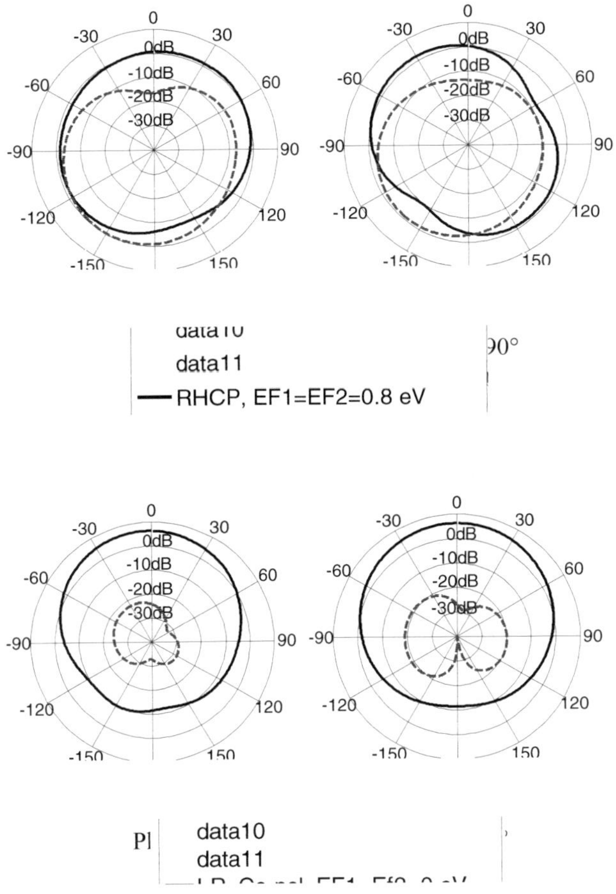

(c)

Fig. 2. (a) $|S_{11}|$, (b) AR, and (c) radiation patterns of the THz antenna at different chemical potentials, left- and right-sides represent Phi=0° and 90°.

V. Conclusion

In summary, a linear to right-handed circularly polarized tunable THz antenna is designed and studied. The employment of graphene switches enables tunable polarization conversion and circular polarization. By electrically shifting the Fermi energy of two graphene switches, the polarization state of the band can be changed without changing the physical geometry. The proposed design has a simple geometry, and gives a strong conversion from linearly to circularly polarized waves at around of 0.55 THz and it can be adjusted to any desired frequency range.

Acknowledgment

This project has received funding from Universidad Carlos III de Madrid and the European Union's Horizon 2020 research and innovation programme under the Marie Sklodowska-Curie Grant 801538.

References

[1] S. Poorgholam-Khanjari, F.B. Zarrabi, "Reconfigurable vivaldi THz antenna based on graphene load as hyperbolic metamaterial for skin cancer spectroscopy," Opt. Comm. Vol. 480, pp. 126482, 2021.

[2] F. Zangeneh-Nejad and R. Safian, "A graphene-based THz ring resonator for label-free sensing," IEEE Sens. J., vol. 16, no. 11, pp. 4338–4344, 2016.

[3] J. P. Fan and Y. Z. Cheng, "Broadband high-efficiency cross-polarization conversion and multi-functional wavefront manipulation based on chiral structure metasurface for terahertz wave," J. Phys. D: Appl. Phys., vol. 53, no. 2, pp. 025109, 2020.

[4] I. A. Gayduchenko, G. E. Fedorov, M. V. Moskotin, D. I. Yagodkin, S. V. Seliverstov, G. N. Goltsman, A. Yu Kuntsevich, M. G. Rybin, E. D. Obraztsova, V. G. Leiman, M. S. Shur, T. Otsuji, and V. I. Ryzhii, "Manifestation of plasmonic response in the detection of sub-terahertz radiation by graphene-based devices," Nanotechnology, vol. 29, no. 24, pp. 245204, 2018.

[5] C. P. Huang, Q. J. Wang, X. G. Yin, Y. Zhang, J. Q. Li, and Y. Y. Zhu, "Break through the limitation of Malus' Law with plasmonic polarizers," Adv. Opt. Mater., vol. 2, no. 8, pp. 723–728, 2014.

[6] H.Lajunen, J.Turunen, J.Tervo, Design of polarization gratings for broadband illumination. Opt. Express, vol. 13, pp. 3055–3067, 2005.

[7] M. M. Fakharian, P. Rezaei, A. A. Orouji "Polarization and radiation pattern reconfigurability of a planar monopole-fed loop antenna for GPS application," Radioengineering, vol. 25, no. 4, pp. 680-686, 2016.

[8] Zhu, H.L.; Cheung, S.W.; Liu, X.H.; Yuk, T.I. Design of polarization reconfigurable antenna using metasurface. IEEE Trans. Antennas Propag., vol. 62, pp. 2891–2898, 2014.

[9] M. M. Fakharian "A graphene-based multi-functional terahertz antenna," Optik, vol. 251, pp. 168431, February 2022.

[10] K.S.Novoselov,A.K.Geim,S.V.Morozov, D.Jiang, Y.Zhang,; S.V.Dubonos,; I.V. Grigorieva, A.A. Firsov, "Electric field effect in atomically thin carbon films," Science, vol. 306, pp. 666–669, 2004

[11] I. Ahmad, S. Ullah, S. Ullah, U. Habib, S. Ahmad, A. Ghaffar,M. Alibakhshikenari, S. Khan, E. Limiti, "Design and analysis of a photonic crystal based planar antenna for THz applications," Electronics, vol. 10, no. 16, pp. 1941, 2021.

[12] G.W. Hanson, "Dyadic Green's functions and guided surface waves for a surface conductivity model of graphene," J. Appl. Phys.,vol. 103, pp. 64302, 2008.

[13] J.S. Gómez-Díaz, J. Perruisseau-Carrier, "Graphene-based plasmonic switches at near infrared frequencies," Opt. Express, vol 21, pp. 15490–15504, 2013.

A broadband and dual-polarization single-layer dichroic filter for applications in Sub-THz range

D. Montofré[1], I. Lapkin[1], D. Meledin[*,1], V. Desmaris[1], L. Helldner[1], M. Fredrixon[1], C. Lopez[1] and V. Belitsky[1]

***Abstract*— In this work we report the design of a single-layer all-metall dichroic filter with an improved spectral response at non-normal beam incidence and nearly equal performance for both polarizations. The dichroic is intended to be employed for dual-frequency receivers. Since the dichroic is produced purely from metal it facilitates its use at cryogenic temperatures. Therefore, the contribution to the system noise is minimized. The dichroic design concept demonstrated a measured transmission of electromagnetic radiation of 85-90% for both polarizations in the range of 35-50 GHz (37% fractional bandwidth). The measurements were performed at room temperature. The spectral properties of the dichroic have been optimized by modeling in 3D FEM simulation software.**

***Keywords*— dichroic, filters, mm-wave**

I. Introduction

Frequency-selective surfaces (FSSs) are widely used as dichroic quasi-optical filters, which allow either dividing a beam into two based on the frequency difference or combining beams from different sources generating different frequencies into a single beam for further processing. The idea of spacially dividing a beam is commonly found in astronomical applications since it has enabled the deployment of multi-band receivers for radio astronomy, multi-pixel systems for current and future facilities on Earth, and stratospheric and space astronomy missions [1,2]. There are numerous benefits of multi-pixel and multi-frequency receiver systems since they allow for increasing the mapping speed and spectral line surveys. Additionally, multi-frequency enhances the phase calibration in high-frequency interferometric observations. Furthermore, the multi-frequency receiver systems have the potential to provide unique information for future Very Large Baseline Interferometry (VLBI) observations [3]. All above, along with the continuous demand for improved instrument sensitivity, explains the increasing interest in developing state-of-the-art dichroic filters.

Traditionally, a dichroic filter employs a sequence of several patterned metallic and dielectric layers, making the devices inherently sensitive to the control of linear dimensions of the pattern elements of each layer and the layers' separation [4]. In fact, the geometrical properties such as the patterns' shapes and sizes, their periodicity, and the thickness of the metal and dielectric layers determine the frequency response of any dichroic. Therefore, complicated configurations are utilized for improving different relevant aspects, such as transmission stability, cross-polarization levels, increasing the bandwidth, or reducing the angular degradation [5].

In this work, we present a novel design and technologically simpler concept for dichroic filters using a single metallic layer. The plate is perforated at an angle of θ=13 degrees with the perforation shape optimized to reduce the insertion loss, frequency response, dual-polarization capability, and operational bandwidth.

Fig. 1 Dichroic array showing an hexagonal layout.

II. Design, simulations, and characterization

We have chosen to confirm the design feasibility and fabrication of the filter in the Q-band (35-50 GHz) as it could be further used in a VLBI Tri-band receiver planned for Onsala Space Observatory 20 m antenna. However, since the single-layer metal structure is used, the FSS filter is perfectly scalable to higher frequency bands and fabrication with, e.g. microfabrication techniques [6]. The proposed design is a single-layer FSS based on a metallic plate with tilted perforations of customized geometry as shown in Fig1. The unit hexagonal cells are depicted in Fig. 1 insert. The spectral response of a dichroic filter is determined by its geometrical parameters, i.e., the shape and size of the apertures, the aperture spacing, the thickness of the metal layer, and how the layers are arranged (in the case of multi-layer systems). The transmissivity and reflectivity of a dichroic filter also depend on the polarization orientation (horizontal or vertical) of the incoming electromagnetic wave and the angle of incidence, θ (AOI). The latter at non-normal beam incidence (e.g. $\theta \neq 0$), affects the dichroic performance in a form known as angular degradation. Typically, this degradation causes a reduction of the RF bandwidth, shifting in the frequency cut-off, resonance spikes in the pass-band, and increased levels of cross-polarization [7]. The AOI of θ = 13 degrees was aimed to be acceptable for the telescope optical systems as well as because of feasibility for the filter's manufacturing that is critical for achieving desired performances. A

[1] Group for Advanced Receiver Development, Chalmers University of Technology, SE412 96, Gothenburg, Sweden.

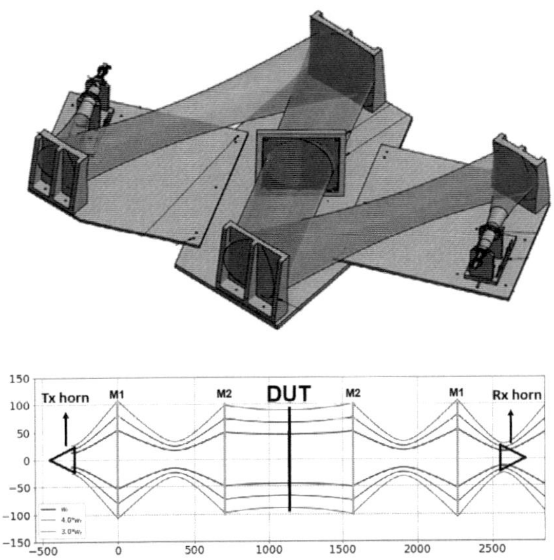

Fig.2. Designed optical system to characerize the dichroic filter. (Top) 3D sketch of the optical assembly, the blue-shaded volume corresponds to the calculated quasi-optical beam. (Bottom) Optical train calculation for a beam at 42.5 GHz..

hexagonal units cell array was chosen for its higher mechanical stability, however, at the expense of having a somewhat reduced bandwidth.

Simulations of the dichroic filter and its optimization were performed using HFSS simulation software. Measurements of transmissivity of the fabricated filters as a function of frequency were carried out using a quasi-optical setup shown in Fig.2. Two spline-profile horns optimized for the Q-band frequency range are connected to a VNA and used as a transmitter (Tx) and a receiver (Rx) for the optical system consisting of 4 active mirrors. During the measurements dichroic filter (DUT) was placed across the collimated beam between the two M2 mirrors (see Fig. 2). The polarizations grids in front of the horns were implemented to improve the quality of the cross-polarization measurements.

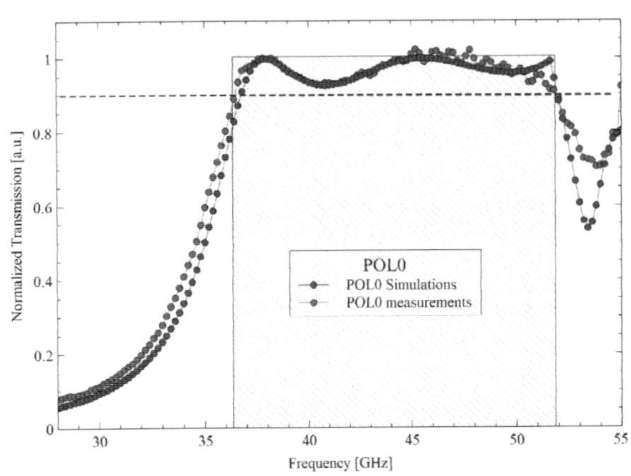

Fig.3 Simulated and measurement performance of the designed dichroic filter at the AOI of 13 degrees for one of polarization with *horizontally* oriented E–field (POL0). Red and blue color circles corresond to transmission .The partly filled region illustrates the frequency range between 37 and 52 GHz where the transmittion is over 90% marked with a blue dashed horizontal line.

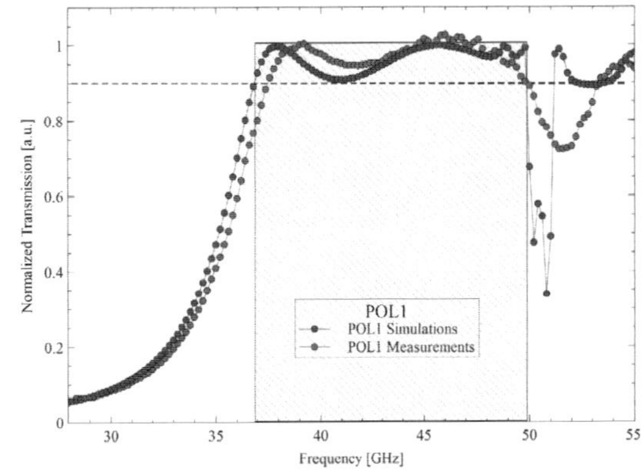

Fig.4 Simulated and measurement performance of the designed dichroic filter at the AOI of 13 degrees for one of polarization with *vertical* oriented E–field (POL1). Red and blue color circles corresond to transmission .The partly filled region illustrates the frequency range between 37 and 50 GHz where the transmittion is over 90% marked with a blue dashed horizontal line.

The simulated and measured performances of the designed dichroic filter at θ=13 degrees are presented in Fig.3 for POL0 and Fig.4 for POL1. Red and blue color circles correspond to co-polar transmission, and the cross-hatched region indicates a frequency range where the transmission is over 90%, the level marked with a blue dashed horizontal line. As could be seen, the results of the measurements are in good agreement with the simulated data, especially for POL0, demonstrating a fractional bandwidth of 37%.

III. Conclusion

We have proposed a novel design for the dichroic filter based on a perforated metal plate. The proposed design can reach 90% of the transmission in at least 37% of the RF bandwidth around 42 GHz central frequency. The measured performance showed a very good agreement with the simulations for both polarization. The presented design is scaleable to higher frequencies and could be employed as a cold dichroic filter for providing simultaneous operation at higher frequencies, for instance, e.g. 230 and 345 GHz channels of the Event Horizont Telescope.

IV. References

[1] I. Mehdi, et al., "Far-infrared heterodyne array receivers", *Bulletin of the American Astronomical Society*, vol. 51, no. 7, 2019.

[2] M. C. Wiedner et al., "A Proposed Heterodyne Receiver for the Origins Space Telescope," *IEEE Transactions on Terahertz Science and Technology*, vol. 8, no. 6, pp. 558-571, Nov. 2018.

[3] M. J. Rioja et al. "The Power of Simultaneous Multifrequency Observations for mm-VLBI: Astrometry up to 130 GHz with the KVN", *The Astronomical Journal*, 150.6 (2015), p. 202.

[4] Xiaolong You, Christophe Fumeaux and Withawat Withayachumnankul, "Tutorial on broadband transmissive metasurfaces for wavefront and polarization control of terahertz waves", J. Appl. Phys. 131, 061101 (2022); https://doi.org/10.1063/5.0077652.

[5] D. Montofre, et al., "Single-Layer Dichroic Filters for Multifrequency Receivers at THz Frequencies," *IEEE Transactions on Terahertz Science and Technology*, vol. 10, no. 6, pp. 690-697, Nov. 2020.

[6] V. Desmaris, D. Meledin, A. Pavolotsky, R. Monje, and V. Belitsky, “All-metal micromachining for the fabrication of sub-millimetre and THz waveguide components and circuits,” *J. Micromech. Microeng.*, vol. 18, 2008, Art. ID 095004 (6 pp).

[7] Chao-Chun Chen, "Transmission of Microwave Through Perforated Flat Plates of Finite Thickness," *IEEE Transactions on Microwave Theory and Techniques*, vol. 21, no. 1, pp. 1-6, Jan. 1973, doi: 10.1109/TMTT.1973.1127906.

Stepped Impedance Metal-Mesh Filters for Terahertz Frequencies

Adhitya B. Sriram[1,2], Cassandra Whitton[1], Yiran Cui[2], Christopher Groppi[1], Georgios Trichopoulos[2] and Philip Mauskopf[1]

***Abstract*—Square metal grids embedded in low-loss dielectrics have been studied and used extensively as quasi-optical (i.e. multi-mode and coupled to free space) low pass filters in ground and space-based astronomical instrumentation applications. Multilayer designs incorporating grids separated by dielectric spacers have been modeled as capacitive shunts or open circuit resonant stubs separated by ideal transmission lines. Typical low pass filter designs with quarter wavelength transmission line spacing to transform alternating capacitive shunts into inductive shunts at the cutoff frequency have been found to have leaks at approximately twice the stub's resonance frequency. Stepped impedance structures manufactured using multilayer dielectric films are one of the most commonly encountered structures for optical wavelengths [1]-[3] and also for microwave filters. In this poster, we describe the design of stepped impedance metal mesh quasi-optical filters using artificial dielectrics for THz frequencies.**

Keywords—Stepped Impedance filters, Metal-Mesh filters, Terahertz.

I. Introduction

In this poster, we discuss a prototype design for a stepped impedance low-pass filter with a wide passband. Stepped impedance filter designs are the most prevalent designs for filters implemented in microstrip where wideband performance is required. Here we demonstrate the simulated performance a low-pass filter where the stepped impedance sections are designed as artificial dielectric layers using square metal grids embedded in polypropylene that virtually behave like an ideal transmission line section with a specific impedance and phase length. The impedance of an array of square metal grids can be calculated as [4]

$$\frac{X_c}{Z} = \frac{1}{n^2}\left[\frac{4g}{\lambda}\ln\left(cosec\left(\frac{\pi a}{g}\right)\right)\right]^{-1}$$

Where n is the refractive index of the artificial dielectric material, a is the gap between two adjacent metal grids and g is the period of the adjacent metal grids. The impedance of capacitive metal grid is completely described by the refractive index, geometrical ratio (a/g) and the wavelength of interest [4]. The effective impedance of the grid can be calculated as,

$$Z_{eff} = \sqrt{\frac{\mu_0}{\varepsilon_0\varepsilon_r + \left(\frac{C_{grid}}{l}\right)}}$$

Fig.2 shows the simulated performance of the prototype stepped impedance low-pass filter.

[1]School of Earth and Space Exploration, Arizona State University, Tempe, 85281, USA ; [2] School of Electrical Engineering, Arizona State Univeristy, Tempe, 85281, USA;

II. Results

The prototype stepped impedance low-pass filter was designed to have a pass-band cutoff at 175 GHz. The stop-band response seen below from the simulated design does not have leakages at twice the desired center frequency.

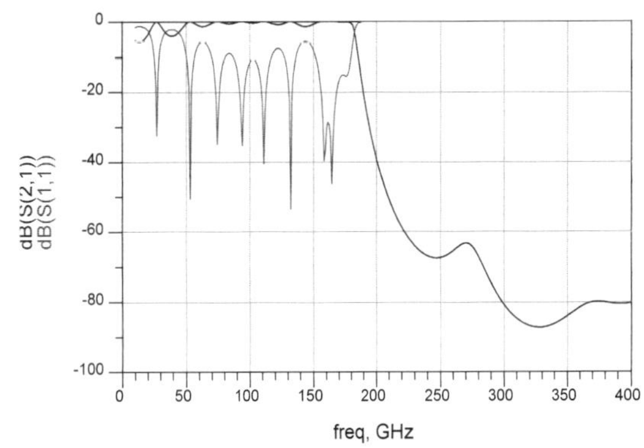

Fig.1 : Simulated Transmission and Reflection Performance of the prototype stepped imedance filter design

The use of a low refractive index anti-reflection coating on the outer sections of the filter is found to further improve the matching and suppress reflections in the passband.

As the next step, this prototype design is planned to be fabricated by spin coating layers of dielectric with embedded square metal grids as opposed to hot-pressing of sections of dielectric. The fabricated filter structure would subsequently be measured to validate the simulated results presented here and would be presented in a future article.

References

[1] Massaro, Alessandro & Pierantoni, Luca & Ciandrini, C. & Rozzi, Tullio. (2005). Analysis of Sensitivity for Low-Pass Multilayer Optical Filters.

[2] Takashi Sato, Takashi Sasaki, Kanji Tsuchida, Kazuo Shiraishi, and Shojiro Kawakami, "Scattering mechanism and reduction of insertion losses in a laminated polarization splitter," Appl. Opt. 33, 6925-6934 (1994).

[3] Rong-Chung Tyan, Pang-Chen Sun, Axel Scherer, and Yeshayahu Fainman, "Polarizing beam splitter based on the anisotropic spectral reflectivity characteristic of form-birefringent multilayer gratings," Opt. Lett. 21, 761-763 (1996).

[4] R. Ulrich, Far-infrared properties of metallic mesh and its complementary structure, Infrared Physics, Volume 7, Issue 1, 1967, Pages 37-55, ISSN 0020-0891, https://doi.org/10.1016/0020-0891(67)90028-0.

A Compact Dielectrically Loaded Quad-ridge Feed Horn for Octave Band Radio Astronomy Application

Sara Salem Hesari*[1], Bruce Veidt[1], Ivan Wevers[1]
[1] National Research Council Canada, Herzberg Astronomy and Astrophysics Research Centre, Victoria, BC, Canada

***Abstract*—This work presents the design and simulation of an octave band quad-ridge feed horn with a dielectric rod installed in the center of the throat to improve the beamwidth, phase stability, and polarization efficiency, which leads to receiving higher aperture efficiency. The proposed dielectrically loaded quad ridge antenna is designed to meet the next generation Very Large Array (ngVLA) optics. This antenna covers a fractional bandwidth of 66% from 25 to 50 GHz with a return loss better than 20 dB and aperture efficiency higher than 91% over the entire operating frequency range.**

***Keywords*— feed horn, quad-ridge flared horn, octave-band horn, ngVLA, high-efficiency reflector antenna**

I. Introduction

Quad-ridge feed horns are an excellent choice for applications that require a feed horn with equal or greater than 2:1 relative bandwidth. Various quad-ridge feed horns have been designed and proposed over the years. Although they are great candidates for ultra-wideband applications, they suffer from asymmetric beamwidth and lower efficiencies than corrugated feed horns, and that is where the dielectrically loaded quad-ridge antenna steps in for performance improvement.

Recently, octave band feed horns are becoming very popular in the radio astronomy field since they can cover a large continuous bandwidth, consequently reducing the number of feed horns required to cover a large frequency band. For large telescopes like Next Generation very large array with an array of 244 antennas each 18m in diameter, supplemented with a short baseline array of 19 antennas each 6m in diameter dishes covering the frequency range of 1.2 to 116 GHz [1] using octave or multi-octave band systems improve the time and cost-effectiveness by making the maintenance cost in the long term more economical than standard waveguide bandwidth.

This paper proposes a dielectrically loaded quad-ridge feed horn designed for the ngVLA optics; a shaped offset Gregorian reflector optimized to receive a feed with a half angle of 55° and edge taper of 16 dB on the secondary. The primary has a projected aperture of 18 m [1].

II. Results

This novel design has excellent performance in frequency bandwidth ratio of 2:1 from 25 to 50 GHz:

- Excellent input return loss across the band >20 dB,
- Stable and high phase efficiency,
- High polarization efficiency ,
- Excellent aperture efficiency > 91%,
- Symmetric radiation patterns in E-, H-, and D-planes,

The feed performance makes it a suitable candidate for radio astronomy applications which are seeking high sensitivity. The proposed antenna consists of three main parts, a quad ridge input waveguide, transition from quad-ridge to open space, and a uniquely designed dielectric rod for performance improvements.

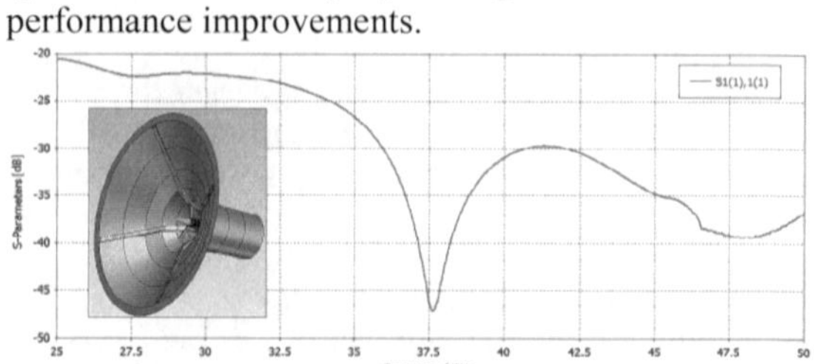

Fig. 1. Feed horn reflection coefficient with a rendering of the horn shown in the inset.

The structure of the compact dielectrically loaded quad-ridge feed horn for the next generation of the ngVLA telescope is shown in Fig. 1. This feed provides a reflection loss of better than 20dB over the band. The ngVLA MTEX optical model is used for the optical analysis using TICRA GRASP. Based on this analysis, the minimum aperture efficiency is 91.3% at 25GHz and the maximum is 96% at 47.5 GHz. The spillover efficiency is better than 93.6% over the entire operating bandwidth.

References

[1] R. Selina et al., "System Reference Design," Nat. Radio Astron. Observatory, NRAO Doc. 020.10.20.00.00-0001-REP-B SYSTEM_REFERENCE_DESIGN, ngVLA Reference Design, Vol. 1, Jul. 2019.

[1]First Affiliation, City, Postal Code, Country; [2]Second Affiliation, City, Postal Code, Country; [3]Third Affiliation, City, Postal Code, Country; [4]Fourth Affiliation, City, Postal Code, Country

NOTES:

A satellite tracking system at 78GHz using the over-moded TE21 ground-station antenna pattern

Hugh Gibson*[1], Ralf Henneberger[2], Axel Tessmann[3], Ingmar Kallfass[4], Laura Manoliu[4] and Benjamin Schoch[4]

***Abstract*—Low power cube-satellites are becoming increasingly popular for earth observations and fast data relay. Their low earth orbit means the available time to download data is only about 10 minutes per fly-by. High frequency (71 to 76 and 81 to 86GHz) radio links have the wide bandwidth required for fast data-transfer and very large antenna gain for a modest diameter (meaning low power can be used) but tracking the fast-moving satellite with a high-gain beam is a problem, requiring a pointing accuracy similar to radio telescopes, but with tracking speed up to 3 degrees a second. This paper outlines a method of automatic servo-tracking of the ground-station antenna to the fast-moving CubeSat (once signal has been acquired) using just the over-moded ground-station linear-polarization TE21 mode to obtain Azimuth/Elevation tracking information from the circular-polarized data signal.**

Keywords—CubeSat, circular-polarization, E-band, monopulse-tracking, TE21, wideband satcom

I. Introduction

Using the over-moded TE21 or TM01 mode of a waveguide-fed antenna to obtain tracking information has been discussed before for low frequencies (around 10GHz) but as far as we are aware, has not been tried at significantly higher frequencies where very high gain antennas with modest diameters are possible as well as very wide signal bandwidth (5GHz). We present a system which couples the fundamental mode TE11 circular-polarization data channel (using a septum polarizer), but also uses a linear-polarization TE21 mode (with 4 off-axis lobes and a deep central null) to additionally sample the circular-polarized TE11 signal when the antenna is slightly off-axis for tracking. This type of system is referred to as "amplitude and phase mono-pulse" from historical use in radar, but in this instance, we are guiding the ground station antenna to a circular-polarized signal being transmitted by the satellite. Traditional dual-waveguide-mode coupling systems are very complex, requiring ultra-precision small component fabrication. This is not possible at higher frequencies so it was necessary to construct a significantly simpler TE21 mode coupler but preserving performance. We also investigated the use of easier-to-manufacture smooth-wall spline horns and corrugated feed horns to support the TE21 mode for use with these systems as well as the main TE11 circular mode.

The beauty of the system, using a circular polarized input beam is that the four TE21 linear-polarized lobes have unique phase characteristics that can be separated: opposite lobes are 180-degree shifted in phase, and (due to the propagating nature of the circular polarization wave) the orthogonal axis is automatically in phase-quadrature. This is a characteristic of intercepting the circular-polarized TE11 main beam with the linear polarized TE21-lobes and we believe this beneficial property has not been fully exploited. If our simulations are correct and the waveguide-lengths are correctly calibrated, we require only ONE additional down-converting mixer for the TE21 tracking mode. AZ/EL steering data can all be unscrambled from the phase/amplitude comparison with the TE11 mixer down-converting the fundamental circular-polarization mode (the mixers share a common local oscillator and the relative waveguide paths are matched). The final phase comparison is done at intermediate frequencies, leading to a relatively simple set-up for the amplitude/phase comparison and AZ/EL steering. Many papers show that the TE21 method provides a linear-enough error signal, to allow servo-tracking of the beam. We also investigated TM01 tracking possibilities which would also work in this way, and have similarly been used before but the TM01 mode converter is harder to implement, and the beam pattern is an annulus rather than 4 beams, so the gain is lower.

We will explain the TE21/TE11 converter and simulations using a variety of horn antennas, with beam patterns and phase plots, and hopefully full measurements.

This design is scalable to higher frequencies as it is fabricated using split block sections and avoids small parts or tight tolerances. Useable frequency bandwidth is about 4%.

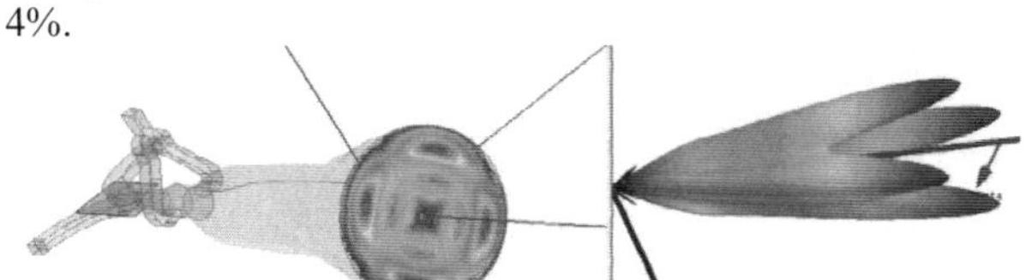

HFSS Simulation of the TE21 mode with a smooth-wall spline horn

Acknowledgement: This work was performed under the project EIVE, supported by the German Federal Ministry of Economics and Technology (BMWi) and the German Aerospace Center (DLR) under grant number 50RK1960

[1]GMD EURL, Antony, 92160, France; [2]RPG Radiometer Physics, Meckenhaim, 53440, Germany; [3]Fraunhofer-Institut für Angewandte Festkörperphysik (IAF) 79108 Freiburg Germany; [4]Institute of Robust Power Semiconductor Systems (ILH) University of Stuttgart, Germany

NOTES:

Diffraction efficiency of reflective metallic gratings operating in the THz range

M. Fernández[1], M.C. Torquemada[1], G. García-Lozano[1], G. Mercant[1], L.M. Sanchez-Brea[2], A. Cuadrado[3], J. Alda[4], L. González[1], T. Belenguer[1]

***Abstract*— Far IR spectrometers have a prominent role to play in future space IR missions devoted to unveiling the obscure universe. Diffraction gratings are the heart of these instruments that, when operating in the range of a few THz, shall cover a wide spectral range, provide the highest resolution and maximize the efficiency, generally for only a single polarization. Therefore, the diffraction grating design and optimization become a determinant exercise where the main grating parameters shall be univocally established and ensured for maximizing the diffraction efficiency for the complete spectral range. This paper describes the optimization of the reflective diffraction efficiency for order *m* = 1 by comparing two different methods. Since the grating period is of the same order of magnitude as the wavelength, it is necessary to solve Maxwell's equations rigorously to obtain the best diffraction efficiencies. Results obtained with the methods based on RCWA (Rigorous Coupled-Wave Analysis) and FEM (Finite Element Method) are compared. Manufacturing errors are a severe limiting factor for performance, so their identification, parameterization, and modeling must also be considered.**

Keywords— Terahertz, Far-IR optics, IR spectrometer, diffraction gratings, diffraction efficiency.

I. Introduction

This work presents the design process of a diffraction grating for a THz spectrometer for space applications operating in the spectral band from 70 to 114 microns. The design of gratings for a THz spectrometer uses the equations of the blazed diffraction grating. Blazed gratings are used to enhance the amount of radiation that is forwarded toward a selected diffraction order. The grating profile was optimized in this case to guide the light toward the $m = 1$ diffraction order. The relation between the angle of incidence, ε, the grating period, d, the diffraction angle θ, and the wavelength of the incident light, λ, is given by the grating equation:

$$d(\sin\varepsilon + \sin\theta) = m\,\lambda\,,$$

where m is the diffraction order [1].

In this case, the efficiency obtained for the ideal geometry of a sawtooth profile for a plane wave TM (transverse magnetic mode) at a fixed incidence angle is included. Two modelization different methods were used, the RCWA and a FEM-based model, analyzing similarities and discrepancies in the diffraction efficiency response [2]. Additionally, the diffraction grating has been fabricated, and morphologically characterized, and its behavior modelized to determine the impact of the manufacturing errors on the diffraction grating performances.

II. Results

A. Diffraction grating design

In the design of a diffraction grating, three fundamental parameters must be considered: period, morphological profile, and spectral range. These parameters have been individually optimized to achieve a diffraction grating with a maximum diffraction efficiency for TM polarization and m=1 order, while the other orders are minimized. In this contribution, FEM is applied using COMSOL-Multiphysics®, and RCWA is realized using GSolver® (Grating Solver Development Company, UT, USA) and S^4 (Stanford Stratified Structure Solver, Stanford University) programs. These approaches solve the Maxwell-Equations for periodic structures quite efficiently. FEM requires an appropriate mesh of the calculation domains and customization of the boundary conditions. Even though RCWA is a mesh-free method, it also needs a fine-tuning of some internal parameters of the calculation (number of internal orders, number of slices, etc.). Fig. 1 shows that under the same conditions, similar efficiencies are observed. Those obtained with the FEM method are higher than those calculated with RCWA. The discrepancies can be attributed to the fact that RCWA generates an inter-order energy distribution that depends on the definition of the internal computational parameters, while FEM only depends on the meshing.

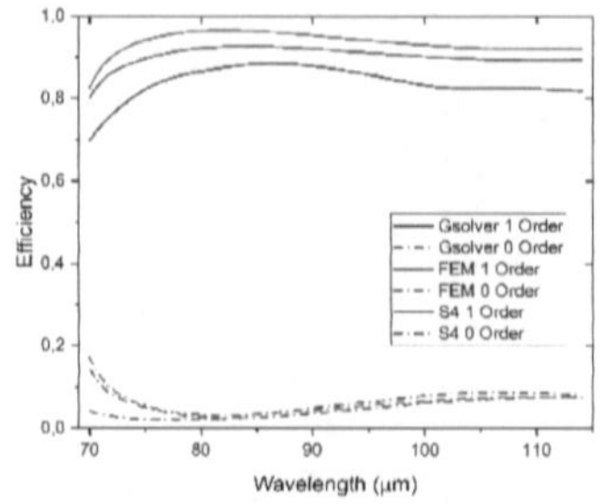

Fig. 1: Comparison of diffractive efficiency obtained by RCWA and FEM methods. m=1 and m=0 orders are shown.

Regarding the diffraction grating efficiency, efforts have been focused on optimizing the period and the blaze angle. As can be seen in Fig. 2, a change in the thickness of the profile produces a variation in the blaze angle. The effects of modifying these variables have been studied by RCWA. The slant angle, the deviation from 90° of the sawtooth profile, was analyzed as an unavoidable manufacturing defect.

[1] Departamento de Óptica Espacial. Instituto Nacional de Técnica Aeroespacial (INTA), Carretera de Ajalvir Km. 4, Torrejón de Ardoz, Madrid, 28850, Spain; [2] Facultad de Ciencias Físicas. Universidad Complutense de Madrid, 28040, Madrid, Spain; [3]Escuela de Ciencias Experimentales y Tecnología, Universidad Rey Juan Carlos, 28933, Madrid, Spain. [4] Facultad de Óptica y Optometría, Universidad Complutense de Madrid, 28037, Madrid, Spain.

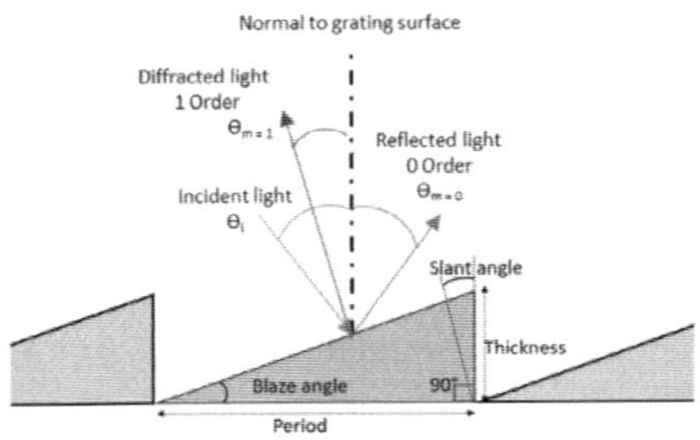

Fig. 2: Sawtooth profile of the blazed diffraction grating.

a) Period: The simulation starts from a period value centered in the wavelength range getting for that period multiple diffraction orders. Cut-off anomalies appear when energy redistributes due to the disappearance of diffraction orders. As can be seen in Fig. 3, this effect can be eliminated by reducing the value of the period. In this case, lowering the period to 76 μm gives a better response for efficiency.

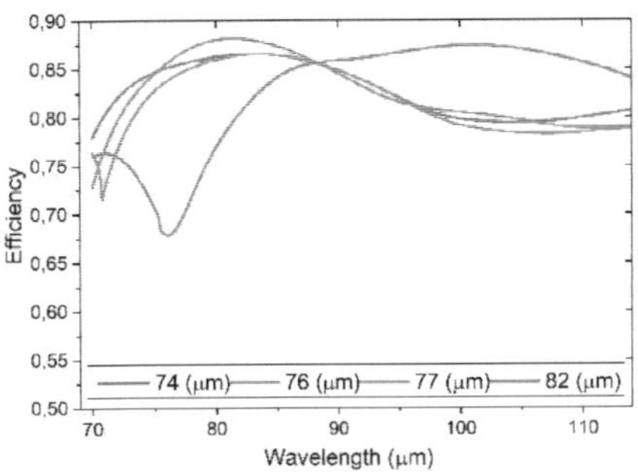

Fig. 3: Diffractive efficiency at different period values.

b) Blaze angle: An increase in this angle produces an increase in the efficiency at longer wavelengths and a decrease in the efficiency at shorter ones as can be observed in Fig. 4.

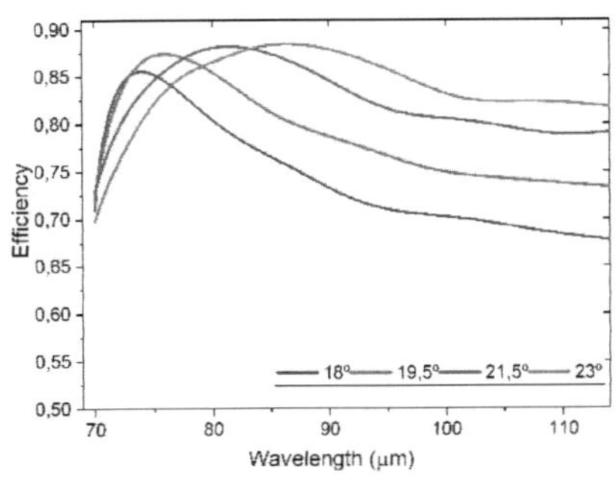

Fig. 4: Diffractive efficiency at different values of blaze angle.

c) Slant angle: When calculating the behavior of the grating as a function of the slant angle from 0 to 30°, for a fixed value of the blaze angle, 24°, and a period a 76μm, we find that the efficiency calculated by the FEM method remains above 90% in the whole spectral range. So, adding a slant angle at the sawtooth profile enhances the diffraction efficiency. The green curve labeled as "Nominal" in Fig. 6 shows the diffraction efficiency for a slant angle of 10°.

B. *Diffraction efficiency analysis of manufactured gratings*

The manufacturing process distorts the geometry of the theoretical design and generates unavoidable discrepancies between the ideal profile and the actual one. In our case, a blazed diffraction grating with sawtooth-shaped geometry and a total area of 54×40 mm^2 has been manufactured by micromachining on a metallic substrate, specifically aluminum 6061T651, which is the most suitable material for space applications due to its mechanical and thermal properties. Fig. 5 shows an image taken by confocal microscopy. It can be observed that the sawtooth profile is not like the ideal one that was modeled and contains fabrication defects.

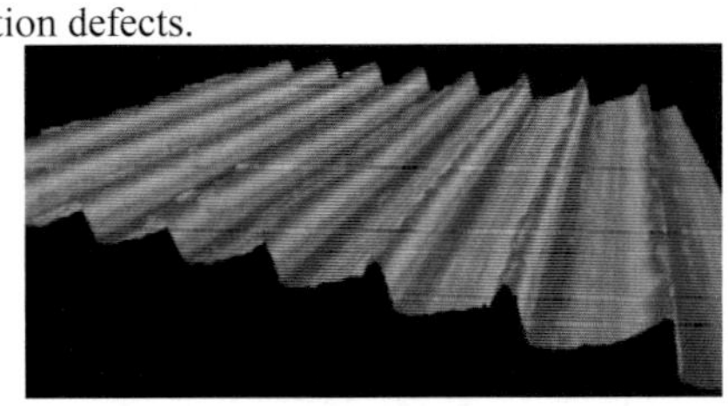

Fig. 5: Confocal image of the micromachined diffraction grating.

The results from the topographic measurements have been used to evaluate the response of the fabricated grating. Fig. 6 shows the diffractive efficiency for the real manufactured profile (red line), and the response to the nominal theoretical ideal profile (green plot). Even though the fabricated topography largely departs from the nominal, the calculated efficiency is above 65% for the whole spectral range.

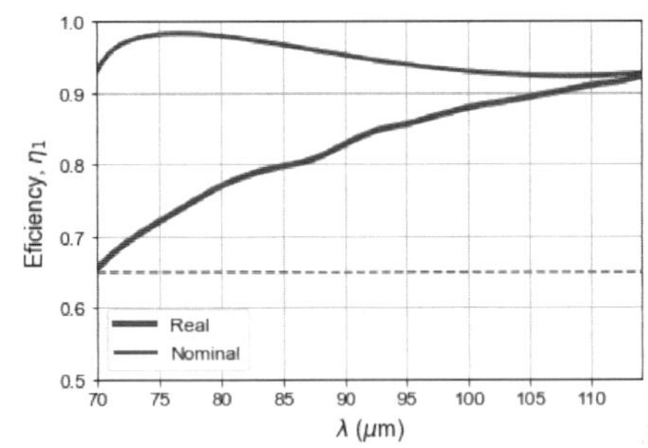

Fig. 6: Diffractive efficiency calculated for the fabricated grating (red) and the nominal profile (green)

III. CONCLUSIONS

A detailed numerical analysis has been made using two main computational electromagnetism methods: the RCWA and the FEM. Both of them produce similar values of efficiency and can be used to optimize the performance of the grating by changing the geometrical parameters. After considering all these issues in the design, we have fabricated an aluminum blazed grating with a period of 76 μm, a blazed angle of 24°, and a slant angle of around 10°. This grating has been fabricated by micromachining. The actual geometry has been also included in the simulations and the diffraction efficiency has been evaluated.

REFERENCES

[1] Loewen, E. G. & Popov, E. "Diffraction gratings and applications". *CRC Press*. https://doi.org/10.1201/9781315214849. January 1997.

[2] Manuel E. Solano, Muhammad Faryad, Akhlesh Lakhtakia, and Peter B. Monk, "Comparison of rigorous coupled-wave approach and finite element method for photovoltaic devices with periodically corrugated metallic backreflector," J. Opt. Soc. Am. A 31, 2275-2284 (2014).

Circularly Polarized Dielectric Resonator Antenna for the Terahertz Band Applications

Isam Eddine Lamri, Sarosh Ahmad, Wael Jaafar, Mohammad Alibakhshikenari, and Halim Yanikomeroglu

***Abstract*—The design of a circularly polarized Dielectric Resonator Antenna (DRA) fed by an aperture coupled technique for THz applications is presented in this paper. The Dielectric Resonator Antenna has a bandwidth of 115 GHz (from 0.925 to 1.04 THz) due to a cross slot aperture and four symmetrically rotated slots in the ground plane. The achieved simulated gain and efficiency values are 8.08 dBi and 83%, respectively. It is proved that the proposed DRA provides an axial ratio (AR) bandwidth covering the whole operating band. Through a parametric study, we optimized the DRA's dimensions to achieve circular polarization and operate efficiently over a large frequency band.**

Keywords—Dielectric Resonator Antenna, terahertz THz, Circular polarization, aperture coupled technique.

I. Introduction

Terahertz (THz) communications use frequencies ranging from 300 GHz to 10 THz, with high spectrum resolution and large bandwidth tunability. Operating in the THz band has the potential to enable several applications such as biomedical imaging, security screening, and high-speed wireless data connections, among other things [1]. Dielectric resonator antennas (DRAs) [2] have recently received a lot of attention due to their compact volume, simple excitation, and adjustable bandwidth. Due to the absence of metallic losses, a high radiation efficiency can be achieved with DRAs.

In this paper, we propose a circularly polarized (CP) DRA, where the cross-slot aperture and four symmetrically rotated slots etched in the ground plane have been addressed, thus yielding to an improvement in the operating and axial ratio bandwidths. Our DRA design has been simulated using CST Microwave Studio. Obtained results demonstrate the high potential of the DRA due to its large THz frequency band, spanning between 0.925 and 1.04 THz, high gain of 8.08 dBi, and enhanced efficiency of 83%. Consequently, our antenna design is very suitable to THz-based applications.

II. Proposed Antenna Design

Fig. 1 shows the proposed single fed broadband circularly polarized DRA, which consists principally of a cylindrical DRA, zigzag-shape strip for feeding, and a perfect electric conductor (PEC) etched ground plane placed just above a silicon substrate of relative permittivity ε_r=11.9, with the antenna's optimal dimensions provided. Moreover, DRA has a relative permittivity ε_r=6.0. An extra axial ratio passband can be induced by truncating a cylinder of radius *R_in* and four orthogonally positioned arms with the same length *L_slot* and width *W_slot* from the DRA, resulting in a wideband CP-radiation.

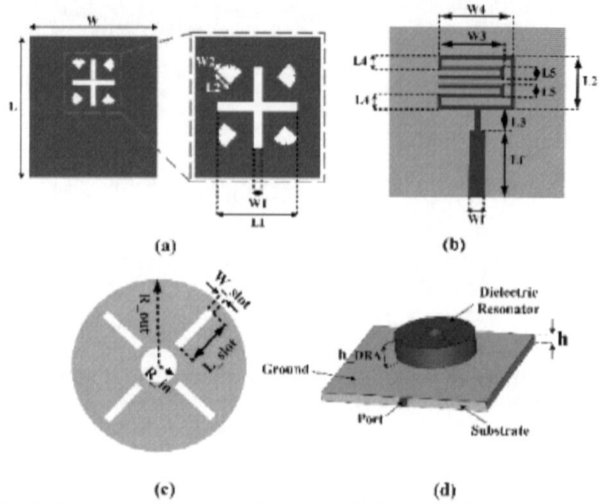

Fig. 1. Geometry of the proposed antenna: (a) Ground (b) Feeding network (c) Dielectric resonator cross-section (d) antenna 3D view.

References

[1] X. Li, W. Yin and S. K. Khamas, "Design of Efficient Photomixer-based Terahertz Dielectric Resonator Antenna," 2020 14th European Conference on Antennas and Propagation (EuCAP), 2020, pp. 1-4.

[2] H. Xu et al., "Single-Fed Dual-Circularly Polarized Stacked Dielectric Resonator Antenna for K/Ka-Band UAV Satellite Communications," in IEEE Transactions on Vehicular Technology, vol. 71, no. 4, pp. 4449-4453, April 2022.

I.E. Lamri is with the laboratory of Electronic Research of Skikda (LRES), 20 August 1955 University, Skikda, Algeria, email: isam.lamri@univ-skikda.dz. S. Ahmad and M. Alibakhshikenari are with the Department of Signal Theory and Communications of Universidad Carlos III de Madrid, Spain, email: {asarosh, malibakh}@ing.uc3m.es. W. Jaafar and H. Yanikomeroglu are with the Systems and Computer Engineering Department of Carleton University, ON, Canada, emails: {waeljaafar, halim}@sce.carleton.ca.}

Design and Analysis of a Spline-Profile Diagonal Horn Antenna at104GHz for Microwave Holography

Daniele Ronso Lima[1,2], Andrey M. Barychev[2], Jacques R. D. Lèpine[1]

***Abstract*— This paper presents the mechanical design and electric simulations, the measurement procedures, beam characterization and Analysis.** (*Abstract*)

Keywords—Antenna Design, Diagonal Antenna, Microwave Holography, Near Field (*key words*)

I. Introduction (*Heading 1*)

Located in the Argentine Andes, at 4800 m of altitude, the Large Latin American Millimeter Array–LLAMA is an astronomical observatory for studying the Universe at sub–millimeter and millimeter wavelengths. Currently under construction and development, LLAMA observatory is composed of a single dish telescope, which has a diameter of 12 m. Once the construction and installation is completed, the telescope must be characterized and validated. The microwave holography technique is a method of evaluating the dish surface profile, which quantifies manufacturing errors, panel alignment and gravity deformation effects among others parameters. The method utilizes the Fourier transform relation between the complex far field radiation pattern of an antenna and the complex aperture distribution [1], which requires a narrow bandwidth and a strong phase uniformity to ensure the accuracy of the usual specifications of sub–mm telescopes surface.

The telescope, as delivered, may have a total error up to 100 μm due to panel misalignment and a total residual sum of squares (RSS) error up to 20 μm, due various contributions. For the second stage of the campaign, the RSS error shall be reduced up to 15 μm by panel alignment, as described by [2]. Those requirements where defined for[1] ALMA observatory microwave holography system and they will be applied at LLAMA observatory. The transmitter of ALMA system has a corrugated horn antenna with a correction lens. However corrugated horn antennas are expensive and impractical for manufacturing when the frequency is increased, due to the high precision requirements applied to small dimensions. The purpose of this study is to validate the design of a spline-profile diagonal Horn antenna, which is suitable for LLAMA telescope microwave holography. The spline-profile was calculated [3] and optimized for 104 GHz. It was machined in a split-block to reduce the manufacturing costs

II. Results

This paper presents the mechanical design and electric simulations, the measurement procedures and the results of the characterization. Furthermore, the results are compared with the specifications of the horn transmitter.

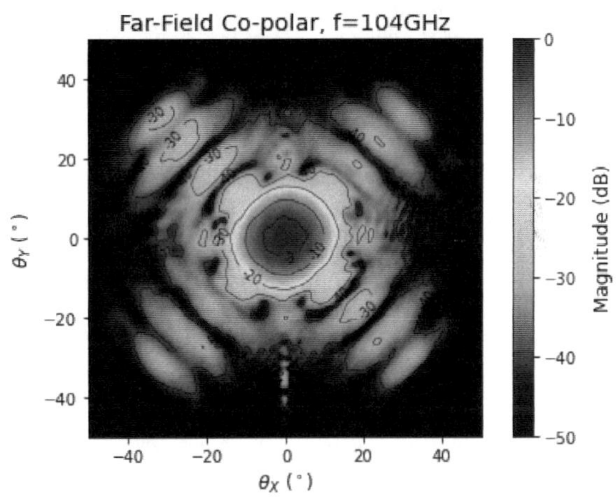

Fig. 1. Far field radiation pattern measured by Near field technique.

References

[1] J. K. Johansson and N. K. Whyborn, "Antennas for sub millimetre wave receivers," in Proc. 2nd Int. Symp. SpaceTHz Technol. (ISSTT 1991), Pasadena, CA, USA, Feb. 26– 28, 1991, pp. 63–81.

[2] L. D'Addario, "ALMA Holography System Overview" ALMA Holography CDR documents. NRAO Tucson, 2000-October-10.

[3] H. J. Gibson, *et al,* "A Novel Spline-Profile Diagonal Horn Suitable for Integration Into THz Split-Block Components" IEEE Transactions on Terahertz Science and Technology, pp 657-663, vol 7, Issue: 6 , Nov. 2017.

[1]University of Sao Paulo, Sao Paulo Brazil; [2]Kaptyen Astronomical Instituto, Groningen, Netherlands; [3]Third Affiliation, City, Postal Code, Country; [4]Fourth Affiliation, City, Postal Code, Country

NOTES:

Design of a 350 GHz Circular Waveguide Superconductor-Insulator-Superconductor Mixer for Array Applications

B-. K. Tan, J. Wenninger, P. Kittara, P. Noptosporn, P. Jaroenjittichai, D. Singwong, I. G. Bernete, C. Chaumont and F. Boussaha

***Abstract*—We present the design of a superconductor-insulator-superconductor (SIS) mixer fed with a 2-probe antenna mounted in a circular waveguide, hence avoiding the need for a rectangular waveguide that is often difficult to machine at high millimetre and sub-millimetre frequencies. The mixer is designed to operate from 275–375 GHz, covering a similar frequency range to the HARP-B receiver of the James Clerk Maxwell Telescope. Each antenna probe is connected to a separate but identical mixer circuit comprising three SIS junctions connected in series to reduce the parasitic capacitance, and the relevant tuning circuits and RF chokes. The down-converted IF power at the output of each mixer branch is expected to be combined using either a microwave Wilkinson power combiner or a 180° hybrid, to recover the full signal strength. In this paper, we present in the detail the electromagnetic simulations of each RF component making up the mixer chip, as well as the performance of the entire 2-probe mixer including the RF and IF performance predicted using SuperMix, a software package developed based on Tucker's theory of quantum mixing. Finally, we show how such circular waveguide SIS mixers can be easily populated onto a simple split-block to form a 16-pixel array.**

***Index Terms*—Superconductor-Insulator-Superconductor (SIS), Heterodyne Mixers, Focal Plane Array, Circular Waveguide, Probe Antenna**

I. Introduction

SINCE the establishment of the National Astronomical Research Institute of Thailand (NARIT), astronomical research and developments in Thailand have expanded rapidly in recent years, including the building of the 2.4 m Thai National Telescope (TNT) and the Thai National Radio Observatory (TNRO) which oversee the operation of the 40 m Thai National Radio Telescope (TNRT) and the 13 m VLBI2010 Global Observing System (VGOS) telescope. NARIT also plays an important role as the regional centre for South East Asian (SEA) countries e.g., hosting the SEA Astronomy Network, the SEA Regional Office of Astronomy for Development and International Training Centre for Astronomy under the auspices of UNESCO, and the International Astronomical Union (IAU) regional office in SEA, which focuses on developing strategic plan to improve the overall education and capital development in SEA via involvement in promoting Science, Technology, Engineering and Mathematics (STEM) subjects in universities & research, children in primary and secondary schools and public outreach.

Although NARIT has been building the infrastructures for instrumentation at radio and optical wavelengths, including the establishment of mechanical/electronic workshops and laboratories for instrument construction, their involvement in millimetre (mm) astronomical receivers is still at the nascent stage. Through the collaboration with the UK, in particular via the support of the UK STFC-NARIT Newton Fund, NARIT aims to accelerate research in astronomy and extend its technological capabilities and infrastructures into the mm wavelengths, one of the modern sciences which utilises the most cutting-edge technologies such as cryogenic engineering, quantum superconducting devices etc. In particular, Thailand has also recently joined the East Asia Observatory (EAO) which operates the James Clerk Maxwell Telescope (JCMT) at Hawai'i, the largest single-dish sub-mm telescope in the world. One of the strategic plans of EAO for the upcoming decade is to upgrade JCMT's existing mm-wave focal plane array (FPA), HARP-B [1], a 16-pixel superconductor-insulator-superconductor (SIS) mixer array operating near 350 GHz. Both EAO and NARIT have already expressed strong interest in collaborating to build such an array receiver.

However, the development of large pixel count FPA at high frequency is challenging, as the dimension of various components comprising the FPA become increasingly smaller and difficult to fabricate to the required accuracy without state-of-the-art machines. For example, fabrication of a high aspect ratio rectangular waveguide [2], [3] to couple the energy from the telescope to the SIS mixer at mm and sub-mm frequency range often requires an investment of an expensive 5- or 6-axis milling machine with high accuracy. Therefore, in this paper, we describe our effort to simplify the construction of the FPA via novel innovation and design of the SIS receiver using planar superconducting circuit technology to ease the build of such FPA. Furthermore, the performance of large format FPA is often boosted with the use of membrane technologies such as silicon-on-insulator technology [4], [5], [6], to ease the mounting, alignment and shaping of mixer chip. Unfortunately,

This project is supported by the UK STFC NARIT Newton Fund (Project No. ST/T007168/1), and the D.Phil. studentship of J. Wenninger is supported by STFC and Lincoln College Kingsgate Scholarship.

B-. K. Tan, J. Wenninger and I. G. Bernete are with the Department of Physics (Astrophysics), University of Oxford, Denys Wilkinson Building, Keble Road, OX1 3RH, Oxford, UK. E-mail: alessandro.traini@physics.ox.ac.uk

P. Kittara and P. Noptosporn are with the Physics Department, Faculty of Science, Mahidol University, Rama VI Rd, Bangkok, 10400 Thailand.

P. Jaroenjittichai and D. Singwong are with the NARIT, Princess Sirindhorn AstroPark, 260, Don Kaew, Mae Rim District, Chiang Mai, 50180 Thailand.

C. Chaumont and F. Boussaha are with GEPI (CNRS UMR 8111), Observatoire de Paris, PSL Université, France.

such technologies are not commonly shared and are only available from a few laboratories worldwide. Therefore, we explore the use of conventional substrate technology to form our mixer chip without compromising the performance of our array. Finally, we also utilise the now widely used smooth-walled horn technology to form the FPA, further simplifying the design of the FPA.

II. Mixer Chip Design

In this section, we describe the design of our SIS mixers which utilise a 2-probe antenna housed within a circular waveguide connected directly to the smooth-walled horn. We utilise the commonly-used quartz substrate technology to form our mixer chips and ensure that we can achieve optimal coupling from the incoming signal to the tunnel junctions without the need for ultra-thin membrane technology. We also utilise three moderately large junctions connected in series to broaden the RF and IF bandwidth of our SIS receiver, and make sure that our mixer design can be easily populated throughout an easily machined mixer array block to form the FPA.

A. 2-Probe Antenna

Most of the SIS mixer chip deployed to dates are housed within a rectangular waveguide, utilising for example a radial probe [7], [8], a unilateral/antipodal finline [9], [10], [11] or a bowtie probe [12], [13] as the waveguide to planar circuit transition. The approach has been highly successful, and such type of mixers has been deployed to numerous millimetre (mm) and sub-mm telescope such as Atacama Large Millimetre/sub-millimetre Array (ALMA), JCMT and Sub-Millimetre Array (SMA). However, this technique often requires a separate rectangular to circular waveguide transition mounted at the back of a feed horn that has a circular waveguide output. This waveguide transition is difficult to machine due to its small size, and the requirement to split the transition piece to fabricate the rectangular section. The situation is even more challenging at high frequencies. Hence, to ease the design of the mixer block, especially for large array applications at short wavelength, we opt to demonstrate a mixer design that can be mounted directly in a circular waveguide, hence removing the requirement for the rectangular waveguide section, to ease operation at sub-mm and supra-terahertz range in the near future.

In our case, we use a 2-probe antenna as our circular waveguide to microstrip transition. The design is similar to a 4-probe orthomode transducer (OMT) [14], [15] that allows the detection of both polarisation states simultaneously with a set of four probe antenna. However, this approach requires a delicate and complicated setup on-chip as the outputs of each opposing probe set need to be combined using additional components with minimal cross-talk [16], [17]. To simplify the mixer chip design, we utilise only one set of opposing probes with two identical SIS mixer branches connected directly to the output of each probe, and we combine the down-converted IF power using a commercially available power combiner such as the Wilkinson combiner or a 180° hybrid. It is worthwhile noting that this approach can also be easily extended to the dual-polarisation operational mode, by simply utilising another IF power combiner to recover the signal strength of the second opposing probe set for the orthogonal polarised state, without the need for an RF crossover. In this paper, we focus on the single polarisation design to demonstrate the feasibility of this new approach.

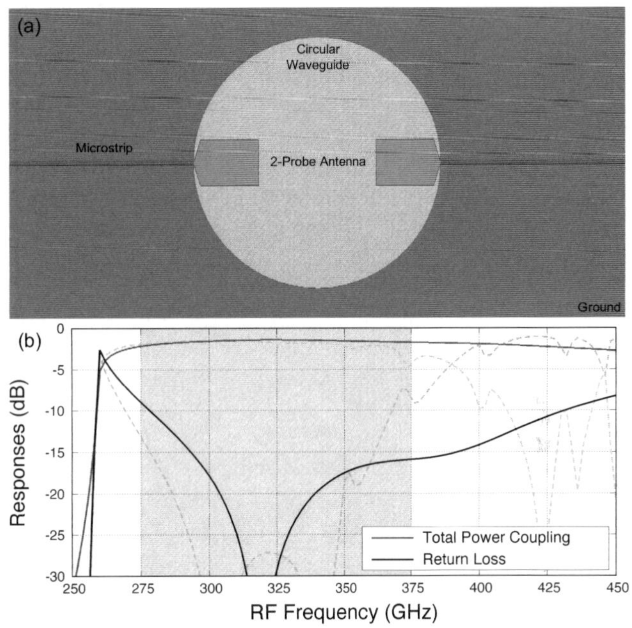

Fig. 1. (a) Image illustrating the layout of the 2-probe antenna mounted across a circular waveguide. The image shown is not-to-scale to show clarity. (b) The total power coupling from the circular waveguide to both the microstrip output ports and the return loss performance was simulated using HFSS. The solid red and black curves indicate the performance of the 2-probe antenna, whereas the dashed pink and grey lines show the inferior performance of the antenna if one of the probes is removed.

Fig. 1 depicts the simple layout of our 2-probe antenna. The set of niobium (Nb) probes is supported on top of a 40 µm thick quartz substrate. For operation near 350 GHz, we set the diameter of the circular waveguide to 0.7 mm with a cutoff frequency of 260 GHz. The antenna chip is mounted along a small opening milled across the circular waveguide with a height of 90 µm, and the antenna is placed approximately 200 µm above the termination of the circular waveguide forming the backshort, without any waveguide choke external to the circular waveguide. The width of the microstrip is set to match the output impedance of the probe. The dimension of the probe, the backshort and the width of the microstrip are then optimised using Ansys® High Frequency Structure Simulator (HFSS) for optimal power coupling and bandwidth performance.

The simulated performance is shown in Fig. 1(b), demonstrating the broadband nature of the antenna, covering a bandwidth from approximately 275–450 GHz. The insertion loss from the circular waveguide to the probe set is about 2 dB, caused mainly by the quartz substrate which is still considerably thick in this case, and more crucially the power leakage through the substrate and the 50 µm gap above the chip required to avoid the planar circuit being shorted by the mixer block. It would be technically very challenging to further thin the quartz beyond 40 µm to reduce substrate loss and minimise power leakage through a narrower slot, but it is possible to improve the total power coupling further by using a thinner membrane such as the use of several µm thick silicon-on-insulator or silicon nitride substrate [18]. However, as mentioned earlier, such membrane technologies are not widely available; hence with conventional low dielectric constant quartz substrate, we believe the presented results are the optimal performance achievable with this technology.

In principle, this approach of utilising direct circular waveguide to planar circuit coupler can be simplified by using only one probe antenna [19] instead of two. In this case, we can eliminate the need for an additional mixer circuit, and more crucially the need for an IF power combiner. Nevertheless, as shown in Fig. 1(b), the performance of a single-probe antenna in a circular waveguide is inherently narrower and tent to excite unwanted modes within the circular waveguide. For example, we can observe from the electromagnetic model that the several notches near 350 GHz, 375 GHz, and above, which are caused by the imbalanced coupling from the other side of the circular waveguide without a probe. Also, without the symmetric arrangement, the bandwidth performance is now only about half of that of the 2-probe antenna design. Therefore, we chose to use the 2-probe antenna for our SIS receiver design to maximise the performance across a much wider band.

B. Mixer Chip Design

Fig. 2(a) shows the configuration of our mixer chip. Apart from the use of the 2-probe antenna, the main feature of our mixer design is that we utilise multiple tunnel junctions connected in series, following the approach presented by [20] and [21], to reduce the total junction capacitance and to improve the RF and IF bandwidth performance. In our case, we use three $1.2\,\mu m^2$ tunnel junctions with a normal resistance of $20\,\Omega/\mu m^2$ and a specific capacitance of $80\,fF/\mu m^2$. The junction arrays are embedded in a microstrip transmission line, formed using a 350 nm top and 200 nm ground Nb film sandwiching a 490 nm thick silicon monoxide (SiO) dielectric layer, to provide convenient access to the required characteristic impedance range. The dielectric is fabricated by stacking a 250 nm SiO on top of a another 240 nm SiO layer, where the upper layer was removed near the junction array to make sure the top Nb is in electrical contact with the first tunnel junction. The series link with the subsequent junctions is provided by using a small coplanar waveguide section, in conjunction with another short microstrip line, as illustrated in Fig. 2(b).

For tuning out the residual parasitic capacitance which would otherwise short the junctions, we utilise a commonly used end-loaded tuner circuit approach [22]. The low impedance stub is shown as the wider microstrip section between the junction array and the probe antenna in Fig. 2(c), and the width and length of the narrower microstrip section connecting the stub and the junction array is adjusted to present an inductive load to the junction array. An impedance transformer section is concatenated before the tuner circuit to bridge the impedance mismatch between the output of the probe antenna and the junction array, and a 5-section high-low impedance RF choke is deployed after the tuner circuit to prevent RF power leakage into the IF chain. The entire planar mixer circuit is then optimised using HFSS, without altering the setup of the antenna and the waveguide. The same mixer circuit is deployed to both the probe antenna forming a symmetric configuration.

Fig. 2(d) shows the predicted performance of our mixer. As

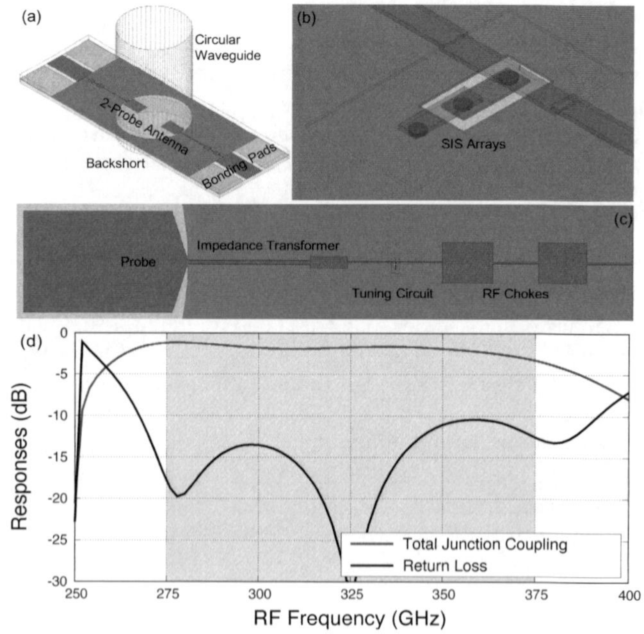

Fig. 2. (a) The layout of the mixer chip mounted across a circular input waveguide with backshort. (b) Zoom-in image of the junction array and (c) the complete mixer circuit connected to one of the probes. (d) HFSS predicted total power coupling from the input waveguide to both the junction array, as well as the return loss performance of the entire setup including the 2-probe antenna.

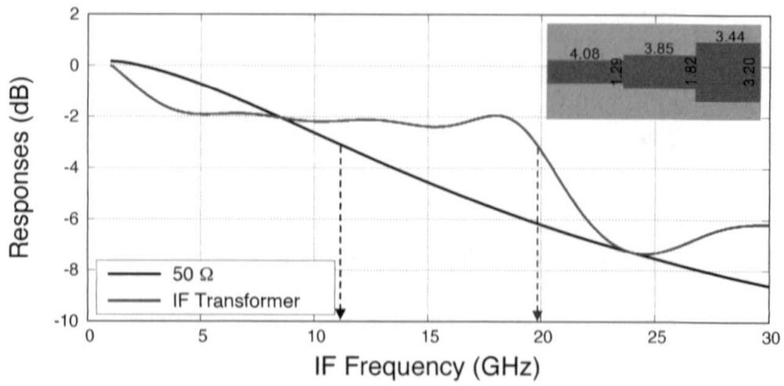

Fig. 3. The power coupling of each set of junctions to the relative IF output port where the black curve shows the performance when the IF is connected directly to a 50 Ω load, while the red curve shows a much broader −3 dB coupling with the use of a 3-sections IF transformer to match the impedance of the tunnel junctions and the 50 Ω load. The inlet shows the 3-sections IF transformer board designed using 20 mil. Roger Duroid 4350b PCB with 17 µm copper on each side forming a microstrip structure. The dimensions shown are in mm.

can be seen, we successfully achieve closed to –3 dB total power coupling from the input circular waveguide to both the junction arrays from 265–375 GHz, slightly wider than the designated operating range. Similar return loss performance is observed over the same frequency range, if we define the bandwidth using the –10 dB cutoff points. Note that the simulated performance includes the effect of the probe antenna which inherently impose a –2 dB insertion loss already across the designated band, which could be further improved if membrane technologies are accessible.

Fig. 3 shows the HFSS predicted IF coupling range of our mixer chip for one of the IF ports, which is identical to the opposite IF port. Using the –3 dB point as the indicator for the operational IF bandwidth, we can see that if we connected these IF ports directly to a 50 Ω environment, inherently we expect an IF bandwidth of approximately 11 GHz. This is considerably wide due to the low total junction capacitance of the 3-junction array. Following the approach we previously presented in [23] to further broaden the IF bandwidth, we opt to install two IF transformers after the mixer chip IF ports to better match the output impedance of the mixer circuit and the subsequent 50 Ω microwave power combiner. Here, we design a simple 3-section transformer using commercially available Roger Duroid® 4350b printed circuit board (PCB). As can be seen from Fig. 3, with such IF impedance transformers, we manage to almost double the IF bandwidth from 11 to 20 GHz, with a very flat and consistent coupling from 4–18 GHz.

C. Heterodyne Performance

Once the mixer chip and the IF transformer designs were finalised using the electromagnetic modelling package to optimise the power coupling between the input circular waveguide to the tunnel junction arrays, as well as from the array to the IF ports in the microwave regime, we checked the heterodyne mixing performance of the SIS receiver using SuperMix [24]. The S-parameter matrixes of the mixer chip and the IF transformers simulated from HFSS were exported to SuperMix to form the full receiver circuit [25]. For this exercise, we make use of an ideal Wilkinson power combiner with the two input ports connected to the respective IF output ports of the mixer chip via the IF transformers to recover the full down-converted signal strength. As shown in Fig. 4, we successfully cover the entire 275–375 GHz range with a noise temperature of less than 30 K and a better than –2.5 dB double sideband (DSB) conversion gain, across the entire bandwidth. Similarly, with the installation of the IF transformer boards, we managed to achieve close to 20 GHz operational bandwidth, if we take the –3 dB point of the DSB gain curve as reference.

III. Mixer Array Block Design

The above-described mixer chip is ready to be extended into an array by populating the individual chips into an array block as shown in Fig. 5. The array block can now be constructed with standard split block technology without much complication. The upper block consist of an array of feed horns which can be directly drilled using a shaped machining tool [26]. In Fig. 6, we show the design of the accompanied

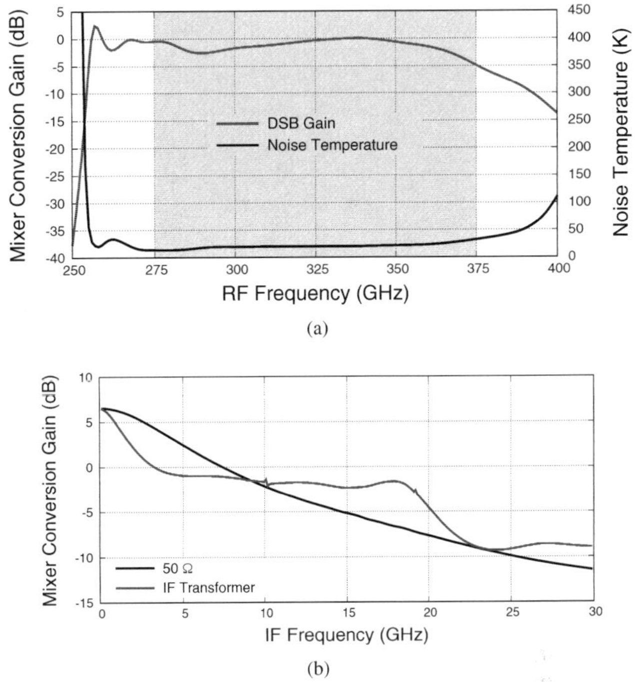

Fig. 4. (a) SuperMix predicted RF noise temperature and DSB conversion gain of the entire SIS receiver, and (b) the IF DSB gain-bandwidth performance.

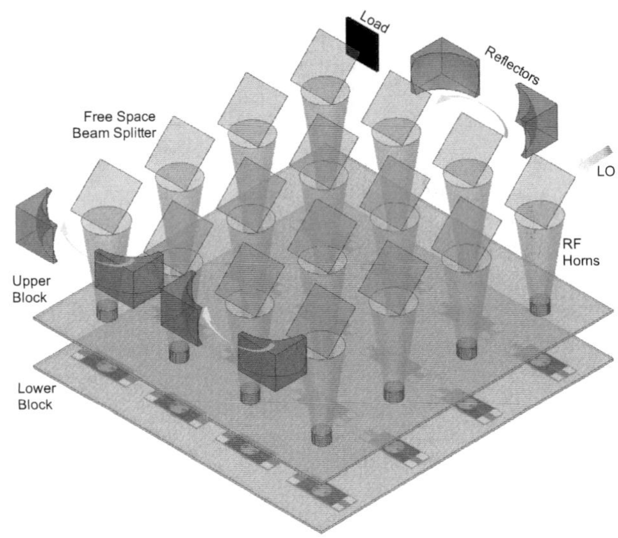

Fig. 5. Illustration of a 4×4 array that can be constructed to house the circular waveguide fed SIS mixer chips, showing the simplicity of the array block. The image shown is not-to-scale to show clarity.

350 GHz 3-section smooth-walled horn that was scaled directly from the design presented in [27], and its corresponding beam patterns at the centre and near the edges of the frequency band. As can be seen, the far field radiation patterns are clean with sidelobes level well below –25 dB. The E- and H-field of the main beam are also identical throughout, except at the lowest frequency point where they start to diverge below the –10 dB level. The design of this smooth-walled horn can be further optimised, if need be, to the specification of the telescope

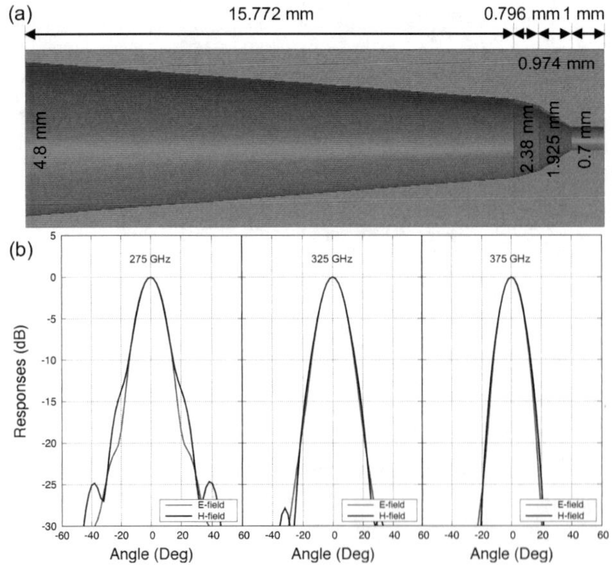

Fig. 6. (a) The design of a 3-section smooth-walled horn for operation from 275–375 GHz, with the dimension given in lengths and diameters of each section, including the output circular waveguide. (b) The HFSS simulated far-field beam patterns of the smooth-walled horn at 275, 325 and 375 GHz.

optics.

The bottom part of the array block would consist of pockets for mounting the SIS mixer chips, as well as the IF transformer PCBs with the corresponding SMA connectors, similar to the design provided in [17]. The mixer chip will be aligned directly underneath the circular waveguide output of the feed horns and wire bonded to the IF board. The outputs of the IF transformer sets will then be guided through the back of the array block via the SMA connectors, which subsequently combine the two down-converted IF signals with a power combiner before feeding to the low noise amplifiers and posterior IF chain for performing characterisation and measurements. The magnetic coil required could be installed adjacent to the feed horn following the design shown in [17] as well. Finally, the LO injection can be provided quasi-optically using the traditional dielectric beam splitter array with corresponding mirror sets to re-focus and meander the LO beam.

IV. Conclusion

We have presented the design of an SIS mixer chip that can be fed directly with a circular waveguide with the use of a 2-probe antenna, eliminating the need for a rectangular waveguide that is more difficult to fabricate than the circular waveguide. Two identical mixer circuits designed to cover the frequency range from 275–375 GHz are concatenated directly to the output of the 2-probe antenna, where the down-converted IF signal from each mixer chain is power combined using a microwave power combiner. Rigorous *em* simulations show that our mixer design can covers the required RF range with good noise temperature performance. With the array of three series-connected tunnel junctions in conjunction with the 3-section IF transformers, we manage to provide an IF coverage close to 20 GHz. In the final section, we show how such a circular waveguide SIS mixer chip can simplify the design of the mixer block array, which can now be fabricated using a standard split block technology without much complication, paving the way for the construction of even larger array in the future, in particular near the supra-THz frequency regime.

Acknowledgment

This research was funded by the STFC-NARIT Newton Grant programme (grant agreement No. ST/T007168/1). For the purpose of Open Access, the author has applied a CC BY public copyright licence to any Author Accepted Manuscript version arising from this submission.

References

[1] H. Smith, R. E. Hills, S. Withington, J. Richer, J. Leech, R. Williamson, H. Gibson, R. Dace, P. Ananthasubramanian, R. W. Barker *et al.*, "HARP-B: a 350-GHz 16-element focal plane array for the James Clerk Maxwell telescope," in *Astronomical Telescopes and Instrumentation*. International Society for Optics and Photonics, 2003, pp. 338–348.

[2] A. Khudchenko, A. M. Baryshev, K. I. Rudakov, P. M. Dmitriev, R. Hesper, L. de Jong, and V. P. Koshelets, "High-gap Nb-AlN-NbN SIS junctions for frequency band 790–950 GHz," *IEEE Transactions on Terahertz Science and Technology*, vol. 6, no. 1, pp. 127–132, 2015.

[3] T. Kojima, M. Kroug, A. Gonzalez, K. Uemizu, K. Kaneko, A. Miyachi, Y. Kozuki, and S. Asayama, "275–500-GHz wideband waveguide SIS mixers," *IEEE Transactions on Terahertz Science and Technology*, vol. 8, no. 6, pp. 638–646, 2018.

[4] A. B. Kaul, B. Bumble, K. A. Lee, H. G. LeDuc, F. Rice, and J. Zmuidzinas, "Fabrication of wide-IF 200–300 GHz superconductor–insulator–superconductor mixers with suspended metal beam leads formed on silicon-on-insulator," *Journal of Vacuum Science & Technology B: Microelectronics and Nanometer Structures Processing, Measurement, and Phenomena*, vol. 22, no. 5, pp. 2417–2422, 2004.

[5] M. Westig, K. Jacobs, J. Stutzki, M. Schultz, M. Justen, and C. Honingh, "Balanced superconductor–insulator–superconductor mixer on a 9 μm silicon membrane," *Superconductor Science and Technology*, vol. 24, no. 8, p. 085012, 2011.

[6] B.-K. Tan, G. Yassin, P. Grimes, and K. Jacobs, "650 GHz SIS mixer fabricated on silicon-on-insulator substrate," *Electronics letters*, vol. 49, no. 20, pp. 1273–1275, 2013.

[7] S. Withington, J. Leech, G. Yassin, K. Isaak, B. Jackson, J. Gao, and T. Klapwijk, "A 350GHz radial-probe SIS mixer for astronomical imaging arrays," *International Journal of Infrared and Millimeter Waves*, vol. 22, no. 9, pp. 1305–1312, 2001.

[8] J. Kooi, G. Chattopadhyay, S. Withington, F. Rice, J. Zmuidzinas, C. Walker, and G. Yassin, "A full-height waveguide to thin-film microstrip transition with exceptional RF bandwidth and coupling efficiency," *International Journal of Infrared and Millimeter Waves*, vol. 24, no. 3, pp. 261–284, 2003.

[9] G. Yassin, S. Withington, M. Buffey, K. Jacobs, and S. Wulff, "A 350-GHz SIS antipodal finline mixer," *IEEE Transactions on microwave theory and techniques*, vol. 48, no. 4, pp. 662–669, 2000.

[10] C. North, G. Yassin, and P. Grimes, "Rigorous analysis and design of finline tapers for high performance millimetre and submillimetre detectors," in *Seventeenth International Symposium on Space Terahertz Technology, held May 10-12, 2006 at Observatoire de Paris, LERMA. Paris, France*. Citeseer, 2006, pp. 284–287.

[11] B.-K. Tan, G. Yassin, P. Grimes, J. Leech, K. Jacobs, and C. Groppi, "A 650 GHz unilateral finline SIS mixer fed by a multiple flare-angle smooth-walled horn," *IEEE Transactions on Terahertz Science and Technology*, vol. 2, no. 1, pp. 40–49, 2011.

[12] W. Shan, S. Shi, T. Matsunaga, M. Takizawa, A. Endo, T. Noguchi, and Y. Uzawa, "Design and development of SIS mixers for ALMA band 10," *IEEE transactions on applied superconductivity*, vol. 17, no. 2, pp. 363–366, 2007.

[13] T. Kojima, M. Kroug, M. Takeda, W. Shan, Y. Fujii, Y. Uzawa, Z. Wang, and S. Shi, "Performance of terahertz waveguide SIS mixers employing epitaxial NbN films and Nb junctions," *IEEE transactions on applied superconductivity*, vol. 19, no. 3, pp. 405–408, 2009.

[14] B.-K. Tan and G. Yassin, "Preliminary design of an ALMA band 10 single-chip dual-polarisation SIS mixer," in *2017 10th UK-Europe-China Workshop on Millimetre Waves and Terahertz Technologies (UCMMT)*. IEEE, 2017, pp. 1–4.
[15] S. Ezaki, W. Shan, S. Asayama, and T. Noguchi, "Fabrication of superconductor integrated circuits of D-band dual-polarization balanced SIS mixers," *IEEE Transactions on Applied Superconductivity*, vol. 29, no. 5, pp. 1–5, 2019.
[16] W. Shan, S. Ezaki, K. Kaneko, A. Miyachi, T. Kojima, and Y. Uzawa, "Experimental study of a planar-integrated dual-polarization balanced SIS mixer," *IEEE Transactions on Terahertz Science and Technology*, vol. 9, no. 6, pp. 549–556, 2019.
[17] J. Wenninger, F. Boussaha, C. Chaumont, B.-K. Tan, and G. Yassin, "Design of a 240 GHz On-Chip Dual-Polarization Receiver for SIS Mixer Arrays," *IoP Superconducting Science and Technology*, submitted.
[18] P. Mauskopf, "Transition edge sensors and kinetic inductance detectors in astronomical instruments," *Publications of the Astronomical Society of the Pacific*, vol. 130, no. 990, p. 082001, 2018.
[19] K.-Y. Liu, M.-J. Wang, C.-T. Li, T.-J. Chen, and S.-C. Shi, "Development of a Dual Polarization SIS Mixer With a Planar Orthomode Transducer at 350 GHz," *IEEE transactions on applied superconductivity*, vol. 23, no. 3, pp. 1 400 705–1 400 705, 2013.
[20] R. Blundell, K. Megerian, J. Stern, S.-K. Pan, M. Pospieszalski *et al.*, "A distributed lumped-element SIS mixer with very wide instantaneous bandwidth," *IEEE transactions on applied superconductivity*, vol. 15, no. 2, pp. 490–494, 2005.
[21] C.-Y. E. Tong, P. Grimes, R. Blundell, M.-J. Wang, and T. Noguchi, "Wideband SIS receivers using series distributed SIS junction array," *IEEE Transactions on Terahertz Science and Technology*, vol. 3, no. 4, pp. 428–432, 2013.
[22] J. Kooi, M. Chan, B. Bumble, H. LeDuc, P. Schaffer, and T. Phillips, "180–425 GHz low noise SIS waveguide receivers employing tuned Nb/AIO x/Nb tunnel junctions," *International journal of infrared and millimeter waves*, vol. 15, no. 5, pp. 783–805, 1994.
[23] B. Tan, G. Yassin, and P. Grimes, "A simple method to widen the IF bandwidth of a high frequency SIS mixer," in *Twenty-Fourth International Symposium on Space Terahertz Technology*, 2013.
[24] J. Ward, F. Rice, G. Chattopadhyay, and J. Zmuidzinas, "SuperMix: A flexible software library for high-frequency circuit simulation, including SIS mixers and superconducting elements," in *Proceedings, Tenth International Symposium on Space Terahertz Technology*, 1999, pp. 269–281.
[25] B.-K. Tan, G. Yassin, and P. Grimes, "Ultra-wide intermediate bandwidth for high-frequency SIS mixers," *IEEE Transactions on Terahertz Science and Technology*, vol. 4, no. 2, pp. 165–170, 2014.
[26] J. Leech, B. K. Tan, G. Yassin, P. Kittara, and S. Wangsuya, "Experimental investigation of a low-cost, high performance focal-plane horn array," *IEEE Transactions on Terahertz Science and Technology*, vol. 2, no. 1, pp. 61–70, 2011.
[27] B.-K. Tan, J. Leech, G. Yassin, P. Kittara, M. Tacon, S. Wangsuya, and C. Groppi, "A high performance 700 GHz feed horn," *Journal of Infrared, Millimeter, and Terahertz Waves*, vol. 33, no. 1, pp. 1–5, 2012.

Design of RF waveguide structure for 2SB SIS mixer at 210-280 GHz

S. Realini*[1], R. Hesper[1], A.M. Baryshev[1]

***Abstract*— We discuss the latest results on the design of a new receiver based on modular sideband separating SIS mixers working in the frequency range 210-280 GHz (approximately corresponding to ALMA Band 6). In particular, we present the design of the components that make up the RF waveguide block, which provides the input to the mixers. We optimized the design taking into account spurious reflections, which have a significant impact on the Image Rejection Ratio. Given the simulated performance of each component, it is also possible to develop a full model of the 2SB SIS receiver, which allows us to verify the influence of each component on the total IRR performance in terms of spurious reflections.**

Keywords*— *sideband-separating mixers, waveguide hybrids, submillimeter receivers

I. Introduction

Sideband separating (2SB) receivers based on SIS (superconductor-insulator-superconductor) mixers are widely used in ground-based observatories such as ALMA, NOEMA and SMA to perform frequency down-conversion and separate the image signal from the signal channel. The use of sideband-separating receivers is highly desirable because it allows the reduction of the atmospheric noise contribution when observing spectral lines.

The receiver we present is based on a modular configuration, which is convenient to characterize all the components separately before assembling. As shown schematically in Fig. 1, the RF input signal is split in two branches by a quadrature hybrid. The local oscillator (LO) power is also divided by a power splitter and coupled to the mixers. The IF outputs of the two mixers are amplified and then combined in the IF quadrature hybrid with the required phase delay. The final outputs are the separated down-converted upper (USB) and lower (LSB) sideband signals.

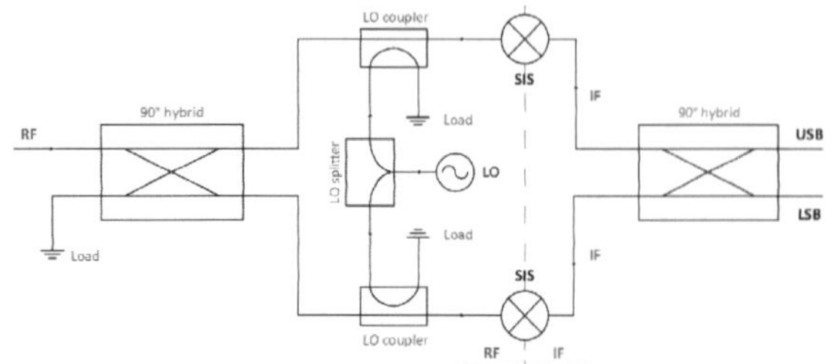

Fig. 1. Schematic design of the modular 2SB SIS receiver.

This parameter should be better than 10 dB, and preferably better than 20 dB, in order to optimize the sensitivity of the receiver and avoid line confusion during observations.

II. RF Waveguide Structure

The waveguide structure is based on a six-branch quadrature hybrid coupler designed to operate in the frequency range 210-280 GHz. The LO signal is equally divided by an E-plane T-splitter and coupled to the SIS mixers with a pair of two-branch directional couplers. The modular design concept is similar to the one presented in [1].

The RF imbalance is one of the key parameter limiting the IRR and its design requires to pay attention to the phase and amplitude balance of the entire RF structure. However, as shown in [2-3], the total RF balance is strongly influenced by reflections within the RF structure. Therefore, we designed the RF hybrid by optimizing the RF matching and isolation by varying slot widths and positions while keeping the phase and amplitude balance within about 0.3° and 0.6 dB, respectively. Due to these imbalances, we predict a contribution to the overall IRR below −25 dB, as shown in Fig. 2.

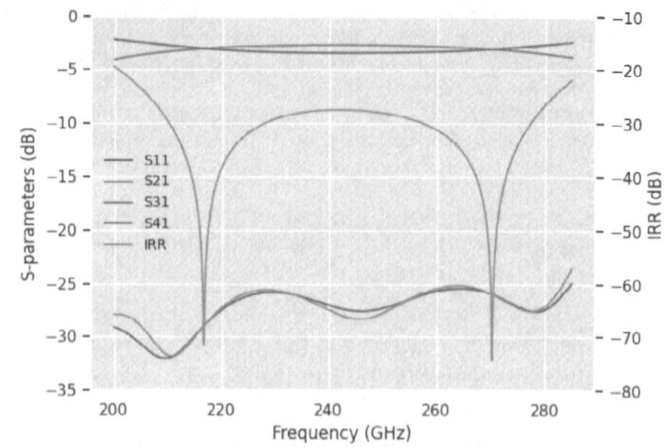

Fig. 2. Simulation of S-parameters of the RF hybrid and its contribution to the IRR.

The LO power is coupled with a two branch directional coupler, which has been optimized to have a low reflection (<−25 dB) and a coupling factor of ≈−16 dB.

The LO is divided by a T splitter with matching sections in the three legs. Since the splitter is a non-dissipative device, it has highly reflective output ports (S_{22}, S_{33}) and/or low isolation between them (S_{23}, S_{32}). Because both affect the standing wave issues, we chose a similar level of -6 dB. The input reflection was minimized to be <-25 dB to avoid further problems with standing waves between the mixer block and the LO itself.

W we performed also a tolerance analysis to understand which parameters affect the most the performance, the accuracy required in the machining process and the effect of possible asymmetries in the arms. Assuming a possible deviation of ±5% around the nominal value, the results of the simulations show that the T-shaped splitter gives a good performance within the tolerance guaranteed by the manufacturing process.

The main loads (hybrid termination and LO dump) are made by terminating the waveguide with an oblique angle into a large block of absorbing material. Simulations show that the transition into the cavity filled with the absorber has

[1] Kapteyn Astronomical Institute, University of Groningen, 9747 AD, Groningen, The Netherlands; *Corresponding author: realini@astro.rug.nl

a reflection of less than -30 dB. Since the simulations are performed assuming room temperature there are still some uncertainties about the expected behavior at 4 K, which will be addressed in the future with experimental tests.

III. Results

Given the results of the electromagnetic design and simulation of each component, we will be able to set up a full circuit simulation of the entire RF structure in order to model the effect of standing waves in the integrated system. A full model of the 2SB SIS receiver allows us to study carefully the influence of each component on the total IRR performance in terms of spurious reflections. The results of electromagnetic simulations are also fundamental to get a comparison with measurements.

The technology developed within this study will be useful for future ground-based telescopes like LLAMA and the Africa Millimetre Telescope (AMT).

References

[1] Hesper, et al., "A new high-performance sideband-separating mixer for 650 GHz", *Proc. SPIE 9914,* Jul. 2016.

[2] Khudchenko et al., "Modular 2SB SIS Receiver for 600-720 GHz: Performance and Characterization Methods", *IEEE Transactions on Terahertz Science and Technology*, Jan. 2017.

[3] Khudchenko, et al., "Comprehensive Description of Sideband Ratio of 2SB SIS Receiver", *44th International Conference on Infrared, Millimeter, and Terahertz Waves*, Sep. 2019.

Design and Performance of the Terahertz Photon Counting System: Detectors and Cryogenics

Hajime Ezawa [*1,2], Hiroshi Matsuo[1,2], Go Fujii[3], Shigetomo Shiki[3], Ayako Niwa[4], Norio Okada[1], Mitsuhiro Fukushima[1]

Abstract— **Development of a Terahertz photon counting system is ongoing. Among various components, this presentation will describe the detector device and the cryogenic system. The detector is designed for lab experiments with relatively narrow bandwidth. Low leakage SIS junction, which forms a PCTJ, is integrated as an antenna coupled photon detector. Two compact helium-4 sorption fridges are integrated into a cryostat with a pulse tube cooler, to provide continuous cooling to 0.8 K by altering their cool-down operation.**

Keywords— Superconducting tunnel junction, Terahertz detector, Photon counting, Cryogenics, Sorption fridge

I. Introduction

Fast detectors sensitive to Terahertz photons may have wide range of applications, including those for observations of astronomy and astrophysics. In general, photon counting detectors will realize high sensitivity by resolving each photon arrivals. Fast photon detectors may also be used for high-precision measurements to study the physical condition of the sources by evaluating the photon statistics.

II. Developments

We are developing a detector system for photon counting experiments at terahertz frequencies. SIS junction (or STJ) is a good candidate as a device for fast photon detection. In order to realize the photon counting capability at Terahertz frequencies, low noise equivalent power (NEP) at the order of $10^{-17}\ \mathrm{W}/\sqrt{\mathrm{Hz}}$ is required. Recently we have developed an antenna coupled detector with SIS junctions (or STJs) of Nb/Al/AlOx/Al/Nb, which exhibit low leakage current of 2 pA at a cryogenic temperature of $T \leq 0.8$ K [1]. The SIS junction is 3 μm × 3 μm in size with critical current density of $J_c = 300\ \mathrm{A/cm^2}$. While wide bandwidth is preferable to realize high sensitivities for astronomical observations, the fabricated detector is designed to have a relatively narrow bandwidth for lab experiment purposes, in order to limit the contribution of the background photons. Two SIS junctions form a PCTJ (parallel connected twin junction) which is designed for 500 GHz, and connected to a twin-slot antenna [2]. The detector was processed at the CRAVITY facility [3] at National Institute of Advanced Industrial Science and Technology (AIST). Detector improvement studies are undergoing; so far we have realized a detector with center frequency of 530 GHz with an improved sensitivity.

The cryogenic system is also being developed. The system is designed to realize a cryogenic environment of 0.8 K, where the detector exhibits the low leakage current. We have developed a compact helium-4 sorption fridge [4], which can operate with less heat load per cool-down cycle to the condensation temperature stage. Two sorption fridges are prepared to alternate their operation, in order to realize continuous cooling to 0.8 K. Recently, the performance of the sorption fridge itself was studied; a single sorption fridge of this design exhibited a hold time of approximately 3 hours, which was shorter than its cycle time of approximately 40 minutes [5]. Following these developments, we are now integrating the sorption fridge into a cryogenic system. The cryostat is cooled with a pulse tube cooler to 3-4 K, which is the condensation temperature of the sorption fridge. So far the initial design of the cryostat has been completed, and integration and evaluation are ongoing.

The design and performance of the detector and cryogenics will be discussed in the presentation. Studies for optics and cryogenic readout circuits are also ongoing, which will be discussed in another presentation.

References

[1] H. Ezawa et al., "Studies on Terahertz Photon Counting Detectors with Low-Leakage SIS Junctions", *J. Low Temp. Phys.* **194**, 426, 2019, DOI: 10.1007/s10909-019-02149-y

[2] H. Ezawa et al., "Optical Performance of SIS Photon Detectors at Terahertz Frequencies", *J. Low Temp. Phys.* **200**, 226, 2020, DOI: 10.1007/s10909-020-02513-3

[3] M. Ukibe et al., "Soft X-ray Detection Performance of Superconducting Tunnel Junction Arrays with Asymmetric Tunnel Junction Layer Structure", *JJAP* **51**, 010115, 2012, DOI: 10.1143/JJAP.51.010115

[4] H. Matsuo et al., "Compact 0.8 K Helium-4 Sorption Cooler", presented at LTD-17, 2017.

[5] A. Niwa, Master thesis, University of Tsukuba, 2022. (in Japanese)

[1]National Astronomical Observatory of Japan, Mitaka, Tokyo, 181-8588, Japan.
[*]h.ezawa@nao.ac.jp
[2]Department of Astronomical Science, The Graduate University for Advanced Studies, SOKENDAI, Mitaka, Tokyo, 181-8588, Japan.
[3]National Institute of Advanced Industrial Science and Technology, Tsukuba, Ibaraki, 385-8568, Japan.
[4]Department of Physics, University of Tsukuba, Tsukuba, Ibaraki, 305-8571, Japan.

NOTES:

Enhanced sensitivity of THz NbN hot electron bolometer mixers

B. Mirzaei[1], J. R. G. Silva[2] W. J. Vreeling[2], W. Laauwen[2] and J. R. Gao[1,2]

***Abstract*—We present an unprecedented low noise temperature of a NbN hot electron bolometer (HEB) mixer at 1.63 THz, which demonstrated the best DSB receiver noise temperature (T_{rec}^{DSB}) of 530 K and DSB mixer noise temperature (T_{mixer}^{DSB}) of 240 K. The latter is 3 times quantum noise at this frequency (hv/k). The new mixer is developed for the proposed OASIS and SALTUS missions and was fabricated using contacts that have a cleaned interface between Au and NbN. The improvement of T_{mixer}^{DSB} is 30 % in comparison with NbN HEB mixers with a superconducting interlayer added to the contacts in our labs. The reduction in T_{mixer}^{DSB} can reduce the integration time of a heterodyne instrument roughly by a factor of 2. The performance is also evaluated at two higher frequencies of 2.52 and 5.25 THz, at which we also obtain better results than our previous devices.**

***Keywords*—terahertz, NbN HEB mixers, low noise**

I. Introduction

NbN HEB mixers have been flown on Herschel [1], SOFIA [2], and STO2[3], and will be flown on GUSTO [4] and ASTHROS [5]. They are also the choice for the FIRSS (or LETO) [6], OASIS [7], and SALTUS [8] space missions, proposed (or to be proposed) to ESA and NASA, respectively. Due to limited lifetime for space missions, a low T_{mixer}^{DSB} is highly demanded to make optimal use of the observation time.

Many years of R&D at different research groups in the world have been devoted to realize low noise NbN HEBs, where the T_{rec}^{DSB} measured at 1.63 THz or extrapolated to this frequency are, for example, 750 K (for HIFI [1]), 690 K (STO2), and 760 K (GUSTO)[9]. Those values have not been reduced in the past decade. The HEBs for STO2 and GUSTO were fabricated at TU Delft using contacts that have a superconducting interlayer (either NbTiN or Nb) between the Au and thin NbN. This interlayer helps to overcome the proximity effect to maintain the NbN underneath the Au to be superconductive. However, this layer may introduce side effects e.g., on RF loss and RF current flow, which can affect T_{mixer}^{DSB}. Here we present the results of a new NbN HEB mixer using contacts that were cleaned and have a direct interface between thick Au and thin NbN. We show that the DSB mixer noise temperature improves by 30 % at 1.63 THz compared to the devices we have previously produced for GUSTO. We also find improvements at higher frequencies, but less significant, which will be briefly explained in section III.

II. HEB and Experimental Setup

HEBs with 400 nm in length and 4 μm in width imbedded in a spiral antenna were fabricated based on a standard thin NbN film [10]. We performed in-situ ion-cleaning of the NbN surface before deposition of thick Au but skipped the superconducting interlayer. Furthermore, we have also merged the two steps of making thin contact pads and thick antenna arms to only one. In this way 200 nm thick spiral arms end up right on both sides of the NbN bridge. This makes the current path from the Au antenna to the NbN bridge straightforward.

It should be mentioned that the spiral antenna of this device is a modified version of the one used in GUSTO [11] to be able to accommodate the two times wider bridge at its center.

The superconducting transition temperature (T_c) of an HEB from this batch is measured to be 8.8 K, which is lower than that of the original NbN film (≥ 9.5 K). This suggests that we indeed have a highly transmissive interface in the contacts [12]. Similar contact structures have been reported earlier by LERMA [13].

T_{rec}^{DSB} of a NbN HEB (OASIS BM2 7B) was measured at an IF of 1.7 GHz and a mixer operating temperature of 4.6 K in a standard setup in air using 1.63, 2.52 and 5.25 THz LOs. The IF chain consists of a circulator between the bias-tee and a low noise SiGe amplifier, which has a noise temperature of 6.5 K. A Si lens with an anti-reflection coating optimized at 1.6 THz was used to couple the radiation to the HEB for all the frequencies.

III. Results

The (un)pumped IV curves at 1.63 THz are shown in Fig. 1, where the optimum operating region, in which the best performance found is indicated.

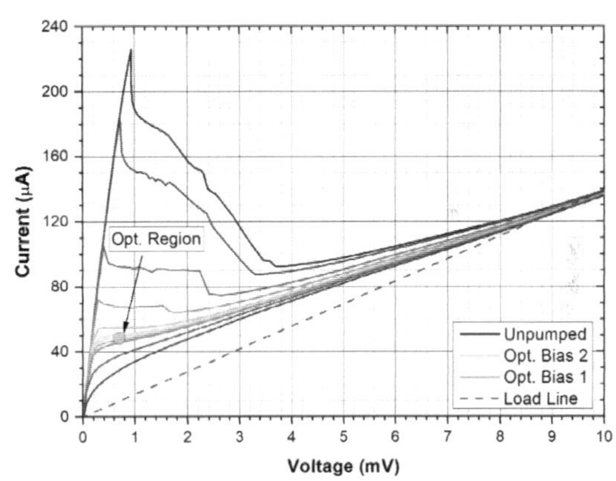

Fig. 1. Pumped IV curves of a NbN HEB at 1.6 THz

Measured T_{rec}^{DSB} together with T_{mixer}^{DSB} (including the contribution of a Si lens ideally antireflection coated for the specific frequency), in unit of hv/k (quantum noise, QN), and DSB mixer conversion gain G_{mixer}^{DSB} are summarized in Table I for all three frequencies. The T_{rec}^{DSB} includes the contributions from the air, a beam splitter, window, heat filter, and the Si lens. The T_{mixer}^{DSB} is about 30 % better than the best device out of 16 GUSTO HEBs at 1.63 THz, which was measured in the exact same setup and has an T_{mixer}^{DSB} of 350 K. We also find a 3-4 dB higher G_{mixer}^{DSB}, which may suggest a lower RF loss due to the contacts.

[1]Department of Imaging Physics, Delft University of Technology, 2628 CJ Delft, The Netherlands; [2]SRON Netherlands Institute for Space Research, Leiden/Groningen, The Netherlands.

At 5.25 THz we also observe an improvement, but it is less significant. This can be partly attributed to the non-optimized spiral antenna used for the higher frequencies since the coupling degraded because of the antenna modification explained in the previous section, and can be partly to the quantum noise, which plays a more role at the higher frequency [14].

Table I: Measured receiver noise temperature T_{rec}^{DSB}, derived mixer noise temperature T_{mixer}^{DSB} (including the contribution of a Si lens ideally antireflection coated for the specific frequency), T_{mixer}^{DSB} in unit of hν/k (QN), and DSB mixer conversion gain (G_{mixer}^{DSB}). All are obtained at an IF of 1.7 GHz and an operating temperature of 4.6 K

LO frequency (THz)	T_{Rec}^{DSB}(K)	T_{mixer}^{DSB}(K)	T_{mixer}^{DSB} in units of hν/k (QN)	G_{mixer}^{DSB}
1.63	530 K	240 K	3.1	-4.35 dB
2.52	640 K	290 K	2.4	-5.45 dB
5.25	2180 K	620 K	2.5	-7.39 dB

In Fig. 2 we show the receiver output power in responding to hot/cold loads as well as the DSB noise temperature at 1.63 THz LO around the optimum bias point. As shown, the low noise performance is present in a relatively wide current range, which gives some freedom in tuning the LO power.

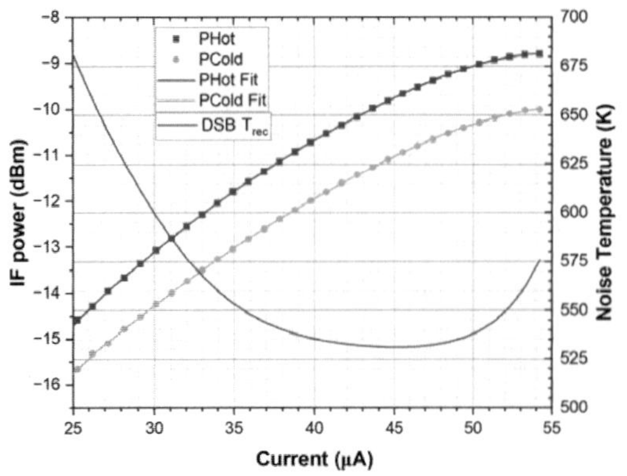

Fig. 2. Measured output power responding to hot and cold loads (left) and resulting DSB receiver noise temperature (right) measured at 1.6 THz and 0.6 mV bias.

This device requires an LO power of ~240 nW at all 3 frequencies, which is a good compromise between availability of the LO source and stability of the receiver.

We acknowledge that this work was supported by NWO (a project number of 614.061.609).

REFERENCES

[1] S. Cherednichenko, V. Drakinskiy, T. Berg, P. Khosropanah and E. Kollberg, "Hot-electron bolometer terahertz mixers for the Herschel Space Observatory", Review of Sci. Instruments, vol. 79, pp. 034501, 2008.

[2] C. Risacher, R. Güsten, J. Stutzki, H.-W. Hübers, R. Aladro, A. Bell, C. Buchbender, D. Büchel, T. Csengeri, C. Duran, U. U. Graf, R. D. Higgins, C. E. Honingh, K. Jacobs, M. Justen, B. Klein, M. Mertens, Y. Okada, A. Parikka, P. Pütz, N. Reyes, H. Richter, O. Ricken, D. Riquelme, N. Rothbart, N. Schneider, R. Simon, M. Wienold, H. Wiesemeyer, M. Ziebart, P. Fusco, S. Rosner and B. Wohler, "The upGREAT Dual Frequency Heterodyne Arrays for SOFIA", Journal of Astronomical Instrumentation, vol. 7, no. 4, pp. 1840014, 2018.

[3] Y. M. Seo, P. F. Goldsmith, C. Walker, D. J. Hollenbach, M. G. Wolfire, C. Kulesa, V. Tolls, P. N.Bernasconi, Ü. Kavak, F. F. S. van der Tak, R. Shipman, J. R. Gao, A. Tielens, M.G. Burton, H. York, E. Young, W. L. Peters, A. Young, C. Groppi, K. Davis, J. L. Pineda, W. D. Langer, J. H. Kawamura, A. Stark, G. Melnick, D. Rebolledo, G. F. Wong, S. Horiuchi and T. B. Kuiper, "Probing ISM Structure in Trumpler 14 and Carina I Using the Stratospheric Terahertz Observatory 2", Astrophys. J., vol. 878, no. 2, 2019.

[4] C. Walker, C. Kulesa, A. Young, W. Verts, J. R. Gao, Q. Hu, J. Silva, B. Mirzaei, W. Laauwen, J. Hesler, C. Groppi, and A. Emrich, Gal/Xgal U/LDB Spectroscopic/ Stratospheric THz Observatory: GUSTO, Proc. SPIE, vol. 12190, 2022, Art. no. 121900E.

[5] Jose V. Siles, Jorge Pineda, Jonathan H. Kawamura, Paul F. Goldsmith, et al., "ASTHROS: The astrophysics stratospheric telescope for high spectral resolution observations at submillimeter-wavelengths," Proc. SPIE 11445, Ground-based and Airborne Telescopes VIII, 114453G (2020).

[6] D. Rigopoulou, C. Pearson, B. Ellison, M. Wiedner, V. Ossenkopf Okada, et al. "The far-infrared spectroscopic surveyor (FIRSS). Experimental Astronomy," Springer Link, 2021, 51 (3), pp.699-728.

[7] C. K. Walker, G. Chin, S. Aalto, C. M. Anderson, J. W. Arenberg, C. Battersby, et al., "Orbiting Astronomical Satellite for Investigating Stellar Systems (OASIS): "Following Water from Galaxies, through Protostellar Systems, to Oceans"," Proc. of SPIE 11820, 11820-26 (2021).

[8] https://live-azs-saltus.pantheonsite.io/sites/default/files/2023-03/SALTUS_Intro_5March23.pdf

[9] José R. G. Silva, Wouter M. Laauwen, Behnam Mirzaei, Nathan Vercruyssen, Matvey Finkel, Menno Westerveld, Nikhil More, Vitor Silva, Abram Young, Craig Kulesa, Christopher Walker, Floris van der Tak and Jian Rong Gao, 4×2 Hot electron bolometer mixer arrays for detection at 1.46, 1.9 and 4.7 THz for a balloon borne terahertz observatory, to be submitted to IEEE TST 2023 (special issue of ISSTT).

[10] The NbN film was provided by TERAHERTZ AND INFRARED PHOTONICS, Moscow, Russia

[11] J. R. G. Silva, M. Finkel, W. M. Laauwen, S. J. C. Yates, B. Mirzaei, N. Vercruyssen, A. Young, C. Kulesa, C. Walker, F. van der Tak and J. R. Gao, "Beam waist properties of spiral antenna coupled HEB mixers at Supra-THz frequencies", IEEE Transactions on Terahertz Sci. and Tech., vol. 13, no. 2, pp. 167 – 177, 2023.

[12] T. M. Klapwijk and A. V. Semenov, "Engineering physics of superconducting hot-electron bolometer mixers", IEEE Trans. Terahertz Sci. Technol. Vol. 7, pp. 627–648, 2017.

[13] R. Lefèvre, Y. Jin, A. Féret, T. Vacelet, W. Miao, L. Pelay, F. Dauplay, M. Ba-Trung, G. Gay, J.-C. Villégier, J. Spatazza, J.-M. Krieg, and Y. Delorme, "Terahertz NbN hot electron bolometer fabrication process with a reduced number of steps", *in Proc. of 23rd ISSTT*, Tokyo, April 2012.

[14] W. Zhang, P. Khosropanah, J. R. Gao, E. L. Kollberg, K. S. Yngvesson, T. Bansal, R. Barends, and T. M. Klapwijk, "Quantum noise in a terahertz hot electron bolometer mixer," Appl. Phys. Lett. 96, 111113 (2010).

10.7 THz HEB Heterodyne Mixer Designs

J. Böhm*[1], K. Jacobs[1], M. Schultz[1], C.E. Honingh[1], and J. Stutzki[1]

Abstract— **We report on the development of a 10.7 THz Hot Electron Bolometer (HEB) mixer. The NbN HEB is fabricated on a 9μm silicon membrane with an integrated gold planar antenna which is attached to the backside of a silicon lens to couple to free space. The mixer designs encompass both narrow-band and broadband antennas. Direct response measurements of two HEB-devices are presented.**

Keywords—HEB mixer, Archimedean spiral antenna, double-slot antenna

I. INTRODUCTION

We are currently developing a Hot Electron Bolometer (HEB) mixer for the high resolution observation of molecular hydrogen at 10.7 THz. The observation of the weak emission feature of the electric quadrupole transition of H_2 at 10.7 THz (28 μm) requires high spectral resolution which only a heterodyne system can supply. The mixer uses a Niobium Nitride (NbN) HEB microbridge as the mixing element. This microbridge is integrated with a gold antenna on a high resistivity silicon substrate that is subsequently glued on a silicon lens. The mixer was initially intended to fly on the Stratospheric Observatory for Infrared Astronomy (SOFIA) [1] whose operational time was recently shortened to the end of September of 2022.

II. DESIGN

The receiver is based on a concept with integrated lens-antenna. A 9μm thick, high resistivity silicon membrane with an integrated planar antenna is directly attached to the flat backside of a silicon (Si) lens to couple to free space. The Si lens consists of an extended hemisphere with a radius of 0.5mm and a cylindrical extension of 164μm. The RF components are integrated in a tellurium copper housing with IF-board (see Fig.1). The front of the block has an about 14° slope to avoid reflections of a possibly not perfectly matched local oscillator (LO) beam, which would lead to standing waves in the LO path negatively influencing the stability of the receiver. The copper mixer housing is the reference towards the further receiver optics.

To enable a very accurate positioning of the antenna relative to the center of the lens and to the copper mixer housing the silicon membrane contains additional markers. With respect to both, the block-assembly and the lens-assembly markers, the antenna is referenced with an accuracy of approximately 1μm.

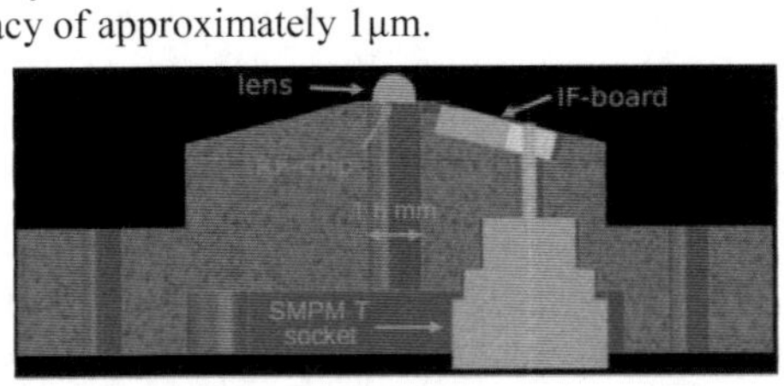

Fig. 1: Sketch trough the cut of the copper receiver housing.

[1]Institute of Physics I , Cologne, 50937, Germany

The broadband and narrow-band RF-circuits presented here are based on a double-slot antenna with RF-choke and an Archimedean spiral antenna. They are composed of a 70nm thick gold layer. A main challenge of the RF-circuit designs is to allow a good power transfer at the HEB-antenna interface within the constraints given by the high HEB layer resistivity and the natural limitation in the dimensions of the RF-structures given by the fabrication technique. The high frequency requires a smaller feed than can be reliable fabricated which leads to a non-vanishing imaginary part in the port-impedance for the Archimedean spiral antenna. The geometric specifications of the spiral antenna are given in Tab.I. The reflection at the HEB-antenna interface is depicted in Fig. 2 for different resistance values for the HEB microbridge in the center of the Archimedean spiral antenna.

Five different versions of a double-slot antenna have been fabricated to account for potential uncertainties for instance given by fabrication tolerances and uncertainties of the material properties (silicon at operating frequency and cryogenic temperatures[2], SiO_2 (bridge substrate), Au resistivity). The full-wavelength slots are between 9 to 10.3μm long, with slot-distances between 5 and 5.6μm. The RF-choke is directly attached to the antenna as shown in Fig.3. It blocks the RF-signal and presents a short circuit at the antenna side at the operating frequency. It consists of 7 lambda-quarter high and low impedance CPW lines. There are fabricated versions where the CPW line at the antenna interface partly is shortened/elongated to study the effect of the filter on the antenna resonance. The effect of the filter on the input impedance at the antenna terminal is exemplarily depicted in Fig.4 for the maximum and minimum value which is assumed to be obtained for the given resistivity and aspect ratio of the HEB microbridge integrated at the antenna terminal.

TABLE I.

Geometrical specifications

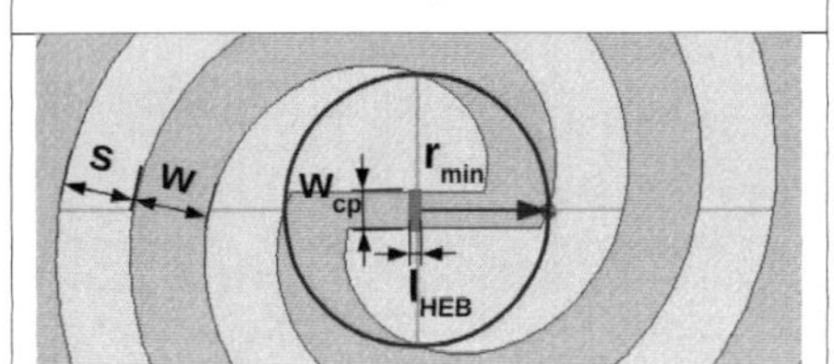

w	1 μm
s	1 μm
w_{cp}	0.5 μm
l_{HEB}	120 nm
r_{min}	1.74 μm
Number of turns	1.75

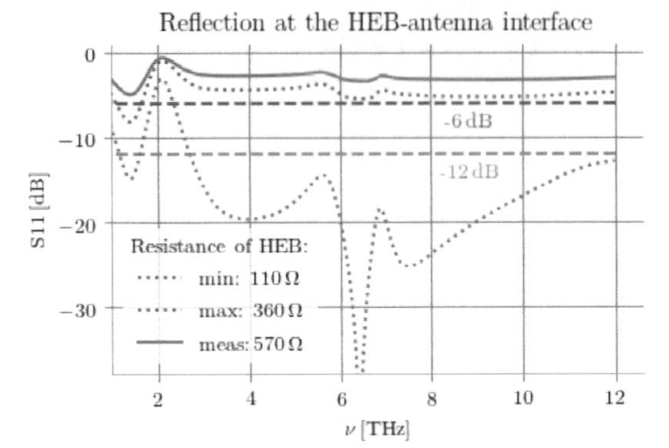

Fig. 2: S11-parameters at the HEB-antenna interface of the Archimedean spiral antenna for different resistance values of the HEB.

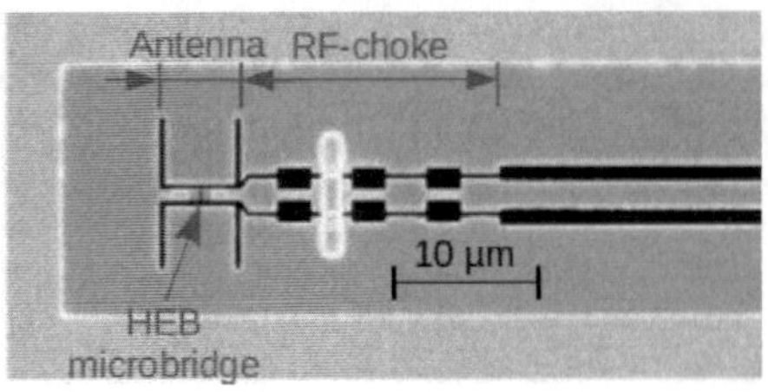

Fig. 3: SEM picture of fabricated double-slot antenna with RF-choke.

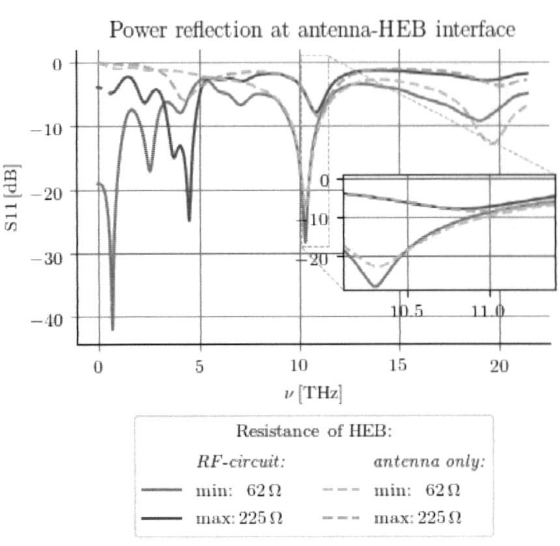

Fig. 4: S11-parameters at the antenna-HEB interface of the double-slot antenna for different resistance values of the HEB.

III. Measurements

First batches with different mixer designs have been fabricated using a combination of E-beam and UV lithography. The local oscillator for the Heterodyne measurement is still pending. Therefore first investigations of the fabricated devices are done using the mixers as direct detectors of a Fourier Transform Spectrometer (FTS) as sketched in Fig.5. The measurement frequency range is restricted by the optical components in the setup to 3 - 12 THz (see Fig.5). The HEB-devices are operated around their critical temperatures and are DC-biased in the constant current mode via a Bias-T, with a stable bias point in the IV curve of the HEB set as indicated as black dot in Fig.6. The spectra have been measured with the maximum FTS resolution of 0.08 cm^{-1} as an average over 5 to 12 spectra.

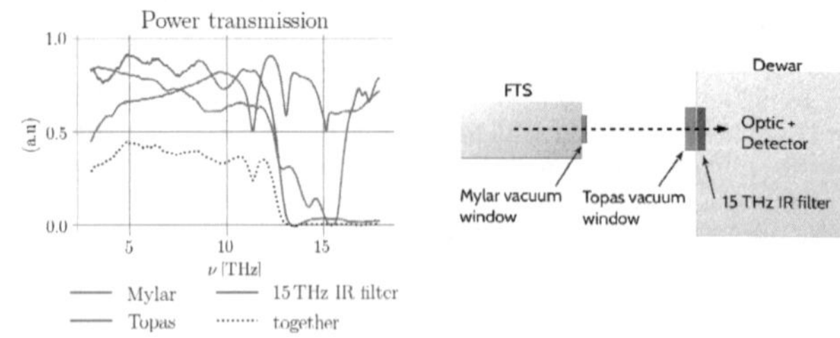

(a) (b)

Fig. 5: (a) Transmission characteristics of different optical components used in the Fourier Transform Spectrometer setup and (b) their location in the FTS setup.

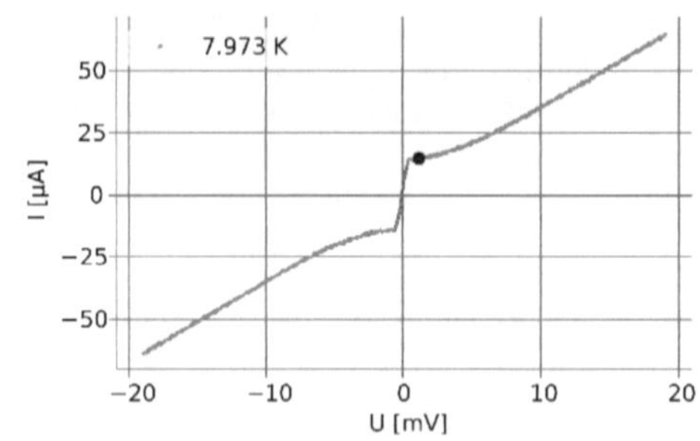

Fig. 6: IV curve of an HEB-device operated around its critical temperature. The black dot indicates the typical location of the bias point chosen for operation.

IV. Results

From a batch with in total 80 RF-chips so far two HEB-devices have been characterized, which are based on one double-slot antenna and one Archimedean spiral antenna, respectively. The raw and calibrated response spectra are depicted in Fig. 7 and Fig .8. The calibration takes into account the transmission characteristic of the FTS and dewar vacuum windows, the IR-filter and the transmission spectrum of the empty FTS, which is measured with the internal detector of the FTS and mainly depicts the spectral transmission of the FTS-beamsplitter. The very large attenuation of the FTS-signal contributed by the sum of all optical components at 11.4 THz does not allow a reasonable assessment of the measured spectrum around that frequency and yields an artefact in the calibrated spectra.

The Archimedean spiral antenna based HEB-device shows a slightly wavy steady response signal over the whole measurement frequency range of the setup as to be expected from simulation results (see Fig .7). For the HEB-device with the double-slot antenna (see Fig. 8) a distinct linearly polarization-dependent response behavior could be shown by means of a grid placed in the optical pathway between FTS and dewar window. The device exhibited maximum response in the dedicated linear polarization direction of the antenna and no response for the respective cross-polarized radiation. The response spectrum (see Fig. 8) exhibits a response peak around 10 THz and one pronounced peak around 7.5 THz. The response peak around 10 THz corresponds to the full-wavelength resonance of the antenna. Based on simulation results carried out with CST Microwave studio[3], the response peak around 7.5 THz likely corresponds to a resonance, where the double slot antenna behaves as four, in-phase driven lambda-quarter slot pieces.

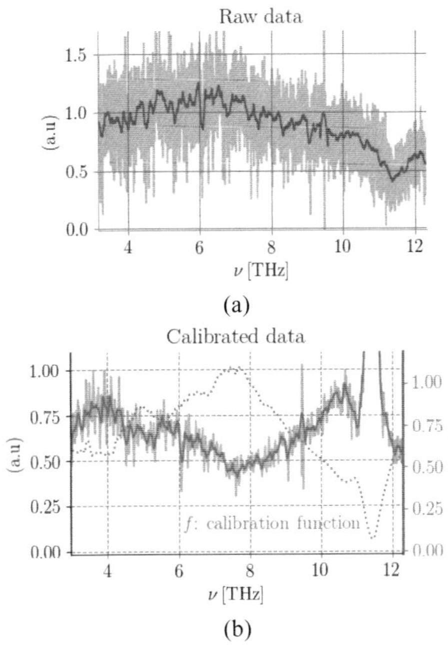

Fig. 7: Spectral direct response ((a) raw and (b) calibrated data) of HEB-device based on the Archimedean spiral antenna.

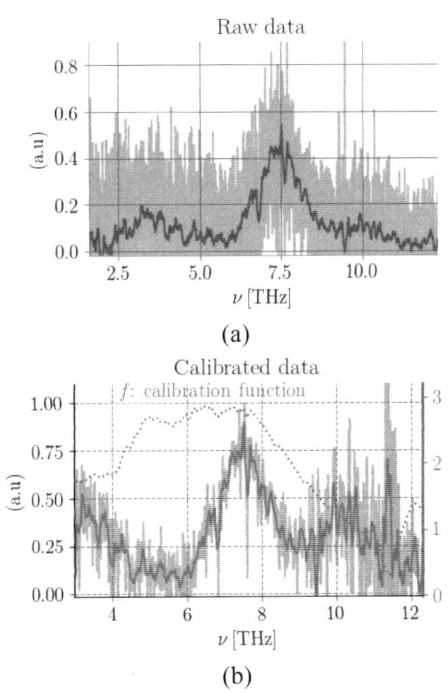

Fig. 8: Spectral direct response ((a) raw and (b) calibrated data) of HEB-device based on the double-slot antenna with RF-choke (*Antenna*: slot-length: 9µm, slot-distance: 5.6µm, slot-width: 0.3µm, *CPW between slots*: inner conductor width: 0.8µm, slot width: 0.3µm).

V. Conslusion

Direct response of HEB-devices based on a narrow-band and a broadband antenna have successfully been measured within a frequency range from 3 - 12 THz. Measurements of further HEB-devices including double-slot antennas, Archimedean spiral antennas and logarithmic spiral antennas will follow soon with an improved FTS-setup with a better signal-to-noise-ratio around 11.4 THz.

References

[1] J. Böhm, et al., "Development of a HEB mixer for the observation of molecular hydrogen on SOFIA." *ISSTT 2019: 30th International Symposium on Space Terahertz Technology*, 2019.

[2] J. Böhm, et al., "Cryogenic temperature-dependence of HRFZ Silicon refractive index from Mid-to Far-IR down to 5 K." *2021 46th International Conference on Infrared, Millimeter and Terahertz Waves (IRMMW-THz). IEEE*, 2021

[3] Computer Simulation Technology GmbH. CST - Computer Simulation Technology.

Noise temperature and *S*-parameter measurements of an SIS-based microwave amplifier

T. Kojima*[1], Y. Uzawa[1], S. Masui[1, 2], Y. Kozuki[1], Y. Fujii[1], K. Makise[1], W. Shan[1]

***Abstract*—In this presentation, we report on characterization of a superconducting microwave amplifier based on superconductor-insulator-superconductor (SIS) mixers, that has comparable performance with conventional cryogenic low-noise amplifiers based on semiconductor devices.**

Keywords—SIS mixer, superconducting amplifier, low noise, microwave

I. Introduction

We have studied a novel low-noise and wideband microwave amplifier operating at a physical temperature of 4 K and a low-power consumption of a few microwatts [1, 2]. The amplifier uses two quasiparticle superconductor-insulator-superconductor (SIS) mixers connected as frequency up- and down-converters in cascade. Since the input and output microwave frequencies can be the same by injecting an identical local oscillator (LO) power at a millimeter wavelength into each mixer, it is possible for this amplifier configuration to have positive gain in total utilizing the bilateral gains in their frequency up and down conversion processes. Our primary target of this study here is to demonstrate the superior performance comparable with conventional cryogenic low-noise amplifiers based on semiconductor devices such as a high electron mobility transistor or heterojunction bipolar transistor.

In this study, we have configured this amplifier using two heterodyne SIS mixer modules as frequency up- and down-converters at W-band (75-110 GHz) developed for the four-beam receiver system on the 45-m telescope (FOREST) [3] and inserting a W-band cryogenic waveguide isolator (2.54 mm x 1.27 mm) between these SIS mixers (see Fig. 1). The W-band cryogenic isolator, produced by Micro Harmonics inc., has an insertion loss less than 1 dB and isolation more than 25 dB across the 75-110 GHz at cryogenic temperature. It minimizes deterioration of amplifier gain and noise temperature and can independently operates the SIS mixers in terms of impedance at the up-converted frequencies. For the independent controllability of LO powers into each SIS mixer, W-band waveguide attenuators have been inserted in the respective LO paths at room temperature. In addition, another simulation indicates that phase rotation of the up-converted signals between the SIS mixers and relative phase difference between two LO powers at the SIS mixers are critical for the amplifier gain. Therefore, a W-band waveguide phase shifter was inserted into one of the LO branches. For the noise temperature measurement setup, we inserted a 20 dB attenuator at the input of this amplifier to avoid the gain saturation and a different low-noise amplifier which covers 0.5-8.0 GHz at the output of the SIS down converter.

The measurement result showed 6 dB gain and 12 K noise temperature up to 5 GHz in typical at an LO frequency of 87.4 GHz. This result is quite comparable with the cryogenic low-noise amplifier based on semiconductor transistors. We also plan to measure *S*-parameters of the amplifier at the microwave frequencies using a vector network analyzer. In this symposium, we will present the detailed characterization, analysis results, and future prospective of this low noise amplifier.

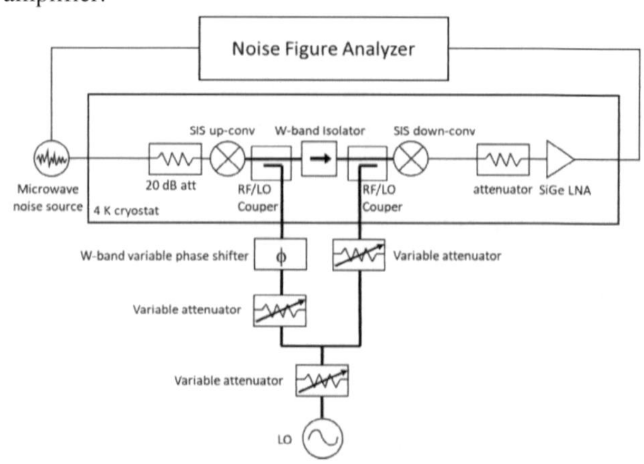

Fig. 1. Schematic of noise temperature measurement setup for the SIS-based amplifier.

Acknowledgements

This work was supported in part by JST [Moonshot R&D] Grant Number [JPMJMS2067] and JSPS KAKENHI Grant Numbers JP18H03881 and JP19H02205.

References

[1] T. Kojima et al., "Microwave amplification based on quasiparticle SIS up and down frequency converters," AIP Advances 8, 025206, 2018

[1]Advanced Technology Center, National Astronomical Observatory of Japan (NAOJ), Mitaka, Tokyo 181-8588, Japan; [2]Graduate School of Science, Osaka Prefecture University, Sakai, Osaka 599-8531 Japan;

NOTES:

[2] Y. Uzawa et al, "An SIS-mixer-based amplifier for multi-pixel heterodyne receivers", *Proc. SPIE*, Dec. 2020.
[3] T. Minamidani, et al., "Development of the new multi-beam 100 GHz band SIS receiver FOREST for the Nobeyama 45-m Telescope," *Proc*. of SPIE, vol. 9914, 99141Z, 2016.

NOTES:

Wideband Permittivity Measurement System in the 67-116 GHz Range

R. Sakai, A. Gonzalez, K. Kaneko, T. Kojima, and Y. Uzawa

***Abstract*—In this study, we present an accurate permittivity measurement system for the 67-116 GHz band and measure dielectric properties of some low-loss polymers. The results obtained were in good agreement with other measurements reported in the literature, demonstrating the validity of the measurement system. The developed free space method allows the measurement of material characteristics with the same system over a large continuous bandwidth, which is key for accurate calibration and results. This is key for the design of high-performance receiver optics components to be used in high-performance application, such as radio astronomy receivers.**

***Keywords*— permittivity measurement, receiver**

I. Introduction

In radio astronomy, it is necessary to measure the permittivity accurately for the design of optical components in order to develop a highly sensitive telescope receiver. Accurate permittivity values are also essential for the optics design of the Atacama Large millimeter/sub-millimeter Array (ALMA) Band 2 receiver, which covers over 67-116 GHz [1].

The free space method has several advantages, such as the possibility of continuous measurement over a wide bandwidth, and the non-contact method, which does not require special jigs or high-precision processing of thin or small samples. In the development stage of an optical system, it is necessary to grasp as quickly as possible a wide variety of materials with different types, lots, and cut-out parts. Therefore, the free space method is suitable because of its ease of sample processing. However, there were few examples of measurements in the wide band of 67-116 GHz without changing the configuration of the equipment, and discontinuities occur in the band, making it difficult to measure accurately. In this study, we developed a system and measured the permittivity of some low-loss polymers in the wide band from 67 to 116 GHz continuously and accurately.

II. Permittivity Extraction

A. Measurement Setup

The measurement setup consisted of a vector network analyzer (VNA), N5225B PNA Series, Keysight Technologies, frequency extender modules, VNAX WR-10+, Virginia Diodes Inc., and an optical system. The detailed optical system design was described in [2]. The optical system was designed with the Gaussian beam theory [3]–[5]. The corrugated horn was designed to have low-reflection loss and high-gaussicity in the operating frequency to make the system works as designed [6]. In the measurement process, *S*-parameters are measured with a metal plate or no object in the beam path. These *S*-parameters are needed to normalize the reflection and transmission coefficients of the sample and to derive the error terms for the 2-port network model.

National Astronomical Observatory of Japan, Mitaka 181-8588, Japan

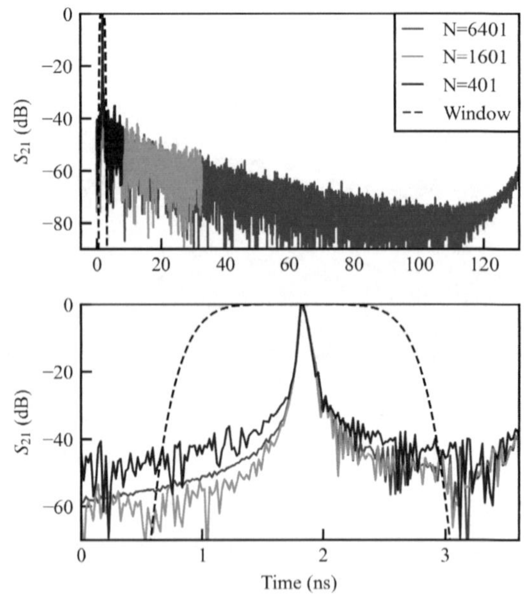

Fig. 1. Measured S_{21} in the time domain for each N.

B. Preprocessing

Before estimating the permittivity by applying the model equation to the measurement data, preprocessing is necessary because the raw data contains reflections due to impedance mismatching at the horn and multiple reflections in the optical system. Therefore, filtering with a window function in the time domain is performed to suppress these effects. First, the measurement data in the frequency domain is transformed into the time domain data using the Discrete Fourier Transform (DFT). After DFT, the time-domain data are multiplied by a window function to minimize the influence of undesired components. When applying DFT, we should carefully select the measurement bandwidth, Δf, and the number of points, N. The measurement range in the time domain is represented by $(N-1)/\Delta f$. Therefore, when Δf is constant, the measurement range in the time domain is proportional to N. According to Perceval's theorem shown in (1), the integral values of the noise power in the time and frequency domains are preserved before and after the Fourier transform. In other words, by selecting a large N, the noise level per unit of time becomes small. By multiplying a window function of appropriate width near the desired signal and performing the DFT, the noise in the frequency domain can be suppressed, resulting in an accurate estimation of the permittivity.

$$\sum_{n=0}^{N-1} |x[n]|^2 = \frac{1}{N} \sum_{k=0}^{N-1} |X[k]|^2 \quad (1)$$

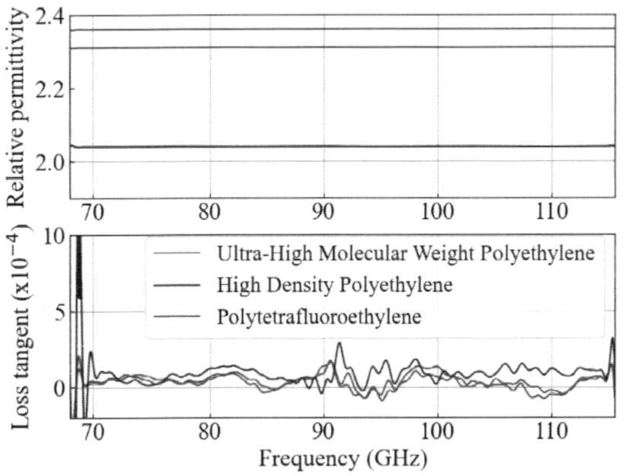

Fig. 2. Extracted relative permittivity (top) and loss tangent (bottom).

where n is the discrete time, k is the discrete angular frequency, $x[n]$ is the discrete-time signal, and $X[k]$ is the DFT of $x[n]$. To investigate the optimal N, we compared the noise level with varying N. Fig. 1 shows the measured S_{21} in the time domain for each N. Since the integrated noise power is preserved in the time domain, the noise level per unit of time decreases as N increases. The larger N, the smaller the noise level in the preprocessed data. However, a larger N increases the frequency sweep time and uncertainty due to the time stability of VNA, so the optimal N was set to 6401 points by balancing them. After preprocessing the above, we estimate the permittivity iteratively.

C. Data Analysis

The model formula used in this study was the 4-parameters method given by the following equations [7].

$$\det S = S_{11}S_{22} - S_{21}S_{12} = \frac{\gamma^2 - \tau^2}{1 - \gamma^2\tau^2} \quad (2)$$

with

$$\gamma = \frac{1/\sqrt{\varepsilon_r} - 1}{1/\sqrt{\varepsilon_r} + 1}, \qquad \tau = \exp(-jk_0 t\sqrt{\varepsilon_r}) \quad (3)$$

where ε_r is the complex relative permittivity, k_0 is the wave number, and t is the thickness of the dielectric slab. The permittivity is estimated by minimizing the difference between the determinant of the measured S-parameter and (2) using an optimization solver.

III. Experimental Results

Fig. 2 shows the measurement results for several types of low-loss polymers, ultra-high weight molecular weight polyethylene (UHMWPE), high density polyethylene (HDPE), and polytetrafluoroethylene (PTFE), which are commonly used for dielectric lenses or vacuum windows. The extracted permittivity of UHMWPE, HDPE, and PTFE are 2.307, 2.360 and 2.045, respectively. The measurement results are consisted with those of reported in [8].

IV. Conclusion

In this study, we measured the dielectric properties of low-loss polymers commonly used for optical components in radio telescopes with the permittivity measurement system in the 67-116 GHz band. The obtained results are in good agreement with previous studies in other bands, demonstrating the validity of the measurement system. This study will enable the high-accurate design of optical components, i.e., dielectric lenses, and vacuum windows, in the 67-116 GHz range.

References

[1] P. Yagoubov *et al.*, "Wideband 67 − 116 GHz receiver development for ALMA Band 2," *Astron. Astrophys.*, vol. 634, 2020.

[2] R. Sakai *et al.*, "Accurate Free-Space Measurement of Complex Permittivity with the Angular Spectrum Method," *IEEE Trans. Terahertz Sci. Technol.* (in press)

[3] Paul F. Goldsmith, *Quasioptical Systems: Gaussian Beam Quasioptical Propogation and Applications*. Wiley-IEEE Press, 1998.

[4] A. Gonzalez, "Frequency Independent LO Injection in Quasi-Optical Heterodyne Systems," *IEEE Trans. Terahertz Sci. Technol.*, vol. 5, no. 4, pp. 546–550, 2015, doi: 10.1109/TTHZ.2015.2434107.

[5] A. Gonzalez, "Frequency Independent Design of Quasi-optical Systems," *J. Infrared, Millimeter, Terahertz Waves*, vol. 37, no. 2, pp. 147–159, 2016, doi: 10.1007/s10762-015-0205-1.

[6] A. Gonzalez, K. Kaneko, and S. Asayama, "Recent work on (Sub-)Mm-wave ultra wideband corrugated horns for radio astronomy," *2017 11th Eur. Conf. Antennas Propagation, EUCAP 2017*, pp. 3327–3331, 2017, doi: 10.23919/EuCAP.2017.7928625.

[7] J. Baker-Jarvis, E. J. Vanzura, and W. A. Kissick, "Improved technique for determining complex permittivity with the transmission/reflection method," *IEEE Trans. Microw. Theory Tech.*, vol. 38, no. 8, pp. 1096–1103, 1990, doi: 10.1109/22.57336.

[8] C. Y. E. Tong, K. Carter, and J. Connors, "Quasi-optical characterization of low-loss polymers at 300 GHz for vacuum window applications," *IEEE Trans. Terahertz Sci. Technol.*, vol. 10, no. 6, pp. 713–720, 2020, doi: 10.1109/TTHZ.2020.3021352.

Applying Energy Absorption Interferometry to THz direct detectors using photomixers

Ian Veenendaal*[1,2], Edgar Castillo Dominguez[1,3], Stephen Yates[1], Bram Lap[1,4], and Willem Jellema[1,4]

Abstract—**The sensitivity and spatial form of the individual optical modes to which a potentially few-moded direct detector is sensitive can be determined by the complex valued coherence matrix from two identical sources as they are scanned through the field of view. We provide experimental verification of this technique using THz photomixers and MKIDs.**

Keywords—Energy Absorption Interferometry, photomixer, partial coherence, verification techniques

I. Introduction

Astronomical imaging and spectroscopy in the THz domain typically involves the study of formation and evolution of planetary, stellar, and galactic systems. Resolution, throughput, and bandwidth requirements for instruments operating at these wavelengths are all key parameters which drive limitations on the size and density of detectors [1, 2]. It is generally the case that these requirements drive the optimum size of detectors to be similar to the wavelengths measured, such that they exhibit a few-moded response to incoming radiation. In such a regime, the detector beam pattern cannot be fully represented by a fully coherent response, as is the case for single-moded detectors. In addition, the system also cannot be fully described using geometric optics and radiative transfer models, such as typically used in the infrared and visible domains. Understanding few-moded power absorbing detectors requires a method to measure the number and relative sensitivities of the natural modes, as well as their spatial forms.

In this work, we applied Energy Absorption Interferometry (EAI) [3], which is a technique by which the coherent properties of any power absorbing structure can be measured, to an end-to-end THz optical system including detectors. In doing so, we demonstrated we were able to recover a complex valued correlation matrix which described sensitivity of the system in terms of individually coherent natural modes.

We introduced two independent source probes in the field of view of an optical system, which while being spatially decoupled, exhibited identical beam profiles, were frequency matched, and the relative phase between the sources could be adjusted. The sources were independently scanned throughout the field of view, and at each pair of positions, the relative phase was varied and an interference fringe was measured. Extracting the phase and amplitude of the fringe allowed us to construct a directly measured representation of the spatial correlation matrix of the complete optical system. Such a matrix could be propagated through the optical system, and decomposed into a weighted set of orthogonal functions which represented the individually coherent optical modes of the system.

II. Theory

Wolf shows that λ_i and ψ_i, which are the eigenvalues and eigenfunctions which describe the natural modes and relative sensitivities of a partially coherent beam, are eigenfunctions and eigenvalues of the following equation [4],

$$\int_D W\,(\mathbf{r}_1, \mathbf{r}_2) \cdot \psi_n(\mathbf{r}_1) d^3\mathbf{r}_1 = \lambda_n \psi_n(\mathbf{r}_2),$$

Where $W(\mathbf{r}_1, \mathbf{r}_2, \nu_0)$ is the cross-spectral density of an optical beam evaluated at a single frequency, ν_0, and $\boldsymbol{r}_i$ are position vectors within the domain D,

Withington et al. show that the power coupling of the system to a source is a projection of the cross-spectral density of the detector beam, $\overline{\overline{D}}(\mathbf{r}_1, \mathbf{r}_2)$, as well as the cross-spectral density of the source field, $\overline{\overline{E}}(\mathbf{r}_1, \mathbf{r}_2)$ evaluated over an arbitrary plane S which forms a cross-section of the beam [3],

$$\langle P \rangle = \iint_{S^2} \overline{\overline{D}}(\mathbf{r}_1, \mathbf{r}_2) \cdot\cdot\, \overline{\overline{E}}(\mathbf{r}_1, \mathbf{r}_2) d^2\mathbf{r}_1 d^2\mathbf{r}_2.$$

By illuminating a system under test with two independent monochromatic point sources, the power measured at the detector can be represented by [5],

$$\langle P(\phi) \rangle = D_{m,m} + D_{m',m'} + 2\,|D_{m,m'}| \cos(\phi + \theta_{m,m'}),$$

Where m, m' are discrete position indices of the two sources, $|D_{m,m'}|$ and $\theta_{m,m'}$ are the amplitude and phase of the discretized version of complex-valued cross-spectral density at coordinates m and m'. If the relative phase difference between the sources, ϕ, can be systematically adjusted, $|D_{m,m'}|$ and $\theta_{m,m'}$ can be measured as the complex amplitude of the resulting fringe. Repeating over all possible pairs of position coordinates with a plane with sufficient sampling and size recreates the full discretized cross-spectral density matrix of the system under test, which we call the detector response function (DRF).

III. Setup

We created two phase-matched, frequency-tunable monochromatic THz source by photomixing two infrared lasers in two identical GaAs photomixers [6]. The two sources are driven with the same pair of lasers, using a 2x2 fiber optic splitter. Phase rotation is applied with fiber stretchers, which modify the differential length between the optical paths [7]. Each source is mounted on a motorized xyz-stage, and the output THz beams are coupled together in free-space with a Mylar beamsplitter.

[1]SRON Netherlands Institute for Space Research, Groningen, the Netherlands; [2]Astronomical Instrumentation Group, Cardiff University, Cardiff, United Kingdom; [3]Department of Astrophysics, University of Oxford, Oxford, United Kingdom; [4]Kapteyn Astronomical Institute, University of Groningen, Groningen, the Netherlands

IV. Results

The DRF was produced by recording the amplitude and phase of the fringe pattern for each source position pair. The amplitude and phase of the complex DRF is shown in Fig. 1. We limited the source position points to include only a single cut of the focal plane, but at a diagonal between the E- and H- planes such that the cut intersected non-zero regions of both the simulated co- and cross- polarization responses of the system.

Using a technique similar to angular plane wave spectrum (APWS) decomposition and propagation [8], we were able to propagate the measured DRF from an arbitrary measurement plane to an image plane of the system under test. This approach also allowed for spatial filtering of the DRF to limit the impact of detector noise and straylight on the recovery of the natural modes.

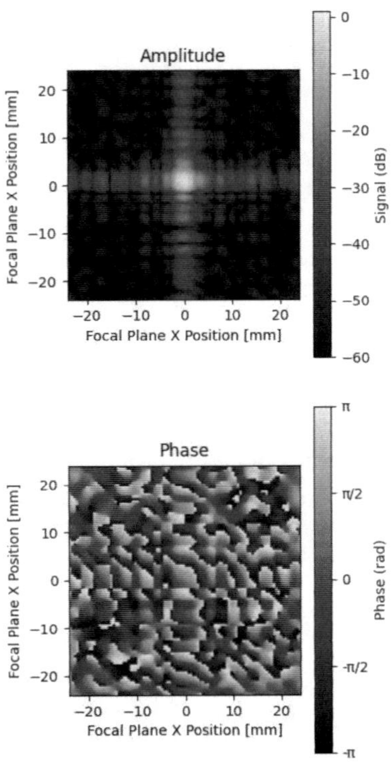

Fig. 1. Amplitude and phase of the complex valued detector response function (DRF) produced by measuring correlations between two point sources over a one-dimensional cut of an instrument beam pattern.

A diagonalization of the DRF matrix, utilizing singular value decomposition (SVD) allowed the extraction of the eigenvalues and eigenvectors, which corresponded to the sensitivity and spatial form of the natural modes (Fig. 2).

The demonstration described in this work proved that this detector could be described as having a single dominant spatial mode and that cross-pol and surface wave stray light can be described as coherent with the main antenna response, at least within measurement accuracy. This would not be expected in a intrinsically multi-moded detector, like a distributed absorber detector. Application of this technique to such detectors will be a topic of future focus, as it has implications for how they couple, particularly for application in wideband instruments such as Herschel/SPIRE and SPICA/SAFARI [9].

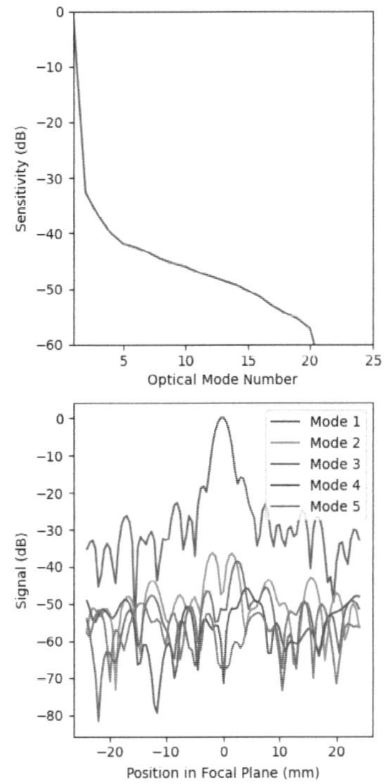

Fig. 2. Extracted sensitivity (top) and spatial form (bottom) of the natural modes extracted from the one-dimensional DRF.

References

[1] R. Wu et al., "Observing extended sources with the Herschel SPIRE Fourier Transform Spectrometer," *A&A*, Aug. 2013.

[2] S. Shu, et al., "Prototype High Angular Resolution LEKIDs for NIKA2," *J Low Temp Phys*, Nov. 2018.

[3] S. Withington et al., "Probing quantum correlation functions through energy-absorption interferometry," *Physical Review A*, Aug. 2017.

[4] E. Wolf, "New theory of partial coherence in the space–frequency domain Part I: spectra and cross spectra of steady-state sources," *J. Opt. Soc. Am*, Mar. 1982.

[5] D. Moinard, et al., "Probing infrared detectors through energy-absorption interferometry," in *SPIE Infrared Sensors, Devices, and Applications VII*, Aug. 2017.

[6] S. Preu, et al., "Tunable, continuous-wave Terahertz photomixer sources and applications," *Journal of Applied Physics*, 2011.

[7] A. Roggenbuck et al., "Using a fiber stretcher as a fast phase modulator in a continuous wave terahertz spectrometer," *Journal of the Optical Society of America B*, 2012.

[8] K. Davis, et al., "Analysis techniques for complex-field radiation pattern measurements," *SPIE Millimeter, Submillimeter, and Far-Infrared Detectors and Instrumentation for Astronomy IX* p. 96, Jul. 2018.

[9] B. Lap, et al., "Modeling the partially coherent behavior of few-mode far-infrared grating spectrometers," *J. Opt. Soc. Am. A*, Jul. 2022.

Space qualification of MMIC Schottky diodes chips for SWI instrument of JUICE mission

J. Valentin*[1], L. Gatilova[1], J. Treuttel[1],
A. Feret[1], S. Caroopen[1], G. Gay[1], T. Vacelet[1], S. Mignoni[1], J.-M. Krieg[1], Y. Jin[2], A. Cavanna[2],
J.-L. Roux[3], C. Goldstein[3], F. Courtade[3], C. Larigauderie[3], D. Moirin[4], J.-L. Espaignol[5],
V. Lavignolle[5] and A. Maestrini[1,6]

***Abstract*—The Sub-millimeter Wave Instrument (SWI) is a heterodyne spectrometer onboard JUpiter ICy moons Explorer (JUICE) mission working in two spectral ranges around 1200GHz and 600GHz with a spectral resolving power of about 10^7. The frequencies conversions in the receivers are based on Schottky diodes integrated on MMIC type chips fabricated on GaAs substrates. The contribution from LERMA[1] to the instrument includes the chips dedicated to the 300GHz doublers of both channels and the 600GHz doubler and 1200GHz subharmonic mixer of the 1200GHz channel. These chips were manufactured in the clean room of C2N-CNRS[2] in the frame of a long-term partnership between LERMA[1] and C2N. These chips have been qualified for integration into SWI for the JUICE mission. The presentation will give an overview of the chip qualification process and results.**

Keywords—Schottky diode, GaAs, heterodyne detection, submillimeter wave, THz, space qualification, SWI, JUICE, Jupiter.

I. Introduction

Interplanetary space missions such as JUpiter ICy moons Explorer (JUICE) implies particularly severe constraints on devices reliability over a time range that includes approximately 8 years cruising to Jupiter and operations in varied environments. In addition to the requirement for high scientific return, high technological demands are required on the instruments. The Sub-millimeter Wave Instrument includes a state-of-the-art heterodyne receiver technology based on Schottky diodes MMIC chips operating at frequencies as high as 1200GHz. These devices are research laboratory products: they are produced on a small scale, unlike industrial components which benefit from mass production capabilities.

The approach chosen for qualifying these parts was based on ESCC and MIL standards, with several adaptations to fit into the scientific program needs and constraints (funds, technical, process). It covered two main objectives, to ensure that : i) the active components perform according to their requirement specifications and in the JUICE mission environment, ii) the assemblies (components bonding, substrate mounting …) withstand the JUICE environment specifications (thermal, vibrations, mechanical shocks, …). We will present the approach methodology and main results obtained during the qualification program.

II. Methodology

The contribution from LERMA to the SWI includes the 300GHz doubler of the 600GHz channel (D300X1 block), and the 300GHz doubler (D300X2 block), the 600GHz doubler (D600 block) and the subharmonic mixer (SHM1200 block) of the 1200GHz channel. All these parts include MMIC GaAs-based Schottky diode chips. These chips were designed by LERMA-Observatoire de Paris and fabricated in the clean room of C2N-CNRS specially for JUICE-SWI instrument. Each block contains 1 chip, except the D300X2 which contains 2 chips per block. The qualification process described below was applied both to the chips and to the blocks themselves.

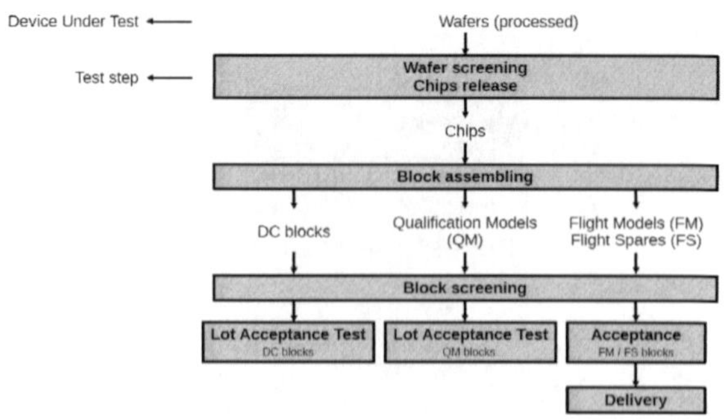

Fig. 1. Overview of qualification methodology.

Fig. 1 summarizes the qualification methodology. The qualification process begins at the end of chips fabrication. Three chip types were fabricated (Fig. 2): 300GHz doubler type (D300 chip), 600GHz doubler type (D600 chip) and 1200GHz subharmonic type (SHM1200 chip). The batches selected for flight provided 140 D300 chips, 140 D600 chips and 119 SHM1200 chips. Systematic measurements of current versus voltage characteristics were performed on all chips by probe testing on wafers before the chips release.

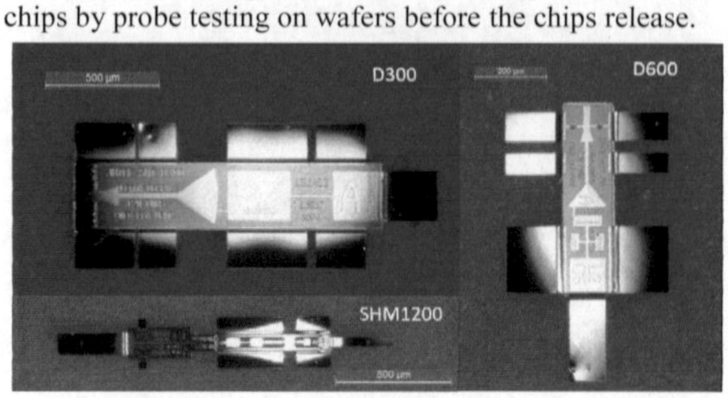

Fig. 2. Photos of D300, D600 and SHM1200 chips after release.

TABLE I. Wafer screening specification for 1 diode

Chip type	Series resistance (Ω)		Ideality factor	
	Min.	Max.	Min.	Max.
D300	5	10	1.12	1.22
D600	15	25	1.15	1.25
SHM1200	35	55	1.35	1.65

[1]Observatoire de Paris, LERMA, Paris, 75014, France; [2]Université Paris-Saclay, CNRS, C2N, Palaiseau, 91120, France; [3]CNES, Toulouse, 31401, France; [4]Hensoldt Space Consulting, Toulouse, 31100, France; [5]Bureau Veritas Group, Toulouse, France ; [6]Jet Propulsion Laboratory, California Institute of Technology, 4800 Oak Grove Driva, Pasadena, CA 91109, USA

Table I shows the specification applied to the wafer screening. The values are related to one single Schottky diode. Following the wafer screening, the chips were released and integrated in various blocks.

Two block types were used during the qualification process: i) the DC test blocks, which had no RF functionality and ii) the Qualification Model (QM) blocks, which were identical to Flight Model (FM) and Flight Spare (FS) blocks. The diagram on Fig. 1 also includes the process followed by the FM and FS blocks as these ones had communal steps with QM blocks.

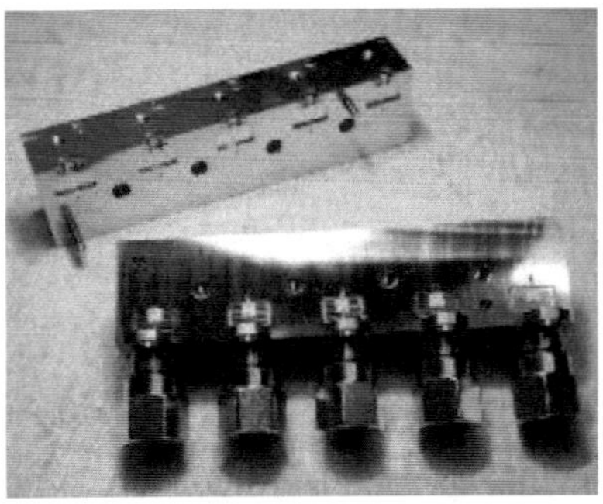

Fig. 3. Photo of a DC test block during the chips integration.

The DC blocks were intended to test a significant number of chips. Each DC block was containing 5 chips of the same type (Fig. 3). Each chip was connected to its individual SMK connector and could be controlled independently in static regime. The qualification program included 3 DC D300 blocks (15 chips), 2 DC D600 blocks (10 chips) and 2 DC SHM1200 blocks (10 chips).

The QM blocks were intended to test the chips under their JUICE operational conditions. Each QM block was first individually submitted to DC and RF measurements in order to check its performance. Then, the D300X2, D600 and SHM1200 blocks were assembled together to form 1200 GHz chains (Fig. 4) and the following tests were performed on assembled chains.

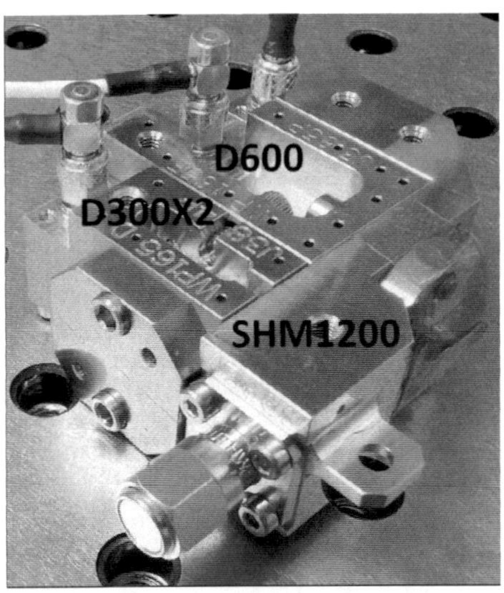

Fig. 4. 300GHz doubler, 600GHz doubler and 1200GHz subharmonic mixer assembled in 1200GHz chain before qualification steps.

All DC, QM and FM/FS blocks were submitted to the block screening tests: thermal cycling (5 cycles), power burn-in and High Temperature Reverse Bias (HTRB) tests. These tests were intended to detect possible early failures on the devices. DC and RF measurements were done before and after each test to check the possible evolution of the ideality, series resistance, doublers output power and receiver sensitivity. These values were compared to the specification.

TABLE II. BLOCK SCREENING CONDITIONS ON QM/FM/FS BLOCKS (BIASING AS SEEN BY EACH SINGLE DIODE)

Test	General conditions	Biasing conditions		
		D300	D600	SHM1200
Thermal cycling	-160°C / +80°C 5 cycles	No bias	No bias	No bias
HTRB	48h 60°C	-5.5V (<3µA)	-4V (<0.5µA)	No bias
Power burn-in	48h 100°C	+2.5mA (<0.8V)	+0.5mA (<0.8V)	+30µA

After the block screening sequence, the DC and QM blocks were submitted to extensive tests (Lot Acceptance Test) intended to model the mission profile at levels higher than the highest planned operational levels. The tests include thermal cycling (50 cycles), humidity, radiation and DC life test for DC blocks and thermal cycling, shocks, vibration, RF life test and destructive physical analysis for QM blocks. The FM/FS blocks were submitted to moderate acceptance tests in view of their integration in the SWI.

III. Screening Results

Fig. 5 shows an example of current versus voltage curves measured during the wafer screening. It should be noticed that the presence of 2 diodes in series on the D300 chip results in a curve slope 2 times lower. Fig. 6 shows ideality factor values obtained from the SHM1200 chips. These values were obtained from current setpoints ranging from 1µA to 10µA using the Shockley equation. On the selected batches, the wafer screening specification yield values were 86%, 93% and 71% for D300, D600 and SHM1200 chips respectively.

Table II shows the conditions applied during the block screening on RF blocks. For each test, the biasing conditions are detailed. These conditions refer to the voltage and current as seen by each single diode of the chip. For chip protection, current or voltage limits were fixed, their values are given in brackets.

It should be noticed, that each QM/FM block includes a filtering circuit after the DC ports, allowing to protect the circuits from potential electrostatic discharges. The drawback is that for the mixer diodes, the DC current cannot be directly measured.

The mixer forward current was measured thanks to refined measurement conditions (see Fig. 8). The figure shows the equivalent circuit in static measurement conditions. The nominal values of R_1 and R_2 were around 600 ohms and 400 ohms. The IF port was kept open during the measurement. At low voltage, an accurate value of R_1+R_2 was measured. At high voltage, the diode current was extracted by using this measured value together with the nominal value of R_2 as given by the relation:

$$I_D = \frac{I(R_1 + R_2) - U}{R_2}$$

In the same way, the diodes voltage is obtained by the relation:

$$V_D = V - R_1 \cdot I$$

Thanks to the classical Shockley diode equation, the ideality factor η can be extracted by calculating the slope between V_D (abscissa) and $\log(I_D)$ (ordinate) and by using the relation:

$$\eta = \frac{q}{slope \cdot k \cdot T}$$

With q the elementary electrostatic charge, k the Boltzmann constant and T the temperature of the diodes. The slope was calculated for $V_D > 1.30$V. This method was used to control the ideality factor value on SHM1200 QM/FM/FS blocks before and after the tests and also to control the 30μA current during power burn-in test.

Fig. 7 shows the ideality factor variation during the block screening sequence on QM/FM/FS blocks. All values, including series resistance, receiver sensitivity or RF output power (D300X1), were within specification.

IV. Conclusion

The qualification tests were performed on LERMA's hardware for the SWI of JUICE ESA mission. The qualification process ended successfully and the hardware was delivered for its integration in the SWI.

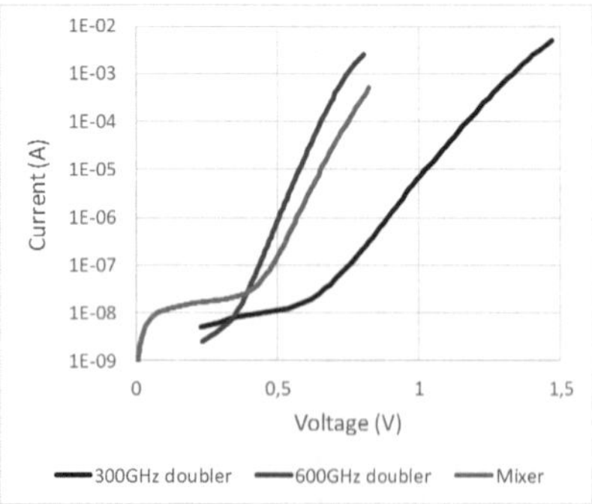

Fig. 5. I(V) curves measurements obtained by probe testing during the wafer screening. The measurement was done on 2 diodes in series on the D300 chip and on 1 diode on D600 and SHM1200 chips.

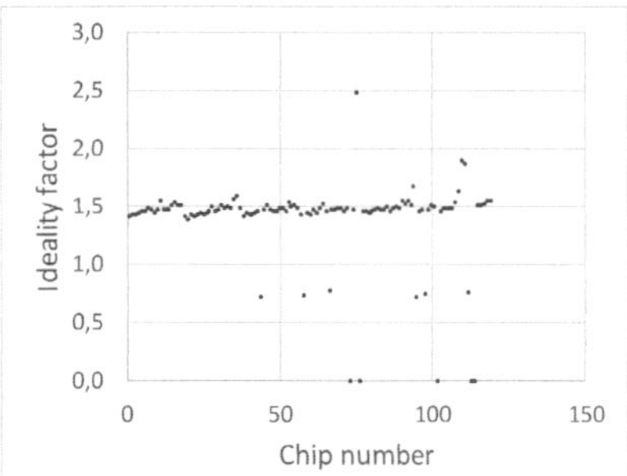

Fig. 6. Ideality factor values obtained on SHM1200 chips from the wafer screening (one point per chip, each chip has two diodes, the value is the mean value of the two diodes).

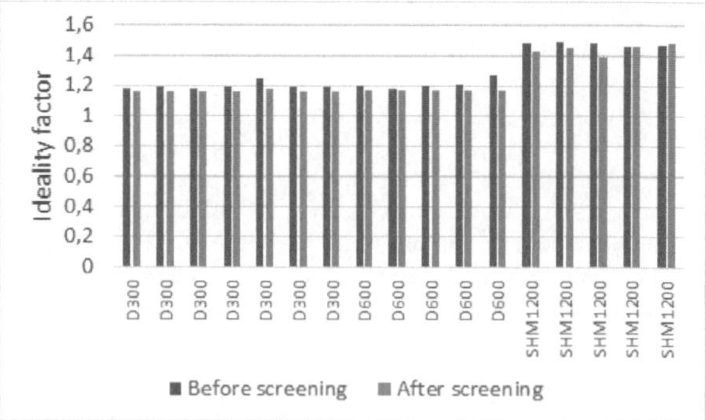

Fig. 7. Ideality factor variation before and after block screening on QM/FM/FS blocks (mean value over the chip).

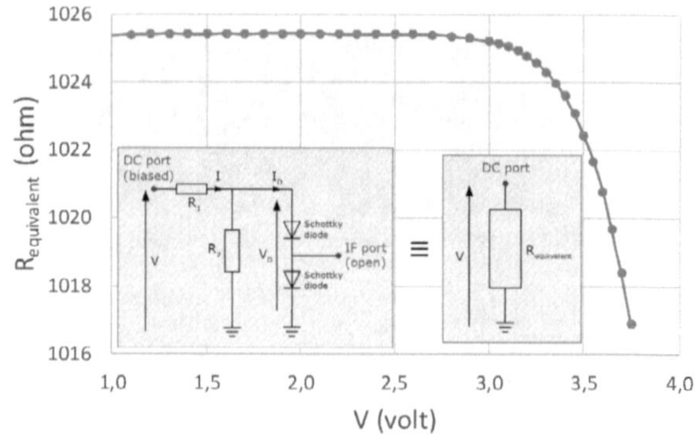

Fig. 8. Equivalent resistance measured from the biasing port on a QM/FM/FS SHM1200 block in static regime. The equivalent simplified circuit is shown as an insert. The IF port should be open (with precaution) so as to include both diodes in the measurement. The current versus voltage diodes characteristic can be deduced (lowest forward current detected around 1μA).

References

[1] J. Treuttel et al, Jupiter Icy Moon Explorer, Submilimeter wave Instrument: Status and performances of the 1200 GHz high spectral resolution receiver front end. *14th European Conference on Antennas and Propagation (EuCAP),* March 2020

[2] J. Treuttel *et al.*, "1200 GHz High Spectral Resolution Receiver Front-end of Submillimeter Wave Instrument for Jupiter Icy Moon Explorer: Part I - RF Performance Optimization for Cryogenic Operation," *IEEE TTST, April 2023, pp. 1-14.*

In situ 1-Port Cryogenic Vacuum Device Calibration

Marko Neric*[1], Sasha Sypkens[1], Farzad Faramarzi[1], and Christopher Groppi[1]

***Abstract*— We present a method for a 1-port vacuum cryogenic in-situ calibration of a Vector Network Analyzer (VNA) using Commercial Off The Shelf (COTS) RF latching electro-mechanical switches. This procedure is being developed for use in a closed-cycle cryogenic test bench to measure the reflection coefficient of a single port connectorized Device Under Test (DUT) which is loaded onto a cold plate alongside calibration standards. The RF switch allows for the calibration and device measurement to be carried out in a single thermal cycle as opposed to a minimum of 4 cycles, which is required for industry standard Open-Short-Load (OSL) device calibration. This test procedure has been carried out on an arbitrary DUT at ~ 3K temperature, over a 6 GHz bandwidth. However, the goal is to develop a setup and procedure for measuring the frequency and temperature dependent complex impedance of superconducting devices such as Hot Electron Bolometer (HEB) mixers, which are used for down converting the signal in the IF chain of astronomy instruments. Characterization of superconducting devices while they are at their operating temperature is challenging using traditional calibration methods. This COTS alternative is less expensive and more efficient in terms of thermal cycles and set up.**

Keywords*— *Astronomy Instrumentation, Cryogenics, Device Characterization

I. Introduction

THz astronomy often makes use of superconducting non-linear devices as mixing elements in receivers to down-convert incident sky signal to an intermediate frequency (IF) by convolving it with a local oscillator (LO) signal. Examples of common astronomy instrumentation mixers include superconductor-insulator-superconductor (SIS) mixers (Zmuidzinas et al. 2004), and hot electron bolometers (HEBs) (Gousev et al. 1994). Because these devices are superconducting they must be operated at cryogenic temperatures, achievable only inside vacuum cryostats. Their complex impedance is dependent on frequency, LO power, and temperature. The impedance is an important factor to characterize because if there is impedance mismatch between mixer output and the amplifiers in the IF system, standing waves will be present, complicating data collection. Vector network analyzers (VNAs) are typically used to characterize devices by measuring their scattering, or s-parameters. The accuracy of the s-parameter measurements are dependent on VNA calibration. Calibration requires a series of measurements using already well characterized objects known as calibration standards. A one-port calibration can be carried out using a short circuit, open, and matched 50 ohm load as reference standards, this is known as an OSL calibration. VNA calibration is difficult when the device under test (DUT) is physically isolated inside a vacuum system. Calibration inside the cryostat may be done manually by measuring the individual standards separately at the operating cryogenic temperature, but this would require a minimum of four cool-down, warm-up cycles to obtain one measurement (Rodriguez-Morales, et al. 2010).

In this work we present a method to calibrate a VNA and measure the DUT all in the same cycle. This setup makes use of commercial-off-the-shelf calibration standards, and cryogenic latching switches. A one port in-situ calibration was used to characterize an RLC circuit in a single cycle, and that result was compared to a manual method with four cryogenic cycles.

II. Set-up

The cryogenic test bench is shown in Fig. 1. The closed cycle system can reach ~ 4 K temperatures, which was below the operating point for most superconducting materials. The VNA was a Rhode & Schwarz ZVA 24. The calibration standards were COTS 85033D DC - 6 GHz, 3.5 mm SMA calibration kit. Each of the calibration standards were heat-sunk to the 4 K test stage using copper clamps so that each component was measured at the same temperature. A temperature sensor was fastened to the DUT heat bracket at the highest point. This calibration scheme was made possible by a single pole six-through (SP6T) cryogenic electromechanical latching DC – 18 GHz Radiall R583423251 switch. Port one of the VNA was connected to the switch through the cryostat. Four channels on the switch were connected to the calibration standards, and DUT via short Mini-Circuits hand formable coaxial UT-047 cables. A supplied 28V DC bias actuated the different switch channels. Outside of the cryogenic system was a hermetic DC bias box that could be manually operated to change the switch channels.

Fig. 1. Cryogenic test bed setup. The calibration standards and DUT were heat-sunk to the test bed using copper clamps and brackets, and are numbered 1-4. The Radiall SP6T switch is in the center with DC power supplied through cryo-wire.

[1] Arizona State University School of Earth and Space Exploration, 85287, Tempe AZ, USA.

At room temperature the SP6T DC bias of 28 V resulted in an actuating current of 61 mA. At ~ 10 K temperature, the switch was actuating with a 2 V bias. By attaching the reference standards, DUT, and VNA to the same switch, one could measure each component in a single cycle. A one port RLC made with COTS surface mount components was the DUT for both a single cycle in-situ measurement and a four cycle manual calibration. For all testing the VNA port power was -20 dBm, the IF bandwidth was 10 Hz, the averaging was x16.

III. Results

The RLC was measured at room temperature using an automated electronic calibration kit, and then using the manual OSL kit. First, this manual calibration was done without the switch. The individual standards were connected one at a time to the first port of the VNA via coax cable. Then the VNA was calibrated with the same standards using the switch. Channel one of the switch was the open, channel two was the short, channel three was the matched load, and channel four was the DUT. These three measurements were in agreement which validated an OSL measurement using the SP6T at 300 K. The S_{11} of the DUT at 300 K can be seen in Fig. 2 as the blue trace. The cryostat was brought to ~ 4 K temperature. The room temperature calibration had drifted due to a change in temperature and conditions inside the cryostat. The switch test was repeated when the system was at 3.3 K. The new result can be seen as the pink trace in Fig. 2. The RLC resonance was recovered by recalibrating at cryogenic temperatures using the SP6T. The result was shifted slightly in frequency. The shift was likely due to objects contracting at cooler temperatures and changing the values of not only the RLC test device, but of some of the calibration standards. The linear coefficient of thermal expansion (CTE) determines how objects will change with temperature for the different materials. The ~ 1.75% shift in frequency indicates a change of ~ 3.50% in the inductance and capacitance LC. The cryogenic calibration was repeated using a manual method where each standard was measured separately at cryo and replaced over multiple cool downs was used to verify that this shift is not coming from the switch. Each cycle changed the measurement environment by changing the thermal load on the test bed, and because the internal coax lines were disturbed by the act of replacing standards and the DUT. The manual method took 2 days, and the results can be seen in Fig. 2 as the brown trace. The two traces are nearly identical with some variation due to the change in environment in the manual method and the lack of temperature control. Without stable temperature control of the test stage the DUT, measurements had a narrow window to be recorded. The final temperature with the bulky SP6T and calibration standards installed together, was 3.3 K. The individual standards and DUT installed alone would reach ~ 2.7 K. This meant averaging had to be started before the objects reached 3.3 K to measure the same conditions.

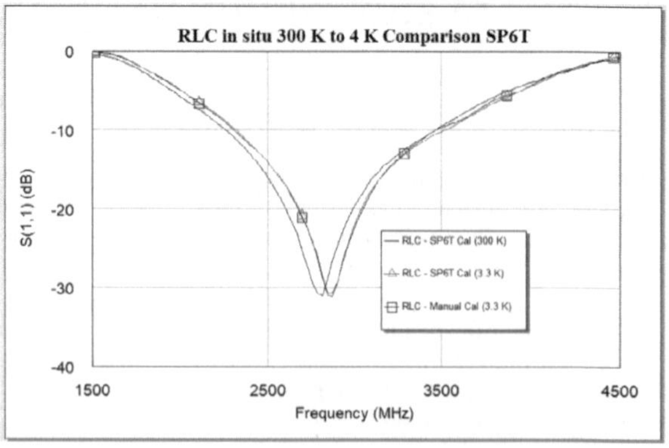

Fig. 2. An RLC DUT test circuit measured at room temperature 300 K (blue), and at 3.3 K using an RF SP6T (pink). The result is compared to a manual method (brown).

IV. Conclusion

The in-situ one-port VNA calibration using an SP6T to switch between reference standards and DUT provides an alternate calibration scheme to a multi-cycle manual error correction. The SP6T method is in good agreement with the manual as shown in Fig. 2. The S_{11} response is nearly identical between both methods. The variations result from the drawbacks of the manual method, namely the inconsistency of environment between measurements, i.e. cable position and exact temperature. The SP6T offers several improvements over the alternatives. The setup is entirely COTS and relatively inexpensive. The technique only requires a single cryogenic cycle to calibrate and test which can save days of measurement time. Due to that fact, the calibration can be restarted in the same cycle if temperature fluctuates before or during measurement with very little time lost. Having characterized a generic RLC DUT, the follow up measurement will be that of an HEB built for THz astronomy instrumentation. With more accurate characterization a matching network may be implemented to improve power transfer from the HEB to the rest of the IF system.

References

[1] Gousev, Y. P., et al., "Broadband ultrafast superconducting NbN detector for electromagnetic radiation", *Journal of Applied Physics*, April. 1994.

[2] Rodriguez-Morales, F K., et al., "Wide-band IF-integrated terahertz HEB Mixers: Modeling and characterization", *IEEE Transactions on Microwave Theory and Techniques*, April. 2010.

[3] Zmuidzinas, J., et al., "Superconducting detectors and mixers for millimeter and submillimeter astrophysics," *Proceedings of the IEEE*, Sept. 2004.

Characterization of a 183 GHz radiometer receiver with heterodyne noise injection calibration system

Tomas Thuroczy[1,2], Alexandre Feret[2], Benoit Collay[2], Jeanne Treuttel[2]

Abstract — **This paper explores the possibility of calibrating a 183 GHz radiometer with noise injection using a heterodyne noise source. The system comprises two mixers with a shared LO source for signal down-conversion and noise up-conversion. A standard 16 dB baseband noise source is used for generating the noise at the IF port before upconversion.**

Keywords—Heterodyne, Noise Source, ENR, Sub-harmonic, Calibration, Sub-millimeter

I. INTRODUCTION

Radiometer systems at submillimeter wavelengths usually employ calibration methods with mechanically moving mirrors, switching alternatively the signal to onboard calibration targets and the scene of observation. For space applications where moving parts are always a risk factor, we can use electronically controlled noise injection calibration method as alternative to bulky systems that are sensitive to ageing. Traditional solution to generate noise signals at the RF frequency consists in using noise diodes [1]. In this work, we propose heterodyne noise generation with the advantage of reusing existing mixer designs and frequency scalability, as previously demonstrated in [2].

II. SYSTEM DESCRIPTION

A block diagram of the radiometer system bench with the integrated heterodyne noise source is shown in the Figure 1 and the implemented test setup in Figure 2. The system is comprised of two sub-harmonic mixers in the 183 GHz band utilizing the United Monolithic Semiconductors (UMS) Schottky diode process [3]. The Sub-Harmonic Mixer (SHM) 2 mixer serves as a down-converter and the SHM 1 as an upconverter of the IF noise signal. A wideband intermediate frequency noise source Agilent 346B with around 16 dB of Excess Noise Ratio (ENR) is used for the IF noise signal. The upconverted noise signal is injected into the receiving system using a 15 dB directional coupler. Due to the same Local Oscillator (LO) feeding both mixers, the noise upconverted by SHM 1 falls into the same frequency band as the receiving band of the SHM 2.

III. MEASUREMENT RESULTS

Prior to the noise injection demonstration, the mixers were first individually characterized to determine the optimal LO power levels that yield the best system noise temperature and conversion loss. The LO power levels can be adjusted by the mechanical variable waveguide attenuators. The measured optimum LO power and noise temperature of SHM 1 and SHM 2 are shown in the Figure 3. The SHM 2 was also characterized in a "dual-LO" configuration when the LO signal was split to feed both subharmonic mixers at the same time.

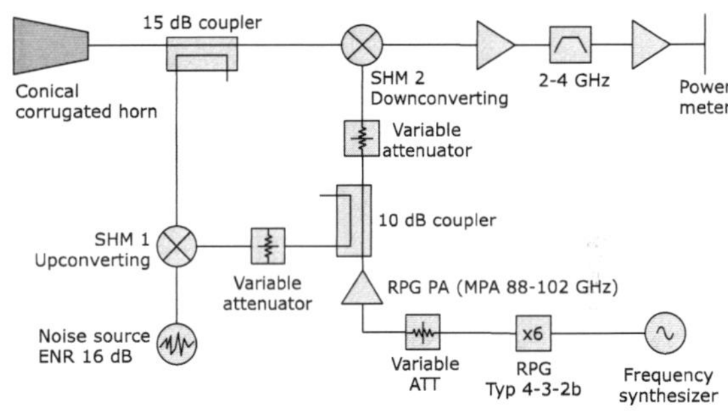

Figure 1 Block diagram of the receiver system

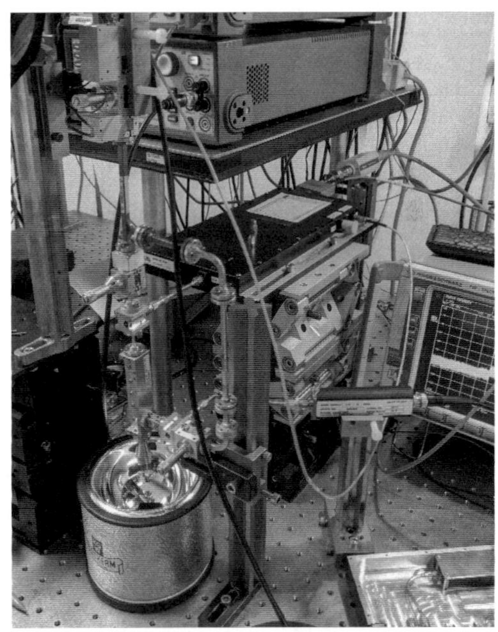

Figure 2 Implemented test setup

[1] Université de Rennes 1, CNRS, IETR, 35000 Rennes, FRANCE
[2] Observatoire de Paris, LERMA, 75014 Paris, FRANCE;

NOTES:

In the subsequent step, we demonstrated a noise injection capability using a heterodyne noise source. For a constant system noise temperature T_{sys} and antenna temperature T_{ant}, injecting the noise can be regarded as if the temperature of the observed target would increase. This manifests as an increase of the detected IF power as well.

Figure 4 shows the measured IF spectra when the heterodyne noise source is ON and OFF at the LO frequency of 183 GHz. The ENR variation of the employed IF noise source in the 2-4 GHz band is 0.09 dB. The measured flatness of the injected noise obtained by subtracting the ON and OFF IF spectra is within the ENR variation of the IF noise source (Figure 5).

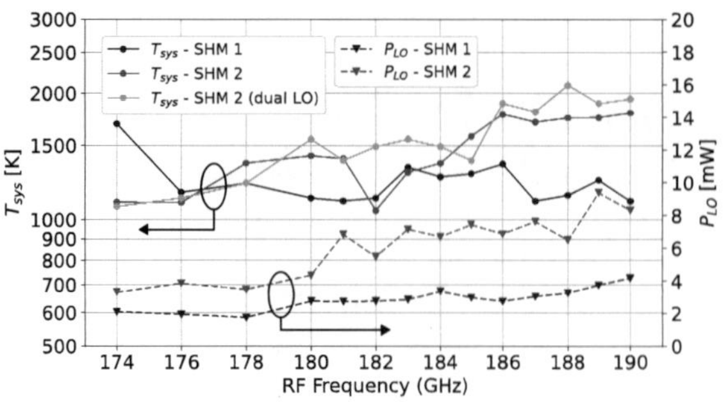

Figure 3 Noise performance of the two subharmonic mixers

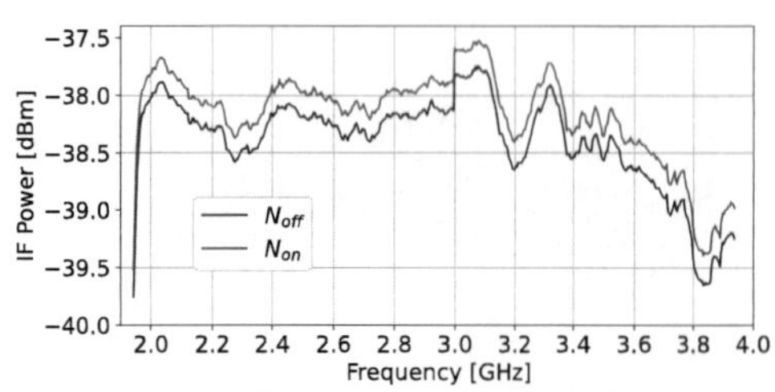

Figure 4 IF frequency spectra with noise source ON and OFF

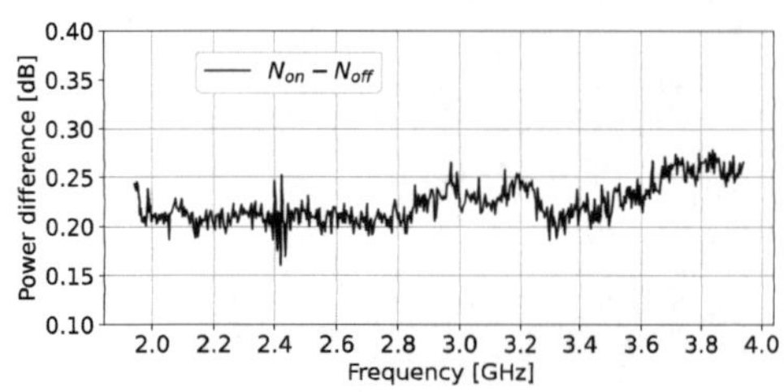

Figure 5 Difference of the ON and OFF IF spectra

IV. CONCLUSION AND FUTURE WORK

We presented measurements of a radiometric receiver in the 183 GHz band with an integrated heterodyne noise source. The employed subharmonic mixers were measured in terms of their noise performance, and we demonstrated a noise injection capability into the receiving system by measuring the IF spectra of the down-converting mixer before and after noise injection. We demonstrated that the flatness of the injected noise was very similar to the flatness of the IF noise source.

In the future, we will focus on characterizing the heterodyne noise source at different RF frequencies and by using IF frequency noise sources with different ENR levels. Furthermore, the stability of the noise source and its adequateness as a calibrator will be determined. Operation at higher millimeter wavelengths will be investigated as well.

ACKNOWLEDGEMENTS

This project has received financial support from the CNRS through the MITI interdisciplinary programs.

REFERENCES

[1] N. Ehsan, et al., "A robust waveguide millimeter-wave noise source," in 2015 European Microwave Conference (EuMC), Paris, France, Sep. 2015, pp. 853–856.

[2] Colin Viegas et al. "Millimeter-Wave Noise Sources using Heterodyne Mixing of Signals". In: 2019 IEEE MTT-S International Microwave and RF Conference (IMARC). Dec. 2019, pp. 1–2.

[3] Hui Wang. Conception et modélisation de circuits monolithiques à diode Schottky sur substrat GaAs aux longueurs d'onde millimétriques et submillimétriques pour des récepteurs hétérodynes multi-pixels embarqués sur satellites et dédiés à l'aéronomie ou la planétologie.. Engineering Sciences [physics]. Observatoire de Paris, 2009. English.

[4] F. Ulaby, Microwave Radar and Radiometric Remote Sensing. Ann Arbor: University of Michigan Press, 2013.

NOTES:

Numerical and theoretical modeling of a heterodyne noise injection radiometer system

Tomas Thuroczy[1,2], Alexandre Feret[2], Jeanne Treuttel[2]

***Abstract* — This paper presents a theoretical and numerical model of a radiometer noise injection system based on a heterodyne generation of the noise signal. The coupled 3D EM and nonlinear circuit simulation is performed at 600 GHz using the subharmonic mixers developed at LERMA and C2N.**

Keywords—Heterodyne, Noise Source, ENR, Sub-harmonic, Calibration, Sub-millimeter

I. Introduction

Noise signals for radiometer calibration are generated directly at the RF frequency using noise diodes up to 220 GHz [1]. At submillimeter wavelength, they can be generated by electrical or optical heterodyning [1], [2]. We present a simulation model of a 600 GHz radiometer noise injection system featuring a sub-harmonic mixer that up-converts the noise signal. The mixers used in the simulation were developed at LERMA and C2N [3].

II. Results

At submillimeter wavelength, radiometer calibration is made using two well calibrated blackbodies at physical temperatures below and above the nominal value of the equivalent brightness temperature of the scene to be detected (typically 4 K and 300 K). In case of noise injection, the injected noise signal effectively increases the T_{sys}. If one can verify that the radiometer output power, during the noise injection, is comparable to its predicted by an equivalent increase of $T_{ant,hot}$, the noise injection system can be used for hot calibration.

The simplified ADS simulation schematic of the system is shown in the Figure 1. The noise source is modelled by an IF resistor at the upconverting mixer with variable physical temperature. The bandwidth of the noise source is limited by an ideal filter. The LO, RF and IF frequencies are 290, 584 and 4 GHz, respectively.

When IF noise with temperature $T_{N,IF}$ is upconverted by a mixer with conversion loss L_{cnv}, the temperature at the output $T_{N,RF}$ is reduced according to the attenuator model – Eq. (1).

$$(1)\quad T_{N,RF} = L_{cnv} T_{N,IF} + (1 - L_{cnv}) \times T_{phys}$$

The input noise temperature of a system is dependent on the $T_{N,RF}$, coupler's loss, noise temperature, coupling factor (G_C, T_C, C) and noise temperature of the receiving mixer T_{mix}. The system noise temperature as a function of $T_{N,RF}$ can be expressed by Eq. (2).

$$(2)\quad T_{sys} = \left[T_C + (T_{N,RF} - 290)\frac{C}{G_C}\right] + \frac{T_{mix}}{G_C}$$

The comparison of the theoretically calculated with Eq. (2) and Keysight Advanced Design System Harmonic Balance (ADS HB) simulated T_{sys} is given in the Figure 2. The coupler is assumed to be lossless and with -10 dB coupling. The T_{sys} value at ENR of 0 dB corresponds to the circuit consisting of a coupler followed by a mixer. As the level of the IF noise increases, the effective system temperature increases too: an increase of the T_{sys} by 300K corresponds to the required ENR at the IF input of the first mixer of around 20 dB. The small discrepancy between the two curves is due to the finite number of harmonics in HB which affects the absolute value of the calculated noise.

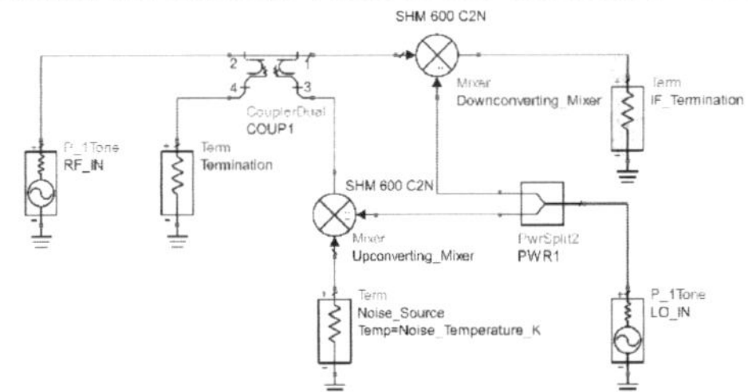

Figure 1 Simplified schematic of the simulated noise-injection system

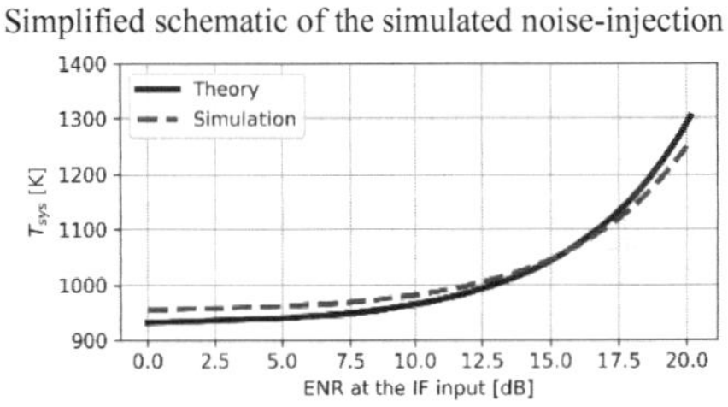

Figure 2, System noise temperature as a function of the IF noise level

References

[1] Colin Viegas et al. "Millimeter-Wave Noise Sources using Heterodyne Mixing of Signals". In: 2019 IEEE MTT-S International Microwave and RF Conference (IMARC). Dec. 2019, pp. 1–2.

[2] W. Liu et al., "Broadband and flat millimeter-wave noise source based on the heterodyne of two Fabry–Perot lasers," Opt. Lett., OL, vol. 47, no. 3, pp. 541–544, Feb. 2022.

[3] J. Treuttel et al., "A 520–620-GHz Schottky Receiver Front-End for Planetary Science and Remote Sensing With 1070 K–1500 K DSB Noise Temperature at Room Temperature," IEEE Trans. THz Sci. Technol., vol. 6, no. 1, pp. 148–155, Jan. 2016.

[1] Université de Rennes 1, CNRS, IETR, 35000 Rennes, FRANCE

[2] Observatoire de Paris, LERMA, 75014 Paris, FRANCE;

NOTES:

Frequency Locking of a 4.7 THz Quantum Cascade Laser using a Delay Line

Sajjad Mahdizadeh[1,*], Urs Graf[1], Matthias Justen[1] and Jürgen Stutzki[1]

***Abstract*—Quantum Cascade Lasers (QCLs) are powerful monochromatic radiation sources at higher THz frequencies and are highly favorable for heterodyne receivers such as the upGREAT of the SOFIA telescope. Apart from their advantages, they can suffer from different line-broadening mechanisms, such as optical feedback, that broaden the effective linewidth and reduce the receiver's spectral resolution. In this work, we present frequency stabilization of a 4.7 THz QCL based on a delay line frequency discriminator. The latter comprises a power divider, a delay line, and a double-balanced mixer that forms a frequency discrimination constant of 91.5 mV/MHz. The QCL emission is down-converted with the help of a Superlattice harmonic generator and mixer, itself pumped with a diode multiplier chain at 182.6 GHz. The QCL line in the intermediate frequency band is then amplified and filtered to feed the frequency discriminator, where its output is used to stabilize the QCL frequency. More than 10 MHz of frequency deviations can be reduced to a static and stable line with an FWHM of 780 kHz.**

***Index Terms*—Quantum Cascade Laser, Local Oscillator, Frequency stabilization, Delay line frequency discriminator, Superlattice harmonic generator and mixer**

I. Introduction

QCLs are the most convenient sources of THz emission at higher THz Frequencies, such as 4.7 THz [1]. Heterodyne receivers operating at these frequencies, therefore, use them as Local Oscillators (LO). UpGREAT of the SOFIA telescope and STO and GUSTO balloon-born missions are examples of such receivers. QCLs usually suffer from different mechanisms that broaden the effective linewidth of the laser. These include current noise, temperature fluctuations, and optical feedback. Since in a heterodyne receiver, the LO line shape convolves with the observed line, frequency stabilization of the LO eventually increases the spectral resolving power of the receiver. In addition, providing an absolute frequency reference for the QCLs will increase the frequency accuracy of those receivers.

II. Delay Line Frequency Discriminator

In order to stabilize the frequency, it is crucial to be able to discriminate it with a frequency detector. This frequency discriminator works as the following. Since this experiment is heterodyne, a small, down-converted signal from the QCL is available in the IF. A power divider divides this signal into two equal amplitude outputs. One is delayed through the delay line cable, and both are mixed with a double-balanced mixer. Both mixer inputs have the same frequency, but the phase difference between them is frequency dependent due to the delay. As a result, a DC is created at the output of the mixer, which is variable with the frequency of the input. Figure 1 shows this DC output during a frequency sweep as a calibration measurement. One can also see this as a Michelson interferometer with unequal arm length. When the frequency of incoming light changes, the brightness of the fringe changes subsequently.

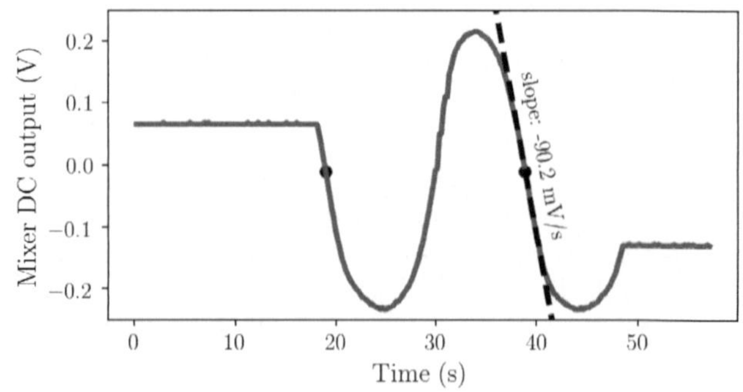

Fig. 1. The output of the delay line frequency discriminator during a frequency sweep with a rate of 1 MHz/s, centered at 400 MHz. The shown slope in this measurement is taken as a calibration of frequency to voltage conversion of this frequency discriminator.

III. Method

Figure 2 shows the schematic of this experiment. The QCL is housed inside a pulse-tube cryostat where the coldhead is thermally coupled and mechanically isolated from the QCL. QCL's emission is coupled to a superlattice harmonic generator and mixer (SLD) [2] using three mirrors. The SLD is pumped with the mm-wave power generated in a VDI diode multiplier chain with the 26th harmonic close to the QCL frequency. The SLD also mixes the two signals, and the difference signal is extracted at the IF port of the SLD. The latter is amplified and band-pass filtered to the optimum level to drive the frequency discriminator. The latter detects the frequency deviations and creates a proportional voltage, which is processed in the control electronics to form the correction signal. This correction signal is then added to the QCL current, and since the QCL frequency is current tunable, the correct settings on the control electronics stabilize the QCL's frequency.

IV. Experimental Setup

Figure 3 shows a photo of the experimental setup. The QCL is running at a current of 160 mA and 32 K bath temperature. The SLD is pumped with the diode multiplier

[1] I. Physikalisches Institut, Universität zu Köln, Zülpicher str. 77, 50937 Köln, Germany
[*] e-mail: mahdizadeh@ph1.uni-koeln.de

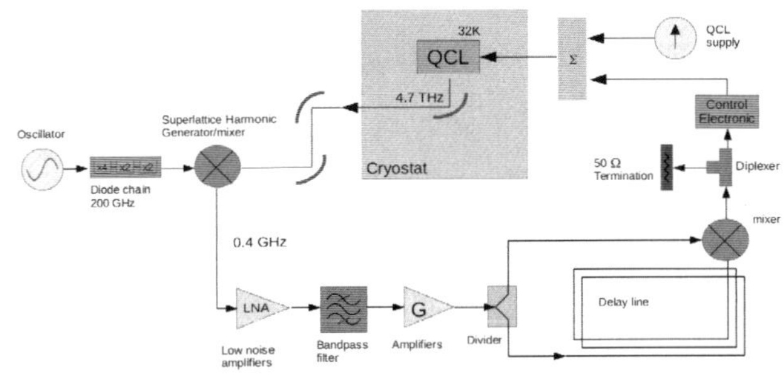

Fig. 2. Schematic of the experiment.

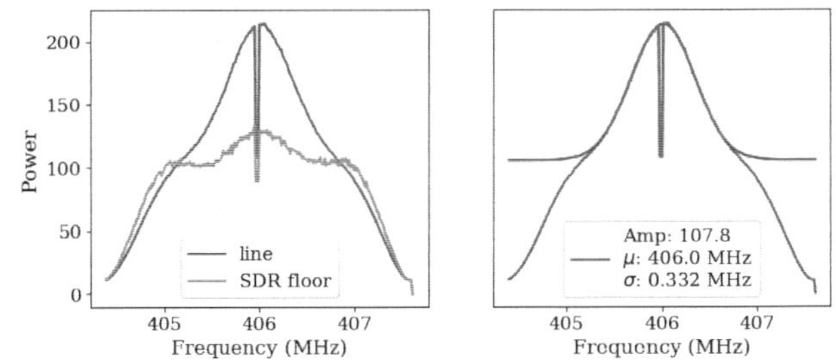

Fig. 4. Left: Example of the integrated SDR noise floor without and with the QCL line in the IF monitor output. Right: A Gaussian function is fitted to the line segment of the same data, with the joint parts used for the fit colored in green.

chain generating 182.6 GHz. The SLD's IF output contains the down-converted QCL signal of 9 dB over the noise floor (Resolution Bandwidth = 3 MHz) at an IF frequency of 400 MHz. While amplifying, it is essential to filter the wide-band noise generated by the SLD. The delay line is a 10-meter coax cable that introduces 50 ns of delay, resulting in a 91.5 mV/MHz frequency discrimination constant. The high-frequency products in the mixer are dumped into a 50-ohm termination load. The low-frequency components are processed in an OPAMP circuit as an amplifier and filter, where its output is the correction signal.

V. Results

Before stabilization, the spectrum of the QCL emission suffered from more than 10 MHz of frequency fluctuations. Figure 4 presents the IF output spectrum with the stabilized QCL. An FWHM of 780 kHz corresponds to the width of the Gaussian function fitted to the QCL line. The narrow dips on the plots correspond to three SDR channels that were not functional, and they were excluded in the Gaussian fit. During the measurement, it was understood that the primary source of the frequency instabilities in this experiment had been the optical feedback to the laser: Inserting attenuators in the beam reduced the strength of frequency modulation. Stabilization could be kept for many hours, and the method showed the needed robustness for use in an actual receiver. Interestingly, stabilization was still possible at small SNRs, such as only 2 dB over the noise floor. This means a significant fraction of the QCL power can be used to pump an array of mixers.

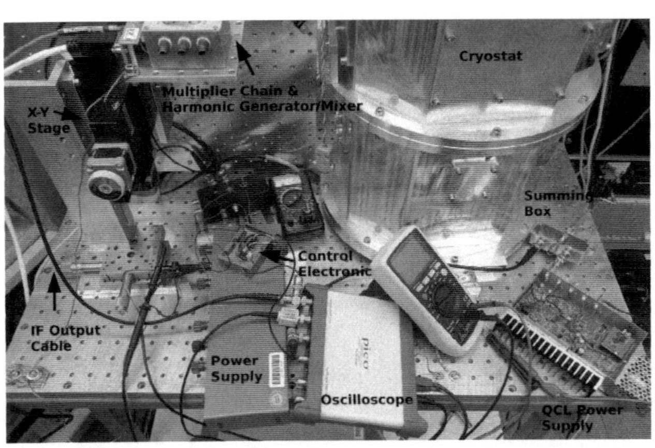

Fig. 3. A large fraction of the experimental setup: Optics bench, cryostat, and a part of the electronics.

VI. Conclusion

In conclusion, we have shown a robust frequency locking of a 4.7 THz QCL using a room-temperature superlattice device. The achieved locked linewidth of 780 kHz is enough for heterodyne astronomical spectrometers to reach a velocity resolution of 50 m/s, which is more than needed. In addition to this work, with a similar setup in the future, we hope to down-convert the SLD emission with an HEB mixer and demonstrate robust phase locking.

Acknowledgment

This research is carried out within the Collaborative Research Center 956, funded by the Deutsche Forschungsgemeinschaft (DFG). The authors acknowledge the effort of colleagues in ETH Zurich to produce QCLs for us and also the help of Dr. Dmitry Paveliev in setting the superlattice device up.

References

[1] L. Bosco, C. Bonzon, K. Ohtani, M. Justen, M. Beck, and J. Faist, "A patch-array antenna single-mode low electrical dissipation continuous wave terahertz quantum cascade laser," *Applied Physics Letters*, vol. 109, no. 20, 11 2016, 201103. [Online]. Available: https://doi.org/10.1063/1.4967836

[2] D. Paveliev, Y. I. Koshurinov, A. S. Ivanov, A. N. Panin, V. L. Vax, V. I. Gavrilenko, A. V. Antonov, V. M. Ustinov, and A. E. Zhukov, "Experimental study of frequency multipliers based on a GaAs/AlAs semiconductor superlattices in the terahertz frequency range," *Semiconductors*, no. 46, pp. 121–125, 2012.

Study on Oscillation Characteristics of G-band Gyrotron Traveling Wave Tubes with a Tapered Interaction Circuit

Wenqi Gao, Yelei Yao*, Ran Yan, Yibin Sun, Haibing Huang, and Yong Luo
School of Electronic Science and Engineering
University of Electronic Science and Technology of China (UESTC)
Chengdu, 611731, China
**yeleiyao@uestc.edu.cn*

***Abstract*— The oscillation characteristics of G-band gyrotron traveling wave tube (gyro-TWT) are analyzed and verified. The contours of oscillation threshold with the changes of the magnetic field, beam current and pitch ratio are plotted. It is found that a tapered circuit with small angle can effectively improve oscillation threshold.**

***Keywords*—G-band, gyro-TWT, interaction circuit, oscillation threshold**

I. Introduction

The gyro-TWT is a fast-wave device featuring high power, high frequency and broad bandwidth for advanced radar and communication applications in millimeter-wave and terahertz wave band. A major obstacle to the development of gyro-TWT has been its susceptibility to oscillations caused by the absolute instability [1]. This paper examines the oscillation characteristics of G-band gyro-TWT, a slowly tapered circuit that can improve oscillation threshold is demonstrated.

II. Results

A. Theoretical Oscillation Threshold

According to the linear theory of gyro-TWTs [2], the oscillation threshold of both the operating mode and backward-wave modes are calculated. Main parameters of the interaction circuit are the beam voltage of 60 kV, the waveguide radius of 0.9 mm and the interaction circuit length of 10 mm, respectively.

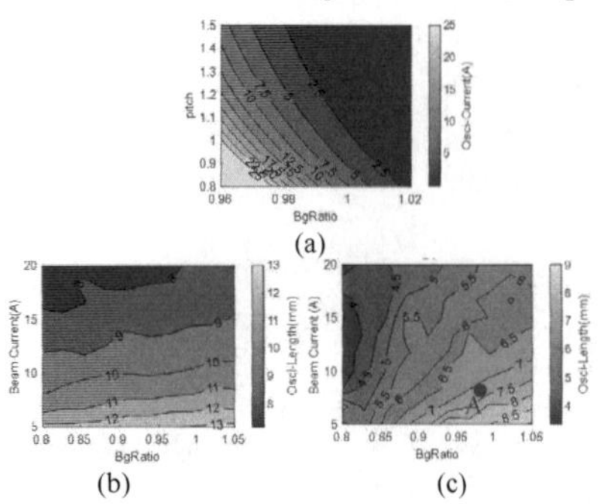

Fig.1. (a) Oscillation current contour of TE_{01} mode with magnetic field and velocity pitch ratio. (b) Oscillation length contour of TE_{11} mode with magnetic field and beam current. (c) Oscillation length contour of TE_{21} mode with magnetic field and beam current.

B. PIC Simulation of Uniform Circuits

In order to verify the theoretical results, point A where the B_gRatio is 0.97 and the beam current is 8 A in the Fig. 1(c) is chosen for simulation verification. According to Fig. 2, the cut-off frequency of TE_{01} mode is 203.1 GHz and the backward intersection frequencies with TE_{11} and TE_{21} mode are 154.6 GHz and 172.2 GHz respectively. As in Fig. 1, the oscillation current of the TE_{01} mode is 9.406 A, the oscillation lengths of the TE_{11} and TE_{21} mode are 10.88 mm and 7.06 mm respectively. In this case, only the TE_{21} mode will oscillate by predicting.

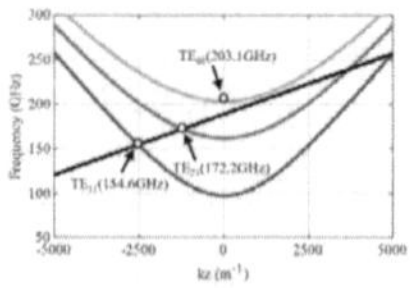

Fig.2 Dispersion diagram of G-band gyro-TWT.

As shown in Fig. 3, clear signal of TE_{21} mode with an oscillation frequency of 175.66 GHz at the output port is observed. No oscillation of the TE_{01} and TE_{11} mode is observed. The theoretical prediction and PIC simulation results show a good agreement.

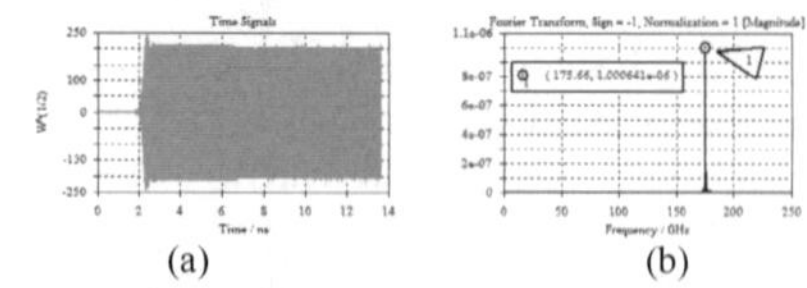

Fig.3. TE_{21} mode (a) input port signal. (b) input port spectrum.

C. PIC Simulation of Tapered Circuits

To decrease the effective interaction length within the whole beam-wave interaction process, a tapered circuit is employed to change the intersection frequency with different circuit positions, in this way the oscillation threshold can get improved. A tapered circuit with small angle is introduced to alleviate the oscillation for design verification with above simulation parameters. As shown in Fig. 4, the oscillation of the TE_{21} mode is suppressed. In addition, no clear oscillation signal is observed for the TE_{01} and TE_{11} mode. It can be seen that a tapered circuit with small angle is beneficial to improve the oscillation threshold.

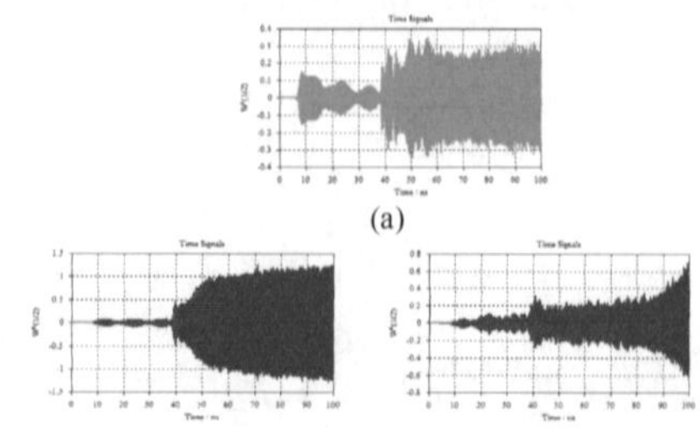

Fig.4. (b) TE_{11} mode input port signal. (c) TE_{01} mode output port signal.
. (a) TE_{11} mode input port signal.

References

[1] W. C. Tsai, et al. "Study of absolute instabilities in the gyrotron traveling-wave amplifier." *IEEE International Conference on Plasma Science IEEE*, 2004.

[2] C. S. Kou, et al. "High-power harmonic gyro-TWT's. I. Linear theory and oscillation study." *IEEE Transactions on Plasma Science* 20.3(1992):155-162.

[3] K. R. Chu and A. T. Lin, "Gain and bandwidth of the gyro-TWT and CARM amplifiers," *IEEE Transactions on Plasma Science*, vol. 16, no. 2, pp. 90-104, April 1988.

Phase-locking THz QC-VECSEL local oscillators using diode mixers and sources

Christopher A. Curwen*[1], Alain Maestrini[1], Anthony Kim[2], Sadhvikas J. Addamane[3], John L. Reno[3], Boris S. Karasik[1], Benjamin S. Williams[2], Jonathan H. Kawamura[1]

***Abstract*—The output frequency and phase of an external cavity quantum-cascade laser is phase-locked to an RF reference via two techniques: 1) down conversion with a diode harmonic mixer and locking to 100 MHz reference using the laser as a voltage tunable oscillator, and 2) directly injection locking the laser using the output of a 2.5 THz diode frequency multiplier source.**

Keywords*—*quantum-cascade laser, phase-locking

Terahertz quantum-cascade lasers (QCLs) are compact, semiconductor sources of coherent, milliwatt level radiation in the 2-5 THz frequency range. They are excellent candidates as local oscillators for heterodyne receiver arrays, and have been used as such in the upGREAT instrument on SOFIA, and the upcoming GUSTO balloon mission; observing neutral oxygen at 4.74 THz in both cases. However, limited beam quality and cryogenic operating temperatures of QCLs can make systems bulky and difficult to align. One design strategy for improving performance is to build the QC gain material into a vertical-external-cavity surface-emitting laser (VECSEL). QC-VECSELs have demonstrated excellent beam patterns, broad frequency tunability, and high continuous wave operating temperatures [1]. However, to achieve high spectral resolution from a heterodyne receiver, the output frequency of the VECSEL must be locked to a more stable frequency reference. Here, we demonstrate phase-locking of the VECSEL to a diode frequency multiplied RF source in two different ways. The VECSEL operates at 77 K and outputs ~1 mW of power.

In the first phase-locking approach, we down convert the output of a 2.5 THz QC-VECSEL to a 2 GHz IF using a harmonic diode mixer [2]. The mixer is pumped with an 840 GHz LO chain, and the 3rd harmonic mixes with the signal from the VECSEL. The IF is further down converted to 100 MHz and fed into a phase-lock loop (PLL) with a quartz reference. The error signal from the PLL controls the bias on the QC material, which acts as a frequency tuning mechanism. The phase noise of the locked VECSEL is shown in Fig. 1(b).

For the second phase-locking approach, we use the same 840 GHz LO chain, but we replace the mixer with a diode tripler source capable of outputting ~15 μW of power at 2.5 THz. This power can directly injection lock the QC-VECSEL with the RF source. The VECSEL is particularly suited to such an approach as it is easier to couple power in the large aperture of the VECSEL cavity than the sub-wavelength sized facet of a ridge-waveguide device. The locked spectrum, is shown in Fig. 1(c). The advantage of the injection locking approach is that the VECSEL output power is more stable than the first approach, but the downside is that it requires a diode frequency multiplier source, which becomes increasingly difficult at higher THz frequencies.

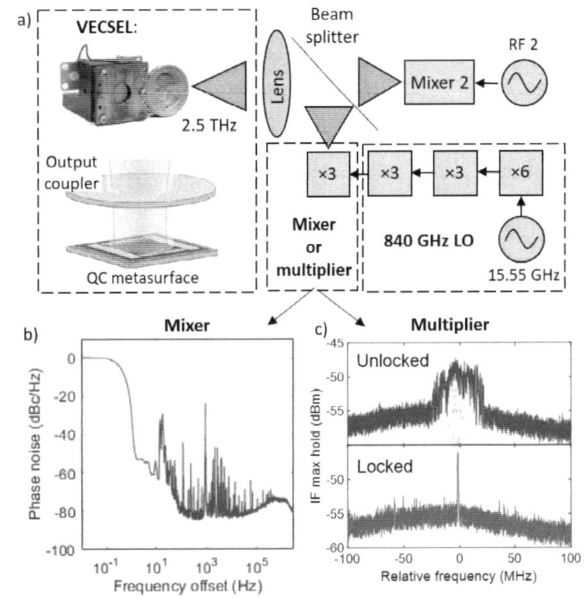

Fig. 1 a) Block diagram of experimental setup. b) Phase-noise when QC-VECSEL is locked to 3rd harmonic mixer. c) Locked vs. unlocked spectrum when injecting power from 2.5 THz tripler.

REFERENCES

[1] C. A. Curwen , *et al.*, "Broadband continuous single-mode tuning of a short-cavity quantum-cascade VECSEL," *Nat. Photon.* (2019).

[2] B. Thomas , *et al.*, "A Broadband 835-900-GHz Fundamental Balanced Mixer Based on Monolithic GaAs Membrane Schottky Diodes," *IEEE Trans. Microw. Theory Thechn.* (2010).

[1]Jet Propulsion Laboratory, Pasadena, CA 91109; [2]SUniversity of California, Los Angeles, CA 90095; [3]Sandia National Laboratories, Albuquerque, NM 87185.

NOTES:

The Submillimeter Wave Instrument on JUICE

Paul Hartogh[1] and the SWI team

JUICE – JUpiter ICy moons Explorer it the first large class mission of ESA's Cosmic Vision 2015 - 2025 program [1-3]. The JUICE satellite is planned to be launched in April 2023 and its Jupiter orbit insertion is foreseen in 2031. The primary mission in the Jupiter system will take about three years. The focus of the mission is Jupiter itself and the Galilean satellites, their internal oceans and potential habitability. Recent ground-based observations of Europa and Ganymede showed water vapor plumes, probably related to geysers on their surfaces. JUICE intends to identify the geysers, monitor their potential activity and molecular and isotopic composition in order to constrain satellite formation models and development processes (of chemical, physical and potentially biologic nature) in the interior of their oceans. Jupiter itself is seen as an archetype of a gas giant. A better understanding of its atmospheric processes will be a baseline for a better understanding of gas giants outside of our solar system. JUICE will characterize the general circulation of Jupiter's atmosphere, its meteorology, chemistry and structure between the upper cloud deck and the ionosphere and magnetosphere.

The Submillimeter Wave Instrument (SWI) on JUICE sounds the middle atmosphere of Jupiter and the exospheres and surfaces of the Galilean satellites with a 290 mm telescope, movable in azimuth and elevation using two heterodyne spectrometers covering the frequency ranges of 530 to 630 GHz and 1065 to 1275 GHz. The heterodyne receivers with tunable solid-state local oscillators and sub-harmonically pumped mixers are passively cooled to about 150 K. The receiver temperatures are about 1000 and 2000 K DSB (600 and 1200 GHz respectively). The intermediate frequency outputs of 4.5 to 8.5 GHz are analyzed by two real-time spectrometer backends per receiver consisting of broadband autocorrelators (4 GHz bandwidth, 4 MHz resolution), high resolution Chirp Transform Spectrometers (1 GHz bandwidth, 100 kHz resolution) and continuum channels. The structure of the Telescope and Receiver Unit (TRU) and the 29 cm mirror are made of AlBeMet 162. The Flight Model (FM) was integrated into the JUICE satellite in September 2021. The main effort of the present work is put on Flight Spare (FS) testing. This presentation will address the science objectives, the technology challenges and details about the instrument and its performance.

[1]Max Planck Institute for Solar System Research (MPS), 37077 Göttingen, Germany

Fig. 1. SWI FM in MPS anechoic chamber during EMC test

REFERENCES

[1] http://sci.esa.int/juice/50073-science-payload/
[2] http://sci.esa.int/juice/50068-science-objectives/
[3] http://sci.esa.int/juice/49837-juice-assessment-study-report-yellow-book/#

NOTES: